Atlas of Bedforms in the Western Mediterranean

Jorge Guillén · Juan Acosta
Francesco Latino Chiocci
Albert Palanques
Editors

Atlas of Bedforms in the Western Mediterranean

Springer

Editors
Jorge Guillén
Marine Geology
Instituto de Ciencias del Mar (CSIC)
Barcelona
Spain

Francesco Latino Chiocci
Dipartimento di Scienze della Terra
Sapienza Università di Roma
Rome
Italy

Juan Acosta
Instituto Español de Oceanografia, IEO
Madrid
Spain

Albert Palanques
Geology-Physics
Instituto de Ciencias del Mar (CSIC)
Barcelona
Spain

ISBN 978-3-319-33938-2 ISBN 978-3-319-33940-5 (eBook)
DOI 10.1007/978-3-319-33940-5

Library of Congress Control Number: 2016944907

Printed on acid-free paper

This Springer imprint is published by Springer Nature
The registered company is Springer International Publishing AG Switzerland

This atlas is dedicated to our colleague and friend, Juan Acosta. Juan studied Geology (Madrid, 1972) and joined the Marine Geology group of the Spanish Institute of Oceanography in 1973. Since then he has accompanied several generations of marine geologists in Spain, combining marine research (his vocation) and resource management (a duty). To highlight just a few things from his extensive curriculum, Juan participated as a Spanish delegate in the Second United Nations Conference on the Law of the Sea (UNCLOS) (Geneva, 1975); in the first Spanish Antarctic campaign (1986); in the study of the fixed link between Morocco and Spain across the Strait of Gibraltar; in many precursor studies on marine protected areas; in the achievement of the 3D bathymetry of the Spanish continental shelf; and in the study of the birth of a new volcanic island (El Hierro, 2011). In the preparation of the FORMED research project, Juan proposed the compilation of this atlas and he has actively collaborated in editing it even after his retirement in 2014. Above all, however, we want to emphasize the pleasure of working with Juan at the office, in oceanographic surveys and in meetings, due to his positive thinking and his friendliness. He has always made things easier for the people around him.

Gracias amigo!

Contents

Part VII Slope and Deep Bedforms

Introduction to the Atlas

Bedforms are depositional morphologies generated by the interaction between a mobile bottom and the force induced by a fluid. The study of bedforms was first undertaken in the field of aeolian and river geomorphology and subsequently has spread to the coastal zone and the marine environment in general. The technological revolution brought about by the use of multibeam echo-sounders, with swath-bathymetric data coverage and previously unimaginable precision, has transformed our small-scale view of the seabed; whereas we initially considered seafloor bottoms to be predominantly flat, we now know that bedforms are ubiquitous throughout marine environments. In parallel to this technological and scientific development, we are changing our perception of the sediment dynamics of the marine environment: significant sediment movement near the bottom was previously thought to be confined to very shallow depths but is now known to occur throughout the continental margin and even in deep basins. This conceptual "revolution" has occurred in a relatively short period of time (30 years) and has led to the incorporation of new terminology for describing bedforms and sediment transport mechanisms and to a great increase in knowledge of the mechanisms that generate and/or maintain bedforms.

Bedforms can be composed of gravel, sand or mud and of siliciclastic or carbonate sediment. They have a wide range of spatial and temporal scales. In shallow water, the common types are ripples, megaripples and nearshore bar systems with a variety of morphological characteristics (shore-parallel, crescentic, etc.). On the continental shelf, the forms described include shoreface-connected ridges, sorted bedforms, sand banks, sand waves, sand ridges, dunes and ribbons. Finally, on prodeltas, on the outer continental shelf and slope, on submarine canyons and in deep areas of the ocean, the presence of large-scale sediment waves and cyclic steps is reported.

The profusion of bedforms has led to various attempts in classifications based on morphological, sedimentological or genetic criteria, although a generic classification that covers all sedimentary environments is still missing. Among the most commonly used classifications are that of Ashley (1990) on large-scale flow-transverse sandy bedforms in fluvial and shallow marine sediment and that of Wynn and Stow (2002) on sediment waves in deep water regions. The basic factors traditionally considered to determine the type of bedform are sediment size and current intensity. Several diagrams containing sediment size, bedform morphology and current intensity have been proposed for both shallow (Allen 1984; Middleton and Southard 1984; Boguchwal and Southard 1990) and deep (Stow et al. 2009) waters. These diagrams are preferably used to interpret the hydrodynamic conditions of relict structures and the geological record, because two parameters allow the third one to be deduced (for example, the intensity of the current can be deduced from the morphology and sediment). However, these diagrams include only certain types of bedforms (usually transverse) and are excessively simple. Additionally, the use of cross-bedding stratification to the classification and interpretation of bedforms provides a useful tool for studying bedforms in the geological record (Rubin 1987). Interpretation of the genesis of bedforms is often complex: in addition to near-bottom currents, other mechanisms such as waves, internal waves and density flows can be involved. A dimensionless parameter combining terms such as speed, height and wave period, sediment density and grainsize, fluid density and viscosity, and bottom slope may well be more appropriate (Fredsoe and Deigaard 1992). In the proper interpretation of bedform formation, an important additional question is which of the observed morphologies are dynamic at present and which are relict features inherited from past environmental conditions.

Interestingly, the morphology of bedforms is basically the same in all marine environments and, except for the effect of gravity waves, which is restricted to the shallower area, the mechanisms proposed to explain the formation of bedforms are quite similar from shallow to deep water areas, suggesting the existence of common processes of interaction between the fluid and the mobile seabed. In an initial approximation, we can consider that the main groups of mechanisms that can generate or maintain bedforms in different marine environments are the following: (1) waves, tides and wind-induced currents (e.g. Swift et al. 1978; Van Dijk 2005; Li and King 2007); (2) thermohaline oceanic currents, contour currents and other specific bottom currents (e.g. Wynn and Stow 2002; Masson et al. 2004); (3) density flows (e.g. cascading) and turbidity currents (e.g. Trincardi and Normark 1988; Lee et al. 2002); and (4) internal waves (e.g. Flood 1988; Puig et al. 2007; Ribó et al. 2016). However, knowledge of the sediment transport processes associated with the generating mechanisms of bedforms is incomplete, and the predictive ability of the models is limited (Davies and Thorne 2008). While we can make appropriate predictions of small-scale bedforms such as ripples based on the size of the sediment and statistics of the current intensity near the bottom (e.g. Nielsen 1992; Wiberg and Harris 1994), these relationships are more ambiguous when it comes to larger bedforms (Gallagher et al. 2005). Currently, the mechanisms that generate shallow bedforms are interpreted as self-organization processes (Coco and Murray 2007).

The study of bedforms is important for the advancement of knowledge of bottom boundary-layer processes, such as sediment transport and variations in the circulation system caused by the changes in the roughness of the seabed. However, bedforms can also be used as indicators of the position of ancient shorelines, to infer the hydrodynamic processes that have taken place in the past, and to interpret the geological record (even on other planets!). Many bedforms have developed above larger sedimentary bodies and they are used as a criterion for inferring whether these bodies are dynamic. The implications of this dynamism on biogeochemical cycles and habitat distribution still deserve focussed research with new technological advances. From the application point of view, bedforms can create specific habitats that must be known in order to optimize the management of the marine environment and, as some of them are often exploited by fishing activities. Some bedforms may also offer exploitable mineral resources (e.g. sand deposits for beach nourishment) or have applications in the oil industry as analogues of possible reservoir rock. Finally, the dynamism of bedforms may have impacts on coastal and offshore structures such as harbours, estuaries and offshore wind farms.

Advances in our knowledge of bedforms have mainly resulted from a huge effort to obtain high-resolution bathymetric records and boundary-layer hydrodynamic and sediment transport measurements over the past decades. We must now collect and analyse in depth the existing information on a regional scale, provide an overview of the morphology of bedforms and identify gaps in our knowledge of the formation mechanisms. This book fits into this goal. The inventory of the bedforms on the Spanish Mediterranean continental margin was one of the objectives of the FORMED research project (CGL2012-33989). We asked our colleagues to collaborate in the preparation of this atlas by providing their expertise on specific topics or examples of bedforms from the western Mediterranean Sea. The aim of this book is to illustrate the characteristics of present-day bedforms, from the shoreline to deep-sea environments, and it also includes short reviews of the main mechanisms that generate such bedforms. The atlas is aimed at the research community, in addition to students, the public at large and companies with interests in the marine environment. The book is divided into seven parts composed of a number of short chapters: (I) Bedform Analysis and Main Physical Processes (eight chapters), (II) Bedforms in the Coastal Zone (eight chapters), (III) Bedforms in Prodeltas and Sorted Bedforms (seven chapters), (IV) Bedforms in the Continental Shelf (six chapters), (V) Bedforms and Benthos (four chapters), (VI) Bedforms in Submarine Canyons (six chapters) and (VII) Slope and Deep-sea Bedforms (seven chapters). We hope that the contents of the book can give the reader a comprehensive, though not exhaustive, view of the diversity of bedforms and associated processes and of the morphological and temporal scales in the enclosed tideless western Mediterranean Sea.

This book would not have been possible without the support and cooperation of all the authors. We especially appreciate the work of reviewers, who greatly improved the original manuscripts.

Jorge Guillén
Juan Acosta
Francesco Latino Chiocci
Albert Palanques

References

Allen, J.R.L. (1984). Principles of Physical Sedimentology. George Allen&Unwin, London, 272 pp.

Ashley, G.M., Chairperson and others. (1990). Classification of large-scale subaqueous bedforms: a new look at an old problem. SEPM bedforms and bedding structures. Journal of Sedimentary Petrology, 60 (1): 160–172.

Boguchwal, L. A. and Southard, J.B. (1990). Bedform configurations in steady unidirectional water flows. Part 1: Scale model study using fine sands. Journal of Sedimentary Petrology 60(5): 649–657.

Coco, G. and Murray, A.B. (2007). Patterns in the sand: From forcing templates to self-organization. Geomorphology, 91: 271–290.

Davies, A.G. and Thorne, P.D. (2008). Advances in the study of moving sediments and evolving seabeds. Surv. Geophys. 29: 1–36.

Flood, R.D. (1988). A lee wave model for deep-sea mudwave activity. Deep-Sea Research, 35(6): 973–983.

Fredsoe, J. and Deigaard, R. (1992). Mechanics of coastal sediment transport. Advanced series on ocean engineering, vol. 3, World Scientific, Singapore, 369 pp.

Gallagher, E.L., Elgar, S., Guza, R.T. and Thornton, E.B. (2005). Estimating nearchore roughness with altimeters. Marine Geology, 216: 51–57.

Lee, H.L., Syvitski, J.P.M., Parker, G., Orange, D., Locat, J., Hutton, E.W.H. and Imran, J. (2002). Distinguishing sediment waves from slope failure deposits: field examples, including the "Humboldt slide", and modelling results. Marine Geology, 192: 79–104.

Li, M.Z. and King, E. (2007). Multibeam bathymetric investigations of the morphology of sand ridges and associated bedforms and their relation to storm processes, Sable Island Bank, Scotian Shelf. Marine Geology, 243: 200–228.

Masson, D.G., Wynn, R.B. and Bett, B.J. (2004). Sedimentary environmet of the Faroe-Shetland and Faroe Bank Channels, north-east Atlantic, and the use of bedforms as indicators of bottom current velocity in the deep ocean. Sedimentology, 51: 1207–1241.

Middleton, G.V and Southard, J.B. (1984). Mechanics of sediment movement. SEPM Short Course No 3, 246pp.

Nielsen, P. (1992). Coastal bottom boundary layers and sediment transport. Advanced series on ocean engineering, vol. 4, World Scientist, Singapore, 324 pp.

Puig, P., Ogston, A.S., Guillén, J., Fain, A.M.V. and Palanques, A. (2007). Sediment transport processes from the topset to the foreset of a crenulated clinoform (Adriatic Sea). Continental Shelf Research, 27: 452–474.

Ribó, M., Puig, P., Muñoz, A., Lo Iacono, C. Masqué, P., Palanques, A., Acosta, J., Guillén, J., Gómez Ballesteros, M. (2016). Morphobathymetric analysis of the large fine-grained sediment waves over the Gulf of Valencia continental slope (NW Mediterranean). Geomorphology, 253: 22–37.

Rubin, D. M. (1987). Cross-Bedding, Bedforms, and Paleocurrents, Volume 1 in a new SEPM series, Concepts in Sedimentology and Paleontology vii + 187 pp. Oklahoma: SEPM.

Stow, D.A.V., Hernández-Molina, F.J., Llave, E., Sayago-Gil, M., Díaz del Río, V. and Branson, A. (2009). Bedform-velocity matrix: The estimation of bottom current velocity from bedform observations. Geology, 37(4): 327–330.

Trincardi, F. and Normark, W.R. (1988). Sediment waves on the Tiber prodelta slope. Geo-Mar. Lett., 8: 149–157.

Swift, D.J.P., Parker, G., Lanfredi, N.W., Perillo, G. and Figge, K. (1978). Shoreface-connected Sand Ridges on American and European Shelves: A comparison. Estuarine and Coastal Marine Science, 7: 257–273.

Van Dijk, T.A.G.P. (2005). Processes controlling the dynamics of compound sand waves in the North Sea, Netherlands. J. Geophys. Res., 10, F04S10.

Wiberg, P.L. and Harris, C. (1994). Ripple geometry in wave-dominated environments. Journal of Geophysical Research, 99, C1: 775–789.

Wynn, R.B. and Stow, D.A.V. (2002). Classification and characterization of deep-water sediment waves. Marine Geology, 192: 7–22.

Part I

Bedform Analysis and Main Physical Processes

Bedform Mapping: Multibeam Data Processing, Metadata and Spatial Data Services

1

S. Diez and J. Sorribas

Abstract

Multibeam data processing is essential for bedform mapping because the resulting bathymetric map is the basis for the characterization of bedforms and for further analysis and interpretations. The existence of a large amount of spatial data in general and swath bathymetry data in particular requires the creation of metadata and spatial data services that should make it possible to find, view and download data in order to share them at various levels, meeting international standards following the INSPIRE Directive (2007).

Keywords

Multibeam • Echo sounder • Bathymetry • Backscatter • Metadata • Spatial data services • Standards • INSPIRE

1.1 Introduction: Multibeam Data Processing

Acoustic mapping systems help us to improve our knowledge of the morphology and nature of the seafloor. Among these acoustic techniques, multibeam echo sounders have undergone a revolution during the last 30 years and are the most significant advance in bedform mapping. This technology provides accurate and extensive measurement of seafloor depth, which, combined with detailed positioning information (acquired through modern GPS navigation systems and motion reference units), can be processed in order to get the swath bathymetry that is essential for characterizing bedforms to a great level of detail.

Multibeam systems produce a swath of soundings covering the seafloor up to 5 times the water depth. To obtain a complete cartography of the seafloor, the echo sounder scans adjacent swaths, drawing up a mosaic of seafloor topography and mapping large areas rapidly and accurately.

The resolution of the systems increases with frequency, but so does the attenuation in the water, so the working depth of the echo sounder is crucial for the resolution of the map. New technologies, such as autonomous underwater vehicles and remotely operated vehicles, can be used to install a multibeam echo sounder working with high frequencies near the seafloor and providing very high-resolution data (centimetres).

Multibeam data need to be processed to get a bathymetry map where bedforms are recognized. The first step of the bathymetry processing flow is to import the raw data into a dedicated software in order to get navigation, depth for each sounding and complementary data such as attitude data (motions of the vessel or platform). After navigation post-processing, which mainly consists of an automatic filtering and interpolation of anomalous values, navigation and sounder depth are merged and a georeferenced sounding file (x, y, z) is generated. A digital terrain model (DTM) is first interpolated from raw soundings to search for errors. Despite accurate data acquisition, some errors persist and it is often difficult to distinguish the source of these errors. The most

S. Diez (✉) · J. Sorribas
Unidad de Tecnología Marina (CSIC), Passeig Marítim de la Barceloneta 37-49, 08030 Barcelona, Spain
e-mail: sdiez@utm.csic.es

J. Guillén et al. (eds.), *Atlas of Bedforms in the Western Mediterranean*,
DOI 10.1007/978-3-319-33940-5_1

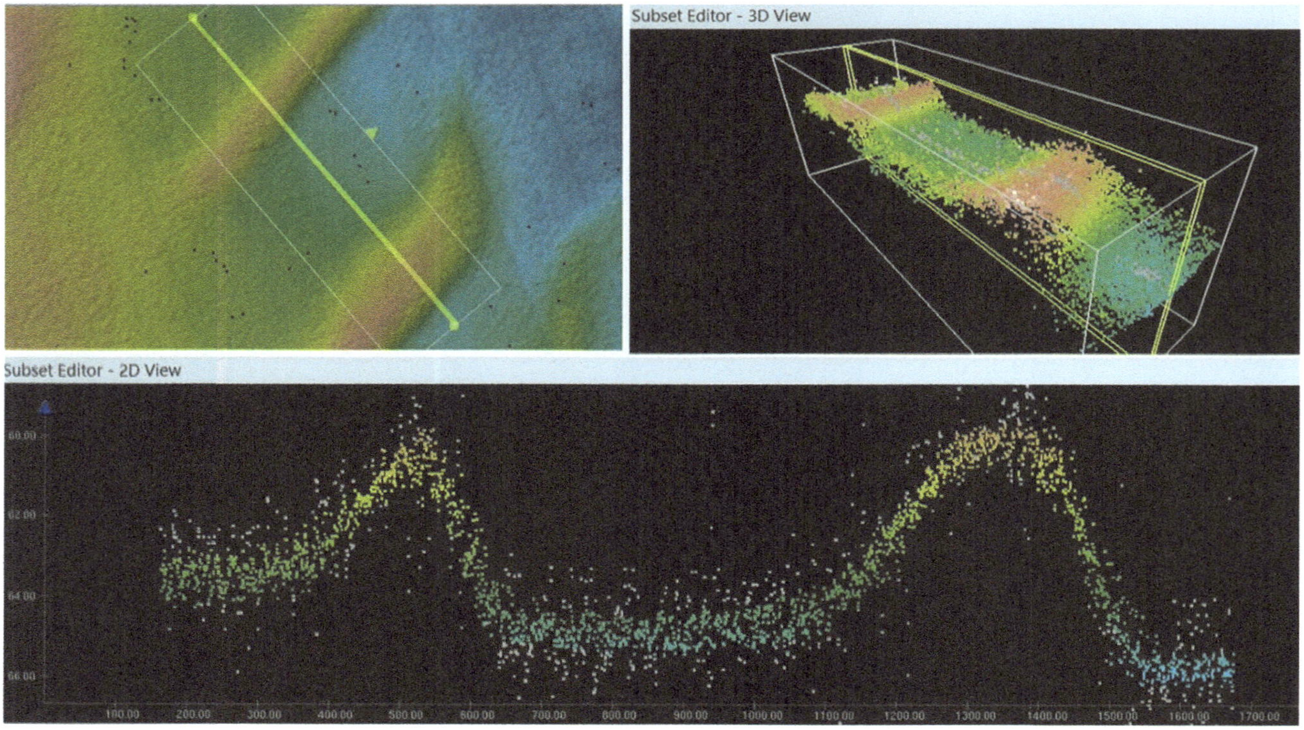

Fig. 1.1 CARIS HIPS&SIPS Subset Editor: the 2D View displays soundings from the subset area created (*white bounding box*) that are within the 2D *yellow slicing box*. Invalidated soundings appear in *white colour*. As the *yellow box* moves through the surface, a manual data cleaning can be performed

common are the effect of tide for shallow waters, variations in sound velocity profiles applied, and calibration values of roll, pitch and yaw. Once the corresponding corrections are introduced, data are filtered and cleaned up using statistical methods and manually cleaning to get more accuracy in the control of depth data (Fig. 1.1). After filtering, bathymetric data are interpolated at nodes of a regular spacing grid in order to get a final DTM. The choice of the grid node spacing takes into account the density of the soundings and the quality of the data and will define the final resolution of the product offered to the users.

The final DTM can be displayed under different cartographic views, such as shaded relief maps and 3D visualizations. Moreover, final products such as isocontour maps, slope maps and profiles can be generated in order to characterize bedforms. In addition, bathymetries that have been acquired in different surveys can be compared to study the temporal evolution of bedforms.

Multibeam echo sounders provide us with other complementary information: the seafloor backscatter, i.e. the amount of acoustic energy being received by the sounder after the acoustic signal is returned by the seafloor. Reflectivity varies according to the angle of incidence, the seabottom local topography and the nature of the seafloor (Blondel and Murton 1997). The main goal of the reflectivity processing is to obtain a digital image mosaic generated from the seabed backscattered strength measured in dB. After data importation and raw data radiometric corrections such as *angle varying gain, gain correction* and *beam pattern correction,* the mosaic is prepared, including bathymetric and navigation values and accurate positioning of the reflectivity pixels (Augustin et al. 1994). The creation of mosaics of reflectivity can be useful for identifying variations in grain size of the sediments along bedforms. The final mosaic can be not only visualized in 2D but also draped on top of three dimensional bathymetry.

An integration of different kinds of data can be done in geographic information system applications, which help us to manage all the information that has been collected and processed. Interoperability between layers (not only multibeam data) is a powerful tool for modelling, analysing and visualizing bedforms.

1.2 Metadata

A large amount of multibeam data is available, and if we want datasets to be reused, integrated with other datasets and even understood, the creation of metadata is crucial.

Metadata are data describing information about a dataset. There are many uses for metadata. Metadata allow us to understand and reuse data, and having access to searchable metadata allows reliable information to be shared. A metadata record for a dataset may include its geographic coverage and time period, the source or person responsible for it, the dataset format, special characteristics, access and usage constraints and distribution information when the dataset is available, the spatial reference, and any relevant supplemental information. Metadata can also provide us with information about the data acquisition, the instrumentation involved and the boundary conditions of the data acquisition process (state of the sea, weather, etc.). This information is crucial to better understand the reliability of the data, going beyond the conception of data quality and providing enough subjective elements to qualify the data sets, but it is also a good way to provide a label of confidence for second-generation products, such as DTMs, that have lost the real values of the seabed that our instrumentation recorded. For example, having a description of the maximum spatial resolution of the echo sounder used to "see" the seafloor could help us to know whether the final map produced using a specific DTM is below or above the expected resolution.

Since such metadata should be compatible and usable in a global community, it is necessary to lay down conventions, vocabularies and standards concerning the metadata elements and terms used to describe the spatial data sets (Commission Regulation 2008). An example of metadata standards, in terms of data, is the one adopted for the European Marine Observation and Data Network (EMODnet) Bathymetry Portal from SeaDataNet metadata standard ISO 19115 (Commission Infrastructure 2009). An example of metadata standards used to describe the data acquisition process and instrumentation involved is the Sensor Web Enablement standard family from the Open Geospatial Consortium (OGC). More specifically, the SensorML standard is very well suited to describing any sensor, instrument and logical system involved in any data acquisition process (Sorribas et al. 2012).

1.3 Spatial Data Services

Spatial data services refer to the operations which may be performed via an internet interface on the spatial data contained in spatial data sets or on their metadata. The main services are discovery, view and download services (Fig. 1.2).

A discovery service consists of a *Catalogue Service for Web*, which makes it possible to quickly and easily search for spatial data based on corresponding metadata stored in a dataset. These services allow users to identify which spatial data are available and who holds the data.

A view service is a *Web Map Service* which provides a simple HTTP interface for requesting map images from a distributed geospatial database. A Web Map Service request defines the geographic layer(s) and area of interest to be processed. The response to the request is a spatially referenced image (returned as TIFF, JPEG, etc.) that can be displayed in a browser application. The interface also supports the ability to specify whether the returned images should have some degree of transparency so that layers from multiple servers can be combined.

Download services provide an interface allowing requests for geographical information. A *Web Feature Service* allows downloading of vectorial data (such as a shapefile containing bedform footprints) and a *Web Coverage Service* enables downloading of coverages, which in the case of bedform mapping may be a DTM.

All these services are developed according to the OGC standards created for use in World Wide Web applications.

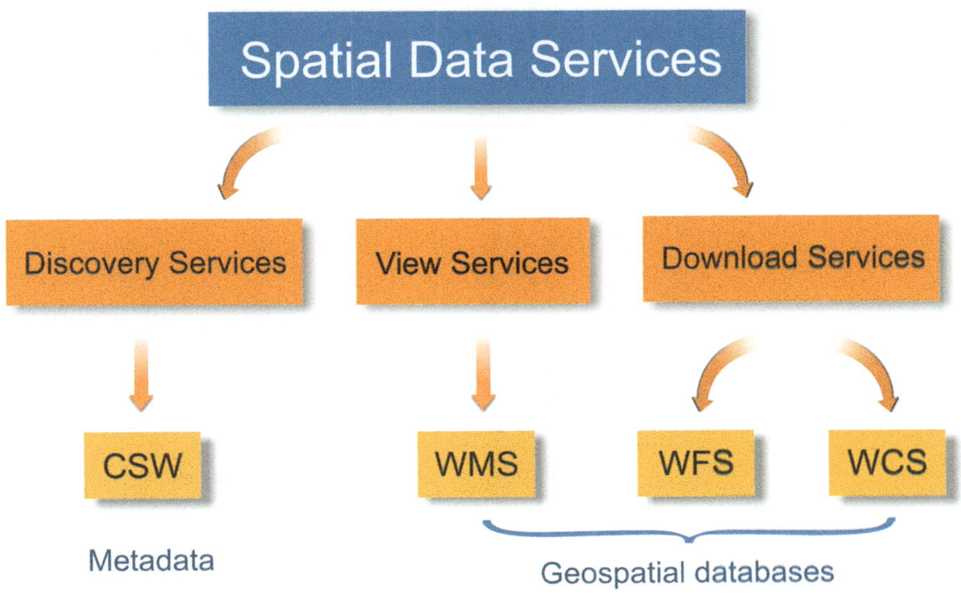

Fig. 1.2 Main spatial data services, in terms of utility, for metadata and spatial data sets: Catalogue Service for Web, Web Map Service, Web Feature Service and Web Coverage Service

1.4 The INSPIRE Directive

Problems regarding the availability, quality, organization, accessibility and sharing of spatial information are usual. Solving these problems requires measures that address exchange, sharing, access and use of interoperable spatial data and spatial data services. An infrastructure for spatial information, INSPIRE, was therefore established to provide the legal framework for the measures (INSPIRE Directive 2007). The INSPIRE Directive aims to create a spatial data infrastructure to enable the sharing of environmental spatial information among public sector organizations and to better facilitate public access to spatial information across Europe. The components of this infrastructure include metadata, spatial data themes (as described in Annexes I, II, III of the Directive), network services and technologies; agreements on data sharing, access and use; and coordination and monitoring mechanisms, processes and procedures (Technical Guidance for INSPIRE 2014).

To ensure that the spatial data infrastructures are compatible and usable in a trans-boundary context, the Directive requires that common implementing rules to be adopted for metadata and spatial data services so that certain standards, such as the OGC standards, are met (OGC Market Report 2012). A significant example of a spatial data infrastructure that meets the INSPIRE Directive is the EMODnet Bathymetry Portal, an infrastructure for discovery, management and delivery of bathymetry data (Commission Infrastructure 2009).

1.5 Conclusions

Bathymetric maps obtained from multibeam data acquisition and processing are essential for the characterization of bedforms. Technological advances in echo sounders, in the platforms where they are installed and in the processing software are leading to more efficient mapping of bedforms (and with higher resolutions).

The creation of spatial data services for metadata and datasets that meet international standards following the INSPIRE Directive enables the sharing of reliable information. This sharing is crucial not only for avoiding data duplication but also for having temporal sequences of data, since the seabed is a dynamic environment, especially in shallow areas of instability or significant currents.

References

Augustin J.M., Edy, C., Savoye, B., Le Drezen, E. (1994). Sonar mosaic computation from multibeam echo sounder. Oceans'94, Vol2, 433-438.

Blondel, Ph. and Murton, B.J. (1997). Handbook of Seafloor Sonar Imagery, PRAXIS-Wiley and Sons: Chichester (UK), 314 pp.

Commission Infrastructure: Building a European marine knowledge infrastructure: Roadmap for a European Marine Observation and Data Network. COMMISSION OF THE EUROPEAN COMMUNITIES, Brussels, 7.4.2009.

Commission Regulation (EC) No 1205/2008 of 3 December 2008 implementing Directive 2007/2/EC of the European Parliament and of the Council as regards metadata (Text with EEA relevance) OJ L 326, 4.12.2008, 12 pp.

INSPIRE DIRECTIVE 2007/2/EC OF THE EUROPEAN PARLIAMENT AND OF THE COUNCIL of 14 March 2007 establishing an Infrastructure for Spatial Information in the European Community (INSPIRE), OJ L 108, 25.4.2007, p. 1.

OGC Market Report: Open Standards and INSPIRE. Open Geospatial Consortium (OGC), 25 April 2012, 32 pp.

Sorribas, J.,Olive, J. Diviacco, P., Bermudez, L. Representing oceanographic vessels by means of sensor web enablement standards. SWE'12 Proceedings of the First ACM SIGSPATIAL Workshop on Sensor Web Enablement.Pages 1-8.ACM New York, NY, USA ©2012. ISBN: 978-1-4503-1701-6.

Technical Guidance for INSPIRE Spatial Data Services and services allowing spatial data services to be invoked. (2014) European Commission, Technical Guidelines, TG_for_INSPIRE_SDS_3.1.-docx 17.12.2014, 100 pp.

Methods for Analysing Bedform Geometry and Dynamics

2

Thaiënne A.G.P. Van Dijk and Roderik C. Lindenbergh

Abstract

Rhythmic bedforms of different spatial scales are common in the marine environment. In shallow seas, the monitoring of bedforms is important because changes in morphology may interfere with offshore infrastructure and navigation. In addition, investigating bedform dynamics improves our understanding of the processes that cause their behaviour. Quantified bedform characterization also contributes to the validation of morphodynamic models. Modern, high-precision and high-resolution bathymetric data enable the detailed analysis of bedforms. Several semi-automated methods (e.g. geostatistical and spectral techniques) have been developed to quantify the geometry (size and shape) and dynamics (growth, change in shape and migration) of subaqueous bedforms. An overview of these different approaches is given and differences in use and potential are described.

Keywords

Bedform analysis • Spectral analysis • Kriging • Bedform geometry • Sand wave dynamics • Review

2.1 Introduction

Rhythmic bedforms are a common feature in marine environments in either contemporary or relict forms (Allen 1980; Ashley 1990; Albarracín et al. 2014). Bedforms of different scales may be superimposed; for example, megaripples (wavelengths *O* tens of metres) may overlie sand waves (wavelengths *O* hundreds of metres), which in turn may overlie sand banks (spacings *O* of kilometres) (see left panel of Fig. 2.1). Each type of bedform changes at a different temporal scale, causing a complex dynamic behaviour of the seabed (Ernstsen et al. 2006; Van Dijk et al. 2008). Determining bedform geometry (size and shape) and dynamics (growth, change in shape and migration) is important for understanding the spatial variation in seabed dynamics. Insight into the changing geometry, growth and migration of bedforms contributes to the understanding of the processes that drive bedform dynamics under local environmental conditions (Van Dijk and Kleinhans 2005; Van Santen et al. 2011) and quantified results are important for validating morphodynamic models. In shallow waters, the study of bedforms is also important for safe navigation and offshore engineering. For instance, the growth of sand waves at critical water depths increases the grounding danger for ships on shallow marine traffic routes (Dorst 2009; Dorst et al. 2011), and sand wave migration may affect the stability of offshore wind turbines and pipelines (Morelissen et al. 2003; Santoro et al. 2004).

Bedform monitoring started with single-beam echo sounding, but determining dynamics was still problematic due to the low precision in horizontal positioning and low

T.A.G.P. Van Dijk (✉)
Department of Applied Geology and Geophysics, Deltares, Utrecht, The Netherlands
e-mail: t.a.g.p.vandijk@utwente.nl

T.A.G.P. Van Dijk
Department of Water Engineering and Management, University of Twente, Enschede, The Netherlands

R.C. Lindenbergh
Department of Geosciences and Remote Sensing, Delft University of Technology, Delft, The Netherlands

J. Guillén et al. (eds.), *Atlas of Bedforms in the Western Mediterranean*,
DOI 10.1007/978-3-319-33940-5_2

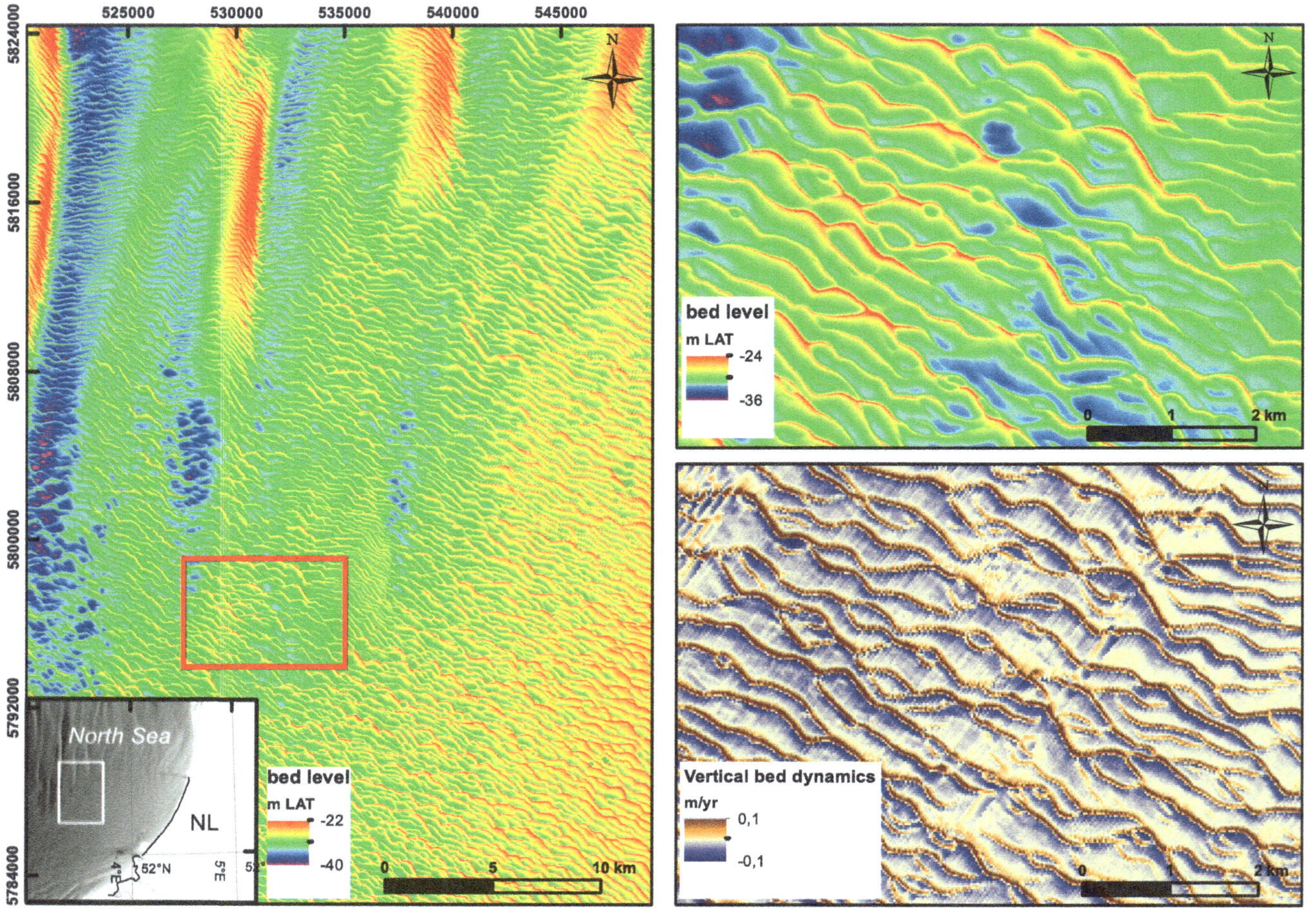

Fig. 2.1 *Left panel* Offshore sand waves superimposed on sand banks on the Netherlands Continental Shelf in the Southern Bight of the North Sea (UTM31N WGS84 coordinates). The map comprises 5-m (where available) and 25-m resolution overlays with bed levels in meter below LAT. *Top right* Zoom of rhythmic bedforms (5-m resolution grid). *Bottom right* Quantified vertical dynamic trends (m/y) per grid node

spatial resolution of the data (McCave 1971; Terwindt 1971; Lanckneus and De Moor 1991). The spatial resolution of modern multibeam echo soundings is up to several tens of observations per m^2 (Lurton 2002). The vertical precision of the echo-sounding data essentially has not increased, still being around 0.1–0.5 m, depending on the applied survey requirements (IHO 2008), the beam angle and the quality of the post-measurement corrections, such as tidal reduction (Hughes Clarke 2012). In shallow and clear water, bedforms can also be monitored using bathymetric LiDAR (Galparsoro et al. 2010) at a comparable quality.

With the progress in survey techniques, the analysis of bedform geometry and migration rates also developed. From manual measurements on plotted maps, present-day methods of bedform analysis have progressed to quantifying digital, high-resolution spatial data in automated or semi-automated ways. The review given in this chapter focuses on methods for analysing marine bedforms (for a review on aeolian dunes, see Hugenholtz et al. 2012). The methods described are applicable to all types of rhythmic bedforms and are not specific to the Mediterranean, being employed in different areas of the world.

2.2 Review of Methods

Methods for analysing bedforms vary from analysing regional morphological patterns to individual bedforms and to the spatial and quantitative analysis of vertical seabed elevation (*dz/dt*) from digital elevation models (DEMs).

2.2.1 Bedform Orientation and Wavelength

The dominant orientations and wavelengths of bedforms in an area can be characterized by a *variability analysis*, in which anisotropic, experimental variograms display the lengths in the *x*- and *y*-directions (Goovaerts 1997; Dorst

2004; Pluymaekers et al. 2007). Basically this means that depth variations in different directions are assessed and averaged. For a regular bedform field, the variability along the crest direction is expected to be low, in contrast to the direction perpendicular to the crests. The variogram analysis allows these directions to be detected in an automated way and allows the length of the bedform to be estimated. Alternatively, a *2-D spectral* (*Fourier*) *analysis* returns the dominant orientations and wavelengths, also in the *x*- and *y*-directions (power-frequency in the wavenumber domain, *kx-ky*) (e.g. Van Dijk et al. 2008; Lefebvre et al. 2011; Cazenave et al. 2013). These methods can be used to infer dominant orientations and wavelengths of distinct groups of bedforms in an area, such as sand banks, sand waves and megaripples. However, the geometries and dynamics of individual bedforms are not revealed.

2.2.2 Superimposed Bedforms and Identification of Crests and Troughs

For the 1-D and 2-D separation of superimposed bedforms into different components that each represent a bedform of particular length scale, Van Dijk et al. (2008) compared two well-known, automated signal-processing methods for bathymetric data derived from multibeam echo soundings. The application of *Factorial Kriging* and a *Fourier decomposition* to two case studies of the North Sea shows that both methods are successful and that the results correspond well. Using Kriging interpolation at an appropriate spatial scale, first a surface is estimated through the observations at the scale of the largest bedforms, in this case sand waves. By subtracting the sand wave surface from the original surface, a residual surface is created that represents bedforms at a smaller scale (e.g. megaripples). Consecutively, crest and trough points of sand waves and megaripples are identified by simple along-profile identification of local extrema in the respective surface.

The 2-D spectral method (discrete Fourier transform, DFT), using densely spaced profiles in two directions, separates bedforms by truncating the Fourier series at frequencies discriminating the different types of bedforms (Van Dijk et al. 2008). Crest, trough and inflection points of the sand wave or megaripple signals are identified semi-automatically, using second derivatives in the 1-D method and curvatures in the 2-D method. These are then used to quantify the geometrics and—when applied in time series—dynamics. The spectral method of Van Dijk et al. (2008) may also identify brink points and toe points, as defined in Allen (1968). Duffy (2012) identified crest and trough points by row-by-row *scanning* of a surface (DEM). The scanning must be done on a rotated DEM, since most depth variation occurs in the direction perpendicular to the sand wave crests. Therefore, this approach is most suitable for 2-D straight-crested sand waves. For the identification of crest points, curvature was also used, and for trough points the steepest descent method was used to overcome less regular cross-sectional shapes of bedforms in the Bay of Fundy, Canada. This method adds to previous methods by correcting locations of crest and trough lines to the unsmoothed bathymetric data, and by quantifying all bedforms in geographical space. The latter allows geographic distribution and morphometric relationships to be presented in histograms and scatter plots.

Cazenave et al. (2013) used a Fourier decomposition to analyse orientation and wavelength, as described by Van Dijk et al. (2008), and added to the method by calculating adjusted heights from the 2-D wavenumber domain, which corrects for underestimation due to spectral leakage in the DFT method, and by comparing the method performance with a synthetic dataset. Both Van Dijk et al. (2008) and Cazenave et al. (2013) present methods for spatial representation of propagated error estimates within the different steps of the spectral analyses.

A 1-D Fourier analysis was also applied by Winter and Ernstsen (2007) to separate superimposed smaller dunes of compound bedforms in a tidal inlet in Denmark. Knaapen et al. (2005) used a low-pass Manning filter to separate bedforms and to map crests and troughs of megaripples. Knaapen et al. (2001) earlier used a Fourier analysis and discovered a new type of bedform: 'long bed waves', with a wavelength and orientation between those of sand banks and sand waves.

Another known method for a frequency analysis of quasi-harmonic forms is a *wavelet transform*, which can deal with more irregular records and has also been applied to bedforms (e.g. Cataño-Lopera and Garcia 2006; Cataño-Lopera et al. 2009; Fraccascia et al. 2011). Although a wavelet transform is less prone to the effects of sample size, its performance relies on the choice of wavelet, depending on the size and shape of the investigated bedform (Cazenave et al. 2013). Moreover, a discrete Fourier transform allows for 2-D analyses, whereas wavelet transform is a 1-D technique, and a Fourier analysis results in good representations of heights and lengths (Cazenave et al. 2013).

2.2.3 Bedform Dynamics

The shift of bedform patterns can be quantified using a spatial *cross-correlation analysis* (e.g. Duffy and Hughes Clarke 2005; Buijsman and Ridderinkhof 2008a, b; Van Dijk and Egberts 2008; Franzetti et al. 2013). This technique calculates the maximum correlation of two 3-D surfaces (DEMs) in time, by shifting the earlier surface one grid cell at a time and adding up the nodal products of the overlapping surfaces for each shift. The correlation coefficients provide the strength of the correlation. This analysis results in a local migration vector, which therefore has a distance and direction of displacement, or a full migration vector field. The differences in the application of this method in the literature largely comprise the choices of (i) the analysed areas and (ii) pre-imposed restrictions, such as limiting the distance or direction of displacement. Duffy and Hughes Clarke (2005) compared three different cross-correlation techniques: that of maximum correlation (free in distance and direction but limited to the grid resolution), weighted centroid (allows for oblique migration) and least distance to regression line (thereby accepting an assumed migration normal to the crests), all applied to sand waves on a banner bank offshore of New Brunswick, Canada. Similarly, Buijsman and Ridderinkhof (2008a, b) created a vector field for sand waves in the Marsdiep, a tidal inlet of the Wadden Sea, Netherlands, using the maximum correlation technique (i.e. method 1 in Duffy and Hughes-Clarke 2005). Van Dijk and Egberts (2008) applied the maximum correlation technique to larger patterns (including several sand waves) in order to increase pattern recognition abilities and to minimize a priori restrictions. Their result was merely one regional migration vector. They combined the cross-correlation technique with a Fourier analysis on profiles in the cross-correlation direction, thereby accurately analysing the maximum migration rates of individual sand waves. Their finding that maximum migration rates in the cross-correlation direction are significantly higher than migration rates normal to the bedform crests implies that the assumption of crest-normal sand wave migration made by most morphodynamic modellers and in the regression method of Duffy and Hughes-Clarke (2005) is invalid. Franzetti et al. (2013) used the weighted centroid method, resulting in a vector field for giant sand waves at the Banc du Four, Brittany, France.

Furthermore, geodetic *deformation analysis* was used to analyse the geometry and dynamics of bedforms (Lindenbergh 2004; Dorst et al. 2009, 2011). A deformation analysis estimates parameters from relatively simple kinematic models, while explicitly incorporating measurement uncertainty and model deficiencies. In this method, the seabed morphology is considered as a composition of different types of bedforms. A drawback is that it works with idealized morphology and kinetics (for example, regularly moving sinusoidal sand waves), although any misfit is reported in the corresponding quality analysis. A comparison of the deformation analysis results with Fourier analysis results (Van Dijk et al. 2011) showed that wavelengths correspond well (heights were not specified by Dorst et al. 2011), but also that migration rates at one location with back-and-forth migration were either estimated to be zero or not detected by the deformation analysis method.

A different approach of analysing seabed dynamics is to calculate nodal vertical dynamics (*dz/dt*). Previously, calculating differences in bed elevations between two DEMs (in metres) was simply done in GIS software. However, to deal with large numbers of surveys of different extents, varying periods of data acquisition and different periods between surveys, a *vertical dynamic trend* (in metres/year) would be a better measure of quantified seabed dynamics (see Fig. 2.1, bottom right panel). This fully automated analysis was recently performed on the Netherlands Continental Shelf (Van Dijk et al. 2011, 2012a, b), now also taking into account areas of human interference, such as dredging (Van Dijk et al. 2014). All datasets from the late 1980s to 2010 for the entire shelf were gridded into DEMs of 25 m resolution. Knowing the periods of data acquisition per record, a rate of change (*dz/dt*) was calculated from the vertical stacks of bed elevations in time at each grid node. Other labels (metadata, precision, etc.) can potentially be attached to the nodal dynamics, taking into account the different sounding methods, precisions and resolutions. In a macro-scale analysis of the German Bight (Winter 2011), although presented as elevation differences in metres, the dynamic pattern corresponds very well to that of the Netherlands Continental Shelf. These methods are not directly aimed at analysing bedform geometry, but the vertical dynamic trends reveal growth or migration of bedforms by aggrading-degrading couplets in the pattern of bedforms.

2.3 Discussion

The methods for the analysis of bedform geometry and dynamics described here all have advantages and disadvantages in performance, dependency on assumptions and input parameters or in the "user-friendliness" of the software and/or procedures. In terms of performance, the geostatistical methods (Kriging, scanning) perform best when bedforms are two-dimensional in shape. The discrete Fourier transform for the separation of bedforms of different length scales performs best when distinct groups of frequencies represent the bedforms. Curvature for the crest, trough and inflection points works well when the cross-sections of bedforms are fairly sinusoidal, e.g. sand waves on continental shelves. However, for less regular cross-sections, e.g. isolated bedforms on a nearly flat bed, the steepest descent method would be more appropriate for troughs. The wavelet transform deals with irregular cross-sections, but is a 1-D method. Most methods (Kriging, DFT and wavelet) are dependent on well-chosen input parameters of wavelength and, in the case of a wavelet transform, on the choice of the wavelet applied. The scanning method also requires a priori insight into the orientation of the bedforms.

The scanning-analysing script of Duffy (2012) is available online from www.ria.ie/publications/journals/irish-journal-of-earth-sciences.aspx. Bathymetric DEMs of the Netherlands Continental Shelf in time series are available at Open Earth of Deltares (https://publicwiki.deltares.nl/display/OET/OpenEarth). The applications (*dz*/*dt* and 1-D DFT-methods) of Van Dijk et al. (2008, 2011) will become available at Open Earth as well.

Computational efficiency is becoming increasingly important for contemporary studies that are dealing with increasing amounts of high-resolution data in time series and more often include macro-scale investigations. Apart from Van Dijk et al. (2008), who specify that for N observations the 2-D DFT needs $O(N \log_2 N)$ calculations and the Kriging method needs a maximum of $O(N^3)$ calculations, not many authors have provided information on computational efficiency. In general, the efficiency increases with lower resolutions and with the use of pre-set limits such as the distance of migration in the cross-correlation technique. The *dz*/*dt* method is fully automated and deals in principle with an unrestricted number of surveys and observations in reasonable calculation times (e.g. just over one day for time series since the late 1980s for the entire Netherlands Continental Shelf, >80 GB), thereby providing several statistics of the time series per grid node (e.g. the number of surveys, minimum and maximum water depths, most recent bathymetric map, vertical dynamic trends and goodness of fit). However, the latter does not result in specific bedform quantification of individual bedforms.

A data-related issue when assessing bedform dynamics from bathymetric data from multiple epochs is the temporal sampling versus the temporal variation in bedform dynamics. Empirical results are net changes, thereby not specifying whether changes occurred during one single event, such as storms (Van Son et al. 2009), or at a continuous rate over time. In extreme cases this may lead to the conclusion that no dynamics took place, whereas in reality large changes during a high-energy event may have been compensated for by regular opposite changes between events. Short-term monitoring data of bedforms are scarce, but have revealed that, during one tidal cycle (Ernstsen et al. 2006), seasons (Buijsman and Ridderinkhof 2008a) and storms (Houthuys et al. 1994), the dynamics of bedforms are variable. Contemporary research (e.g. Van Dijk et al. 2014; SMARTSEA-project, 2014–2019) investigating the impact of storms on sand wave dynamics, therein combining modelling and empirical approaches, relates dredging to bedform dynamics and will try to incorporate precision estimates into spatial analyses supporting resurvey policies.

For quantified morphometrics and dynamics of bedforms to serve the validation of morphodynamic models, the same entities need to be analysed. For example, modellers often work with spacings between crests (horizontal equivalents), whereas marine geologists often define bedform length as the real distance between two troughs. The methods described in this chapter (e.g. Kriging, DFT and scanning) should be able to define these different values effortlessly.

To our insight, the next step in the spatial analyses of bedforms would be to add precision estimates. For example, propagated errors per grid node for the entire path from measurements to the last step in the method, including temporal issues of the surveys, would provide the horizontal and vertical precision of the results. Furthermore, combining morphometrics and dynamics with local environmental conditions is expected to increase our understanding of the processes and temporal variations that control the behaviour of marine bedforms. Specific monitoring for short-term variations is crucial in explaining event-driven bedform dynamics.

2.4 Conclusions

Present-day methods for quantifying the geometry and dynamics of marine bedforms may analyse nearly unrestricted amounts of data in space and in time at a data-mining level and covering country-wide continental shelves. A spatial cross-correlation technique allows for the determination of bed pattern shifts, resulting in a migration vector field of a bedform field. Bedforms of different length scales can be separated, in order to determine the geometries and dynamics of the different types of bedforms by methods such as factorial Kriging, geographical "scanning" and 2-D spectral analyses. Semi-automated identification of crest, trough and inflection points allows for the quantification of geometry and dynamics of individual bedforms. A deformation analysis works the other way around, building up a seabed morphology from different components. The spatial analysis of vertical seabed dynamics reveals the vertical dynamic trends per grid node (in metres/year). Improvements to the existing methods include good estimates of precision (propagated errors) that include all steps in the method, from data acquisition to the effects of the analysing methods themselves.

Acknowledgments This chapter was improved by the comments of Paul P.J. Egberts (TNO, Geological Survey of the Netherlands).

References

Albarracín, S., J. Alcántara-Carrió, I. Montoya-Montes, Á. Fontán-Bouzas, L. Somoza, C.L. Amos and J. Rey Salgado (2014). Relict sand waves in the continental shelf of the Gulf of Valencia (western Mediterranean). *Journal of Sea Research* **93**(): 33–46.

Allen, J.R.L. (1968). *Current ripples, their relation to patterns of water and sediment motion*. North-Holland publishing company, Amsterdam, 433 pp.

Allen, J.R.L. (1980). Sand waves: a model of origin and internal structure. *Sedimentary Geology* **26**: 281–328.

Ashley, G.M. (1990). Classification of large-scale subaqueous bedforms: a new look on an old problem. *Journal of Sedimentary Petrology* **60**(1): 160–172.

Buijsman, M.C. and H. Ridderinkhof (2008a). Long-term evolution of sand waves in the Marsdiep inlet, II: relation to hydrodynamics. *Continental Shelf Research* **28**(9): 1202–1215.

Buijsman, M.C. and H. Ridderinkhof (2008b). Long term evolution of sand waves in the Marsdiep inlet, I: high-resolution observations. *Continental Shelf Research* **28**(9): 1190–1201.

Cataño-Lopera, Y.A., J.D. Abad and M.H. Garcia (2009). Characterization of bed form morphology generated under combined flows and currents using wavelet analysis. *Ocean Engineering* **36**(9–10): 617–632.

Cataño-Lopera, Y.A. and M.H. Garcia (2006). Geometry and migration characteristics of bedforms under waves and currents Part 2: ripples superimposed on sandwaves. *Coastal Engineering* **53**: 781–792.

Cazenave, P.W., J.K. Dix, D.O. Lambkin and L.C. McNeill (2013). A method for semi-automated objective quantification of linear bedforms from multi-scale digital elevation models. *Earth Surface Processes and Landforms* **38**: 221–236.

Dorst, L.L. (2004). Survey plan improvement by detecting sea floor dynamics in archived echo sounder survey. *International Hydrographic Review* **5**(2): 49–63.

Dorst, L.L. (2009). *Estimating sea floor dynamics in the southern North Sea to improve bathymetric survey planning*. Published Ph.D. Thesis, University of Twente, 218 pp.

Dorst, L.L., P.C. Roos and S.J.M.H. Hulscher (2011). Spatial differences in sand wave dynamics between the Amsterdam and the Rotterdam region in the Southern North Sea. *Continental Shelf Research* **31**: 1096–1105.

Dorst, L.L., P.C. Roos, S.J.M.H. Hulscher and R.C. Lindenbergh (2009). The estimation of sea floor dynamics from bathymetric surveys of a sand wave area. *Journal of Applied Geodesy* **3**(2): 97–120.

Duffy, G.P. (2012). Patterns of morphometric parameters in a large bedform field: developement and application of a tool for automated bedform morphometry. *Irish Journal of Earth Sciences* **30**: 31–39.

Duffy, G.P. and J.E. Hughes Clarke (2005). Application of spatial cross correlation to detection of migration of submarine sand dunes. *Journal of Geophysical Research* **110**(F04S12).

Ernstsen, V.B., R. Noormets, C. Winter, D. Hebbeln, A. Bartholomä, B.W. Flemming and J. Bartholdy (2006). Quantification of dune dynamics during a tidal cycle in an inlet channel of the Danish Wadden Sea. *Geo-Marine Letters* **26**: 151–163.

Fraccascia, S., C. Winter, V.B. Ernstsen and D. Hebbeln (2011). Bedform evolution: in a tidal inlet referred from wavelet analysis. *Journal of Coastal Research* **64**(Special Issue): 751–755.

Franzetti, M., P. Le Roy, C. Delacourt, T. Garlan, R. Cancouët, A. Sukhovich and A. Deschamps (2013). Giant dune morphologies and dynamics in a deep continental shelf environment: example of the Banc du Four (Western Brittany, France). *Marine Geology* **346**: 17–30.

Galparsoro, I., A. Borja, I. Legorburu, C. Hernández, G. Chust, P. Liria and A. Uriarte (2010). Morphological characteristics of the Basque continental shelf (Bay of Biscay, northern Spain); their implications for integral coastal zone management. *Geomorphology* **118**(3–4): 314–329.

Goovaerts, P. (1997). *Geostatistics for Natural Resources Evaluation*. Oxford University Press, Oxford, UK pp.

Houthuys, R., A. Trentesaux and P. De Wolf (1994). Storm influences on a tidal sandbank's surface (Middelkerke Bank, southern North Sea). *Marine Geology* **121**: 23–41.

Hugenholtz, C.H., N. Levin, T.E. Barchyn and M.C. Baddock (2012). Remote sensing and spatial analysis of aeolian sand dunes: A review and outlook. *Earth Science Reviews* **111**: 319–334.

Hughes Clarke, J.E. (2012). Optimal use of multibeam technology in the study of shelf morphodynamics. *Sediments, Morphology and Sedimentary Processes on Continental Shelves: Advances in technologies, research and applications.*, International Association of Sedimentology. Special Publication: 1–28.

IHO (2008). *IHO Standards for Hydrographic Surveys*. Monaco, International Hydrographic Bureau. **Special publication No. 44**.

Knaapen, M.A.F., S.J.M.H. Hulscher, H.J. de Vriend and A. Stolk (2001). A new type of sea bed waves. *Geophysical Research Letters* **28**: 1323–1326.

Knaapen, M.A.F., C.N. Van Bergen Henegouw and Y.Y. Hu (2005). Quantifying bedform migration using multi-beam sonar. *Geo-Marine Letters*.

Lanckneus, J. and G. De Moor (1991). Present-day evolution of sand waves on a sandy shelf bank. Oceanologica Acta, Proceedings of the International Colloqium on the environment of epicontinental seas, Lille, France. **11**: 123–127.

Lefebvre, A., V.B. Ernstsen and C. Winter (2011). Bedform characterisation through 2D spectral analysis. Journal of Coastal Research. **Special Issue 64 (Conf. Proc. 11th International Coastal Symposium)**: 781–785.

Lindenbergh, R.C. (2004). Parameter estimation and deformation analysis of sand waves and mega ripples. 2nd International Workshop on Marine sandwave and river dune dynamics (MARID2004), University of Twente, Enschede, Netherlands. 192–199.

Lurton, X. (2002). *An introduction to underwater acoustics*. Springer, 347 pp.

McCave, I.N. (1971). Sand waves in the North Sea off the coast of Holland. *Marine Geology* **10**: 199–225.

Morelissen, R., S.J.M.H. Hulscher, M.A.F. Knaapen, A.A. Németh and R. Bijker (2003). Mathematical modelling of sand wave migration and the interaction with pipelines. *Coastal Engineering* **48**: 197–209.

Pluymaekers, S., R. Lindenbergh, D. Simons and J.d. Ronde (2007). A deformation analysis of a dynamic estuary using two-weekly MBES surveying. Oceans '07, Aberdeen, UK. IEEE.

Santoro, V.C., E. Amore, L. Cavallaro and M. De Lauro (2004). Evolution of sand waves in the Messina Strait, Italy. *Ocean Dynamics* **54**(3–4): 392–398.

Terwindt, J.H.J. (1971). Sand waves in the Southern Bight of the North Sea. *Marine Geology* **10**: 51–67.

Van Dijk, T.A.G.P. and P.J.P. Egberts (2008). The variability of sand wave migration in the North Sea. 3rd International Workshop on Marine and river dune dynamics (MARID2008), University of Leeds, Leeds, UK. 63–68.

Van Dijk, T.A.G.P. and M.G. Kleinhans (2005). Processes controlling the dynamics of compound sand waves in the North Sea, Netherlands. *Journal of Geophysical Research* **110**(F04S10).

Van Dijk, T.A.G.P., M.H.P. Kleuskens, L.L. Dorst, C. Van der Tak, P. J. Doornenbal, A.J.F. Van der Spek, R.M. Hoogendoorn, D. Rodriguez Aguilera, P.J. Menninga and R.P. Noorlandt (2012a). Quantified and applied sea-bed dynamics of the Netherlands Continental Shelf and the Wadden Sea NCK-days 2012: Crossing borders in coastal research, Enschede, Netherlands. 223–227.

Van Dijk, T.A.G.P., R.C. Lindenbergh and P.J.P. Egberts (2008). Separating bathymetric data representing multi-scale rhythmic bedforms: a geostatistical and spectral method compared. *Journal of Geophysical Research* **113**(F04017).

Van Dijk, T.A.G.P., C. Van der Tak, W.P. De Boer, M.H.P. Kleuskens, P. J. Doornenbal, R.P. Noorlandt and V.C. Marges (2011). *The scientific validation of the hydrographic survey policy of the Netherlands Hydrographic Office, Royal Netherlands Navy*. Deltares, Report 1201907-000-BGS-0008: 165 pp. http://kennisonline.deltares.nl/3/m/search/products.html?q=hydrographic&qtype=1.

Van Dijk, T.A.G.P., S. Van Heteren, M.H.P. Kleuskens, L.M. Vonhögen, P.J. Doornenbal, A.J.F. Van der Spek, R.M.

Hoogendoorn, L.L. Dorst and D. Rodriguez Aguilera (2012b). Quantified sea-bed dynamics of the Netherlands Continental Shelf and the Wadden Sea: a morphological and sedimentological approach. Hydro12 - Taking care of the sea, SS Rotterdam, Rotterdam, Netherlands. Hydrographic Society Benelux: 356.

Van Dijk, T.A.G.P., T. Vermaas and M.P. Hijma (2014). *KPP Onderzoek Bodemdynamiek 2014: effect van baggeren op bodemdynamiek locatie Maasgeul & pilot koppeling Kust en Zee*. Deltares, Report 1209377-010-ZKS-0001: 42 pp. http://kennisonline.deltares.nl/3/m/search/products.html?q=bodemdynamiek&qtype=1.

Van Santen, R.B., H.E. De Swart and T.A.G.P. Van Dijk (2011). Sensitivity of tidal sand wavelength to environmental parameters: A combined data analysis and modelling approach. *Continental Shelf Research* **31**(9): 966–978.

Van Son, S.T.J., R.C. Lindenbergh, M.A. De Schipper, S. De Vries and K. Duijnmayer (2009). Using a personal watercraft for monitoring bathymetric changes at storm scale. *Proc.* Hydro9, Cape Town, South Africa.

Winter, C. (2011). Macro scale morphodynamics of the German North Sea coast. *Journal of Coastal Research* **SI 64**(Proceedings of the ICS2011, Poland): 706–710.

Winter, C. and V.B. Ernstsen (2007). Spectral analysis of compound dunes. 5th IAHR Symposium on River, Coastal and Estuarine Morphodynamics (RCEM 2007), University of Twente, Enschede, The Netherlands Taylor & Francis, **2**: 907–911.

Principles of Physical Modelling of Bedforms Under Waves and Currents

3

G. Simarro and Á. Galán

Abstract
Physical modelling in laboratory is a powerful tool to understand the main mechanisms that drive some processes observed in the nature. In this chapter we provide a brief but general overview of the principles of physical modelling, using the bedform dynamics as an example of application.

Keywords
Laboratory experiments • Physical modelling • Scale effects • Wave-current • Bedforms • Wave-induced ripples

3.1 Introduction

Physical modelling is a powerful tool for understanding the behaviour of real-size problems using small-scale models in the laboratory. This technique has been applied to a wide variety of problems, including river hydraulics and the design of bridges, ships and aircraft, to mention but a few. It has also been applied to the analysis of bedforms under waves and currents, which is the topic of this chapter (Fig. 3.1).

3.2 Physical Modelling in Three Steps

In the study of bedforms under waves and currents, the main aim is to predict their height and wavelength, or their migration rate, under given hydrodynamic conditions.

The **first** step in designing a physical model is to identify the variables that play an important role in the process under consideration. In order to focus on the influence on bedforms of a few variables (wave height and period, current velocity, etc.), the experiments are most often carried out in idealized conditions: horizontal beds, monochromatic waves and uniform sediments are usual assumptions. Hence, the variables of interest usually considered are the gravitational acceleration (g), the sediment properties (density ρ_s and size d_{50}), the water properties (density ρ and kinematic viscosity ν) and the ones describing the hydrodynamics (water depth h, current velocity u_c, wave orbital velocity u_w, wave period T and angle between waves and currents θ).

Some of the above variables can be substituted by others. For example, the wave period T can be replaced by the "orbital diameter", $d_o = u_w T/\pi$, since u_w has also been included in the above list. Less evident replacements are also possible. One can write, for example, for the bedform height η (the same arguments below hold for the bedform wavelength λ),

$$\eta = f(g, \rho_s, d_{50}, \rho, \nu, h, u_c, u_w, d_o, \theta). \tag{3.1}$$

G. Simarro (✉)
Institut de Ciències del Mar, CSIC, Passeig Maritim de la Barceloneta 37-49, 08830 Barcelona, Spain
e-mail: simarro@icm.csic.es

Á. Galán
Universidad de Castilla-la Mancha, Avda/Camilo José Cela s/n, 13071 Ciudad Real, Spain
e-mail: alvaro.galan@uclm.es

J. Guillén et al. (eds.), *Atlas of Bedforms in the Western Mediterranean*,
DOI 10.1007/978-3-319-33940-5_3

Fig. 3.1 Dune field generated by a current in the laboratory. Experiment performed at the Hydraulics Laboratory of the UCLM

In a **second** step, the above expression is written in dimensionless form using Vaschy-Buckingham's theorem (see, e.g., Novak et al. 2010) to get, e.g.,

$$\frac{\eta}{d_{50}} = f\left(\Delta \equiv \frac{\rho_s}{\rho} - 1, d_* \equiv \sqrt[3]{\frac{g\Delta d_{50}^3}{\nu^2}}, \frac{h}{d_{50}}, \frac{u_c}{\sqrt{g\Delta d_{50}}}, \frac{u_w}{\sqrt{g\Delta d_{50}}}, \frac{d_o}{d_{50}}, \theta\right), \tag{3.2}$$

where the function f in expression (3.2) is different than that in Eq. (3.1). The above expression is made up of dimensionless "groups": for example, Δ is the submerged relative density. Of note, the submerged density appears in other groups, and always in combination with g, implicitly indicating that the gravitational acceleration and sediment density play their role only as they contribute to the submerged weight of the particles (e.g., Lança et al. 2015).

Writing the equation in dimensionless form has two main advantages. First, it reduces the number of variables: the function f in Eq. (3.2) depends on fewer variables than that in Eq. (3.1). Second, it ensures dimensional homogeneity of the expression obtained from the experimental results.

Expression (3.2) is just one out of the many possible (but equivalent) expressions that Vaschy-Buckingham's theorem yields starting from Expression (3.1). Furthermore, recalling that some of the variables in Expression (3.1) could be replaced by others (as it has been done above with d_o and T), the amount of equivalent dimensionless expressions is enormous. For instance, it can be seen that expression (3.2) is equivalent to

$$\frac{\eta}{d_{50}} = f\left(\Delta, d_*, \frac{h}{d_{50}}, \frac{u_{*c}}{u_{*0}}, \frac{u_{*w}}{u_{*0}}, \frac{d_o}{d_{50}}, \theta\right), \tag{3.3}$$

where u_{*c} and u_{*w} are the shear velocities corresponding to the current and wave respectively, and u_{*0} is the critical shear velocity for the sediment entrainment (Soulsby 1997). The critical shear velocity is $u_{*0} = \sqrt{g\Delta d_{50}\tau_{*0}}$ with τ_{*0} the Shields parameter, related to d_* through the Shields diagram.

Remarkably, it is not always straightforward to determine whether two different dimensionless expressions rely on the same initial hypotheses, and the choice of the dimensionless groups is not simply an academic issue, since some combinations of groups allow a simpler explanation of the problem than others.

The **third** step in the physical modelling is to ensure that the dimensionless groups on the right hand side of Eqs. (3.2) or (3.3) take the same values in the physical model as those in the real-size problem (or prototype), i.e., $\xi^m = \xi^p$ for all the groups ξ on the right hand sides, where "m" stands for model and "p" for prototype. In doing so, the left hand sides will automatically be the same in both the model and the prototype, and therefore $\eta^p = d_{50}^p \eta^m / d_{50}^m$, i.e., one will obtain the bedform height in prototype using physical model results.

However, in general it is not possible to get $\xi^m = \xi^p$ for all the groups on the right hand side unless the physical model has the same size as the prototype (Novak et al. 2010). First note that if $g^m = g^p$, $\rho^m = \rho^p$, $\rho_s^m = \rho_s^p$ and $\nu^m = \nu^p$ (as usual, since laboratory experiments are run with water and sand on Earth) then, imposing $d_*^m = d_*^p$ requires $d_{50}^m = d_{50}^p$. Now, imposing $(h/d_{50})^m = (h/d_{50})^p$ requires $h^m = h^p$, i.e., the model should have the same water depth as the real-size problem.

To make the model smaller than the prototype, the influence of at least one dimensionless group has to be neglected. Insofar as these influences are not actually negligible, there will appear "scale effects", i.e., errors due to the fact that the physical model is not able to properly represent the real-size problem.

3.3 Two Simple Cases

Two particular cases (pure waves and pure currents) are considered below for illustrative purposes. Saving exceptional cases, $\Delta^m = \Delta^p \approx 1.65$, and it will be ignored below. Further, since only waves or currents are considered, the angle θ will also be dropped.

Pure waves. If there is no current velocity, Expression (3.3) reduces to

$$\frac{\eta}{d_{50}} = f\left(d_*, \frac{h}{d_{50}}, \frac{u_{*w}}{u_{*0}}, \frac{d_o}{d_{50}}\right). \tag{3.4}$$

As stated, it is in practice unviable to impose $(h/d_{50})^m = (h/d_{50})^p$, as the above expression (3.4) requires. However, particularly in pure wave cases, the influence of h/d_{50} can be disregarded as long as it is large enough, i.e.,

$$\frac{\eta}{d_{50}} = f\left(d_*, \frac{u_{*w}}{u_{*0}}, \frac{d_o}{d_{50}}\right). \tag{3.5}$$

Most studies in the literature (Wiberg and Harris 1994; Donoghue et al. 2006; Soulsby et al. 2012, to mention a few) use expressions consistent with Eq. (3.5) or similar. Because it is not the aim of this chapter to review the expressions in the literature, but to introduce the physical modelling principles in bedform studies, we only mention that the classification of wave generated by Wiberg and Harris (1994) includes "anorbital" ripples (that scale with d_{50}, i.e., η/d_{50} = constant, and occur for d_o/d_{50} "small"), "suborbital" ripples and "orbital" ripples (that scale with d_o and occur for d_o/d_{50} "large").

Pure currents. Ignoring again Δ and θ, if there is no wave velocity Expression (3.3) reduces to

$$\frac{\eta}{d_{50}} = f\left(d_*, \frac{h}{d_{50}}, \frac{u_{*c}}{u_{*0}}\right), \tag{3.6}$$

which is the one found in van Rijn (1984). This author, using mainly laboratory results, obtained

$$\frac{\eta}{d_{50}} = 0.11\left(\frac{d_{50}}{h}\right)^{0.3}\left(1 - e^{-0.5t}\right)(25 - t), \tag{3.7}$$

where $t \equiv (u_{*c}/u_{*0})^2 - 1$ is the transport stage parameter. Comparing the above expression to Eq. (3.6), the dimensionless diameter d_*, the group including the viscosity, is missed: i.e., viscous effects are being neglected in Expression (3.7). Also of mention, Vanoni (1974) provided a classification of bedforms under pure currents. Assuming constant Δ, his classification relies on d_*, h/d_{50} and $u/\sqrt{gh}$, which happens to be consistent with Expression (3.2) if the waves are ignored.

3.4 Concluding Remarks

Physical modeling is a very useful tool for analyzing some sediment dynamic processes that are difficult to isolate in field experiments, and it has extensively been used to find expressions to predict bedform shapes under waves and currents. However, scale effects are almost always present to some extent; understanding and limiting their influence is critical to obtain useful expressions.

Acknowledgments This research was supported by the project BUS2 (CGL2011-22964). G. Simarro is supported by the Spanish Government through the Ramón y Cajal programme.

References

Donoghue, T. O., Doucette, J. S., van der Werf, J. J., Ribberink, J. S. (2006). The dimensions of sand ripples in full-scale oscillatory flows. Coastal Engineering 53, 997–1012.

Lança, R., Simarro, G., Fael, C. M. S., Cardoso, A. H. (2015). Effect of Viscosity on the Equilibrium Scour Depth at Single Cylindrical Piers. Journal of Hydraulic Engineering 06015022.

Novak, P., Guinot, V., Jeffrey, A., Reeve, D. E. (2010). Hydraulic modelling – an introduction: principles, methods and applications. Spon Press, London and New York.

Soulsby, R. L. (1997). Dynamics of Marine Sands: a manual for practical applications. Thomas Telford, London, ISBN 0-7277-2584-X.

Soulsby, R. L., Whitehouse, R. J. S., Marten, K. V. (2012). Prediction of time evolving sand ripples in shelf seas. Continental Shelf Research 38, 47–62.

van Rijn, L. C. (1984). Sediment transport, part iii: bedforms and alluvial roughness. Journal Hydraulic Engineering 110(12), 1733–1754.

Vanoni, V. (1974). Factors determining bedforms of alluvial streams. Journal of Hydraulic Division HY3.

Wiberg, P. L., Harris, C. K. (1994). Ripple geometry in wave dominated environments. Journal of Geophysical Research 99 (C1), 775–789.

Modelling Bedforms on the Continental Shelf

4

Daniel Calvete

Abstract

A number of morphodynamic models have been developed to simulate the dynamics of large-scale bedform patterns on the continental shelf. Most of the models are based on linear stability analysis and therefore only describe the characteristics of the bedforms at their formation. Apart from studying the principal formation mechanism, the models have taken into account various processes (three-dimensional flows, sediment sorting, wave stirring, tidal constituents, etc) to describe the characteristics of the bedforms at specific locations. Exploration of the finite amplitude regime have also been made with simplified models. Future model developments are needed to explore long-term dynamics.

Keywords

Morphodynamic modelling • Continental shelf • Tidal sand waves • Tidal sand banks • Long bed waves • Shoreface-connected ridges

4.1 Introduction

The bottom of many continental shelves is covered with regular large-scale patterns of elongated bedforms. Their wavelengths range from several hundreds of metres to a few kilometres. Their amplitudes are of the order of metres and can migrate up to several metres per year, depending on the hydrodynamic conditions (see Dyer and Huntley 1999; Belderson 1986 for a general overview). According to their morphological characteristics, four types of large-scale bedforms can be distinguished: tidal sand waves, tidal sand banks, long bed waves and shoreface-connected ridges. Superimposed on these bedforms, smaller bedforms such as ripples, dunes and antidunes are usually found. Here the focus is on the large-scale features because they are specific to the continental shelves.

Large-scale bedforms are of interest to both scientists and engineers. Rather than being geological relics, most of these features are currently active due to tides or storm-generated currents. Understanding their dynamics furthers our knowledge of the sedimentary processes of the continental shelf. In addition, navigation channels, pipelines, communication connections and other offshore infrastructures can be affected by active bedforms. They are also used as a source of sand and may protect the shore from the wave action during severe storms. A number of models (e.g. Idier et al. 2010; Roos and Hulsher 2002) have been developed to further our understanding of the dynamics of these features and to provide a theoretical background for human interventions.

The modelling of bedforms in the continental shelf is constrained by the scales of the features: they extend over large domains (several kilometres) and develop during long periods (hundreds of years). To deal with these constraints, simplified models have been developed for each specific type of bedform. In the first step, the models are used to understand the formation mechanism. These models study the growth of small perturbations of the sea bottom (linear stability analysis) to determine the particular processes that lead to the development of a specific bedform. This kind of model is simple but allows for exhaustive exploration of the

D. Calvete (✉)
Universitat Politècnica de Catalunya—BarcelonaTech,
Jordi Girona 1-3, 08023 Barcelona, Spain
e-mail: daniel.calvete@upc.edu

J. Guillén et al. (eds.), *Atlas of Bedforms in the Western Mediterranean*,
DOI 10.1007/978-3-319-33940-5_4

parameters and provides an initial approximation of the characteristics and dynamics of the bedforms. Because the models are based on linear stability, they only provide information related to the initial formation of the bedforms. To obtain better predictions, the models for the initial growth have been complemented with processes (e.g. sediment sorting, tide components, elliptical tides and waves) that improve the physical description of the processes. For a review of the use of linear stability analysis on coastal morphodynamics, see Dodd et al. (2003), Blondeaux (2001) and Seminara (1998).

A second step, which has not been reached for all bedforms, is the development of nonlinear models. By including nonlinear interactions, these models are able to describe the long-term dynamics: finite amplitude behaviour, merging and splitting of bedforms, effects of sea level rise and climate change, human interferences, etc. Since these experiments are computationally expensive, several techniques have been used to reduce cost, such as weakly nonlinear analysis and expanding the solutions in terms of the modes obtained with the linear stability models (Dodd et al. 2003). Recently, taking advantage of increasing computing power, full nonlinear models have been developed.

A description of each bedform and the models used to study them is presented below.

4.2 Tidal Sand Waves

Sand waves (see Fig. 4.1) are characterized by having their crests perpendicular to the main direction of the tidal current (Stride 1982). The distance between successive crests is of the order of hundreds of metres (200–800 m) and they

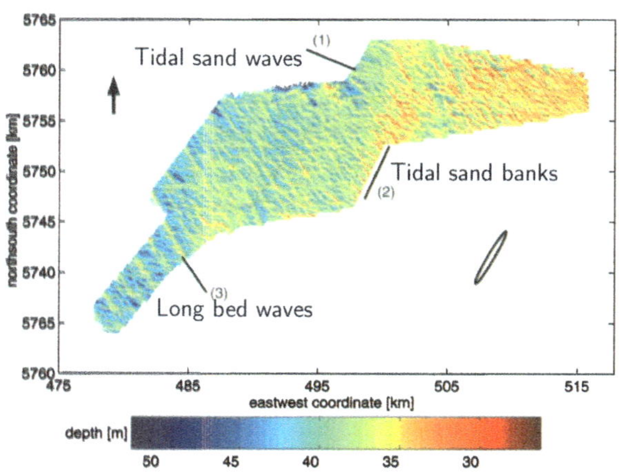

Fig. 4.1 High-resolution bathymetry of the North Sea at 52°N and 3° E. The smaller patterns correspond to tidal sand waves. In the southwest of the area long bed waves are present. Tidal sand banks can also be identified. Adapted from Knaapen et al. (2001), with permission from John Wiley & Sons

migrate up to a few metres a year. Their profile, with an amplitude up to 5 m, is symmetric except if the tide is asymmetric or there are significant residual currents. Under such conditions the migration of the sand waves is enhanced.

The first model for sand wave formation was proposed by Hulscher (1996), who performed a linear stability analysis of bottom perturbations coupled to a tidal current. The model is based on shallow water equations without averaging along the vertical direction. Resolving the vertical structure of the flow is crucial for describing the formation of sand waves. This model showed that their formation is a consequence of the vertical residual recirculation cells that result from the interaction between an undulating sea bed and the tidal current. The residual flow produces a net sediment flux towards the crests and therefore a positive feedback between the growth of the sand banks and the development of recirculation cells. This formation mechanism is similar to that of sea ripples forced by wind waves or river dunes.

The reference model of Hulscher (1996) has been subsequently modified and improved to account for the effects of the tidal constituents, the sediment sorting (Van Oyen and Blondeaux 2009a, b) and the three-dimensional flows (Besio et al. 2006; Blondeaux and Vittori 2011), among others. Thanks to the improved tidal description, the migration of sand waves can be described more accurately. It is found that sand waves migrate upstream/downstream depending on the ratio of the strength of the residual current to the amplitude of the quarter-diurnal tide constituent and on the phase shift between the semi-diurnal and quarter-diurnal tide constituents (Besio et al. 2004). The orientation of the sand waves appears to be related to the rotation of the velocity vectors induced by the tides (Besio et al. 2006). However, the presence of different tidal components is not crucial for the formation of sand waves. The wavelengths of these features may be affected by the modulation of the tidal currents during the spring-neap cycle (Blondeaux and Vittori 2010). Predictions from linear models have been verified with field observations in many studies (Van Santen et al. 2011). For an extended overview of sand wave modelling, see Besio et al. (2008).

At present, there are already a few models to simulate finite amplitude sand waves (e.g. van den Berg et al. 2012). These models validate results from the linear stability analysis. Additionally, it has been found that the saturation in the growth of sand waves is caused by the balance between the shear stress at the seabed and the tendency of the sediment to move downhill (Németh et al. 2007). Weakly nonlinear interaction (Blondeaux and Vittori 2009) also shows that the observed three-dimensional bottom pattern may be a result of resonant interactions between harmonic components of the bottom perturbations. For the moment, nonlinear models take into account fewer physical processes than linear models do.

4.3 Tidal Sand Banks

Tidal sand banks (see Figs. 4.1 and 4.2) have their crest rotated 10°–30° counter-clockwise (in the Northern Hemisphere) with respect to the direction of the tidal current. Their spacing is of the order of a few kilometres, they have an amplitude that is a large fraction of the water depth (up to 20 m), and they hardly migrate. The length of their crest can be up to 80 km.

The formation and orientation of this bedform is a consequence of the horizontal residual circulation around the sand banks that developed as result of the interaction between the bedform, the tidal currents and the Coriolis force (Huthnance 1973). Frictional torques and the Coriolis force enhance the residual circulation on a bank oriented counter-clockwise (in the Northern Hemisphere). As a result, the currents, and therefore the sediment transport fluxes, on the upstream side of the banks are larger than on the downstream side and a net deposition of sediment occurs at the sand bank crest (Zimmerman 1981).

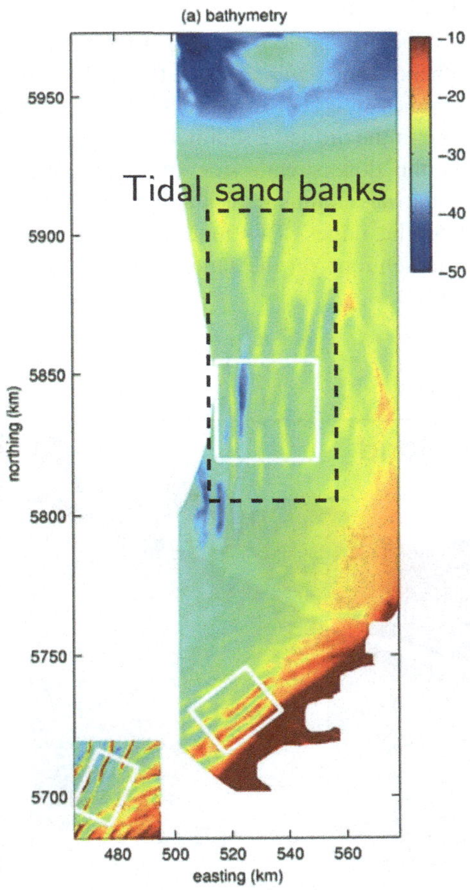

Fig. 4.2 Tidal sand banks in the Southern Bight of the North Sea. Adapted from Roos et al. (2004), with permission from John Wiley & Sons

An initial numerical model for the formation of tidal sand banks was presented by Huthnance (1982), who predicted the formation of linear sand banks with their crest oriented anticlockwise (in the Northern Hemisphere) with respect to the tidal current. The water motion is modelled by the depth-averaged shallow water equations and a bed load sediment transport formulation is used. By means of a linear stability analysis, this model shows that tidal sand banks may develop as a result of morphodynamic instability between a unidirectional tidal current and an erodible sea bottom. The growth rate and wavelength of the preferred growing bedform predicted by the model agree well with observations. Consequently, this model confirmed the mechanism proposed in the conceptual model of Zimmerman (1981). The model of Huthnance (1982) was first improved by Hulscher et al. (1993), who included circular tides, and was subsequently complemented with 3D effects by de Swart and Hulscher (1995). By including circular tides, a finite wavelength growing mode is found in critical conditions that allow a weakly nonlinear model to be set up. Conversely, vertical current has a minor effect on the initial growth of tidal sand banks.

Further modelling improvements have been the incorporation of the effects of steady currents, higher components of tidal currents (Walgreen et al. 2002), and grain-size sorting (Walgreen et al. 2004; Roos et al. 2007). It was found that tidal asymmetry enhances migration velocities while growth rates decrease. The presence of sediment mixtures increases the growth rates and migration velocities of tidal ridges. Symmetric tidal currents cause a grain-size distribution that is in phase with the ridge topography. On the other hand, an asymmetric tide or the presence of steady currents results in a coarser landward flank. As in the case of tidal sand waves, if 3D effects are taken into account when non-unidirectional tides are present, the orientation of the tidal sand banks is related to the residual velocities induced by the tides (Besio et al. 2006).

The finite amplitude behaviour of tidal sand banks was initially studied within the framework of the weakly nonlinear theory. The saturation process was found to be mostly due to the hydrodynamic processes (Idier and Astruc 2003). By using a more complex model that allows for full nonlinear interaction and incorporates the effect of wind wave stirring, Roos et al. (2004) found that the shape of finite amplitude banks depends on the dominant sediment transport mode and the hydrodynamic conditions. Settling lag of suspended sediment and wind wave stirring tend to smooth and lower the height of the banks, while tidal asymmetry causes asymmetric profiles. The saturation in the growth of the banks results from the balance between fluid drag and the downslope transport (Tambroni and Blondeaux 2008). These authors also found that to properly capture the physics of finite amplitude tidal sand banks, it is essential to fully resolve the dynamics on both the fast (tidal) and slow timescales. The role of locally generated residual flow and overtides was studied by Sanay et al. (2007), who concluded that the former is always larger than the latter.

4.4 Long Bed Waves

Long sand bed waves (see Fig. 4.1) were discovered by Knaapen et al. (2001) using high-resolution bathymetries. In addition to the tidal sand waves and tidal sand banks, they found a third harmonic mode with wavelengths of about 1.6 km, three times larger than those of the tidal sand waves, and with the crest rotated 60° anticlockwise (in the Northern Hemisphere) with respect to the tidal current. Blondeaux et al. (2009) showed that these bedforms can be modelled with a traditional depth-averaged model for tidal bedforms on the continental shelf (e.g. Huthnance 1982), supplemented with a sediment transport that includes a critical shear stress for the erosion.

4.5 Shoreface-Connected Sand Ridges

Shoreface-connected sand ridges (see Fig. 4.3) are observed on the inner part of some continental shelves. These ridges have wavelengths of 4–10 km and their growth, orientation and along-shore migration is related to the mean storm-driven currents. They migrate downstream with velocities of 1–10 m/year and the seaward end of the ridges is shifted upstream with respect to their shoreface attachment (see van de Meene et al. 1996).

The formation of these ridges was first simulated by Trowbridge (1995) by applying linear stability analysis to a morphodynamic model based on a depth-averaged storm-driven flow over a sloping inner shelf. Realistic predictions for the growth and migrations of the ridges were modelled by Calvete et al. (2001) and accounted for the dynamics of bedload and suspended load. These authors found that an increase in the concentration of suspended sediment towards the shore is an essential element of the models for properly reproducing the characteristics of the ridges, while the tidal currents hardly affect the dynamics of the ridges on the inner shelf (Walgreen et al. 2002). Using a similar morphodynamics model, grain-size sorting was studied by Walgreen et al. (2003), who were able to model the observed pattern of grain sizes on a field of sand ridges (see also Vis-Star et al. 2009). Since the ridges develop under storm weather conditions (Calvete et al. 2001), the role of wave stirring has been examined with several models. Vis-Star et al. (2007) showed that perturbations in the wave stirring significantly enhance the growth of the ridges.

Finite amplitude dynamics (nonlinear regimen) of the ridges has also been studied with a model for the amplitude of the different bottom modes by Calvete et al. (2002), Calvete and de Swart (2003) and Vis-Star et al. (2008), among others. These authors found that the growth of the ridges becomes saturated in a timescale of centuries due to the increase in the diffusive sediment transport caused by the slopes on the downstream side of the ridges. They also modelled the development of small-scale bedforms superimposed on the ridges. This model was also used to study human interventions on the inner shelf (e.g. de Swart and Calvete 2003).

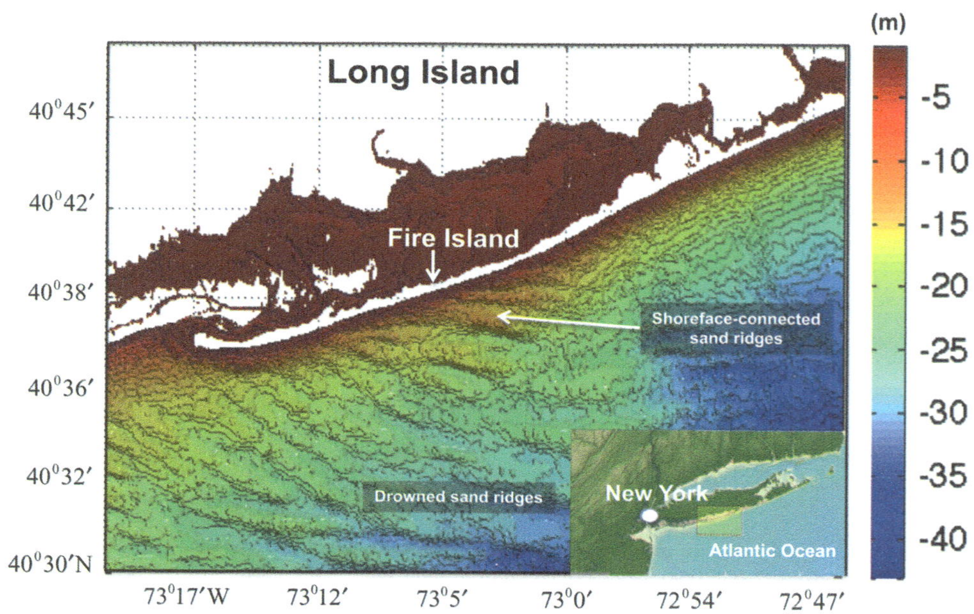

Fig. 4.3 Shoreface-connected sand ridges on the Long Island continental shelf. Reprinted from Nnafie et al. (2014b), with permission from Elsevier

Recently, Nnafie et al. (2014a, b, 2015) applied a full nonlinear model, in a finite differences scheme, to model shoreface ridges of finite amplitude, corroborating the results of previous studies. They also studied how sea level rise and the evolution of the shelf affect the dynamics of the ridges. They conclude that for moderate rates of sea level rise the growth of the ridges can be enhanced. However, rapid sea level rises in comparison with the timescale of ridge formation limit the growth of the ridges and can even allow the ridges to become a moribund feature. Nnafie et al. (2014b) also model human interventions on the inner shelf.

4.6 Future Perspectives

Several fundamental issues still remain to be addressed, although a large number of models exist. In the few models for finite amplitude patterns, the physics is simplified. The effects of three-dimensional flows, sediment sorting, wave stirring and different tidal constituents, among others, have not yet been explored in the nonlinear regime. Additionally, processes occurring at the long-term scale, such as evolution of the shelf morphology, sea level rise and changes in the wave climate and hydrodynamic conditions, also need to be covered. Despite the interest of these outstanding issues, problems of a numerical nature and limitations due to computational requirements may temporarily delay the extensive use of nonlinear models.

Acknowledgements D. Calvete gratefully acknowledges the support through the proyect CTM2015-66225-C2-1-P (MINECO/FEDER).

References

Belderson, R. H. (1986). Offshore tidal and non-tidal sand ridges and sheets: differences in morphology and hydrodynamic setting. In R. J. Knight & J. R. McLean (Eds.), *Shelf Sands and Sandstones* (pp. 293-301). Can. Soc. Petr. Geol.

Blondeaux, P. (2001). Mechanics of Coastal Forms. *Annual Review of Fluid Mechanics*, *33*(1), 339-370.

Blondeaux, P., de Swart, H. E., & Vittori, G. (2009). Long bed waves in tidal seas: an idealized model. *Journal of Fluid Mechanics*, *636*, 485

Blondeaux, P., & Vittori, G. (2009). Three-dimensional tidal sand waves. *Journal of Fluid Mechanics*, *618*, 1-11.

Blondeaux, P., & Vittori, G. (2010). Formation of tidal sand waves: Effects of the spring-neap cycle. *Journal of Geophysical Research: Oceans*, *115*(10), 1-7.

Blondeaux, P., & Vittori, G. (2011). The formation of tidal sand waves: Fully three-dimensional versus shallow water approaches. *Continental Shelf Research*, *31*(9), 990-996.

Besio, G., Blondeaux, P., Brocchini, M., & Vittori, G. (2004). On the modeling of sand wave migration. *Journal of Geophysical Research C: Oceans*, *109*(4), C04018.

Besio, G., Blondeaux, P., Brocchini, M., Hulscher, S. J. M. H., Idier, D., Knaapen, M. A. F., Németh, A. A. (2008). The morphodynamics of tidal sand waves: A model overview. *Coastal Engineering*, *55*(7-8), 657-670.

Besio, G., Blondeaux, P., & Vittori, G. (2006). On the formation of sand waves and sand banks. *Journal of Fluid Mechanics*, *557*, 1-27.

Calvete, D., & de Swart, H. E. (2003). A nonlinear model study on the long-term behavior of shore face–connected sand ridges. *Journal of Geophysical Research*, *108*(C5), 3169.

Calvete, D., De Swart, H. E., & Falqués, A. (2002). Effect of depth-dependent wave stirring on the final amplitude of shoreface-connected sand ridges. *Continental Shelf Research*, *22*, 2763-2776.

Calvete, D., Falqués, a, de Swart, H. E., & Walgreen, M. (2001). Modelling the formation of shoreface-connected sand ridges on storm-dominated inner shelves. *J. Fluid Mech.*, *441*, 169-193.

de Swart, H. E., & Calvete, D. (2003). Non-linear response of shoreface-connected sand ridges to interventions. *Ocean Dynamics*. Vol. 53, pp. 270-277.

de Swart, H. E., Walgreen, M., Calvete, D., & Vis-Star, N. C. (2008). Nonlinear modelling of shoreface-connected ridges; Impact of grain sorting and interventions. *Coastal Engineering*, *55*(7-8), 642-656.

de Swart, H. E., & Hulscher, S. J. M. H. (1995). Dynamics of large-scale bedforms in coastal seas. In A. Doelman & A. van Harten (Eds.), *Nonlinear Dynamics and Pattern Formation in the Natural Environment* (pp. 315-331). Addison–Wesley–Longman.

Dodd, N., Falqués, A., Hulscher, S. J. M. H., Rózynski, G., Vittori, G., Blondeaux, P., Calvete, D., and de Swart, H. E. (2003). Understanding Coastal Morphodynamics Using Stability Methods. *Journal of Coastal Research*, *19*, 849-865.

Dyer, K. R., & Huntley, D. A. (1999). The origin, classification and modelling of sand banks and ridges. *Continental Shelf Research*, *19* (10), 1285-1330.

Hulscher, S. J. M. H. (1996). Tidal-induced large-scale regular bed form patterns in a three-dimensional shallow water model. Journal of Geophysical Research, 101(96), 20727.

Hulscher, S. J. M. H., de Swart, H. E., & Vriend, H. J. de. (1993). The generation of offshore tidal sand banks and sand waves. *Continental Shelf Research*, *13*(11), 1183-1204.

Huthnance, J. M. (1973). Tidal current asymmetries over the Norfolk Sandbanks. Estuarine and Coastal Marine Science, 1, 89–99.

Huthnance, J. M. (1982). On one mechanism forming linear sand banks. Estuarine, Coastal and Shelf Science, 14(1), 79-99.

Idier, D., & Astruc, D. (2003). Analytical and numerical modeling of sandbanks dynamics. *Journal of Geophysical Research C: Oceans*, *108*(3), 5-15.

Idier, D., Hommes, S., Briere, C., Roos, P. C., Walstra, D., & Knaapen, M. A. F. (2010). Morphodynamic Models Used to Study the Impact of Offshore Aggregate Extraction: a Review. *Journal of Coastal Reseacrh*, SI 51, 39-52.

Knaapen, M. A. F., Hulscher, S. J. M. H., De Vriend, H. J., & Stolk, A. (2001). A new type of sea bed waves. *Geophysical Research Letters*, *28*(7), 1323-1326.

Meene, J. W. H. van de, Boersma, J. R., & Terwindt, J. H. J. (1996). Sedimentary structures of combined flow deposits from the shoreface-connected ridges along the central Dutch coast. *Marine Geology*, *131*(3-4), 151-175.

Nnafie, A., de Swart, H. E., Calvete, D., & Garnier, R. (2014a). Modeling the response of shoreface-connected sand ridges to sand extraction on an inner shelf. *Ocean Dynamics*, *64*(5), 723-740.

Nnafie, A., de Swart, H. E., Calvete, D., & Garnier, R. (2014b). Effects of sea level rise on the formation and drowning of shoreface-connected sand ridges, a model study. *Continental Shelf Research*. 80, 32-48

Nnafie, A., de Swart, H. E., Garnier, R., & Calvete, D. (2015). Dynamics of shoreface-connected and inactive sand ridges on a shelf, Part 2: The role of sea level rise and associated changes in shelf geometry. *Continental Shelf Research*., vol. 104, pàgs. 63-75

Németh, A. A., Hulscher, S. J. M. H., & Van Damme, R. M. J. (2007). Modelling offshore sand wave evolution. *Continental Shelf Research, 27*(5), 713-728.

Roos, P. C., & Hulscher, S. J. M. H. (2002). Formation of offshore tidal sand banks triggered by a gasmined bed subsidence. *Continental Shelf Research,* Vol. 22, pp. 2807-2818.

Roos, P. C., Hulscher, S. J. M., Knaapen, M. A. F., & Damme, R. M. J. V. (2004). The cross-sectional shape of tidal sandbanks: modeling and observations. *J. Geophys. Res., 109*(F02003), F02003.

Roos, P. C., Wemmenhove, R., Hulscher, S. J. M. H., Hoeijmakers, H. W. M., & Kruyt, N. P. (2007). Modeling the effect of nonuniform sediment on the dynamics of offshore tidal sandbanks. *Journal of Geophysical Research: Earth Surface, 112*(2).

Sanay, R., Voulgaris, G., & Warner, J. C. (2007). Tidal asymmetry and residual circulation over linear sandbanks and their implication on sediment transport: A process-oriented numerical study. *Journal of Geophysical Research: Oceans, 112*(12).

Seminara, G. (1998). Stability and Morphodynamics. *Meccanica, 33,* 59-99.

Stride, A. H. (1982). *Offshore tidal sands: processes and deposits.* Chapman & Hall.

Tambroni, N., & Blondeaux, P. (2008). Sand banks of finite amplitude. *Journal of Geophysical Research, 113*(C10), C10028.

Trowbridge, J. H. (1995). A mechanism for the formation and maintenance of shore-oblique sand ridges on storm-dominated shelves. *Journal of Geophysical Research, 100*(C8), 16071.

Van den Berg, J. H., Sterlini, F., Hulscher, S. J. M. H., & Van Damme, R. M. J. (2012). Non-linear process based modelling of offshore sand waves. *Continental Shelf Research, 37,* 26-35.

Van Oyen, T., & Blondeaux, P. (2009a). Grain sorting effects on the formation of tidal sand waves. *Journal of Fluid Mechanics,* 629, 311-342.

Van Oyen, T., & Blondeaux, P. (2009b). Tidal sand wave formation: Influence of graded suspended sediment transport. *Journal of Geophysical Research, 114*(C7), C07004.

Van Santen, R. B., de Swart, H. E., & Dijk, T. a G. P. van. (2011). Sensitivity of tidal sand wavelength to environmental parameters: A combined data analysis and modelling approach. *Continental Shelf Research, 31*(9), 966-978.

Vis-Star, N. C., de Swart, H. E., & Calvete, D. (2009). Effect of wave-bedform feedbacks on the formation of, and grain sorting over shoreface-connected sand ridges. *Ocean Dynamics, 59*(5), 731-749.

Vis-Star, N. C., de Swart, H. E., & Calvete, D. (2008). Patch behaviour and predictability properties of modelled finite-amplitude sand ridges on the inner shelf. *Nonlinear Processes in Geophysics, 15*(6), 943-955.

Vis-Star, N. C., de Swart, H. E., & Calvete, D. (2007). Effect of wave-topography interactions on the formation of sand ridges on the shelf. *Journal of Geophysical Research: Oceans, 112*(6)

Walgreen, M., de Swart, H. E., & Calvete, D., (2004). A model for grain-size sorting over tidal sand ridges. In *Ocean Dynamics.* Vol. 54, pp. 374-384.

Walgreen, M., Calvete, D., & de Swart, H. E. (2002). Growth of large-scale bed forms due to storm-driven and tidal currents: A model approach. *Continental Shelf Research,* Vol. 22, pp. 2777-2793.

Walgreen, M., de Swart, H. E., & Calvete, D., (2003). Effect of grain size sorting on the formation of shoreface-connected sand ridges. *Journal of Geophysical Research*, 108(C3), p. 3063.

Zimmerman, J. T. F. (1981). Dynamics, diffusion and geomorphological significance of tidal residual eddies. *Nature, 290*(5807), 549-555.

Internal Waves and Bedforms

5

Hans van Haren

Abstract

The physics of internal waves in the density-stratified deep sea is reviewed with the aim of understanding the waves' potential effects on undular bedforms, 'sediment waves', at the seafloor. Such bedforms occur mainly on continental slopes. Sloping topography is also a prerequisite for internal wave breaking, which is the dominant process for sediment resuspension in the deep sea. Internal and sediment waves have common horizontal length scales. They differ in vertical length scale and, foremost, in propagation velocity and age.

Keywords

Internal waves • Sediment waves • Sloping topography • Turbulent bores • Inertial motions

5.1 Introduction

As seas and oceans are basically heated from above, they are stably stratified in density because warm water is less dense than cold water. The heat stored as potential energy near the surface is transported into the deep sea by means of mechanical turbulent mixing. The main supply of kinetic energy comes from tides, and from inertial motions following the passage of atmospheric disturbances on the rotating Earth. The former dominate in most oceans, the latter in the Mediterranean Sea. The fact that as little as 1/30,000 of the potential energy is needed in kinetic energy to provide sufficient mixing to keep the deep-sea stratified (Munk and Wunsch 1998) shows the importance of ocean tides and inertial motions.

The same density stratification supports freely propagating 'internal waves' at frequencies between inertial frequency $f = 2\Omega \sin\varphi$ at latitude φ, where Ω is the Earth's angular velocity, and buoyancy frequency $N = (-g/\rho \cdot d\rho/dz)^{1/2}$ plus

H. van Haren (✉)
Royal Netherlands Institute for Sea Research (NIOZ), P.O. Box 591790 AB Den Burg, the Netherlands
e-mail: hans.van.haren@nioz.nl

J. Guillén et al. (eds.), *Atlas of Bedforms in the Western Mediterranean*,
DOI 10.1007/978-3-319-33940-5_5

compressibility correction, where g is the acceleration of gravity and ρ the density. In contrast with surface (wind) waves, internal waves are essentially three-dimensional, as only N-waves propagate strictly along density interfaces with their velocity directed vertically, while pure inertial motions are horizontal. As a result, waves at all frequencies (σ) in the range $f < \sigma < N$ propagate at a particular angle to the vertical (e.g., LeBlond and Mysak 1978). Thus, depending on local values of f and N, internal waves are generated and break at particular slope sections of the bottom topography. This fact associates particular sections of underwater seamounts and continental slopes with relatively strong turbulent mixing.

Undulating bedforms rather than sliding structures are also observed uniquely above particular slopes, mostly seaward of river outflow. Examples of such 'sediment waves' in prodeltas of river outflows above continental slope sections are given in Urgeles et al. (2011) and in various chapters of this book. Although the mechanisms of their formation, development or maintenance are not well understood, it is tempting to compare physical processes of internal wave generation and, more importantly, wave breaking, with the formation of sediment waves. In this chapter, properties of physical and geological processes are reviewed and some dedicated observational setups for the investigation of internal wave impact on sediment wave formation are discussed.

5.2 Internal Wave Impact on Sediment

While in shallow seas—say less deep than 50 m—the sediment is mainly reworked by the impact of surface wind waves and (tidal) currents through bottom friction in the lower 5–10 m, at greater depths these effects are less important. Above the continental slope, boundary currents like the Northern Current in the NW Mediterranean may be responsible for some friction when they reach the bottom. However, most resuspension of sediment is observed to occur during the relatively short moments of breaking of shoaling internal waves (Hosegood et al. 2004). Thereby, the internal waves deform nonlinearly from their sinusoidal shape in the interior to create an upslope-moving, vigorously turbulent bore (Vlasenko and Hutter 2002; Klymak and Moum 2003). Including their pre-frontal sharpening, the bores account for 60 % of the turbulence generated in a tidal cycle within half an hour (van Haren and Gostiaux 2012). The sediment is swept up by 0.05–0.15 m s^{-1} vertical currents of the 10- to 60-m-high bores (examples in Figs. 5.1 and 5.2), as in an atmospheric dust storm. This phenomenon results in nepheloid layers dispersed along isopycnals in the ocean interior (e.g., Armi 1978; Cacchione and Drake 1986; McPhee-Shaw and Kunze 2002).

It has been suggested that nepheloid layers are associated with bottom slopes (β) that 'critically' match the internal

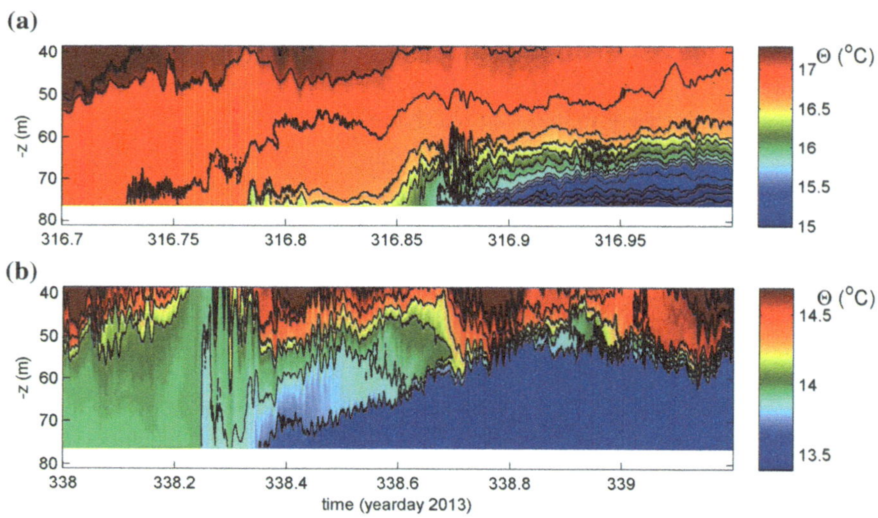

Fig. 5.1 Two examples of high-frequency internal waves in the Llobregat prodelta (Barcelona harbour entrance buoy), observed using high-resolution NIOZ temperature sensors during the autumn of 2013 (van Haren, Guillén, Puig, unpublished results). Contour interval is 0.2 °C. **a** 7 h time-window showing a near-bottom bore moving upslope, while it mixes with overlying water (e.g. day 316.87 between 60 and 75 m) and generates high-frequency internal waves. **b** Three weeks later, note the cooler temperatures and the time-window of 28 h, also showing a cold-water bore but with high-frequency internal waves at an interface 20–30 m above the bottom

wave slope $\alpha = \arcsin(\sigma^2 - f^2)/(N^2 - f^2) \equiv \beta$ (see e.g. Lamb 2014 for geometry of super- and sub-critical bottom slopes for internal wave beams). This association may be related to the generation site of, e.g., internal tides at which the near-bottom (<5 m) shear appears to be high as modelled by Zhang et al. (2008). In their model, internal wave breaking is not resolved. However, ocean observations confirm that most vigorous internal wave breaking, and associated sediment resuspension transported much higher up to about 50 m above the bottom, does not occur above critical slopes, but rather on steep slopes that are supercritical $\beta > \alpha$ for semidiurnal lunar tides (Hosegood et al. 2004; van Haren et al. 2015). At such slopes internal tides reflect back into the interior (Lamb 2014) and higher frequency internal waves may be critical. There, the average turbulent kinetic energy dissipation is found to be 100 times greater than at subcritical slopes and in the ocean interior. Furthermore, turbulent bores have been associated with tidal, inertial and sub-inertial 'carrier' waves. The latter two are not considered freely propagating internal waves in the traditional approximation because, e.g., at f, $\alpha = 0$ in theory. In practice, the inertial band has a finite width of approximately [0.95f, 1.05f] and near-inertial internal waves in modulation with the Northern Current are observed to occasionally create bores in the Gulf of Valencia (van Haren et al. 2013). In general in the Mediterranean, tides are very weak with a few exceptions, e.g., near Gibraltar and in the northern Adriatic.

Although to my knowledge a precise relation between bedform occurrence and slope angle is presently lacking, it seems that sediment waves mainly occur on the deep-sea side of the steepest part of a continental slope (see examples in Verdicchio and Trincardi 2006; Puig et al. 2007; Urgeles et al. 2011; Ribó et al. 2016). Typical amplitudes are 1–10 m in the vertical, with 100- to 1000-m horizontal wave lengths. The former are one to two orders of magnitude smaller than those of internal wave (bores). The latter, however, match the length scales of internal waves, which vary between several tens of metres at N and about 1 km at f (and tidal frequencies) (LeBlond and Mysak 1978).

5.3 Observational Setup

Variations in internal wave breaking and sediment resuspension are best studied using underwater moored instrumentation that is left unattended for at least a month to build-up some statistics with the varying low-frequency motions. Sediment traps and optical backscatter devices may be attached to the mooring line to monitor variations in suspended materials. Preferably, multiple instruments should be attached to the typically 100-m longline in the vertical, together with multiple current meters. To investigate internal wave breaking in detail and to quantify the turbulence dissipation, high-resolution temperature sensors may be used, provided the temperature-density relationship is known, e.g. from additional shipborne CTD profiles (van Haren and Gostiaux 2012). The sensors are typically spaced at 1-m intervals over the entire 100-m-long mooring line and sample at a rate of 1 Hz. Thus, high-frequency internal waves (examples in Fig. 5.1 from the Llobregat prodelta), and especially also turbulent overturning (examples in Fig. 5.2 from the Ebro prodelta), can be detailed.

The time-scale of the largest overturns in Fig. 5.2a, b is 1000–1500 s. This is nearly equal to the buoyancy period observed in very thin (<2 m) stratified layers that border the

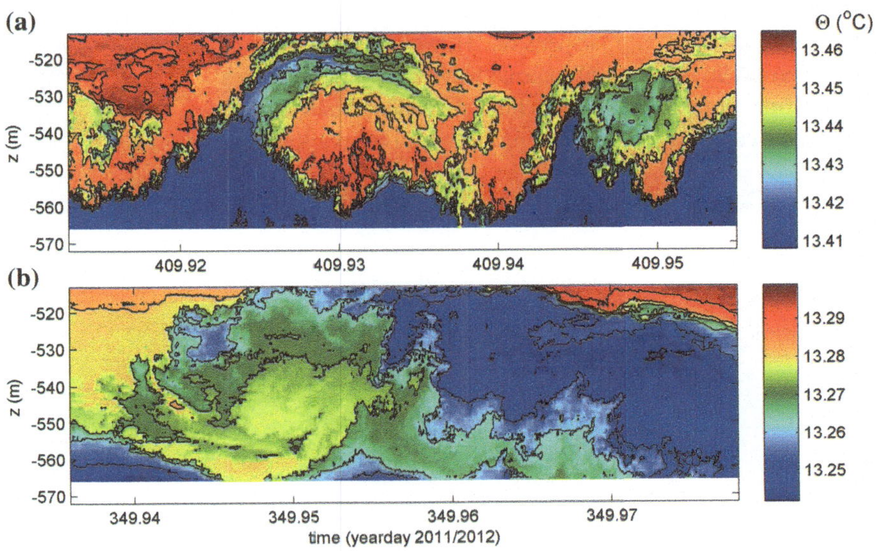

Fig. 5.2 One-hour examples of depth-time series of ~50 m vertical overturns observed using high-resolution temperature sensors above the continental slope in the Gulf of Valencia (after van Haren et al. 2013). *Both panels* have a temperature range of 0.062 °C and contours every 0.01 °C, but they differ in absolute values. **a** During strong downslope flow. **b** During upslope flow

overturns. The shear that drives these overturns varies locally on the buoyancy scale (reaching 9×10^{-3} s^{-1} over 20-m intervals for the period in Fig. 5.2a). It is superposed on a larger-scale shear of about 2×10^{-3} s^{-1} created by inertial motions and the sub-inertial Northern Current (van Haren et al. 2013).

From a single mooring with multiple instruments, one can measure the vertical internal wavelength $2\pi/m$, where m is the wavenumber, from the phase shift in a depth-time diagram. The horizontal wavelength $2\pi/k_h$, where k_h is the wavenumber, is found through the dispersion relation (e.g., LeBlond and Mysak 1978): $\sigma^2 = N^2k_h^2/(k_h^2 + m^2 + (N^2/2g)^2)$, where g denotes the acceleration of gravity.

5.4 Discussion

Despite the similarity in horizontal wavelength between sediment and internal waves, the need for scepticism is suggested by the fact that sediment waves take such a long time to grow compared with the time scales of internal waves. However, other frictional processes such as surface waves and the processes of near-inertial, tidal and boundary currents also have time scales that are far shorter than those of sand-ripple and sediment wave growth. In the deep sea, at depths greater than about 100 m above the steepest part of continental slopes, these frictional processes are less important than those induced by internal wave motions, in particular breaking. This is because observed overturns are not associated with friction on a rotating Earth as they do not comply with Ekman dynamics above sloping bottoms, which predicts greater turbulence for a downwelling-favourable along slope flow like the Northern Current in the NW Mediterranean.

When turbulent bores move up the slope, their peak currents far exceed those of other processes (both frictionally at the bottom and in the interior) that are important for sweeping material into suspension. However, a more comprehensive classification of internal wave breaking processes in relation to bedforms is still highly needed. Such a classification can only be made using detailed observations.

References

Armi, L. (1978). Some evidence for boundary mixing in the deep ocean, *J. Geophys. Res., 83*, 1971–1979.

Cacchione, D.A., and D.E. Drake (1986). Nepheloid layers and internal waves over continental shelves and slopes, *Geo-Mar. Lett., 16*, 147–152.

Hosegood, P., J. Bonnin, and H. van Haren (2004). Solibore-induced sediment resuspension in the Faeroe-Shetland Channel, *Geophys. Res. Lett., 31*, L09301, doi:10.1029/2004GL019544.

Klymak, J.M., and J.N. Moum (2003). Internal solitary waves of elevation advancing on a shoaling shelf, *Geophys. Res. Lett., 30*, 2045, doi:10.1029/2003GL017706.

Lamb, K.G. (2014). Internal wave breaking and dissipation mechanisms on the continental slope/shelf, *Ann. Rev. Fluid Mech., 46*, 231–254.

LeBlond, P.H., and L.A. Mysak (1978). *Waves in the Ocean*, 602 pp., Elsevier, New York.

McPhee-Shaw, E.E., and E. Kunze (2002). Boundary-layer intrusions from a sloping bottom: A mechanism for generating intermediate nepheloid layers, *J. Geophys. Res. 107*, doi:10.1029/2001JC000801.

Munk, W., and C. Wunsch (1998). Abyssal recipes II: Energetics of tidal and wind mixing, *Deep-Sea Res. I, 45*, 1977–2010.

Puig, P., A.S. Ogston, J. Guillén, A.M.V. Fain and A. Palanques (2007). Sediment transport processes from the topset to the foreset of a crenulated clinoform (Adriatic Sea), *Cont. Shelf Res., 27*, 452–474.

Ribó, M., et al. (2016). Large fine-grained sediment waves over the Valencia slope, *This book.*

Urgeles, R., et al. (2011). A review of undulated sediment features on Mediterranean prodeltas: distinguishing sediment transport structures from sediment deformation, *Mar. Geophys. Res., 32*, 49–69.

van Haren, H., and L. Gostiaux (2012). Detailed internal wave mixing observed above a deep-ocean slope, *J. Mar. Res., 70*, 173–197.

van Haren, H., M. Ribó, and P. Puig (2013). (Sub-)inertial wave boundary turbulence in the Gulf of Valencia, *J. Geophys. Res., 118*, 2067–2073, doi:10.1002/jgrc.20168.

van Haren, H., A. Cimatoribus, and L. Gostiaux (2015). Where large deep-ocean waves break, *Geophys. Res. Lett., 42*, 2351–2357, doi:10.1002/2015GL063329.

Verdicchio, G., and F. Trincardi (2006). Short-distance variability in slope bed-forms along the Southwestern Adriatic Margin (Central Mediterranean), *Mar. Geol., 234*, 271–292.

Vlasenko, V., and K. Hutter (2002). Numerical experiments on the breaking of solitary internal waves over a slope-shelf topography, *J. Phys. Oceanogr., 32*, 1779–1793.

Zhang, H.P., B. King and H.L. Swinney (2008). Resonant generation of internal wave on a model continental slope, *Phys. Rev. Lett., 100*, 244504.

Turbidity Current Bedforms

6

Matthieu J.B. Cartigny and George Postma

Abstract

Turbidity currents in the submarine seascape are what river flows are in terrestrial landscapes. While rivers transport sediment from the mountains through valleys towards the sea, turbidity currents transport sediment from the shallow marine realms through canyons towards the deeper abyssal plains. The large scale architecture of both systems is remarkably similar. Yet, there are some fundamental differences between rivers and turbidity currents, the most fundamental one being their density difference; the density of river water is thousand times denser than its surrounding air, while the density of a turbidity current can never be more than twice as dense as its ambient water. In addition, rivers do not depend on their sediment load to flow, while turbidity (density) currents do need the sediment derived excess density to flow. These physical differences change their morphodynamics on the bedform scale. Present day high-resolution seafloor observations show that turbidity current path ways are covered with bedforms that are fundamentally different from those that occur in river channels. In this chapter we point out these differences and present a 3D bedform stability diagram for turbidites.

Keywords

Bedforms • Bedform stability diagram • Turbidity currents • Antidunes • Cyclic steps

6.1 Turbidity Current Bedforms

Compared with the well-studied range of bedforms occurring in river systems, very little is known about the equivalent range of bedforms that are observed in turbidity current systems. Most of our present day knowledge of turbidity current bedforms is derived from seafloor surveys (e.g. Normark et al. 1980; Migeon et al. 2001; Zhong et al. 2015) and from studies of bedforms preserved in the geological record through outcrop studies (e.g. Walker 1967; Skipper 1971; Mutti 1992; Postma et al. 2014; Ventra et al. 2015). Recent years have seen an advance in bedform knowledge due to upcoming high-resolution seafloor surveys with Autonomous Underwater Vehicles (i.e. Paull et al. 2011) and detailed short-period-repeat mapping of turbidity current systems (i.e. Hughes Clarke et al. 2012). These studies have shown that turbidity current bedforms are typically long and characterized by low stoss- and leeside inclinations, while their crests migrate against the flow direction. Their dynamic morphology is thus in great contrast to river bedforms, such as dunes, which have steep lee-sides and typically migrate down-current. Given that the seafloor sediments are similar to those in river beds, any of these differences must be related to physical differences between river and turbidity currents associated with their dynamic interaction with the bed.

The main physical difference between river flow and a turbidity current is centred on their difference in relative density. The excess density of turbidity currents, in relation

M.J.B. Cartigny (✉)
National Oceanography Centre, University of Southampton
Waterfront Campus, Southampton, UK
e-mail: m.cartigny@noc.ac.uk

G. Postma
Faculty of Geosciences, Utrecht University,
Utrecht, The Netherlands

J. Guillén et al. (eds.), *Atlas of Bedforms in the Western Mediterranean*,
DOI 10.1007/978-3-319-33940-5_6

to the ambient sea water, is formed by the amount of sediment that it is carrying, which limits the density of the flow to a maximum of twice that of its surrounding water. This is in contrast with the excess density of river water, which is around a thousand times heavier than the surrounding air and thereby relatively insensitive to the extra weight of suspended sediment.

The low relative density of turbidity currents impact their morphodynamics in two ways. Firstly, waves propagating on the top of the flow are slowed down, as these are driven by gravity which in turbidity current is strongly reduced due to buoyancy forces. Early experimental bedform work of Guy et al. (1966) has shown that currents with velocities below the surface wave propagation velocity results in a range of down-current migrating bedforms, like ripples and dunes. In contrast, currents with flow velocities (U_{flow}) that exceed the wave propagation velocity (U_{wave}) lead to up-current migration bedforms, such as antidunes and cyclic steps (Taki and Parker 2005). The ratio of flow velocity over wave propagation velocity is expressed by the densimetric Froude number ($Fr' = U_{flow}/U_{wave} = U_{flow}/\sqrt{(\rho_{flow} - \rho_w)gH/\rho_{flow}}$), where ρ_{flow}, ρ_w are the densities of the flow and the water, respectively, g the acceleration of gravity and H is the flow depth. Supercritical flows have a flow velocity that exceeds the wave propagation velocity ($Fr' > 1$) while for subcritical flows the opposite holds. Prevailing supercritical conditions in turbidity currents are confirmed by: (1) direct measurements of turbidity currents (Xu 2010), (2) experimental and numerical modelling (Middleton 1966; Hand 1974; Kostic and Parker 2006), (3) monitoring up-current migrating bedforms (Migeon et al. 2001; Smith et al. 2005; Hughes Clarke et al. 2012), and (4) observations of supercritical-flow bedforms in successions of turbidity current deposits in the rock record (Walker 1967; Skipper 1971; Postma et al. 2014; Ventra et al. 2015).

Secondly, the direct connection between the sediment-based relative density and the flow velocity make turbidity currents far more unstable than river flows. If turbidity currents entrain more sediment, their driving force goes up, which in turn speeds up the flow and further enhances sediment entrainment (so called auto-ignition, Parker et al. 1986). Running the same cycle in the opposite direction (waning flow) leads to positively enforced loops of deceleration and sediment fall-out. In such waning flow conditions sediment collects in high concentrations at the base of the flow before slowly coming to a complete standstill (Postma et al. 1988; Sohn 1997; Cartigny et al. 2013; Sumner and Paull 2014). Such dense basal layers will have their own densimetric Froude numbers, as waves are often seen to form on the top interface of dense basal layers (Cartigny et al. 2013; Sumner and Paull 2014). This has led to the idea that dense basal layers will have their own bedform dynamics that is largely independent of the main flow (Postma et al. 2009; Postma and Cartigny 2014). Such dense basal layer bedforms are only thought to form under supercritical-flow conditions, as all subcritical bedform are dependent on turbulence, which is severely suppressed in dense basal layers. Dense basal layers flowing subcritically are thought to form long stretches of virtually flat basin floor (Postma and Cartigny 2014). In the rock record these deposits are characterized by centimetre-thick planar laminations of spaced and crude laminations (Hiscott 1994; Talling et al. 2012), which are thought to reflect the high basal sediment concentrations (Cartigny et al. 2013).

6.2 Bedform Stability Diagram for Turbidity Currents

The tendencies of turbidity currents to be supercritical and to form dense basal layers are now taken into account and introduced in a bedform stability diagram originally developed for river flows (Cartigny et al. 2014). Five examples of turbidity currents with increasing basal sediment concentrations, will now be discussed in more detail.

In panel I of Fig. 6.1b the bedform stability diagram is drawn for a dilute, 10 m thick turbidity current without a dense basal layer. The vertical concentration profile and turbulence characteristics are indicated for each panel in Fig. 6.1a. On the horizontal axes there is a dimensionless grain size, called particle parameter, $D^* = D[(\rho_s - \rho_w)g/(\rho_w v^2)]^{1/3}$, where D is the particle diameter, ρ_s is the density of the sediment and v is the viscosity of the water. On the vertical axes there is a dimensionless shear stress, called mobility parameter, $\theta' = \rho_w U^2/[(\rho_s - \rho_w)C'^2 D]$), where C' is the Chezy parameter ($C' = 18\log(1.25H/D)$), see Cartigny et al. (2014) for more details on these parameters. The vertical position on this panel and thereby the type of bedform that will develop is determined by the bed shear stress, which is dominantly controlled by the flow velocity. If the mobility parameter drops below the Shields curve, then the bed shear stresses are no longer sufficient to transport sediment and bedforms will not develop (Shields 1936). The critical densimetric Froude line separates the subcritical bedforms such as ripples and dunes (Guy et al. 1966) from supercritical bedforms such as antidunes and cyclic steps (Cartigny et al. 2014). The indicated length scale of the antidunes is calculated using the work of Hand (1974) and that of the cyclic step is based on numerical simulations of Kostic (2011), Cartigny et al. (2011).

Panel II in Fig. 6.1b shows the case for a slightly higher basal sediment concentration, as indicated on the third axes of the bedform stability diagram, but still lacks a clear basal

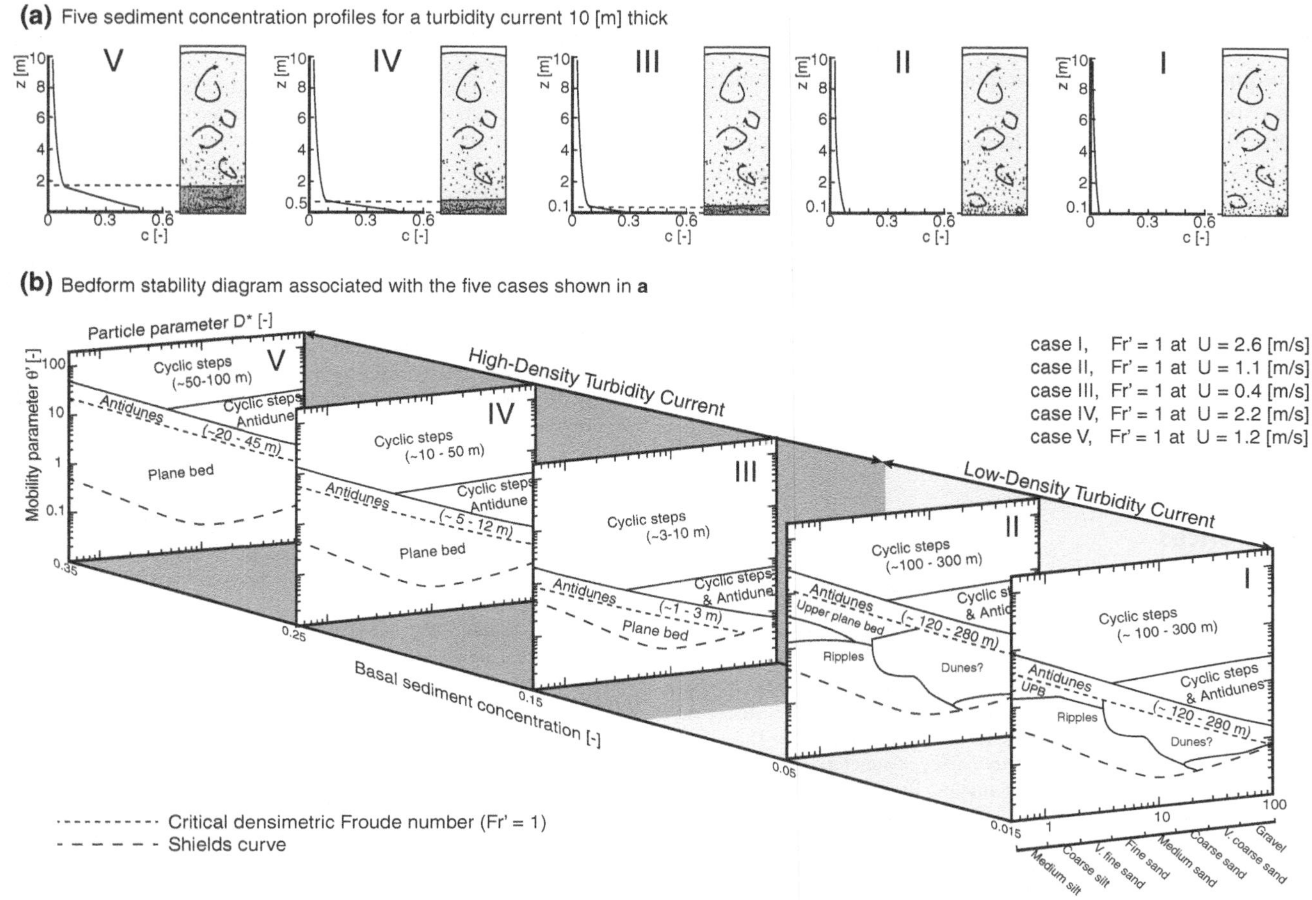

Fig. 6.1 **a** Five 10-m thick turbidity currents with different concentration profiles and resulting dense basal layers; **b** bedform stability diagrams associated with the five flow stratification types depicted in **a** (modified from Cartigny 2012)

density stratification. The stability diagram is very similar to that of case I, but the $Fr' = 1$ line has shifted upwards as a result of the increased sediment concentration.

Panel III is similar to case II, but now bedform formation is governed by a thin dense basal layer (0.1 m) at the base of the flow, which buffers the bed from the dilute overriding flow. Its limited thickness makes it prone to supercritical flow and causes the downward expansion of the supercritical flow domain in the diagram. Furthermore, the high sediment concentrations near the base of the flow are now thought to suppress the turbulence and thereby preventing the formation of any ripples and dunes.

Panel IV is characterized by a thicker (0.5 m) dense basal layer, as in case III, its thickness making it less prone to supercritical flow, thereby shifting the supercritical flow domain back up again in the diagram.

Panel V shows a very thick (2 m), dense basal layer, which again shifts the supercritical flow field further upwards.

6.3 Discussion

As only a few experimental and numerical studies have been done on turbidity current bedforms (Hand 1974; Kostic and Parker 2006; Spinewine et al. 2009; Sequeiros et al. 2010), it is emphasised that the bedform stability diagram presented here is based on many assumptions, and should be seen as a first attempt to estimate the impact of differences in flow dynamics of rivers and turbidity currents on bedform formation.

Although the existence of denser basal layers in turbidity currents is now becoming more widely accepted after they were directly observed from a submersible caught in a turbidity current (Sumner and Paull 2014), the morphodynamics of these basal layers still remains virtually unstudied. Here we assume that a dense basal layer is able to form supercritical bedforms such as antidunes and cyclic steps, as flow instabilities and their associated processes, like standing waves and hydraulic jumps, do not require a specific turbulence structure (Karcz and Kersey 1980; Weirich 1988).

Furthermore, turbidity currents are also likely to carry considerable amounts of clay in suspension, which have a profound effect on their morphodynamics, as demonstrated by experimental studies (Wan et al. 1994; Baas and Best 2008; Baas et al. 2009). These studies suggested that any bedform stability diagram for turbidity currents should also have a fourth dimension, in particular if the amount of transported clay is more than 6 % (see Baas et al. 2009).

The last limitation mentioned here is that the diagram as presented here is only intended for flows that are 10 m thick. Smaller or larger flow depth will have an impact on the position of the line representing critical densimetric Froude numbers, and hence for larger flows the supercritical-flow bedforms will move up in the diagram, while the opposite holds for thinner flows.

References

Baas, J.H. and Best, J.L. (2008). The dynamics of turbulent, transitional and laminar clay-laden flow over a fixed current ripple: Sedimentology, v. 55, p. 635–666.

Baas, J.H., Best, J.L., Peakall, J., and Wang, M. (2009). A phase diagram for turbulent transitional and laminar clay suspension flows: Journal of Sedimentary Research, v. 79, p. 162–183.

Cartigny, M.J.B. (2012). Morphodynamics of supercritical high-density turbidity currents. Utrecht Studies in Earth Sciences, PhD thesis, 153 pp.

Cartigny, M.J.B., Postma, G., Van Den Berg, J.H. and Mastbergen D. R. (2011). A comparative study of sediment waves and cyclic steps based on geometries, internal structures and numerical modeling. *Mar. Geol.*, 280, 40–56.

Cartigny, M.J.B., Eggenhuisen, J.T., Hansen, E.W.M. and Postma, G. (2013). Concentration-dependent flow stratification in experimental high-density turbidity currents and their relevance to turbidite facies models. Journal of Sedimentary Research, 83, 1046–1064, doi:10.2110/jsr.2013.71.

Cartigny, M.J.B., Ventra D., Postma, G. and Van Den Berg, J.H. (2014). Morphodynamics and internal structures of bedforms under supercritical-flow conditions. Sedimentology, doi:10.1111/sed.12076.

Guy, H. P., Simons, D. B., & Richardson, E. V. (1966). Summary of alluvial channel data from flume experiments, 1956–61 (No. 462-I).

Hand, B.M. (1974). Supercritical flow in density currents. *J. Sed. Res.*, 44(3), 637–648.

Hiscott, R. N. (1994). Traction-carpet stratification in turbidites-fact or fiction?. *Journal of Sedimentary Research*, *64*(2).

Hughes Clarke, J.E., Brucker, S., Muggah, J., Hamilton, T., Cartwright, D., Church, I. and Kuus, P. (2012). Temporal progression and spatial extent of mass wasting events on the Squamish prodelta slope, *in* Eberhardt, E., et al., eds., Landslides and engineered slopes: Protecting society through improved understanding: London, Taylor & Francis Group, p. 1091–1096.

Karcz, I. and Kersey, D. (1980). Experimental study of free-surface flow instability and bedforms in shallow flows. *Sedimentary Geology*, *27*(4), 263–300.

Kostic, S. (2011). Modeling of submarine cyclic steps: Controls on their formation, migration, and architecture. *Geosphere*, *7*(2), 294–304.

Kostic, S. and Parker, G. (2006). The response of turbidity currents to a canyon–fan transition: internal hydraulic jumps and depositional signatures. *Journal of Hydraulic Research*, *44*(5), 631–653.

Middleton, G. V. (1966). Experiments on density and turbidity currents: II. Uniform flow of density currents. *Canadian Journal of Earth Sciences*, *3*(5), 627–637.

Migeon, S., Savoye, B., Zanella, E., Mulder, T., Faugères, J.C. and Weber, O. (2001). Detailed seismic-reflection and sedimentary study of turbidite sediment waves on the Var sedimentary ridge (SE France): Significance for sediment transport and deposition and for the mechanisms of sediment-wave construction: Marine and Petroleum Geology, v. 18, p. 179–208, doi:10.1016/S0264-8172(00)00060-X.

Normark, W. R., Hess, G. R., Stow, D. A. V. and Bowen, A. J. (1980). Sediment waves on the Monterey Fan levee: a preliminary physical interpretation. *Marine Geology*, *37*(1), 1–18.

Mutti, E. (1992). Turbidite Sandstones. Agip Spec. Publ, Instituto de geologia, Univ. di Parma, Agip sp. A. 275 pp.

Parker, G., Fukushima, Y. and Pantin, H. M. (1986). Self-accelerating turbidity currents. *Journal of Fluid Mechanics*, *171*, 145–181.

Paull, C. K., Caress, D. W., Ussler, W., Lundsten, E., and Meiner-Johnson, M. (2011). High-resolution bathymetry of the axial channels within Monterey and Soquel submarine canyons, offshore central California. Geosphere, 7, 1077–1101.

Postma, G. and Cartigny, M. (2014). Supercritical and subcritical turbidity currents and their deposits- A synthesis, *Geology* 42, 987–990.

Postma, G., Nemec, W., and Kleinspehn, K. L. (1988). Large floating clasts in turbidites: a mechanism for their emplacement. *Sed. Geol*, 58, 47–61.

Postma, G., Cartigny, M. and Kleverlaan, K. (2009). Structureless, coarse-tail graded Bouma Ta formed by internal hydraulic jump of the turbidity current? Sedimentary Geology 219, 1–6.

Postma, G., Kleverlaan, K. and Cartigny, M. (2014). Recognition of cyclic steps in sandy and gravelly turbidite sequences and consequences for the Bouma facies model. Sedimentology 61: 2268–2290, DOI:10.1111/sed.12135.

Shields, A. (1936). *Anwendung der Aehnlichkeitsmechanik und der Turbulenzforschung auf die Geschiebebewegung*. Preussischen Versuchsanstalt für Wasserbau.

Skipper, K. (1971). Antidune cross-stratification in a turbidite sequence, Cloridorme Formation, Gaspé, Québec. *Sedimentology*, *17*(1-2), 51–68.

Smith, D. P., Ruiz, G., Kvitek, R. and Iampietro, P. J. (2005). Semiannual patterns of erosion and deposition in upper Monterey Canyon from serial multibeam bathymetry. *Geological Society of America Bulletin*, *117*(9–10), 1123-1133.

Sohn, Y.K. (1997). On traction-carpet sedimentation. *J. Sed. Res.*, 67 (3), 502–509. Southard and Boguchwal 1990.

Spinewine, B., Sequeiros, O. E., Garcia, M. H., Beaubouef, R. T., Sun, T., Savoye, B. and Parker, G. (2009). Experiments on wedge-shaped deep sea sedimentary deposits in minibasins and/or on channel levees emplaced by turbidity currents. Part II. Morphodynamic evolution of the wedge and of the associated bedforms. *Journal of Sedimentary Research*, *79*(8), 608–628.

Sequeiros, O. E., Spinewine, B., Beaubouef, R. T., Sun, T., García, M. H. and Parker, G. (2010). Characteristics of velocity and excess density profiles of saline underflows and turbidity currents flowing over a mobile bed. *Journal of Hydraulic Engineering*, *136*(7), 412–433.

Sumner, E. J. and Paull, C. K. (2014). Swept away by a turbidity current in Mendocino submarine canyon, California. *Geophysical Research Letters*, *41*(21), 7611–7618.

Talling, P. J., Masson, D. G., Sumner, E. J. and Malgesini, G. (2012). Subaqueous sediment density flows: Depositional processes and deposit types. *Sedimentology*, *59*(7), 1937–2003.

Taki, K. and Parker, G. (2005). Transportational cyclic steps created by flow over an erodible bed. Part 1. Experiments. *Journal of Hydraulic Research*, *43*(5), 488–501.

Ventra, D., Cartigny, M.J.B., Bijkerk, J. & Acikalin, S. (2015). Supercritical-flow structures on a Late Carboniferous delta front: sedimentologic and paleoclimatic significance. *Geology*. in press.

Walker, R. G. (1967). Upper flow regime bed forms in turbidites of the Hatch Formation, Devonian of New York State. *Journal of Sedimentary Research*, *37*(4).

Wan, Z., Wang, Z. and Julien, P. Y. (1994). Hyperconcentrated flow. *Journal of Hydraulic Engineering*, *120*(10), 1234–1234.

Weirich, F. H. (1988). Field evidence for hydraulic jumps in subaqueous sediment gravity flows. Nature, 332, 626–629.

Xu, J. P. (2010). Normalized velocity profiles of field-measured turbidity currents. *Geology*, *38*(6), 563–566.

Zhong, G., Cartigny, M. J., Kuang, Z. and Wang, L. (2015). Cyclic steps along the South Taiwan Shoal and West Penghu submarine canyons on the northeastern continental slope of the South China Sea. *Geological Society of America Bulletin*, *127*(5–6), 804–824.

7 Dense Shelf Water Cascading and Associated Bedforms

Pere Puig

Abstract

Sea-atmosphere interactions play an important role in oceanographic processes at various spatial and temporal scales. In the world oceans, several regions are key spots of intense air-sea interactions which considerably affect the heat and water budgets. An example of this is the formation of dense waters by cooling, evaporation or freezing over continental shelf regions. Dense shelf waters can eventually cascade to greater depths over the continental slope, being occasionally channelized through submarine canyons, until they reach an equilibrium depth of equal density. The enhanced current velocities associated with this oceanographic process are capable of generating erosive and depositional bedforms along the dense water pathways.

Keywords

Dense shelf water • Cascading • Sediment waves • Sedimentary furrows

7.1 Introduction

Dense shelf water cascading (DSWC) is a global atmospheric-driven oceanographic phenomenon common not only on high latitude continental margins but also on mid-latitude and tropical margins (Killworth 1983; Ivanov et al. 2004). DSWC is a specific type of buoyancy-driven current, in which dense water formed by cooling, evaporation or freezing in the surface layer over the continental shelf descends down the continental slope to greater depths. The general DSWC concept was formulated by Nansen (1906), who made the first direct measurements over the Rockall Bank in the North Atlantic Ocean (Nansen 1913). The term "cascading" was introduced by Cooper and Vaux (1949), but

P. Puig (✉)
Institute of Marine Sciences (ICM-CSIC), Passeig Marítim de la Barceloneta, 37-49, 08003 Barcelona, Spain
e-mail: ppuig@icm.csic.es

J. Guillén et al. (eds.), *Atlas of Bedforms in the Western Mediterranean*,
DOI 10.1007/978-3-319-33940-5_7

the same phenomenon is also referred to as "shelf/slope convection".

The dense water plumes cascade both along- and across-slope in an oblique angle under the influence of gravity, Earth rotation, friction and mixing, until their density reaches the hydrostatic equilibrium depth (Shapiro and Hill 1997; Shapiro et al. 2003). By incising the shelf edge, submarine canyons can channelize dense shelf waters and be preferential cross-margin pathways for such gravity flows (Canals et al. 2006; Allen and Durrieu de Madron 2009). This intermittent process contributes to the ventilation of intermediate and deep waters of the oceans and has a significant impact on biogeochemical cycles, entraining significant quantities of both dissolved and particulate minerals and organic material (Hill et al. 1998; Niemann et al. 2004).

Cascades of dense shelf waters can last for several weeks and the associated strong currents can induce erosion and resuspension of surface sediments on the outer shelf/upper slope and generate bottom nepheloid layers (i.e. layers of water that contains significant amounts of suspended sediment). These layers can be detached at intermediate levels when the density of the mixture of water and particles reaches its equilibrium depth or—if the density is great enough—evolve into a thick bottom nepheloid layer that can reach the lower continental slope and basin (Fig. 7.1).

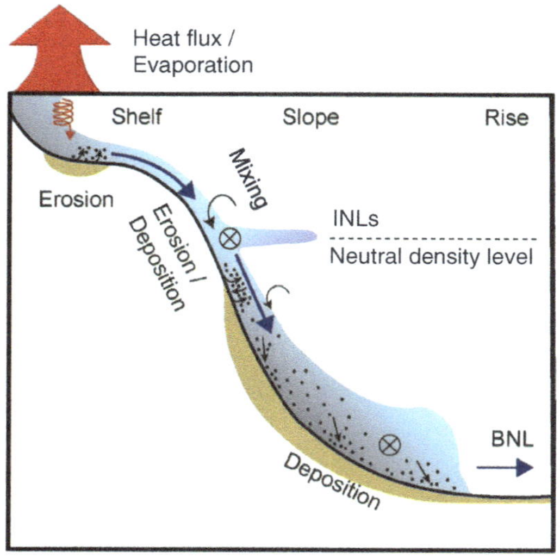

Fig. 7.1 Schematics of the DSWC mechanism illustrating the formation of intermediate nepheloid layers when water and particle detachments occur at the equilibrium depth of equal density, and of a thick bottom nepheloid layer when dense shelf waters reach the basin. Adapted from Fohrmann et al. (1998)

Additionally, the interaction with the seafloor can generate various types of erosional and depositional bedforms along the outer continental shelf and the upper continental slope. The SW Adriatic margin is one of the regions where the seafloor morphologies associated with DSWC have been best documented. This area receives the seasonal arrival of North Adriatic Dense Water (NAdDW), the densest water mass of the whole Mediterranean (T ~ 11 °C, S ~ 38.5, σt ~ 29.5–29.6) (Zore-Armanda 1963; Vilibič and Supic 2005), which is generated in the broad and shallow North Adriatic shelf through intense cooling and evaporation in January and February. NAdDW flows southward along the Italian coast and reaches the shelf edge of the southern Adriatic pit over a prolonged interval (typically several weeks) approximately 2 months later (Zoccolotti and Salusti 1987). Swath bathymetry and side-scan sonar (TOBI) images of the SW Adriatic margin, accompanied by a dense grid of Chirp sonar profiles, have defined an area of extreme seafloor complexity characterized by a variety of sediment waves, erosional scours, longitudinal furrows and giant comet-marks (Verdicchio and Trincardi 2006; Verdicchio et al. 2007; Trincardi et al. 2007). These distinctive bottom-current features are not randomly distributed but appear genetically linked with the off-shelf paths of NAdDW and have a consistent down-current arrangement obliquely to the slope, also reflecting the interaction with the geostrophic flow along the upper slope (Trincardi et al. 2007). Five distinct fields of muddy sediment waves occur at variable water depth along the SW Adriatic margin, down to the basin floor. All these fields are located downslope of large erosional areas. They have sediment waves of similar size and a common upcurrent (and upslope) migration pattern. Data regarding the distribution, morphology, orientation and stratigraphy of these patchy fields of mud waves are interpreted to record the long-term (millennial-scale) average of the seasonally variable direction and intensity of the NAdDW cascading currents. Recently, a broader mapping coverage of the SW Adriatic was reported by Foglini et al. (2015), providing a complete and more comprehensive picture of the morphologies resulting from this oceanographic process.

7.2 Dense Shelf Water Cascading in the Western Mediterranean

The Gulf of Lions is one of the principal areas in the western Mediterranean where massive dense water formation occurs because of cooling and evaporation of surface waters during wintertime. Concurrent with the well-known open-sea convection process over the Provençal-Ligurian subbasin (MEDOC group 1970), winter heat losses and evaporation induced by persistent, cold and dry northerly winds cause densification and mixing of coastal waters. Despite the buoyancy gain induced by freshwater inputs over the shelf, once, surface waters become denser than surrounding waters over the shelf sink, overflow the shelf edge and cascade downslope until they reach their equilibrium depth, which changes from year to year (Millot 1990; Durrieu de Madron et al. 2005).

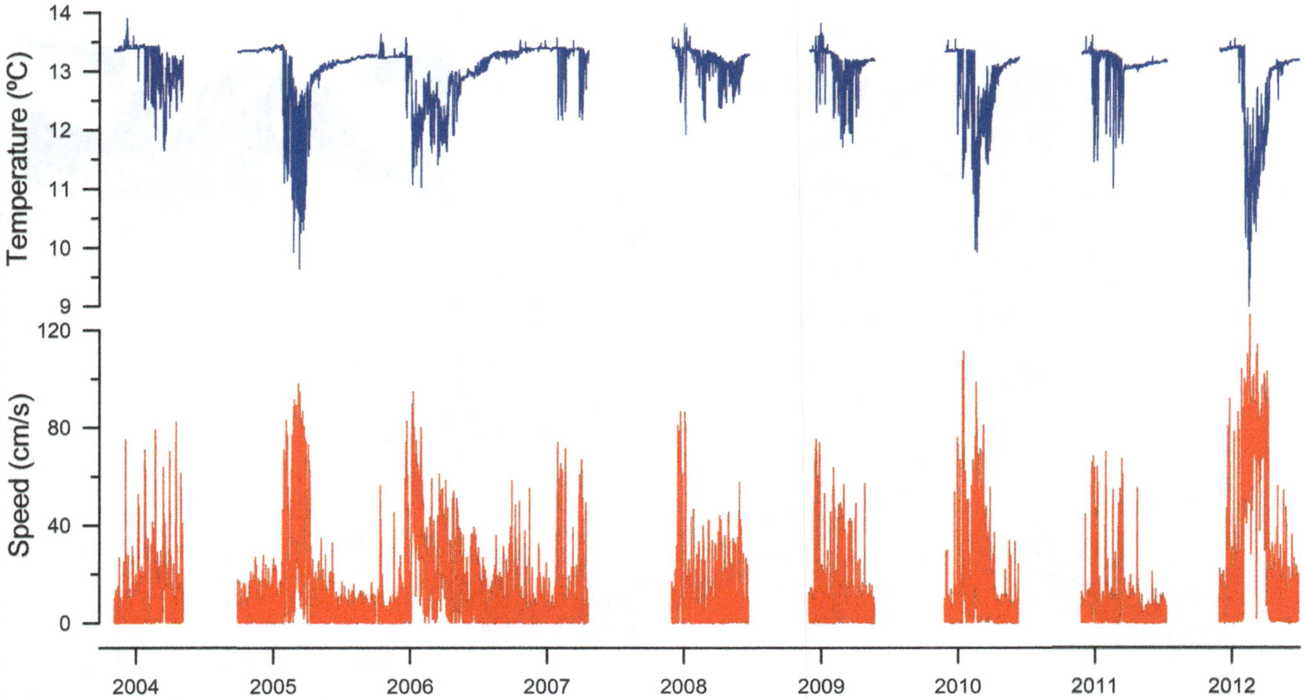

Fig. 7.2 Time series observations of near-bottom (5 mab) temperature and current speed measured at the head of the Cap de Creus submarine canyon (300–500 m water depth), illustrating the occurrence of DSWC events during the winters of 2004–2012

The first evidence of DSWC in the Gulf of Lions was found in 1953 by Bougis and Ruivo (1954), who identified the formation of dense waters along the southwestern shelf and their cascading into the Lacaze-Duthiers submarine canyon. DSWC events in the same area were also observed in 1969 by Fieux (1974) and in 1971 by Person (1974), who traced dense shelf waters down to 800 and 350 m depth, respectively. Further evidence was gained at the same location in winter 1995 by Lapouyade and Durrieu de Madron (2001), who observed a cascade at its final stage with a large tongue of cold water escaping the shelf and reaching its neutral density level around 170 m depth. More recently, a fine spatial resolution hydrographic survey performed in 2004 along the axis of the Cap de Creus Canyon (Durrieu de Madron et al. 2005) showed a cold and less salty filament attached to the seabed down to 350 m.

In all cases, temperature is the sole driver of cascading. At the initial stages, when dense waters escape the shelf and spread around the shelf break depth (150–200 m), they contribute to the formation of Western (previously known as Winter) Intermediate Water (WIW), whose typical properties are T = 12.5 °C, S = 38.0 (Dufau-Julliand et al. 2004). When dense shelf waters become colder, they cascade to deeper levels and mix with the warmer and saltier Levantine Intermediate Water (LIW) layer (T = 13.2 °C, S = 38.5) that extends between 200 and 1000 m depth or eventually with the underlying Western Mediterranean Deep Water (WMDW).

The monitoring of temperature, current and downward particle fluxes conducted since 1993 in the lower part of the Planier and Lacaze-Duthiers submarine canyons revealed an inter-annual variation of the DSWC intensity in the Gulf of Lions (see Heussner et al. 2006; Canals et al. 2006 for details). From these time series, Béthoux et al. (2002) inferred that, during the abnormally cold winter of 1999, the intense shelf cascading episode arrived at 1000 m depth on the continental slope, with downslope velocities of up to 60 cm/s, and contributed to the renewal of the bottom waters of the western Mediterranean basin at depths greater than 2000 m. Further time series observations in the Gulf of Lions were conducted in the framework of the EuroSTRATAFORM project, during which seven submarine canyon heads were monitored simultaneously in winter 2004. These results revealed that the preferential cyclonic circulation of the coastal current and the narrowing of the shelf at the southwestern end of the Gulf cause most of the water and sediment transport during DSWC events to occur through the Cap de Creus submarine canyon (Palanques et al. 2006). Since then, the occurrence and effects of dense shelf water cascades off the Gulf of Lions have been continuously observed at the Cap de Creus submarine canyon head (300–500 m depth) in the framework of several research projects (Fig. 7.2). Cascading events are easily recognizable since they are characterized by abrupt decreases in water temperature (<12 °C) associated with increases in current speed

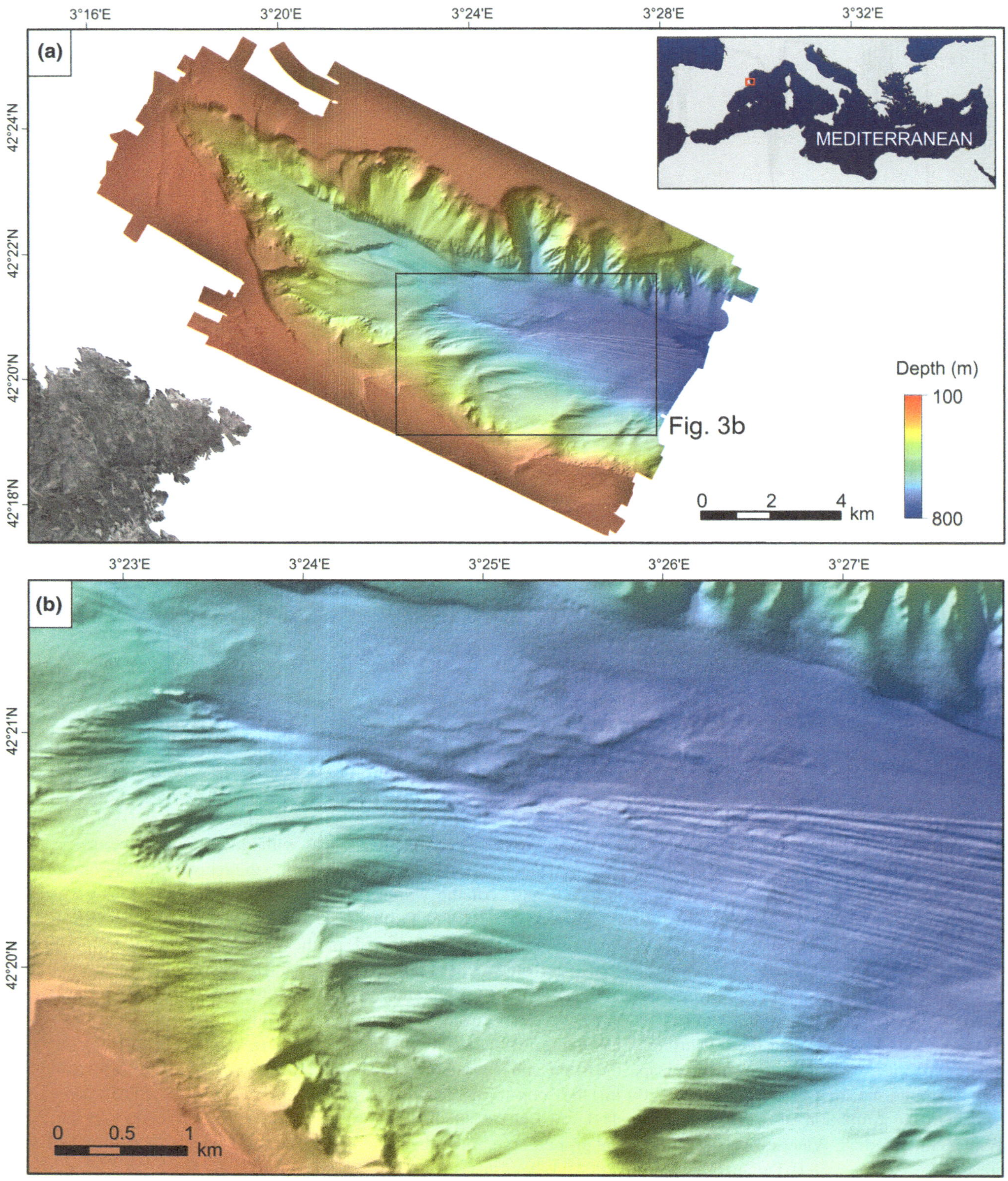

Fig. 7.3 **a** Shaded relief bathymetric map from the Cap de Creus submarine canyon head derived from a multibeam survey (10 × 10 m grid) acquired and processed by Fugro Survey Ltd. and AOA Geophysics Inc. **b** Detailed map of the canyon region where furrows start to develop

(>80 cm/s) and with increases in suspended sediment concentration (data not shown). Though this oceanographic process affects the upper canyon and continental slope region annually, the winters of 2005–06 and 2012–13 were characterized by major cascading events able to reach the basin seafloor. Under these circumstances, dense shelf waters propagated along and across the continental slope, reaching depths of more than 2000 m, where they merged with dense waters formed off-shelf by a typical open-sea convection process (Font et al. 2007; Durrieu de Madron et al. 2013). Previous severe winters with anomalous dense water formation were identified by Béthoux et al. (2002) after the analysis of historical hydrographic data. These events presumably took place at sub-decadal intervals in the winters of 1972, 1981, 1988 and in the previously mentioned winter of 1999.

Several seafloor morphologies identified in this region have been associated with this oceanographic phenomenon. The most noteworthy one is a large field of erosive sedimentary furrows that develop inside the Cap the Creus submarine canyon (Canals et al. 2006; Lastras et al. 2007; Puig et al. 2008; García-García et al. 2012; Fig. 7.3). The furrows field is observed down to 1500 m water depth, as long as the deep DSW cascades are confined within the canyon flanks, and starts to vanish where the canyon widens and the gravity current decreases its speed, losing its eroding capability.

Dense waters can also bypass the Cap de Creus canyon head and spread along the adjacent open slope, off the Gulf of Roses, where large muddy sediment waves genetically related to this process develop (see Ribó et al., this issue). Finally, a large field of muddy sediment waves found at the northern Catalan continental rise south of the Cap de Creus and La Fonera submarine canyons was also interpreted by Jallet and Giresse (2005) to be created by the occurrence of major (sub-decadal) cascading events able to reach the basin seafloor.

References

Allen, S.E., Durrieu de Madron, X. (2009). A review of the role of submarine canyons in deep-ocean exchange with the shelf. *Ocean Sci.*, 5: 607–620.

Béthoux, J.P., Durrieu de Madron, X., Nyffeler, F., Tailliez, D. (2002). Deep water in the western Mediterranean: peculiar 1999 and 2000 characteristics, shelf formation hypothesis, variability since 1970 and geochemical inferences, *J. Marine Syst.*, 33–34: 117–131.

Bougis, P., Ruivo, M., (1954). Sur une descente des eaux superficielles en profondeur (cascading) dans le sud du Golfe du Lion. *Bulletin d'Information du Comité Central d'Océanographie et d'Etude des Côtes*, 6, 147–154.

Canals, M., Puig, P., Durrieu de Madron, X., Heussner, S., Palanques, A., Fabrés, J. (2006). Flushing submarine canyons. *Nature*, 444 (7117): 354–357.

Cooper, L.N.H., Vaux, D. (1949). Cascading over the continental slope of water from the Celtic Sea. *J. Mar. Biol. Assoc. UK*, 28: 719–750.

Dufau-Julliand, C., Marsaleix, P., Petrenko, A., Dekeyser, I. (2004). 3D Modelling of the Gulf of Lion's hydrodynamics (NW Med.) during January 1999 (MOOGLI3 Experiment) and late winter 1999: WIW formation and cascading over the shelf break. *J. Geophys. Res.*, 109, C11002.

Durrieu de Madron, X., Zervakis, V., Theocharis, A., Georgopoulos, D. (2005). Comments on "Cascades of dense water around the world ocean". *Prog. Oceanogr.*, 64: 83–90.

Durrieu de Madron, X., Houpert, L., Puig, P., Sanchez-Vidal, A., Testor, P., Bosse, A., Estournel, C., Somot, S., Bourrin, F., Bouin, M.N., Beauverger, M., Beguery, L., Calafat, A., Canals, M., Cassous, C., Coppola, L., Dausse, D., D'Ortenzio, F., Font, J., Heussner, S., Kunesch, S., Lefevre, D., Le Goff, H., Martín, J., Mortier, L., Palanques, A., Raimbault, P. (2013). Interaction of dense shelf water cascading and open-sea convection in the Northwestern Mediterranean during winter 2012. *Geophys. Res. Lett.*, 40: 1379–1385.

Fieux, M. (1974). Formation d'eau dense sur le plateau du golfe du Lion. In: Processus de formation des eaux profondes. *Colloques Internationaux du CNRS*, 215: 165–174.

Foglini, F., Campiani, E., Trincardi, F. (2015). The reshaping of the South West Adriatic Margin by cascading of dense shelf waters. *Mar. Geol.*, in press.

Font, J., Puig, P., Salat, J., Palanques, A., Emelianov, M. (2007). Sequence of hydrographic changes in the NW Mediterranean deep water due to the exceptional winter 2005. *Sci. Mar.*, 72: 339–346.

Fohrmann, H., Backhaus, J.O., Blaume, F., Rumohr, J. (1998). Sediments in bottom-arrested gravity plumes: Numerical case studies. *J. Phys. Oceanogr.*, 28: 2250–2274.

García-García, A., Schoolmeester, T., Orange, D., Calafat, A., Fabres, J., Grossman, E., Field, M., Lorenson, T.D., Levey, M., Sansoucy, M. (2012). Recent sedimentary processes in the Cap de Creus canyon head and adjacent continental shelf, NE spain: evidence from multibeam bathymetry, sub-bottom profiles and coring. *Int. Assoc. Sedimentol. Spec. Publ.*, 44: 71–98.

Heussner, S., Durrieu de Madron, X., Calafat, A., Canals, M., Carbonne, J., Delsaut, N., Saragoni, G. (2006). Spatial and temporal variability of downward particle fluxes on a continental slope: Lessons from an 8-yr experiment in the Gulf of Lions (NW Mediterranean). *Mar. Geol.*, 234: 63–92.

Hill, A.E., Souza, A.J., Jones, K., Simpson, J.H., Shapiro, G.I., McCandliss, R., Wilson, H., Leftley, J. (1998). The Malin cascade in winter 1996. *J. Mar. Res.*, 56: 87–106.

Ivanov, V.V., Shapiro, G.I., Huthnance, J.M., Aleynik, D.L. Golovin, P.N. (2004). Cascades of dense water around the world ocean. *Prog. Oceanogr.*, 60: 47–98.

Jallet, L., Giresse, P. (2005). Construction of the Pyreneo-Languedocian Sedimentary Ridge and associated sediment waves in the deep western Gulf of Lions (Western Mediterranean). *Mar. Petrol. Geol.*, 22: 865–888.

Killworth P. (1983). Deep convection in the world Oceans. *Rev. Geophys. Space Phys.* 21: 1–26.

Lapouyade, A., Durrieu de Madron, X. (2001). Seasonal variability of the advective transport of particulate matter and organic carbon in the Gulf of Lion (NW Mediterranean). *Oceanol. Acta*, 24: 295–312.

Lastras, G., Canals, M., Urgeles, R., Amblas, D., Ivanov, M., Droz, L., Dennielou, B., Fabrés, J., Schoolmeester, T., Akhmetzhanov, A., Orange, D., García-García, A. (2007). A walk down the Cap de Creus canyon, Northwestern Mediterranean Sea: Recent processes inferred from morphology and sediment bedforms. *Mar. Geol.*, 246: 176–192.

MEDOC group. (1970). Observation of formation of deep water in the Mediterranean Sea, 1969. *Nature*, 227: 1037–1040.

Millot, C.A. (1990). The Gulf of Lions' hydrodynamic. *Cont. Shelf Res.*, 10, 885–894.

Nansen, F. (1906). Northern Waters. Captain Roald Amundsen's oceanographic observations in the Arctic Seas in 1901. *In:* Vid-selskap. Skrifter I, Mat.-Naturv. kl. 1(3), Dybvad, Christiania, 145 p.

Nansen, F. (1913). The waters of the north-eastern North Atlantic, *In:* Internationale Revue der Gesamten Hydrobiologie und Hydrographie. Suppl. to Bd.4, 139 p.

Niemann, H., Richter, C., Jonkers, H.M., Badran, M.I. (2004). Red Sea gravity currents cascade near-reef phytoplankton to the twilight zone, *Mar. Ecol. Prog. Ser.*, 269: 91–99.

Palanques, A., Durrieu de Madron, X., Puig, P., Fabrés, J., Guillén, J., Calafat, A., Canals, M., Heussner, S., Bonnin, J. (2006). Suspended sediment fluxes and transport processes in the Gulf of Lions submarine canyons. The role of storms and dense water cascading. *Mar. Geol.*, 234: 43–61.

Person, R. (1974). Un exemple de descente des eaux superficielles du plateau continental dans un canyon du Golfe du Lion. *Colloques Internationaux du CNRS*, 215: 175–189.

Puig, P., Palanques, A., Orange, D.L., Lastras, G., Canals, M. (2008). Dense shelf water cascades and sedimentary furrow formation in the Cap de Creus Canyon, northwestern Mediterranean Sea. *Cont. Shelf Res.*, 28: 2017–2030.

Ribó, M., Durán, R., Puig, P., Van Rooij, D., Guillén, J. (2016). Large sediment waves over the Gulf of Roses continental slope (NW Mediterranean), *this issue*.

Shapiro, G.I., Hill, A.E. (1997). Dynamics of Dense Water Cascades at the Shelf Edge. *J. Phys. Ocean.*, 27: 2381–2394.

Shapiro, G.I., Huthnance, J.M., Ivanov, V.V. (2003). Dense Water Cascading off the Continental Shelf. *J. Geophys. Res.*, 108. C12, 3390.

Trincardi F., Verdicchio G., Miserocchi S. (2007). Sea-floor evidence for the interaction between Cascading and along-slope bottom-water masses. *J. Geophys. Res.*, 112, F03011.

Verdicchio, G., Trincardi, F. (2006). Short-distance variability in slope bed-forms along the Southwestern Adriatic Margin (Central Mediterranean). *Mar. Geol.*, 234: 271–292.

Verdicchio, G., Trincardi, F., Asioli, A. (2007). Mediterranean bottom-current deposits: an example from the Southwestern Adriatic Margin. In: Viana, A.R. and Rebesco, M. (eds.) Economic and Palaeoceanographic Significance of Contourite Deposits. *Geol. Soc. SP*, 276: 183–204.

Vilibič, I., Supic, N. (2005). Dense water generation on a shelf: the case of the Adriatic Sea. *Ocean. Dynam.* 55: 403–415.

Zoccolotti, L., Salusti, E. (1987). Observations of a vein of very dense marine water in the southern Adriatic Sea. *Cont. Shelf Res.*, 7: 535–551.

Zore-Armanda, M. (1963). Les masses d'eau de la mer Adriatique. *Acta Adriatica*, 10: 5–88.

Currents in the Western Mediterranean Basin

8

E. García-Ladona

Abstract

The main circulation patterns in the western Mediterranean are revisited and updated. The schemes are based on the previous literature, adapted and reviewed. The update concerns in particular the circulation at intermediate depths, exemplified by the Levantine Intermediate Water circulation, and includes deep water cascading and recent observations from drifting floats and mooring recordings.

Keywords

Ocean currents • Western Mediterranean • Circulation patterns

8.1 Introduction

One of the primary constraints of ocean circulation is the bottom topography. Compared with the atmosphere, which can freely evolve over the Earth, ocean circulation is limited by the lateral boundaries of continents and the bottom relief of the basins. Both global and basin-scale circulation patterns, such as sub-tropical gyres, western intensified boundary currents (e.g. the Gulf Stream and the Kuroshio) and upwelling systems (e.g. California, Peru and Benguela), are subject to the interplay of large-scale forcing due to wind stress and differential heating and geometrical constraints (basin shape and boundaries).

Currents near the bottom and along coastal boundaries transport nutrients, organic matter and sedimentary material, which fertilize benthic communities. Erosion, advection and sedimentation of suspended matter through currents contribute over long periods to the configuration of the boundary and bottom morphology. Strong bottom boundary currents, such as gravity and turbidity flows (hyperpycnal flows) can cause drastic changes in the seafloor, in particular in shelf/slope areas and submarine canyons (Canals et al. 2006; Trincardi et al. 2007). These gravity flows may be generated either by spontaneous instabilities of sedimentary deposits (submarine landslides) or by cascading of dense waters from the surface (Ivanov et al. 2004; Durrieu de Madron et al. 2005). Storms surges, tidal flows and internal waves are also known to set up strong resuspension events (Van Rijn 1993). Further, anthropogenic activities related to intensive trawling fisheries are now highlighted as being responsible for systematic resuspension events and gravity flows in exploited areas (Palanques et al. 2006; Puig et al. 2012). All these phenomena are very energetic, often appearing suddenly and acting in a very short time in which potential energy is rapidly converted to kinetic energy that causes a fast vertical transport of properties.

There are, however, other kinds of flows that may shape the structure of the bottom boundary topography associated with changes over much longer times. These flows are driven by pressure and thermohaline changes which are strongly constrained by lateral basin boundaries, often intensified or maintained by density differences between fresh coastal waters and more salty open ocean waters. These flows, known as frontal currents, tend to be aligned along the shelf/slope bathymetry, with a strong dynamic tendency to follow the isobaths and a great persistence over time, leading trapped particles to be transported over great distances. Additionally, the arrangement of forces in the bottom boundary layers at scales where rotation is important may produce a systematic transport of material at a certain angle to the main current

E. García-Ladona (✉)
DOFT, Institut de Ciències del Mar (CSIC), 08039 Barcelona, Catalunya, Spain
e-mail: emilio@icm.csic.es

J. Guillén et al. (eds.), *Atlas of Bedforms in the Western Mediterranean*,
DOI 10.1007/978-3-319-33940-5_8

direction (e.g. Shapiro et al. 2004), thus favouring the detachment of fine material from the bottom, where the front impinges the seafloor (Garret et al. 1993), or enhancing across-shelf exchanges near the submarine canyons (Allen and Durrieu de Madron 2009). However, frontal currents are in general less suitable for resuspending matter because the usual stresses at the boundary layer may not reach the critical values needed to produce vigorous uplift of material. We can therefore consider conceptually that suspended material is supplied by fresh water discharges and the aforementioned mechanisms of resuspension. The main large-scale currents (especially shelf/slope frontal currents) then capture, transport and redistribute matter far away.

8.2 The Mediterranean Basin

The Mediterranean basin is a semi-enclosed basin in which most permanent currents are organized around the coastal boundaries. The first Mediterranean circulation scheme goes back to Nielsen in 1912 based on data collected during the 1908–1909 THOR hydrographic cruises (Nielsen 1912). In the 1960s, I.M. Ochinnikov drew up new schemes of basin scale geostrophic currents incorporating all the previous hydrographic measurements, which were then based on inverse thermometers and water bottle samples (Ovchinnikov 1966). Through subsequent observations using modern equipment (e.g. CTD) and remote sensing measurements (SST, altimetry, etc.), general circulation schemes have been progressively improved, offering greater detail and seasonal variability (see Fig. 2 in Millot and Taupier-Letage 2005a and references therein). While there is some consensus on the circulation scheme in the western basin, there are still some controversial points about the circulation in the eastern basin (e.g. Malanotte et al. 2014). The Mediterranean circulation is mostly determined by thermohaline forcing, as the basin has a positive buoyancy balance between evaporation minus precipitation and river inflows. The deep water formation events in winter in the eastern and western basins and in the Adriatic and Aegean Seas, the permanent supply of fresh waters from rivers, and the Atlantic water inflow through the Gibraltar Strait are the main elements that drive the thermohaline character of the whole circulation basin.

In the western Mediterranean basin the vertical structure of water masses is characterized by three main characteristic layers: mode waters generically called Atlantic Waters in the upper part, Levantine Intermediate Waters (LIW), and Deep Mediterranean waters. Other mode waters, such as Western Intermediate Water, result from relatively conventive regimes in winter and are often found between surface and LIW waters. The alongslope surface frontal flows often reach depths of 400–500 m on the shelf-slope, just over the LIW layer, being strongly constrained by the local topography (e.g. Millot and Taupier-Letage 2005a). Though kinetic energy of frontal flows at deeper layers is generally lower than that of intermittent downslope gravity currents and deep water convection events, these contouring flows are almost permanently active and are therefore a mechanism for modulating the seafloor. From model simulations of the Iberian and western Mediterranean basins, Hernández-Molina et al. (2011) showed the high correlation between some well-known contouring forms and the resulting circulation. Alonso et al. (2014) studied how the detailed topographic features of the Alboran Sea, close to the Gibraltar Strait, can be explained by the expected currents over the depth associated with the recirculation of surface and deep waters in the area occupied by the Western Alboran Gyre.

8.3 Methods and Observations

As mentioned above, the main sources of data leading to the known schemes of surface circulation have been hydrographic measurements, satellite infrared images, altimetry, surface moorings and drifters. The circulation patterns of deeper layers are much more difficult to access than those of surface layers, and mainly rely on direct measurements collected from records through moored current meters (e.g., Schott and Leaman 1991; Flexas et al. 2002; Testor et al. 2005; Palanques et al. 2012). Although they do not cover the entire basin, they have provided us with quantitative values of deep current speeds and many general figures have led to the construction of general circulation schemes (panels B-C, e.g. see Millot 1999; Millot and Taupier-Letage 2005a). Presently, a more complete view of deep circulation currents can be derived for larger areas by tracking consecutive positions of autonomous profilers. In Testor and Gascard (2005) and Testor et al. (2005) the circulation of LIW in the central Algerian basin is described and discussed from the analysis of RAFOS floats (floats that remain with neutral buoyancy at a given density level or isobaric at some pressure level).

Presently, as part of the international Argo programme, the MedArgo programme (Poulain et al. 2007) has deployed autonomous profilers that self-adjust their buoyancy to remain at a deep layer (parking depth) for some time and go up and down, making a vertical profile every 5–10 days. The whole paths of these floats reveal in fact how deep currents are configured at basin scales and velocities can be roughly estimated by the analysis of contiguous locations of vertical profiles (Park et al. 2005; Menna and Poulain 2010).

Here previously proposed and widely accepted circulation schemes have been adapted and complemented with recent sources of information and published work. In particular, more details have been added concerning the Tyrrhenian basin, the Algerian basin and the Catalano-Balearic Basin,

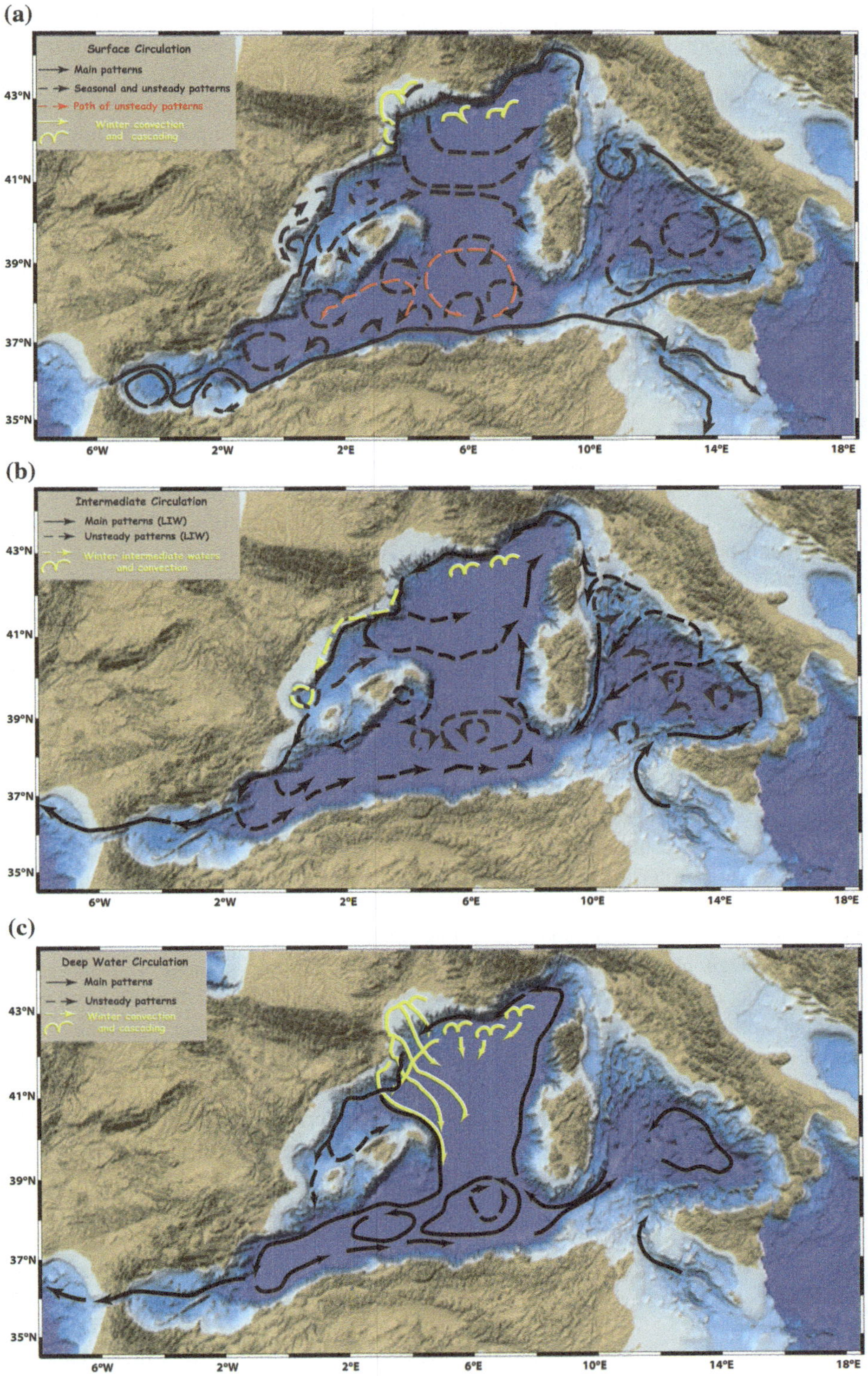

Fig. 8.1 Schemes of the main surface (**a**), intermediate (**b**) and deep (**c**) current patterns in the western Mediterranean basin. *Black solid lines* represent relatively well-established steady patterns. *Black dashed lines* correspond to less steady or seasonal patterns. *Yellow lines* represent patterns associated with deep water flows formed by deep water formation and cascading processes from the shelf area. The drawn features are based on and/or adapted from the schemes and references cited in the text. The background topography is based on the GEBCO one-minute interval grid v. 2.0 (2008)

mainly associated with the circulation of deep and intermediate waters. The panels of Fig. 8.1 represent schematically the circulation in the western basin for each layer.

In terms of magnitude, surface currents in the Mediterranean are generally moderate (0–1 m/s) compared with most of the world ocean current systems, such as the Gulf Stream, the Kuroshio current and the Antartic Circumpolar Current (1–2 m/s). Nevertheless, the Gibraltar Strait at the entrance of the Alboran Sea is a place where strong currents can reach peak speeds of 1.5–2 m/s under specific conditions of internal bores propagating during the spring tide period (Sotillo et al. 2015). The Algerian current is perhaps the most intense surface current in the western Mediterranean, with observed speeds close to 1 m/s. In the north, the Northern current is relatively strong, with surface velocities of around 0.5–0.75 cm/s particularly along the Italian-French coast and slower values in the southern sector along the northwestern Spanish coast (0.4–0.5 cm/s). Obviously, these values are general figures that correspond to currents at the surface when the wind forcing is not the main driver of the circulation. Although intermediate and deep circulation is less intense than surface circulation, it may reach high intensities, as measurements have revealed in several locations. At the LIW levels and deeper along the northern slope of Sicily, Sparnochia et al. (1999) reported values of about 20–40 cm/s. In very deep zones in the central part of the Algerian basin, average speeds of 0–5 cm/s have been observed (Millot 1999) but maximum values can be around 15 cm/s at 1800 and 2700 m (Testor et al. 2005).

In the northern part of the western Mediterranean basin, where the flow of the Northern current almost follows the basin topography, intensive measurements have been undertaken with modern current meters close to the bottom relief. There, submarine canyons are a characteristic features that constrain the flow according to their width; in narrow canyons the flow is polarized by the canyon axis, while in wider canyons the flow can adjust to the geometry of the canyon walls. Measurements in several canyons along the northwestern Mediterranean area indicate low mean velocities but with peaks values that can be in the range 10–15 cm/s inside the canyon and 5–20 cm/s on the adjacent slope at 30 m above the bottom (e.g. Puig et al. 2000; Palanques et al. 2005). While these values are very common in many other places, flows close to the bottom relief are often disrupted due to exceptional events during deep water formation and cascading of dense water downslope in this area of the western Mediterranean basin. Intense currents with values of up to 80 cm/s at 500 m depth over the slope and over 20–40 cm/s in the central basin at depths as great as 1830 and 2130 and at 15 m above the bottom have been recorded (Font et al. 2007; Salat et al. 2010). These events may last from a few days to 2 to 3 months during the winter season.

8.4 Surface Circulation

Surface waters tend to flow in a cyclonic sense around the basin geometry. The two main features are the input flow of Atlantic water through the Gibraltar Strait and frontal flows associated with the contrast between fresher waters over the shelves along the basin boundaries and saltier water over the deep basin. In the Alboran Sea the surface waters tend to recirculate and form two quasi-permanent gyres: the western (WAG) and the eastern (EAG) Alboran gyres. The WAG is more permanent and sometimes appears decoupled from the input of Atlantic water that progresses around its boundary to the eastern Alboran. The EAG is less stable and often disappears completely or becomes strongly deformed. Further east, the Atlantic flow impinges the African coast at 1° E and continues to flow along the coast until the Sardinia channel, constituting the so-called Algerian current (AC). Along its path it develops instabilities that often detach and evolve into large long-lived eddies (Algerian eddies, AE), which are mostly anticyclonic and recirculate within the Algerian basin (e.g. Puillat et al. 2002). By applying the Okubo-Weiss technique to detect eddies from sea level anomaly maps, Isern-Fontanet et al. (2006) found the main circuit followed by AE in the central part of the Algerian basin. The analysis of the 1992–1999 time series of these maps shows two main circuits (see the dashed red arrows in panel A): a closed central path embedded between the Balearic islands the Sardinia island and the African coast; and a secondary one starting at 4° E, where AE are constrained to move westward along the Balearic Islands slope (Isern-Fontanet et al. 2006).

Once the AC arrives at 9° E, south of Sardinia, it mainly goes towards the eastern basin but bifurcates partially, entering the Tyrrhenian sub-basin. Inside the Tyrrhenian basin the circulation is essentially cyclonic along the Italian slope until the Corsica channel. A branch recirculates southwards along the Corsica-Sardinia slope, while in the centre many subgyres, bifurcations and mesoscale eddies have been observed (Vetrano et al. 2010). Whether these structures, such as the Bonifacio gyre between 41° and 42° N in front of the Bonifacio Strait, are persistent enough or vary seasonally is yet to be determined.

From the Tyrrhenian basin and through the Corsica channel, the surface circulation continues to flow inside the Ligurian Sea along the French-Italian coast, reaching the external part of the Gulf of Lions and going southwards along the shelf break of the northwestern Spanish coast. This constitutes the Liguro-Provençal-Catalan current system, also known as the Northern current (NC, Millot 1999), which goes as far as the Ibiza channel. From Genoa (°9 E) to Toulon-Marseille (°6 E) the shelf is practically non-existent or extremely narrow and the flow follows the coastline as a fairly narrow vein. From 6° E the NC usually continues southwestward, following the shelf break of the Gulf of Lions up to the northwestern Spanish coast at cap of Creus (around 42°15″ N, 3°18″ E) and then continues southward along the shelf break. From the Toulon-Marseille area, the NC flow often intrudes the Gulf of Lions, developing a meandering structure and forming eddies (Flexas et al. 2002; Rubio et al. 2009). Along the Italian-French-Spanish coast several river outflows (e.g. Rhône, Ebro) reinforce the frontal character of this current, which exhibits great seasonal variability. However, some eastward recirculation of the NC has been evidenced from the combined analysis of surface drifter tracks and satellite altimetry (Poulain et al. 2011, 2012, 2013). A recirculation of the NC, first to the south near the Rhône deep-sea fan area and then eastward to the west coast of Corsica, has been observed. Sometimes the recirculation is observed at a lower latitude (42° N, Rubio et al. 2009) and sometimes the NC can return within the Catalan-Balearic sub-basin. The atmospheric forcing (especially in winter), the north-south extent of the Balearic front and the conditions in the Ibiza channel may perhaps influence the latitude at which the NC recirculates (López García et al. 1994).

8.5 Intermediate Circulation

\In panel B we have schematically represented the circulation at the LIW level from the first schemes provided by Millot (1999) and updated with the information from Lagrangian measurements (Testor and Gascard 2005; Testor et al. 2005; Menna and Poulain 2010). LIW enters the western basin by the Sicily Strait and progresses around the Tyrrhenian basin in an anticlockwise sense, also following the Sardinia-Corsica shelf/slope from the Sardinia channel to the Ligurian Sea, quite similarly to the general surface circulation pattern. Several sub-gyres and bifurcations have also been observed to form in the central part of the Tyrrhenian basin. The Bonifacio gyre has been detected as a characteristic pattern at the LIW level (400 m), indicating the barotropic character of this structure (Vetrano et al. 2010). From numerical and inverse techniques it has been deduced that part of the LIW flow reaches the Liguro-Provençal basin through the Corsica channel, and the remainder goes to the Algerian basin, following the Corsica-Sardinia coast induced by the basin topography and cyclonic flow separations from the Italian slope. In the southwestern corner of Sardinia the flow can detach and enter the central part of the Algerian basin, either performing a large elliptic cell or being attached and entrained by Algerian eddies (e.g. Testor and Gascard 2005; Millot and Taupier-Letage 2005b). In the north, the LIW progresses along the shelf break until it reaches the Ibiza channel. At some places, detachments from this circuit have been observed from Argo floats, suggesting a deflection along the northern side of the Balearic Island. Then, it turns southwards following the eastern Balearic shelf break (Menna and Poulain 2010). Mooring measurements along the western slope of the Balearic Island indicate that LIW may flow along this side, probably detached from the Spanish shelf break or recirculating from the Gulf of Valencia (Amores and Montserrat 2014). There, an anticyclonic structure has been observed to form, favoured by the shape of the Gulf of Valencia. The path of LIW in the south continues towards the deep region of Cape Gata, where it penetrates the Alboran Sea facilitated by a channel in the middle and leaves the Mediterranean through the Strait of Gibraltar, thus forming the major source of salinity for the world ocean. A recirculation along the African slope is also inferred from the Argo data (Menna and Poulain 2010).

8.6 Deep Circulation

In panel C, a rough view of the circulation in the deep layers is represented, although it cannot be assigned to a specific level. Once again, the major sources of information are the same as for the intermediate layers. The main figures are closely correlated with the LIW circulations patterns and with the basin topography. The bottom circulation in the western Mediterranean basin was analysed by Álvarez et al. (1994), who accounted for eddy-topography interactions (Neptune effect). They showed circulation patterns, even for surface layers, clearly impacted by the Neptune effect, suggesting a strong influence of the sea bed in configuring the main currents. The circulation in deeper layers is dynamically constrained by f/H contours (f is the Coriolis frequency and H is the water depth). Because the range of latitude values that confines the Mediterranean basin is narrow, the circulation is thus forced to follow the topographic contours. This phenomenon has been observed in the Algerian basin, where two elliptic cells have been evidenced from trajectories of deep floats constrained by the local gentle topography (Testor et al. 2005). The trajectories of deep Argo floats in the Tyrrhenian Sea also exhibit similar patterns, adjusting their shape to the irregular configuration of the bottom basin. In the western part of the basin, from the north to the Alboran Sea, the patterns are very similar to those of intermediate circulation. In the Balearic Basin detachments similar to those found at intermediate depths can be observed from Argo trajectories and mooring data (Amoros and Montserrat 2014). Finally, observational evidence has shown that Mediterranean deep waters, similarly to LIW water mass, may escape the Mediterranean basin through the Strait of Gibraltar (Kinder and Parrilla 1987; Millot 2008).

The existence of deep water formation areas in the northern part of the western basin deserves particular attention (see yellow arrows in panel 1a–c). Dry cold winds going down the Rhône valley and along the Pyrenees (the so-called *mistral* and *tramontana* winds) are able to create pools of dense water at the surface that convect into the bottom layer. The penetration depends on the buoyancy of dense plumes and occurs at small scales in the northern area in front of the Gulf of Lion in the Liguro-Provenzal basin (e.g. Schott and Leaman 1991; Schott et al. 1996). Over the shelf, inside the Gulf of Lion and at the northern part of the Spanish coast, a similar phenomenon occurs in such a way that dense water may form and recirculate through the shelf, cascading mainly through the submarine canyons that incise the shelf/slope in that area (Durrieu de Madron et al. 2005; Canals et al. 2006; Puig et al. 2013). This cascading produces a supply of particulate matter to the deep seafloor, which, depending on the intensity and amount of available dense water, has been shown to spread over the deepest part of the Algerian basin, as is represented in panel 1c (adapted from Puig et al. 2013). If the densification process is not so strong, Western Intermediate Water is formed, remaining relatively shallow and being found over the LIW layer in the Balearic Basin (Pinot and Ganachaud 1999). This Western Intermediate Water is assumed to be advected downstream reaching the Gulf of Valencia and also to form some anticyclonic structures there (see panel 1b). Depending on the amount of water formed and the circulation in the Ibiza channel, this layer blocks the output of surface waters, forcing them to recirculate through the Balearic Basin (Pinot et al. 2002).

Acknowledgments This is a contribution of FORMED project (CGL2012-33989) funded by the Spanish Ministry of Economy and Competitiveness (MINECO). The author wishes to thank M. Farran and F. Estrada for their technical assistance in preparing the panels and reviewers for helping to improve the manuscript.

References

Álvarez A., Tintoré J., Holloway G., Eby M. and J.M. Beckers. (1994). Effect of topographic stress on the circulation in the western Mediterranean, Journal of Geophysical Research, 99, 16053–16064.

Alonso B., Ercilla G., García M., Vázquez J.T., Juan C., Casas D., Estrada F., D'Acremont E., Gorini C., El Moumni B. and M. Farran. (2014). Quaternary Mass-Transport Deposits on the North-Eastern Alboran Seamounts (SW Mediterranean Sea). In: Submarine Mass Movements and Their Consequences, Advances in Natural and Technological Hazards Research 37. Ed. S. Krastel et al. Springer. Chap. 50. 561–570. DOI:10.1007/978-3-319-00972-8_50, ISBN: 978-3-319-00972-8.

Allen S.E. and X. Durrieu de Madron. (2009). A review of the role of submarine canyons in deep-ocean exchange with the shelf, Ocean Sciences, 5, 607–620.

Amores A. and S. Montserrat. (2014). Hydrodynamic comparison between the north and south of Mallorca Island. Journal of Marine Systems, 138, 40–50.

Canals M., P., Durrieu de Madron X., Heussner S., Palanques A. and J. Fabres. (2006). Flushing submarine canyons, Nature, 444, 354–357.

Durrieu de Madron X., Zervakis V., Theocharis A. and D. Georgopoulos. (2005). Comments on Cascades of dense water around the world ocean. Progress in Oceanography, 64, 83–90.

Font J., P., Salat J., Palanques A., Emelianov M. (2007). Sequence of hydrographic changes in NW Mediterranean deep water due to the exceptional winter of 2005. Scientia Marina 71(2), 339–346.

Flexas M.M., Durrieu de Madron X., García M.A., Canals M. and P. Arnau. (2002). Flow variability in the Gulf of Lions during the MATER HFF experiment (March-May 1997). Journal of Marine Systems, 33–34, 197–214.

López García, M.J., Millot C., Font J. and E. García-Ladona. (1994). Surface circulation variability in the Balearic Basin. Journal Of Geophysical Research, 99, C2, 3285–3296.

Garret C., MacCready P. and P. Rhines. (1993). Boundary mixing and arrested Ekman Layers: rotating stratified flow near a sloping boundary. Annual Review of Fluid Mechanics, 25: 291–323.

Hernández-Molina F.J., Serra N., Stow D.A.V., Llave E., Ercilla G., Van Rooij D. (2011). Along-slope oceanographic processes and sedimentary products around the Iberian margin, Geo-Mar Lett (2011) 31:315–341. DOI: 10.1007/s00367-011-0242-2

Ivanov V.V., Shapiro G.I., Huthnance J.M., Aleynik D.L. and P.N. Golovin. (2004). Cascades of dense water around the world ocean. Progress in Oceanography, 60, 47–98.

Isern-Fontanet J., García-Ladona E. and J. Font (2006). Vortices of the Mediterranean Sea: An Altimetric Perspective. Journal of Physical Oceanography, 36, 87–103. doi:http://dx.doi.org/10.1175/JPO2826.1.

Kinder T.H. and G. Parrilla. (1987). Yes, Some of the Mediterranena Outflow does come from great depth. Journal of Geophysicical Research, 92, C3, 2901–2906.

Malanotte-Rizzoli P., Artale V., Borzelli-Eusebi G.L., Brenner S., Crise A., Gacic M., Kress N., Marullo S., Ribera d'Alcalá M., Sofianos S., Tanhua T., Theocharis A., Álvarez M., Ashkenazy Y., Bergamasco A., Cardin V., Carniel S., Civitarese G., D'Ortenzio F., Font J., García-Ladona E., García-Lafuente J. M., Gogou A., Gregoire M., Hainbucher D., Kontoyannis H., Kovacevic V., Kraskapoulou E., Kroskos G., Incarbona A., Mazzocchi M. G., Orlic M., Ozsoy E., Pascual A., Poulain P.-M., Roether W., Rubino A., Schroeder K., Siokou-Frangou J., Souvermezoglou E., Sprovieri M., Tintoré J. and G. Triantafyllou. (2014). Physical forcing and physical/biochemical variability of the Mediterranean Sea: a review of unresolved issues and directions for future research. Ocean Science, 10, 281–322. DOI:10.5194/os-10-281-2014.

Menna M. and P.M. Poulain. (2010). Mediterranean intermediate circulation estimated from Argo data in 2003–2010. Ocean Sciences, 6, 331–343.

Millot C. (1999). Circulation in the Western Mediterranean Sea, Journal of Marine Systems, 20, 423–442.

Millot C. (2008). Another description of the Mediterranean Sea outflow. Progress in Oceanography, 82, 101–124.

Millot C. (2013). Levantine Intermediate Water characteristics: an astounding general misunderstanding! Scientia Marina, 77(2), 217–232.

Millot C. and A. Monaco. (1984). Deep Strong Currents and Sediment Transport in the Northwestern Mediterranean Sea. Geo-Marine Letters, 4:13–17.

Millot C. and I. Taupier-Letage. (2005a). Circulation in the Mediterranean Sea. Hdb Env Chem, Springer-Verlag Berlin Heidelberg, Vol. 5, Part K: 29–66, DOI:10.1007/b107143.

Millot C. and I. Taupier-Letage. (2005b). Additional evidence of LIW entrainment across the Algerian subbasin by mesoscale eddies and not by a permanent westward flow. Progress in Oceanography, 66, 231–250.

Nielsen, J. N. (1912). Hydrography of the Mediterranean and adjacent waters, Rep. Dan. Oceanogr. Exp. Medit., 1, 77–192.

Ovchinnikov I.M. (1966). Circulation in the surface and intermediate layers of the Mediterranean. Oceanology, 6: 48–59.

Palanques A., García-Ladona E., Gomis D., Martín J., Marcos M., Pascual A., Puig P., Gili J.M., Emelianov M., Montserrat S., Guillén J., Tintoré J., Segura M., Jordi A., Ruiz S., Basterretxea G., Font J.,

Blasco D. and F. Pagès. (2005). General patterns of circulation, sediment fluxes and ecology of the Palamós (La Fonera) submarine canyon, northwestern Mediterranean Progress in Oceanography, 66, 89–119. doi:10.1016/j.pocean.2004.07.016

Palanques A., Martín J., Puig P., Guillén J., Company J.B. and Sardá F. (2006). Evidence of sediment gravity flows induced by trawling in the Palamós (Fonera) submarine canyon (northwestern Mediterranean). Deep-Sea Res. 53, 201–214.

Palanques A., P., Durrieu de Madron X., Sanchez-Vidal A., Pasqual C., Martín J., Calafat A., S. Heussner and M. Canals. (2012). Sediment transport to the deep canyons and open-slope of the western Gulf of Lions during the 2006 intense cascading and open-sea convection period, Progress in Oceanography, vol 106, pages 1–15.

Park J.J., Kim, K., King, B A. and S.C. Riser. (2005). An Advanced Method to Estimate Deep Currents from Profiling Floats, Journal Atmospheric Oceanic Technology, 22(8), 1294–1304.

Pinot J.M., López-Jurado J.L. and M. Riera. (2002). The CANALES experiment (1996–1998). Interannual, seasonal, and mesoscale variability of the circulation in the Balearic Channels. Progress in Oceanography, vol 55, issues 3-4, 335–370.

Pinot J.M. and A. Ganachaud. (1999). The role of winter intermediate waters in the spring-summer circulation of the Balearic Sea. 1. Hydrography and inverse box modeling. Journal of Geophysical Research, vol 104, C12, 29843–29864.

Poulain, P.-M., Barbant R., Font, J., Cruzado A., Millot C., Gertman I., Griffa A., Molcard, A., Rupolo, V., Le Bras S. and L. Petit de la Villeon. (2007). MedArgo: a drifting profiler program in the Mediterranean Sea. Ocean Science, 3 (3), pp. 379–395.

Poulain, P.-M., Menna M. and E. Mauri. (2011). Surface Geostrophic Circulation of the Mediterranean Sea Derived from Drifter and Satellite Altimeter Data, Journal Physical Oceanography, 42, 973–990. doi:http://dx.doi.org/10.1175/JPO-D-11-0159.1.

Poulain, P.-M., Gerin R., Rixen M., Zanasca P., Teixeira J., Griffa A., Molcard A., de Marte M. and N. Pinardi. (2012). Aspects of the surface circulation in the Liguro-Provençal basin and Gulf of Lion as observed by satellite-tracked drifters (2007–2009). Bolletino di Geofisica Teorica ed Applicata, 53. 2, 261–279.

Poulain P.-M., Bussani A., Gerin R., Jungwirth R., Mauri, E., Menna M. and G. Notarstefano. (2013). Mediterranean surface currents measured with drifters: From basin to subinertial scales. Oceanography 26(1):38–47, doi:10.5670/oceanog.2013.03.

Puig P., Palanques A., Guillén J. and E. García-Ladona. (2000). Deep slope currents and suspended sediment fluxes in and around the Foix submarine canyon (NW Mediterranean), Deep Sea Research I, 47, (3), 343–366.

Puig P., Canals M., Company J. B., Martín J. B., Amblas D., Lastras G., Palanques A. and A.M. Calafat. (2012). Ploughing the deep sea floor, Nature, 489, 286–289. DOI:10.1038/nature11410.

Puig P., Durrieu de Madron X., Salat J., Schroeder K., Martín, J., Karageorgis A.P., Palanques A., Roullier F., Lopez-Jurado J.L., Emelianov M., Mouting T. and L. Houper. (2013). Thick bottom nepheloid layers in the western Mediterranean generated by deep dense shelf water cascading, Progress in Oceanography, 111, 1–23.

Puillat, I., Taupier-Letage, I. and C. Millot, C. (2002). Algerian Eddies Lifetime can near 3 years. Journal of Marine Systems, 31 (4), 245–259.

Rubio A., Taillandier V. and P. Garreau. (2009). Reconstruction of the Mediterranean northern current variability and associated cross-shelf transport in the Gulf of Lions from satellite-tracked drifters and model outputs, Journal of Marine Systems, 78, S63–S78.

Salat A., P. Puig and M. Latasa. (2010). Violent storms whithin the Seaa: dense water formation episodes in the NW Mediterranean. Advances in Geosciences, 26, 53–59.

Shapiro G.I., van der Molen J. and H.E. Swart. (2004). The effect of velocity veering on sand transport in a shallow sea, Ocean Dynamics, 54: 415–423.

Schott F., Visbeck M., Send U., J. Fischer, L. Stramma and Y. Desaubies. (1996). Observations of deep Convection in the Gulf of Lions, Northern Mediterranean, during the winter of 1991/92. Journal of Physical Oceanography, 26, 505–524.

Schott F. and K.D. Leaman. (1991). Observations of deep Convection in the Gulf of Lions, Northern Mediterranean, during the winter of 1991/92. Journal of Physical Oceanography, 21, 558–574.

Sotillo M. G., Garcia-Ladona E., Orfila A., Rodríguez-Rubio P., Maraver J.C., Conti D., Padorno E., Jiménez J.A., Capó E., Pérez F., Sayol J.M., de los Santos J., Amo A., Rietz A., Troupin C., Tintore J. and E. Álvarez-Fanjul. (2015). The MEDESS-GIB database: tracking the Atlantic water inflow. Earth Syst. Sci. Data Discuss., 8, 863–887, doi:10.5194/essdd-8-863-2015.

Sparnocchia, S., Gasparini, G.P., Astraldi, M., Borghini, M., Pistek, P. (1999). Dynamics and mixing of the Eastern Mediterranean Outflow in the Tyrrhenian Basin. Journal of Marine Systems 20, 301–317.

Testor P. and J.C. Gascard. (2005). Large scale flow separation and mesoscale eddy formation in the Algerian Basin. Progress in Oceanography, 66, 211–230.

Testor P., Send U., Gascard J.C., Millot C., Taupier-Letage I. and K. Béranger. (2005). The mean circulation of the southwestern Mediterranean Sea: Algerian Gyres. Journal of Geophysical Research, 110, C11017, doi:10.1029/2004JC002861.

Trincardi F., Verdicchio G. and S. Miserocchi. (2007). Seafloor evidence for the interaction between cascading and along-slope bottom water masses. Journal of Geophysical Research, vol. 112, F03011, doi:10.1029/2006JF000620.

Van Rijn L.C. (1993). Principles of sediment transport in rivers, estuaries and coastal seas. Aqua Publications, Amsterdam, the Netherlands. ISBN: 9080035629.

Vetrano A., Napolitano E., Iacono R., Schroeder K. and G. P. Gasparini. (2010). Tyrrhenian Sea circulation and water mass fluxes in spring 2004: Observations and model results, Journal of Geophysical Research, 115, C06023, doi:10.1029/2009JC005680.

Part II

Bedforms in the Coastal Zone

Bedforms as Self-organized Patterns

9

Giovanni Coco

Abstract

The formation and evolution of regular bedform patterns occurs in response to complex interactions between flow, sediment transport and evolving morphology. The assumption that the emergence of regular patterns is governed by non-linear dissipative feedbacks (the self-organization theory) has helped to provide simple and insightful explanations for the development of rhythmic bedforms and provides an alternative mode of analysis for patterns that could not be explained otherwise.

Keywords

Pattern formation • Self-organization • Emergence • Morphodynamics

9.1 Introduction

The nearshore region puzzles scientists because of its extraordinarily complicated dynamics. This region has "forcing drivers" which operate over different spatial and temporal scales (from waves to tides) and that are intrinsically variable (a sea state ranges from calm to stormy). The nearshore also involves phenomena characterized by strongly nonlinear behaviour such as wave breaking or the non-linear relationship linking hydrodynamics and sediment response (usually parameterized by relating the sediment transport fluxes to flow velocity taken to a power greater than one). Finally, nearshore processes involve dissipation of the energy inputted into the system via bed friction or work done on sediment transport. Despite these complications, scientists and casual observers can only admire in awe the beauty and regularity of bedform patterns shaping beaches and the underwater seabed. Regular bedform patterns (Blondeaux 2001; Ribas et al. 2015) can be characterized by length scales that can range from a few centimetres in the case of ripples to the order of kilometres in the case of large scale cuspate features (Fig. 9.1). Between these two extremes a variety of morphological patterns can be observed, such as linear and lunate megaripples, beach cusps, crescentic bars, systems of multiple alongshore bars and transverse bars, sand waves and shoreface-connected ridges.

Understanding the formation, evolution and disappearance of bedform patterns goes beyond the pure scientific spirit of discovery and has important practical ramifications. How much do we understand about the nearshore? What can we predict, and with what level of accuracy? These questions are usually asked with regard, for example, to shoreline erosion, coastal inundation, transport of pollutants and beach ecology, but they are also valid when applied to bedform patterns, with one important caveat: how can we expect to successfully predict beach evolution or beach ecology if we are not able to understand if, when, why, and how bedform patterns of striking regularity appear?

9.2 Forced or Self-organized Response?

Clearly, our understanding of the nearshore is still limited and it seems unintuitive that regular patterns appear where so many complicated processes are operating and closely intertwined. In fact, shorelines are rarely straight for long distances rather they often look like an extremely irregular

G. Coco (✉)
School of Environment, University of Auckland, Auckland, New Zealand
e-mail: g.coco@auckland.ac.nz

J. Guillén et al. (eds.), *Atlas of Bedforms in the Western Mediterranean*,
DOI 10.1007/978-3-319-33940-5_9

Fig. 9.1 Nearshore patterns: ripples, beach cusps and crescentic sandbars (from *left* to *right*, photos courtesy of G. Barajas, T. Dolphin and A. Short)

series of forms in which no periodicity can be clearly detected. On the other hand, the same reasons that cause irregularity may drive the formation of bedform patterns. In fact, the seabed typically responds to hydrodynamic forcing in one of two ways: either it simply mirrors the forcing conditions (i.e. a beach erodes under intense storms), or the system is governed by non-linear dissipative feedbacks whereby regular autogenic patterns can emerge.

Since our understanding of nearshore processes is still limited, it is not surprising that both approaches have been considered and that the debate on the relative merits of each approach has divided the scientific community for more than a decade (Coco and Murray 2007). The first approach implies that bedform patterns are simply a mirror of a pattern in the forcing conditions. Just as big waves are likely to result in big erosion, specific wave patterns can impose their presence on the hydrodynamics and shape the seabed accordingly. The success of this approach was measured by its ability to provide an explanation for the appearance of patterns such as beach cusps (Guza and Inman 1975), crescentic sandbars and rip channels (Bowen and Inman 1971), multiple sandbars (Short 1975) and bedforms on the inner shelf (Cacchione et al. 1984).

Despite its initial success in explaining formation and geometry of bedform patterns, this approach neglected some obvious features of nearshore processes (Coco and Murray 2007). In particular, it failed to account for the role of feedbacks between hydrodynamics, sediment transport and evolving morphology (Fig. 9.2). An intuitive example of how feedback acts can be seen by considering flow over a mobile seafloor. The flow will cause fluxes of sediment and changes in the seabed morphology that will in turn affect the flow. The feedback loop, characteristic of coastal evolution, involves an interaction between hydro- and morphodynamics in such a way that the flow is affected by topographic changes and vice versa. The factor coupling the two systems is sediment transport, which is sometimes modelled and measured to be proportional to a high-power of the fluid velocity (this is an obvious simplification given the complicated variety of drivers that result in sediment transport, e.g., waves and currents, and even the different possible modes of transport (e.g., seabed or bedload)).

It is also useful to distinguish between positive and negative feedback. Positive feedback is the mechanism responsible for enhancing and amplifying any perturbation of the system, whereas negative feedback is the mechanism that prevents the system from moving away from an equilibrium state and damps the growth of perturbations. If a system is driven by internal feedbacks rather than by external forcing, it might be

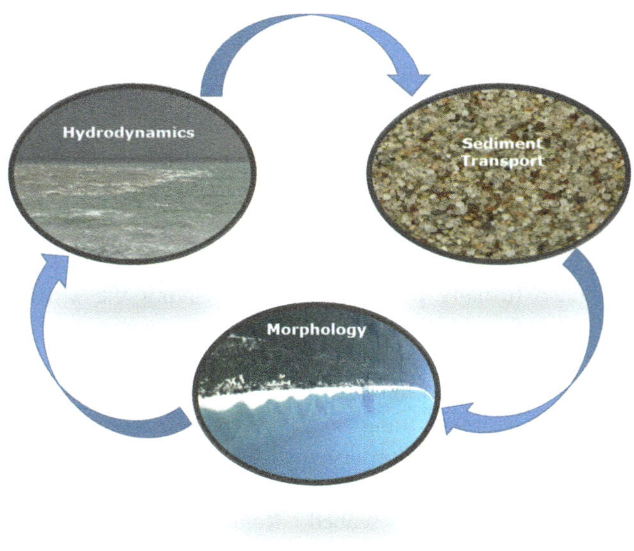

Fig. 9.2 The morphodynamic loop detailing feedbacks between hydrodynamics, sediment transport and evolving morphology

attracted towards a particular, ordered configuration that emerges without any external control and that, at a certain scale, also minimizes the degrees of freedom of the system.

Studies of this kind of behaviour have already been applied to various fields of natural sciences providing evidence for the primary role of self-organization (see for example Werner and Hallet 1993; Werner and Fink 1993; Ashton et al. 2001; Ribas et al. 2015). Overall, it appears that self-organization is a common property of natural systems characterized by interactions between flow and sediment. Hence, self-organization theory is capable of providing simple explanations for the development of rhythmic bedforms and provides an alternative mode of analysis for patterns that could not be explained otherwise.

It is easy to predict that studies on bedform patterns will continue over the next few decades. We are only beginning to unravel the secrets behind the formation of these beautiful features, whose presence affects physical processes over a range of spatial and temporal scales. Importantly, bedform patterns will continue to challenge our understanding and prediction skills and, because of their regularity in a “disordered” environment, will also continue to amaze scientists and observers.

References

Ashton, A., Murray, A.B., Arnault, O. (2001). Formation of coastlines features by large-scale instabilities induced by high-angle waves. Nature, 414, 296–300.

Blondeaux, P. (2001). Mechanics of coastal forms, Annu. Rev. Fluid Mech., 33, 339–370.

Bowen, A.J., and Inman, D.L. (1971). Edge waves and crescentic bars, J. Geophys. Res., 76, 8662–8671.

Cacchione, D.A., Drake, D.E., Grant, W.D., Tate, G.B. (1984). Rippled scour depressions on the inner continental shelf off Central California. J. Sedim. Petrol., 54(4), 1280–1291.

Coco, G., and Murray, A.B. (2007). Patterns in the sand: from forcing templates to self-organization, Geomorphology, 91(3–4), 271-290.

Guza, R.T. and Inman, D.L. (1975). Edge waves and beach cusps, J. Geophys. Res., 80(21), 2997–3012.

Ribas, F., A. Falqués, H. E. de Swart, N. Dodd, R. Garnier, and D. Calvete (2015). Understanding coastal morphodynamic patterns from depth-averaged sediment concentration, Rev. Geophys., 53, doi:10.1002/2014RG000457.

Short, A.D. (1975). Multiple offshore bars and standing waves, J. Geophys. Res., 80(27), 3838–3840.

Werner, B.T., and Fink, T.M. (1993). Beach cusps as self-organised patterns, Science, 260, 968–971.

Werner, B.T., and Hallet, B. (1993). Numerical simulation of self-organized stone stripes, Nature, 361,142–145.

Beach Cusps

10

Giovanni Coco

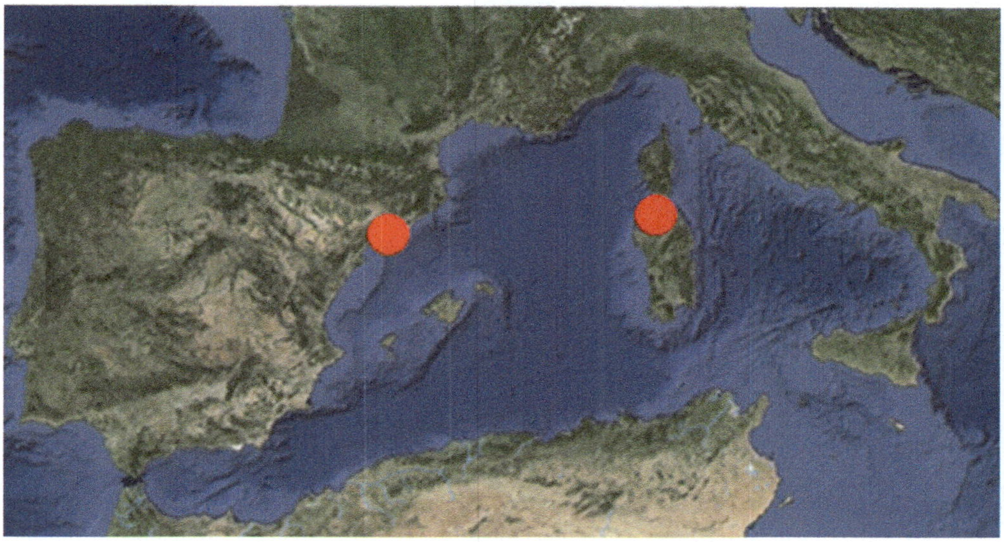

Abstract

Beach cusps are one of the most commonly observed rhythmic features in the nearshore. Their formation has been related amongst other processes to the presence of standing waves or self-organization processes. This chapter provides a short introduction to the modelling of beach cusp formation and the main feed-back mechanisms between flow, sediment transport processes and morphodynamics.

Keywords

Pattern • Beach cusps • Swash • Run-up

10.1 Introduction

Beach cusps are one of the most commonly observed rhythmic features in the nearshore. They appear at the edge of the swash zone, the part of the beach where waves have collapsed into a thin bore that runs up and down the shore. A beach cusp pattern is characterized by a series of uniformly spaced ridges of sediment trending at right angles to the shoreline (Fig. 10.1). When waves running up the shore encounter a beach cusp field, the flow, which is usually close to normally-incident when beach cusps are present, diverges at the horns and converges into the bays. Narrow, concentrated and offshore-directed flows are a typical feature of beach cusps. Beach cusps have been observed along the coastlines of many beaches worldwide (examples from the western Mediterranean can be found in Bowman et al. 2007) and seem to appear under a wide variety of hydrodynamic conditions (from calm to almost stormy conditions) and sediments sizes (from fine sand to gravel). More often, they have been associated with steeper beaches with coarse-grained sediments, normally incident waves and spacing (defined as the distance between consecutive cusp horns) below 100 m. Because of their striking regularity,

G. Coco (✉)
School of Environment, University of Auckland, Auckland, New Zealand
e-mail: g.coco@auckland.ac.nz

J. Guillén et al. (eds.), *Atlas of Bedforms in the Western Mediterranean*,
DOI 10.1007/978-3-319-33940-5_10

Fig. 10.1 Formation of beach cusps at Marinedda, Italy (*left panel*, courtesy of A. Ruju) and at Sitges, Catalunya (*right panel*, courtesy of Y. Ciriano)

beach cusps have been the object of much speculation and several theories have been proposed for their formation. Most of these theories have been ignored due to the poor agreement with field observations, so hypotheses of beach cusp formation centre on linking cusps either (1) to a pattern in the hydrodynamics due to the presence of standing edge waves (Guza and Inman 1975) or (2) to self-organization through sediment-flow feedbacks (Werner and Fink 1993). The two theories propose different explanations and discussions within the scientific community have characterized a still ongoing debate on the causative mechanism for beach cusps formation with even more complicated interactions being hypothesized (Dodd et al. 2008). Herein, in view of the increasing evidence that self-organization processes shape nearshore patterns, focus will be given to this approach.

10.2 Modelling Beach Cusp Formation

On the basis of the model initially proposed by Werner and Fink (1993), a numerical code based on cellular automata was implemented. A detailed description of the model and its sensitivity especially towards sediment transport is given in Coco et al. (2000) so here only a brief description will be provided. In this very simplified model, water motions in the swash zone are abstracted and described through the use of a discrete number of water particles. In the cross-shore direction the domain is divided into two parts representing the surf and the swash zone. In the surf zone water particles are characterized by a linear constant trajectory while inside the swash zone the trajectory deviates from linear under the gravity force resulting from local bed morphology. In addition, friction, percolation and interaction between successive swash cycles are neglected (later studies have shown the importance of including these processes for detailed comparisons with field measurements). For sediment transport parameterization a simplified formulation has been adopted with the sediment flux being proportional to a high power of the water particle velocity. Once the net amount of erosion/deposition has been evaluated a smoothing function operates and spreads the sediment over a fixed area keeping account of the pre-existing topography. The importance of sediment transport in the physical process leading to beach cusp formation is quite evident because once an area of lower elevations starts to develop flow will start to be directed towards it, accelerate and enhance erosion. In fact, because of the non-linear relation with sediment transport, such low area will be characterized by greater sediment divergence than the surroundings leading to a positive feedback that increases the relief of the feature. The opposite behaviour can be observed in depositional areas where a slowing flow will cause further deposition of sediment and flow diversion. The switch from positive to negative feedback occurs when the amounts of net sediment erosion or deposition are considerably reduced because of the beach cusp shape which, once a certain ratio between spacing and amplitude is reached, reduces deceleration and acceleration effects. Finally, the topography does not experience further strong changes and a steady state can be reached. At the beginning of each swash cycle, water particles are assigned an alongshore and cross-shore velocity (plus a random component) and then allowed to move over the grid reproducing the typical sequence of swash run-up and run-down. After a number of cycles, small perturbations begin to appear and after a series of morphological changes, which include merging and separation of incipient features a regular pattern of beach cusps emerges (Fig. 10.2).

Numerical simulations for a range of differing beach slopes and initial swash velocities show a surprisingly good agreement with observations of beach cusp spacing collected in the field and in the laboratory, especially if one considers

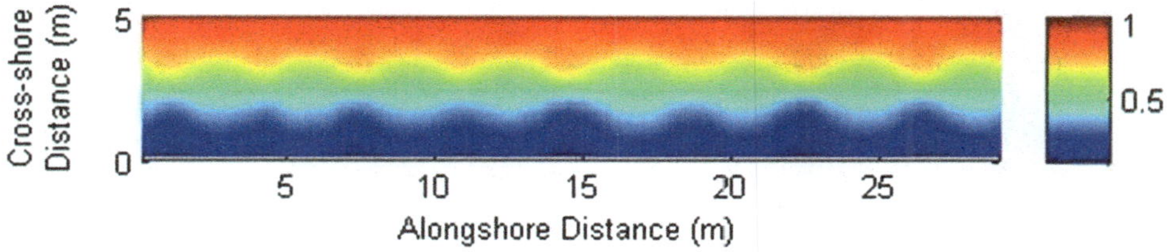

Fig. 10.2 Numerical simulation of beach cusp formation. *Contour lines* show the development of a set of beach cusps with regular alongshore spacing and steep horns. The initial beach slope is around 0.17 and the average initial cross-shore velocity of water particles is around 2.5 m/s resulting in a swash excursion of around 2 m. The swash zone is limited to the higher part of the domain (between 2 and 5 m in the cross-shore direction). Offshore of the swash zone (between 0 and 2 m in the cross-shore direction), in agreement with field observations, bathymetry 'mirrors' the swash features. Average cusp spacing is around 3.5 m

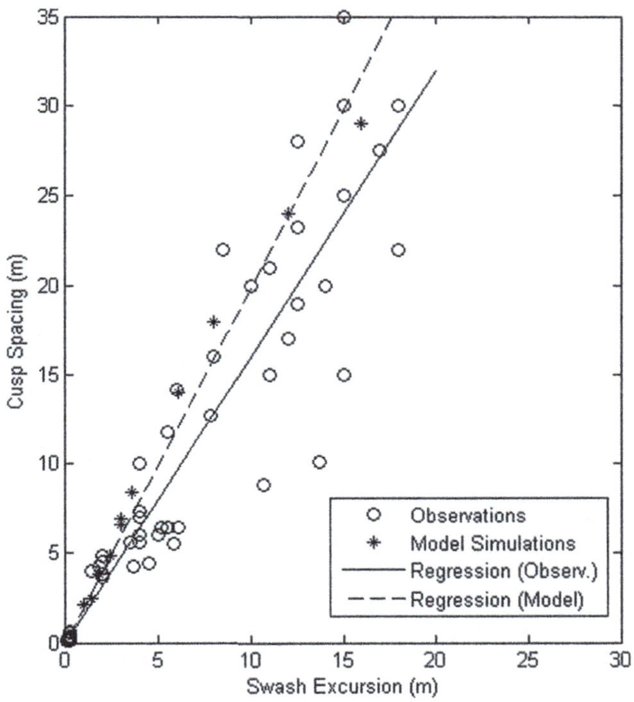

Fig. 10.3 Beach cusp spacing as a function of swash excursion for measurements (field and laboratory) and numerical simulations

the simplifications in the numerical model. The linear regression obtained through the use of most of the available field and laboratory data (details are presented in Coco et al. (1999)), suggests a proportionality coefficient between swash excursion and cusp spacing of around 1.6, while the numerical simulations indicate a slightly larger value of around 1.9 (Fig. 10.3). In agreement with field observations, this numerical code has also simulated the destruction of beach cusps under obliquely approaching waves, as well as the dependence of beach cusp formation on the regularity of the incoming wave field (Coco et al. 2000). Abstraction and inclusion of other processes (infiltration and tides) that clearly affect the development of beach cusps has also allowed more detailed comparisons with field observations (Coco et al. 2004). More recent numerical modelling (Dodd et al. 2008) is now allowing to study with unprecedented level of detail the complicated interactions between hydro- and morphodynamics. Despite their limited importance in terms of coastal hazards or socio-economics, beach cusps remains an enigma that continues to challenge and puzzle scientists and observers worldwide.

Bibliography

Bowman, D., Ferri, S., and Pranzini, E. (2007). Efficacy of beach dewatering—Alassio, Italy. Coastal Engineering, 54(11), 791–800.

Coco, G., O'Hare, T.J., and Huntley, D.A. (1999). Beach cusps: a comparison of data and theories for their formation, J. Coastal Res., 15(3), 741–749.

Coco, G., Huntley, D.A., and O'Hare, T.J. (2000). Investigation of a self-organisation model for beach cusp formation and development, J. Geophys. Res., 105(C9), 21991–22002.

Coco, G., Burnet, T.K., Werner, B.T., and Elgar, S. (2004). The role of tides in beach cusp development, J. Geophys. Res., 109(C4), C04011, doi:10.1029/2003JC002154.

Dodd, N., Stoker, A. M., Calvete, D., and Sriariyawat, A. (2008). On beach cusp formation. Journal of Fluid Mechanics, 597, 145–169.

Guza, R.T. and Inman, D.L. (1975). Edge waves and beach cusps, J. Geophys. Res., 80(21), 2997–3012.

Werner, B.T., and Fink, T.M. (1993). Beach cusps as self-organised patterns, Science, 260, 968–971.

Km-Scale Shoreline Sand Waves

11

Albert Falqués and Francesca Ribas

Abstract

Km-scale shoreline undulations that are linked to similar undulations in the bathymetry up to certain depth are here described. The hypothesis that they may emerge from feedback mechanisms between the wave field and the morphology through the sediment transport is discussed. Basically, waves undergo refraction and shoaling when they approach the shore. On an undulating coastline, differences in refraction and shoaling arise between the updrift and the downdrift sides of an undulation. This causes gradients in alongshore wave-driven sediment transport from which accretional/erosional coastal stretches arise feeding back into the wave field.

Keywords

Shoreline undulations • Wave refraction • Wave shoaling • Alongshore sediment transport

11.1 Shoreline Undulations

Rather than showing an irregular shape in plan view, sandy shorelines sometimes exhibit a series of undulations with a nearly recurrent spacing, L, between crests (or embayments). Sometimes there are only 2 undulations, sometimes more than 10. Their spacing can range from $L \sim 1$ m up to $L \sim 20$ km or more (Magaña et al. 2014). Given the high variety in their morphology, a rigorous classification is difficult. A rough classification can be set as follows.

(A) *Beach cusps* Lunate embayments at the swash zone separated by relatively narrow shoals or horns, the apices of which point seaward. The typical spacing is $L \sim$ 1–50 m.

(B) *Megacusps* Shoreline undulations that are linked to rhythmic surf zone bars and rip channel systems. The typical spacing is $L \sim$ 100–1000 m.

(C) *Km-scale shoreline sand waves* The spacing is at least one order of magnitude larger than the surf zone width and they are not linked to surf zone rhythmic bars and rip channels. Rather, they are typically associated with bathymetric undulations extending well beyond the surf zone. The typical spacing in open ocean coasts is $L \sim$ 1–10 km. This chapter is devoted to this type of shoreline undulations.

Km-scale shoreline sand waves are less known than beach cusps and megacusps, probably because their large length and time scales make them difficult to detect and monitor. Moreover, their spacing on low-energy beaches can be relatively small, $L \sim$ 100–200 m, overlapping the typical spacing of megacups on open ocean coasts, so the two types of undulation can be easily confused.

11.2 Km-Scale Shoreline Sand Waves

The increasingly widespread use of aerial and satellite imagery shows that there are many sites around the world that display km-scale shoreline sand waves (Fig. 11.1). For example, analysis of satellite images shows that they are extremely frequent along the Atlantic sandy shores of Africa (NW coast, Western Sahara and Mauritania; SW coast,

A. Falqués (✉) · F. Ribas
Physics Department, Universitat Politècnica de Catalunya, Barcelona, Catalonia, Spain
e-mail: albert.falques@upc.edu

J. Guillén et al. (eds.), *Atlas of Bedforms in the Western Mediterranean*,
DOI 10.1007/978-3-319-33940-5_11

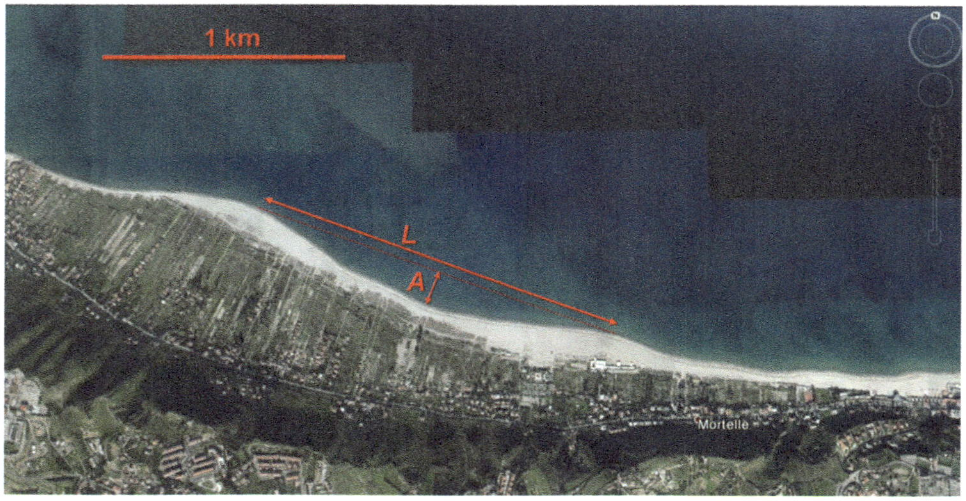

Fig. 11.1 km-scale shoreline sand waves near Mortelle (NE Sicily). Definition of the wavelength, L, and the amplitude, A. *Source* Google Earth, image from 2016 TerraMetrics, map data from SIO, NOAA, US Navy, NGA, GEBCO

Namibia and Angola), where they are clearly distinct from megacusps (van den Berg 2012; Idier and Falqués 2014). Table 11.1 presents a list of coasts where km-scale shoreline sand waves have been observed and measured. Very remarkable is the case of the Dutch coast, where an analysis of long-term (150 year) measurements of coastline position reveals the systematic presence of shoreline sand waves (Verhagen 1989; Ruessink and Jeuken 2002). Also, detailed measurements of shoreline position and bathymetric surveys during a 4-year period has shown the occurrence of such sand waves along the west coast of Denmark (Kaergaard et al. 2012). In the case of low-energy beaches, sand waves with substantially smaller wavelengths (although much larger than the surf zone width) can exist, as illustrated by the analysis of video images of Medellín et al. (2008) from Santander Bay.

Table 11.1 Documented km-scale shoreline sand waves around the world

Site location	Coast type	Number of crests	L (km)	A (km)	V (km/year)	References
West Danish coast	Open	3	5	0.1	0.37	Kaergaard et al. (2012)
Dutch coast	Open	~30	3–10	0.02–0.5	0–0.2	Ruessink and Jeuken (2002), Verhagen (1989)
Santander bay, Cantabria, Spain	Estuary	3	0.13–0.15	0.015	0	Medellín et al. (2008)
Gulf of Finland, Russia	Relatively low-energy	4	0.1–0.9	0.015–0.1	0.015–0.02	Ryabchuk et al. (2011)
Nile Delta, Egypt	Open	Not reported	2.5–5	0.05–0.1	0.5–1	Inman et al. (1992)
Calella beach, Catalonia, Spain	Open	4–6	0.6–1.4	0.05	0	Caballeria et al. (2011)
SW coast of Africa: Angola, Namibia	Open	>50	2–10	0.1–0.7	Not measured	Van den Berg et al. (2012), Idier and Falqués (2014)
NW coast of Africa: Sahara, Mauritania	Open	>50	2–18	0.1–0.5	Not measured	Idier and Falqués (2014)
Lake Erie, Canada	Open	8	0.35–1.5	0.05	0.1–0.3	Davidson-Arnott and van Heyningen (2003)
Southampton beach, Long Island, NY, USA	Open	11	0.6–1.4	0.04	0.35	Thevenot and Kraus (1995)
Fire island, Long Island, NY, USA	Open	9	1–2	0.026	0	Gravens (1999)
Carchuna, Andalucia, Spain	Open	5	0.48–1.7	0.05–0.1	0	Ortega-Sánchez et al. (2003)
Chelem, Yucatan, Mexico	Open	3	2.2	0.09	0	Arriaga et al. (2014)
Port Stephens estuary, Australia	Estuary	1	1.4	0.25	0.07	Vila-Concejo et al. (2009)

The alongshore spacing between crests or embayments is L, the cross-shore displacement of the crests with respect to the embayments is A and the average (yearly) alongshore translation velocity is V

Detailed geological and sedimentary studies also show the occurrence and evolution of such sand waves, e.g. from Lake Erie (Davidson-Arnott and van Heyningen 2003) and from the Gulf of Finland (Ryabchuk et al. 2011). Quite special is the case of the Nile delta shores, where the eastward propagation of the sand blanket episodically impinging on the coast causes erosion/accretion waves (Inman et al. 1992). A detailed study of aerial photographs and topographic surveys has also shown the occurrence of sand waves along Long Island beaches (Thevenot and Kraus 1995; Gravens 1999). Shoreline sand waves can exist typically as wave trains, ranging from a few waves to 10 or even more. Furthermore, several wave trains may occur on a long stretch of coast, as in Africa, where more than 50 crests in total can be seen from satellite images on both the NW and the SW coasts. Sometimes, however, a single sand wave can form and evolve over decades (Vila-Concejo et al. 2009). Finally, other known examples are the Yucatan coast (Arriaga et al. 2014) and the Catalan coast (Caballeria et al. 2011).

Shoreline sand waves can be stationary or can translate downdrift along the coast. This is typically the case when there is a dominant sediment transport direction. Also, they can be relatively symmetrical but in cases of dominant transport direction they tend to be asymmetrical, the distance between crest and embayment on the downdrift flank being smaller than the corresponding distance on the updrift flank.

11.3 Self-organized Origin and Evolution

Km-scale shoreline undulations occurring on wave-dominated coasts can have a number of causes. First, they can be forced by offshore bathymetric anomalies, as for example, in the case of the undulations in Carchuna beach, which, according to Ortega-Sánchez et al. (2003), are primarily due to the modulation of wave energy by offshore drowned fluvial valleys. Second, the periodical opening of river inlets or the sediment carried to the sea by floods can generate sand deposits in the nearshore that can subsequently evolve and migrate under the action of waves and currents, thus forming shoreline sand waves (Thevenot and Kraus 1995). However, in many cases there is neither in the wave forcing nor in the boundary conditions any particular pattern that is similar to the morphology of the observed sand waves. In this case, the alongshore spacing, the migration celerity and the particular shape of the sand waves emerge from feedback mechanisms between the wave field and the morphology through the sediment transport. This situation is referred to as self-organized dynamics (Coco and Murray 2007). The procedure for studying self-organized processes is classical stability analysis: (i) assume a reference situation where the sand waves are absent, (ii) superimpose a small arbitrary perturbation, and (iii) study the coupled dynamics (water-morphology) to see whether the perturbation decays (stable situation, sand waves do not form) or it grows and eventually forms a sand wave train (unstable situation) (Dodd et al. 2003). Notice that even if sand waves have been caused by a template in the forcing, the self-organization mechanisms can always affect their morphology, maintenance and alongshore migration.

We now briefly explain the feedback mechanisms between waves and morphology that can lead to the self-organized formation of shoreline sand waves. Under oblique wave incidence on sandy shores, the dynamics of the shoreline at km-scale is governed by the alongshore gradients in the total transport rate, Q, giving the total sand volume crossing any cross-shore section by time unit (m^3/s). Where there is divergence of Q (sediment loss), the shoreline retreats; where there is convergence (sediment accumulation), it progrades. A widely accepted semi-empirical evaluation of sediment transport is the CERC formula:

$$Q = kH_b^{5/2} \sin 2\alpha_b \tag{11.1}$$

where H_b is the wave height at breaking, α_b is the angle between wave crests and shoreline at breaking and k is an empirical coefficient (Komar 1998). Assume an alongshore uniform wave input in deep water (before refraction and shoaling). If the shoreline and the bathymetric contours are rectilinear and parallel, say to the x axis, both H_b and α_b are constant alongshore and so is Q, with the result that the coastline remains stationary. Under similar conditions, let us assume now an undulating shoreline. Now the relative wave angle, i.e. the angle between wave crests and local shoreline, will be $\alpha_b = \theta_b - \phi$, where θ_b is the wave angle with respect to x and ϕ is the local angle of the perturbed shoreline with respect to x. Assume first that H_b and θ_b remain unchanged along the coast, as in the rectilinear case. Due to the undulating shape of the shoreline, ϕ varies so α_b is maximum on the downdrift flank of each crest (at the inflexion point of the sinusoidal shoreline) and minimum on the updrift flank. Because of Eq. (11.1), Q will be maximum (minimum) at the downdrift (updrift) inflexion point of the shoreline. This implies that there will be divergence of sediment flux around the crests (and convergence at the embayments), making the sand wave decay. This is the *shoreline shape stabilizing effect* illustrated in Fig. 11.2a. Notice that due to wave refraction α_b is typically smaller than 45°. Otherwise, an increase in angle at breaking could cause a decrease in Q.

Thus, if there was no feedback from the shoreline sand wave into the wave field, the rectilinear shoreline would be stable and sand waves would not form. But associated with the shoreline undulation there can be similar undulations in

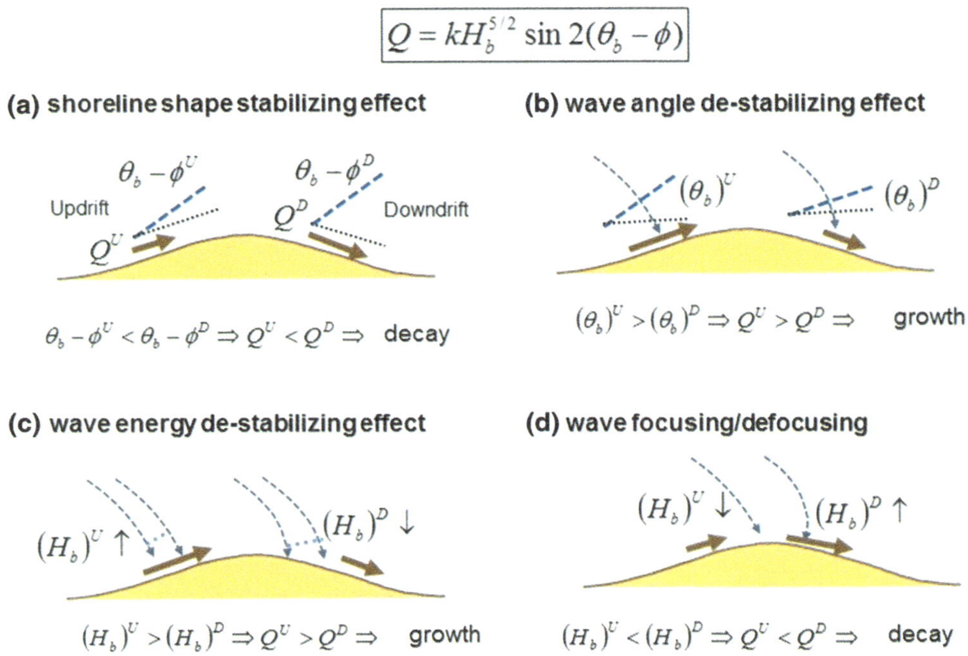

Fig. 11.2 The four basic wave effects controlling alongshore transport on an undulating sandy shoreline and sand wave growth/decay. **a** The relative wave angle is larger downdrift (*D*) than updrift (*U*) of a crest, **b** due to refraction, the absolute wave angle is larger updrift than downdrift, **c** the refractive wave crest stretching is less updrift than downdrift of a crest, and **d** refractive wave focusing by a crest causes maximum wave energy slightly downdrift of the crest (the opposite at embayments)

the nearshore bathymetry up to a certain water depth, D_c. This undulating bathymetry causes modifications in wave refraction and shoaling, so H_b and θ_b are no longer alongshore uniform. This provides the feedback from the sand wave into the wave field which is essential to shoreline instability (Falqués et al. 2011).

Due to the different orientation of the bathymetric contours, refraction is stronger on the downdrift flank of the crest than on the updrift flank, so $(\theta_b)^D < (\theta_b)^U$, causing a decrease in Q_D and an increase in Q_U. This weakens the stabilizing effect of the shoreline shape and, if it is strong enough, it can even reverse it, bringing the maximum in Q onto the updrift side. In this case there is convergence of transport at the crest (divergence at the embayments) and the sand wave grows. This is the *wave-angle destabilizing effect* illustrated in Fig. 11.2b. Wave refraction decreases wave energy density due to *refractive wave crest stretching*. This effect is maximum at the downdrift side, where refraction is maximum, causing a decrease in wave energy (and thereby in H_b) on the downdrift side with respect to the updrift side of a crest. Similarly to the wave-angle destabilizing effect, this tends to bring the maximum in Q onto the updrift flank, contributing to sand wave growth. We call this the *wave-energy destabilizing effect* and it is shown in Fig. 11.2c. The fourth important process is the *wave focusing stabilizing effect*. Due to differences in wave refraction between the updrift and the downdrift flanks, there is a wave energy focusing (defocusing) at the crests (embayments), actually somewhat downdrift of them (it would be exactly at the crests for frontal wave incidence). This tends to bring the maximum in wave energy and hence in Q somewhat downdrift of the crests causing decay of the sand waves. It is illustrated in Fig. 11.2d (Falqués et al. 2015).

Whether the feedback between an incipient sand wave and the wave field will be positive or not depends on the interplay of the four factors described above. Their relative intensity depends on the cross-shore bathymetric profile of the shoreface, on the depth of closure, D_c, i.e. the depth up to which the potential shoreline undulations can affect the bathymetry, and on the wave height, angle and period at D_c. In most conditions the primary competition is between the shoreline shape stabilizing effect and the wave-energy destabilizing effect. For low incidence angles the former is dominant and the shoreline is stable, whereas for high angles the latter is dominant and sand waves grow. The threshold angle may vary but is commonly about 45°. This is called *high-angle wave shoreline instability* (HAWI) (Ashton et al. 2001; Falqués and Calvete 2005; Van den Berg et al. 2012; Kaergaard et al. 2013). It turns out that the wavelength selection is controlled by the wave focusing stabilizing effect. Indeed, the intensity of the wave-energy destabilizing effect increases for decreasing wavelength, *L*, but the wave focusing takes over for very short wavelengths so there is a minimum *L* for sand wave growth, L_m, and a dominant *L*, L_M, i.e. an *L* of faster growth (Van den Berg et al. 2014). In most conditions, the wave-angle destabilizing effect is not very great. It just weakens the decay for low angles and contributes to the growth for high angles. However, if D_c is quite large, the foreshore steep and the shoreface relatively shallow, this effect can overcome the shoreline shape stabilizing effect and cause sand wave growth even for wave angles below the HAWI threshold (Idier et al. 2011; Falqués et al. 2015).

The downdrift migration of shoreline sand waves can be due to a number of causes, but if their dynamics is dominated by the gradients in Q, it is necessarily due to the perturbations of the wave field by the bathymetric undulations associated with the sand waves. Indeed, if these perturbations are disregarded, the maximum in Q is located exactly at the downdrift inflexion point (shoreline shape stabilizing effect) and this implies that the erosion area is centred on the crest, with the result that the sand waves decay in place without moving. Only if the maximum in Q is shifted updrift of the inflexion point does the sand wave translate downdrift while decaying. Similarly, a maximum in Q located downdrift of the updrift inflexion point causes downdrift migration while growing. This shift is necessarily the result of the alongshore gradients in H_b and θ_b due to a combination of the two destabilizing effects and the wave focusing/defocusing.

Acknowledgments This research has been funded by the Spanish government through the research projects CTM2012-35398 and CTM2015-66225-C2-1-P (co-funded by FEDER, U.E.).

References

Ashton, A. D., Murray, A. B. and Arnault, O. (2001). Formation of coastline features by large-scale instabilities induced by high-angle waves, Nature, 414, 296–300.

Arriaga, J., Ribas, F., Mariño, I and Falqués, A. (2014). Km-scale shoreline sand waves: numerical modelling and observations. Coastal Eng. 2014, in press.

Caballeria, M., A. Falqués, and N. van den Berg (2011). Potential instabilities of Catalan coastline induced by high-angle waves, in Proceeding of the 7th IAHR Symposium on River, Coastal and Estuarine Morphodynamics, pp. 2133–2143, Tsinghua Univ. Press, Beijing, China. [CD-ROM].

Coco, G., and A. B. Murray (2007). Patterns in the sand: From forcing templates to self-organization, Geomorphology, 91 (271–290).

Davidson-Arnott, R. G. D. and van Heyningen, A. (2003). Migration and sedimentology of longshore sand waves, Long Point, Lake Erie, Canada, Sedimentology, 50, 1123–1137.

Dodd, N., P. Blondeaux, D. Calvete, H. E. de Swart, A. Falqués, S. J. M. H. Hulscher, G. R Rozynski, and G. Vittori (2003), The use of stability methods in understanding the morphodynamical behavior of coastal systems, J. Coastal Res., 19 (4), 849–865.

Falqués, A. and Calvete, D.(2005). Large-scale dynamics of sandy coastlines: Diffusivity and instability, J. Geophys. Res., 110, C03007, doi:10.1029/2004JC002587.

Falqués, A., Calvete, and Ribas, F. (2011). Shoreline Instability due to Very Oblique Wave Incidence: Some remarks on the Physics. J. Coastal Res. 27(2), 291–295.

Falqués, A., Ribas, F., Idier, D. And arriaga, J. (2015). Formation mechanisms for km-scale shoreline sand waves. Geophys. Res. Lett., submitted.

Gravens, M. B. (1999). Periodic shoreline morphology, Fire Island, New York, in Coastal Sediments '99, edited by N. Kraus and W. McDougal, pp. 1613–1626, Am. Soc. Civ. Eng., Reston, Va.

Idier, D., Falqués, A., Ruessink, B.G. and Garnier, R. (2011). Shoreline instability under low-angle wave incidence, J. Geophys. Res., 116, F04031, doi:10.1029/2010JF001894.

Idier, D. and Falqués, A. (2014). How kilometric sandy shoreline undulations correlate with wave and morphology characteristics: preliminary analysis on the Atlantic coast of Africa. Adv. Geosci., 39, 55–60. www.adv-geosci.net/39/55/2014/doi:10.5194/adgeo-39-55-2014

Inman, D.L., Elwany, M. H. S., Khafagy, A.A. and Golik, A. (1992). Nile Delta Profiles and Migrating Sand Blankets. Coastal Eng. 1992, 3273–3284.

Kaergaard, K., Fredsoe, J., and Knudsen, S. B. (2012). Coastline undulations on the West Coast of Denmark: Offshore extent, relation to breaker bars and transported sediment volume, Coast. Eng., 60, 109–122, 2012.

Kaergaard, K., and Fredsoe, J. (2013). Numerical modeling of shoreline undulations. Part 1: Constant wave climate, Coastal Eng., 75, 64–76.

Komar, P. D. (1998), Beach Processes and Sedimentation, 2nd ed., Prentice Hall, Englewood Cliffs, N. J.

Magaña, P., Lopez-Ruiz, A., Lira, A., Ortega-Sanchez, M. and Losada, M.A. (2014). A public, open Western Europe database of shoreline undulations based on imagery. Applied Geography, 55, 278–291.

Medellín, G., R. Medina, A. Falqués, and M. González (2008). Coastline sand waves on a low-energy beach at "El Puntal" spit, Spain, Mar. Geol., 250, 143–156.

Ortega-Sánchez, M., Losada, M.A., and Baquerizo, A. (2003). On the development of large-scale cuspate features on a semi-reflective beach: Carchuna beach, Southern Spain. Marine Geology 198 (2003) 209–223.

Ruessink, B. G. and Jeuken, M. C. J. L. (2002). Dunefoot dynamics along the dutch coast, Earth Surf. Proc. Land., 27, 1043–1056, doi:10.1002/esp.391.

Ryabchuk, D., I. Leont'yev, A. Sergeev, E. Nesterova, L. Sukhacheva, and V. Zhamoida (2011). The morphology of sand spits and the genesis of longshore sand waves on the coast of the eastern Gulf of Finland, Baltica, 24(1), 13–24.

Thevenot, M. M., and N. C. Kraus (1995), Longshore sandwaves at Southampton Beach, New York: Observations and numerical simulation of their movement, Mar. Geol., 126, 249–269.

Van den Berg, N., Falqués, A. and Ribas, F. (2012). Modeling large scale shoreline sand waves under oblique wave incidence. J. Geophys. Res., 117, F03019, doi:10.1029/2011JF002177.

Van den Berg, N., Falqués, A., Ribas, F. and Caballeria, M. (2014). On the mechanism of wavelength selection of self-organized shoreline sand waves. J. Geophys. Res. Earth Surf., 119, 665–681, doi:10.1002/2013JF002751.

Verhagen, H. J. (1989). Sand waves along the dutch coast, Coastal Eng., 13, 129–147.

Vila-Concejo, A., Short, A.D., Hughes, M.G. and Ranashinghe, R. (2009). Formation and evolution of a sandwave on an estuarine beach, J. Coast.Res., SI56, 153–157.

KM-Scale Shoreline Sand Waves In The Western Mediterranean Sea

12

Albert Falqués and Francesca Ribas

Abstract

Available satellite images of the coastlines of the western Mediterranean Sea are examined to identify km-scale shoreline sand waves. The sites are classified as open coasts, deltas or sheltered shores. We find a significant number of shoreline undulations along the Spanish coast (Andalucia and Catalonia). The Rhône delta, Sicily and the Nador lagoon in Marocco also show coastal stretches with such undulations. We present a number of pictures and a table summarizing the results of the study.

Keywords

Shoreline undulations • Satellite images • River deltas • Sheltered shores

12.1 Introduction

Km-scale shoreline sand waves (KSSW) are shoreline undulations with an alongshore wavelength, L, at least one order of magnitude larger than the surf zone width. They are not linked to surf zone rhythmic bars and rip channels but rather with bathymetric undulations extending well beyond the surf zone. The typical spacing on open ocean coasts is $L \sim 1$–10 km (see more details in chapter 'Km-scale shoreline sand waves') and cross-shore amplitudes, A, are of the order of tens to hundreds of metres. On low energy coasts, where the surf zone is narrower, their spacing can be smaller, of hundreds of metres, but they still bear a morphological resemblance to KSSW on open coasts. There are very few published studies on KSSW in the western Mediterranean Sea (WMS). Magaña et al. (2014) provided a database of

A. Falqués (✉) · F. Ribas
Physics Department, Universitat Politècnica de Catalunya, Barcelona, Catalonia, Spain
e-mail: albert.falques@upc.edu

J. Guillén et al. (eds.), *Atlas of Bedforms in the Western Mediterranean*,
DOI 10.1007/978-3-319-33940-5_12

shoreline undulations at a wide range of scales, covering alongshore wavelengths ranging from 50 m to several km (see Table 12.1). Some of their undulations can be associated with KSSW. According to Falqués and Ribas (this atlas), there are only two site-specific studies in the WMS, one referring to Carchuna beach, Andalucía, Spain (Ortega-Sánchez et al. 2003) and one to Calella beach, Catalonia, Spain (Caballeria et al. 2011).

We used satellite imagery from Google Earth and aerial photos from the Institut Cartogràfic i Geològic de Catalunya (Catalonia, Spain) to find shoreline undulations in the WMS that could be associated with KSSW (see Table 12.1). The association is straightforward in case of an alongshore spacing well in the range 1–10 km but, as stated in Falqués and Ribas (this atlas), relatively short sand waves may be mistaken for megacusps. In the absence of bathymetric surveys to determine the offshore extent of the bathymetric undulations or to discern whether the shoreline undulations are linked to crescentic bars and rip channels, it is sometimes difficult to distinguish between them. Table 12.1 includes the results of the analysis and also the degree of confidence that the observed features are km-scale shoreline sand waves. A high degree of confidence (A) was assigned when either the spacing is much larger than the surf zone width (say at least 1 km on the open coast) or it is clear that the undulations are not linked to surf zone rhythmic bars. Otherwise, a low degree of confidence (B) was assigned. Some of the undulations occur along river delta shores or along sheltered shores in basins (closed or semi-enclosed). Other undulations occur along ordinary open shores. In the following three sections, the observed KSSW on these three types of coast are described and discussed and images with examples are provided.

12.2 Open Coasts

Shoreline sand waves on open western Mediterranean coasts are not common. The first two examples occur in Sicily. Shoreline undulations with a spacing ranging from 0.4 to 2.3 km are found (see Fig. 12.1a) along the western shores of the Messina Strait, SW of the city of Messina.

Although it is a very urbanized area and the shoreline undulations are well integrated with the human development, some of them seem clearly natural since places where the sandy beach is wide (crests) alternate with locations where there is almost no beach at all (embayments). The geographical characteristics of this coast suggest a fetch-limited wave incidence with a wave climate dominated by high-angle waves from the south (see Falqués and Ribas this atlas). Some of the sand waves are asymmetric, indicating a dominant littoral drift from the S-SW. The sand waves occur in several wave trains and some crests are rather subtle but a total of about 12 crests can be seen.

Looking at the historical images, some morphological changes can be traced back. Moving further north, the NE coast of Sicily shows many shoreline undulations but for many of them it is not clear whether the shore is sandy, and others are altered by groynes or breakwaters. However, in the eastern part, near Mortelle, there are two clear natural sand waves with a spacing $L \approx 1.5$ km and a width $A \approx 0.1$ km (see Fig. 12.1b). Further east, there is a series of four subtler undulations with $L \approx 0.4$ km (image from 2015). Looking at images taken at different moments (e.g. 2006), one can see differences from those from 2015, indicating that the sand waves along this coast are quite dynamic.

Another open coast where KSSW exist is that of Andalucía, Spain, where we found four sites (Table 12.1). The case of Carchuna beach (Fig. 12.2a) is well documented by Ortega-Sánchez et al. (2003, 2004, 2008). This is a semi-reflective beach made of coarse and poorly sorted sediment where a series of five horns with a spacing ranging from 0.48 to 1.7 km is observed. The amplitude of the horns may change over time but their alongshore position remains stationary. According to Ortega-Sánchez (2003), the shoreline undulations are essentially forced by the modulation of wave energy by offshore drowned fluvial valleys. However, a number of processes, such as feedbacks between water motions and morphology (that may involve trapped infragravity edge waves), also play a role in their evolution and maintenance. Very striking are the two large shoreline undulations observed at Albuñol, with a spacing of about 1.8 km (Fig. 12.2b). According to historical aerial photos they are the deltas of two seasonal streams. The position of the apices is shifted with respect to the present positions of the mouths of these streams because they have been artificially moved.

Moving eastward, near Almerimar, there is a series of relatively irregular undulations with alongshore wavelengths ranging from 0.9 to 1.5 km (Fig. 12.3a). There are six major crests but some smaller/subtler undulations with a spacing of about 0.3 km can also be seen. This is a fairly unspoiled area and the beaches are made of loose sand. Historical images show some small differences, so these sand waves are dynamic. Moving further east, near Almeria city there is another series of undulations (Fig. 12.3b). The type of sediment is similar to that at Almerimar: the sediment inshore is fixed by vegetation but the sand at the shoreline is unconsolidated. The shape of the undulations is quite different:

Table 12.1 km-scale shoreline sand waves in the western Mediterranean Sea. The number of crests is sometimes not very well defined and may vary over time. The existence of a dominant littoral drift and the dynamic character of the sand waves are indicated only if they are clearly observed. The direction of the dominant drift is defined as the direction of the net sediment transport indicated by, e.g., sand wave asymmetry or the side of sediment accumulation at groynes. Regarding the degree of confidence, A is high and B is low. The geographic coordinates refer to a pointer shown in the corresponding figures

Situation	Coordinates	Coast type	Number of crests	L (km)	Dominat littoral drift?	Dynamic behaviour?	Degree of confidence	Figure
Mesina Strait, Sicily	38° 8′ 56″ N 15° 32′ 45″ E	open	12	0.4–2.3	Yes	Yes	A	1a
Mortelle, Sicily	38° 16′ 52″ N 15° 36′ 38″ E	open	6	0.37–1.5		Yes	A	1b
Carchuna, Andalucía	36° 41′ 19″ N 3° 26′ 36″ W	open	5	0.48–1.7			A	2a
Albuñol, Andalucía	36° 44′ 29″ N 3° 9′ 12″ W	open	2	1.8			B	2b
Almerimar, Andalucía	36° 39′ 58″ N 2° 43′ 8″ W	open	6	0.9–1.5		Yes	A	3a
Almeria, Andalucía	36° 49′ 18″ N 2° 19′ 30″ W	open	7	0.4–1.5			A	3b
Calella, Catalonia	41° 36′ 40″ N 2° 41′ 2″ E	open	6-Apr	0.4–1.3		Yes	B	4
El Fangar, Ebro Delta	40° 47′ 12″ N 0° 45′ 32″ E	delta plain	8-Mar	0.4–1	Yes	Yes	B	5
La Banya, Ebro Delta	40° 33′ 44″ N 0° 35′ 34″ E	delta plain	6-Feb	0.5–0.8	Yes	Yes	B	6a
Beauduc, Rhône Delta	43° 22′ 45″ N 4° 32′ 45″ E	delta plain	7	0.2–0.54	Yes	Yes	B	6b
Cabanes du Levant, Rhône Delta	43° 22′ 54″ N 4° 55′ 25″ E	delta plain	8	0.17–0.43	Yes	Yes	B	
Trabucador (inner shore), Ebro Delta	40° 36′ 27″ N 0° 43′ 33″ E	lagoon	2–14	0.06–0.25		Yes	A	7a
Nador, Morocco	35° 9′ 0″ N 2° 46′ 11″ E	lagoon	3	2.7–3.6	Yes	Yes	A	7b

they are asymmetrical, indicating a dominant drift to the east, and their spacing increases to the east from $L \approx 0.4$ to 1.5 km.

Much further to the northeast, Calella beach, in Catalonia, Spain also displays KSSW (Fig. 12.4), which were described by Caballeria et al. (2011). The shoreline undulations are quite subtle and their spacing ranges from 0.4 to 1.3 km. According to the images from Google Earth, from 2004 to 2014 they were dynamic, so the shoreline could at times display 4, 5 or 6 undulations. There are no detailed bathymetric surveys at a suitable space scale and that study relied on aerial photos combined with a fairly clear signal in the 5 m isobath (from nautical charts) at roughly the same alongshore spacing as the shoreline undulations. Caballeria et al. (2011) investigated the possible connection of these shoreline sand waves and shoreline instability triggered by the bimodal wave climate dominated by high-angle waves from the east and southwest on the Catalan coast. Interestingly, when the instability conditions were artificially strengthened, the range of the emergent wavelength was consistent with that observed at Calella. However, under the real wave climate the beach was at the threshold for instability.

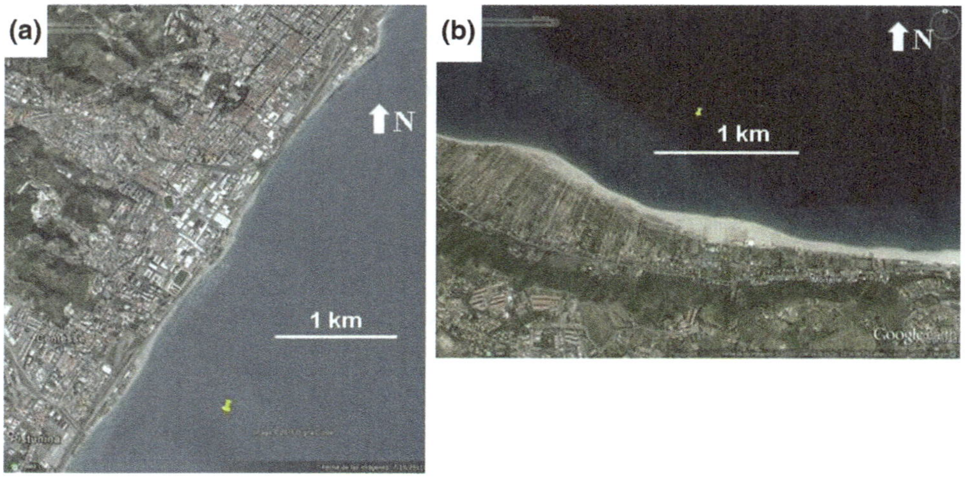

Fig. 12.1 Shoreline undulations in Sicily (Italy): **a** at the Messina strait (*Source* Google Earth, image from 2015 Google and 2015 TerraMetric, taken on 07/10/2011) and **b** near Mortelle (NE coast) (*Source* Google Earth, image from 2015 Google and 2015 TerraMetric, taken on 3/05/2015)

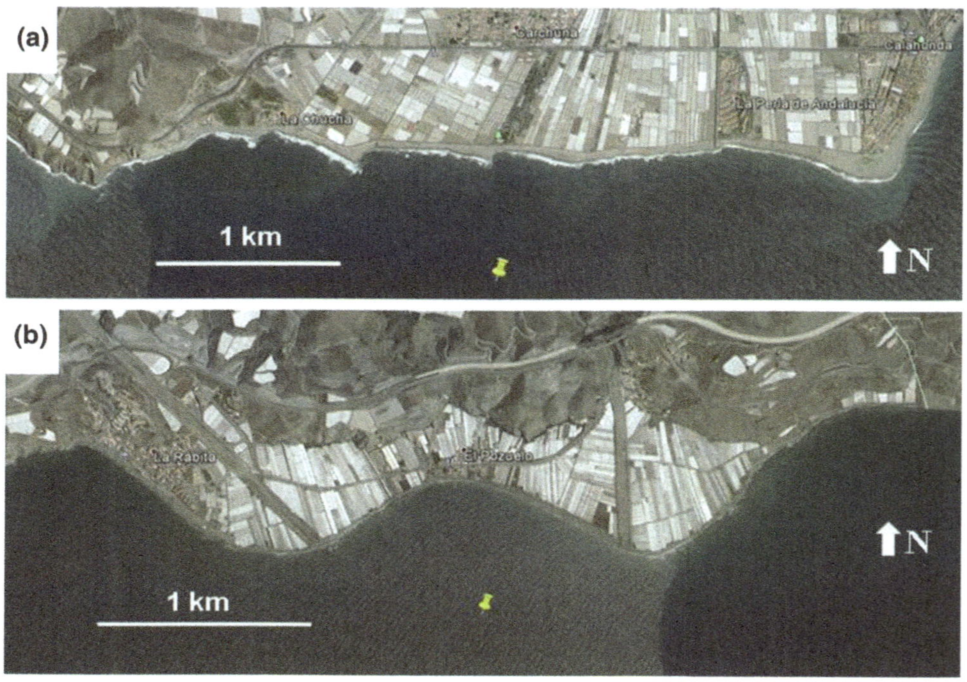

Fig. 12.2 Shoreline undulations in Andalucía (Spain): **a** at Carchuna (*Source* Google Earth, image from 2015 Google, taken on 5/08/2012) and **b** at Albuñol (*Source* Google Earth, image from 2015 Google, taken on 29/06/2013)

12.3 Deltas

The Ebro delta in Catalonia, Spain and the Rhône delta in France provide sandy shores with an abundant supply of sediment that are strongly shaped by wave action and are almost not constrained by human settlement. They are therefore very dynamic and shoreline undulations are frequent. At both sides of the Ebro River mouth there are two spits, la Banya spit to the southwest and El Fangar spit to the northeast. Both spits are very dynamic and they almost always have shoreline undulations (Figs. 12.5a, b and 12.6a). The number of crests ranges from 2 to 8 at each spit and their spacing ranges from 0.4 to 1 km. Although bathymetric surveys are not available for the present study, the relatively fast dynamics of these undulations suggests that their impact on the bathymetry cannot reach too far offshore. In fact, some of them seem to be linked to surf zone bars. Sometimes they are nearly symmetrical (e.g.

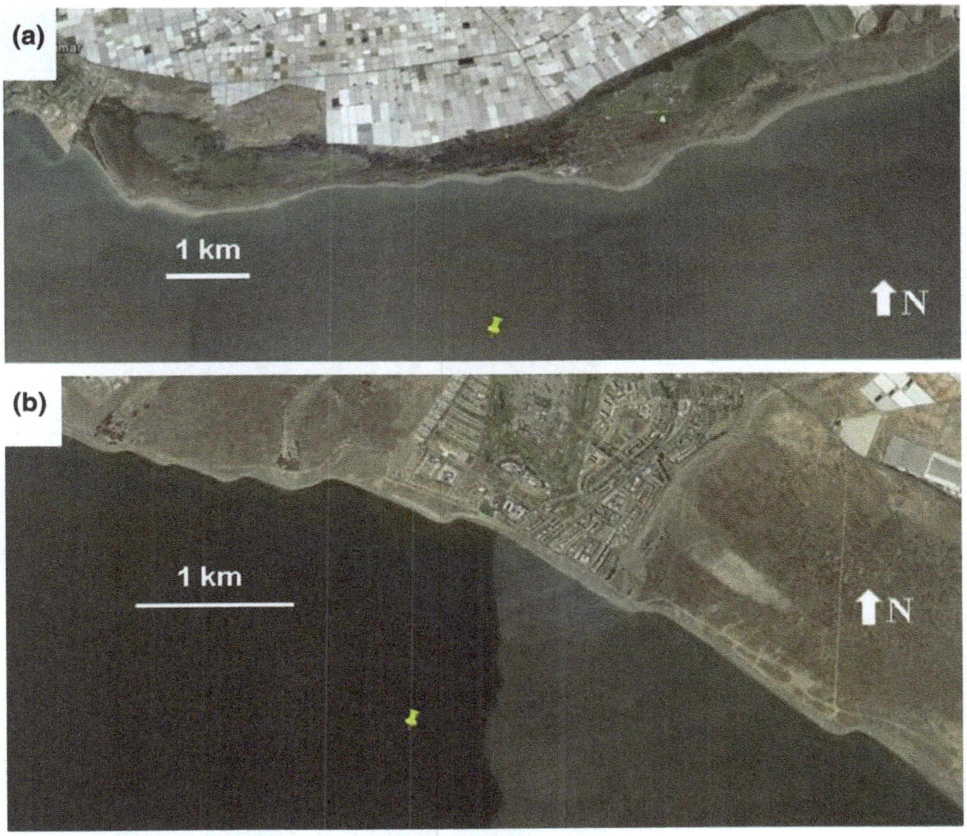

Fig. 12.3 Shoreline undulations in Andalucía (Spain): **a** at Almerimar (*Source* Google Earth, image from 2015 Google, taken on 29/06/2013) and **b** near Almeria (*Source* Google Earth, image from 2015 TerraMetrics and 2015 Google, taken on 29/06/2013)

Fig. 12.4 Shoreline undulations at Calella, Catalonia, Spain (*Source* Google Earth, image from 2015 Digital Globe and 2015 Institut Cartogràfic i Geològic de Catalunya, taken on 27/04/2005)

Figure 12.6a) but they are more frequently asymmetrical, indicating a littoral drift away from the river mouth to the northeast at La Banya spit and to the west at El Fangar spit.

Quite similar shoreline undulations are also common along the shoreline near Beauduc (Fig. 12.6b) at the west side of the Rhône Delta. Seven undulations can be seen in an image taken in 2010 with a spacing of 0.2–0.54 km. They are asymmetric, indicating a littoral drift to the north, away from the river mouth. Further east, there is a spit trending northeast, where six to eight shoreline undulations also appear. They are, however, quite small, with a spacing of 0.17–0.43 km and they could very well be related to surf zone bars. In fact, this comment applies to all the shoreline undulations at both the Ebro delta and the Rhône delta. In

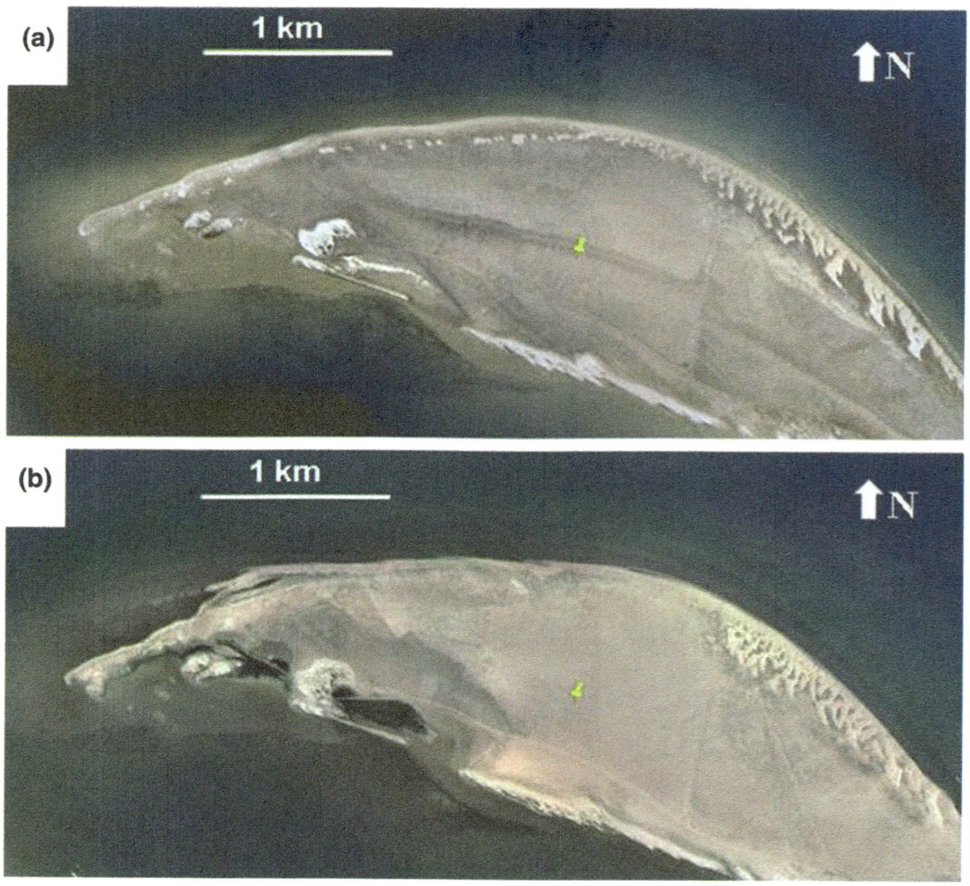

Fig. 12.5 Shoreline undulations along El Fangar spit (northwest Ebro Delta; Catalonia, Spain): **a** in 2005 (*Source* Google Earth, image from 2015 Digital Globe and 2015 Institut Cartogràfic i Geològic de Catalunya, taken on 1/01/2005) and **b** in 2008 (*Source* Google Earth, image from 2015 Institut Cartogràfic i Geològic de Catalunya, taken on 1/01/2008)

the absence of a comparison with bathymetric surveys, and given the relatively short alongshore wavelength, it is not clear whether or not these undulations are linked to surf zone rhythmic bars. Therefore, they cannot be associated with KSSW with certainty. They could be related to pulses in alongshore sediment transport triggered by fluctuating weather conditions.

12.4 Sheltered Shores

Complex morphological patterns and rhythmic features are quite common along sandy shores that are protected from open-sea wave energy. This is the case of enclosed or semi-enclosed basins, bays and lagoons. An interesting example is the Alfacs bay on the southwest side of the Ebro delta, which is bounded from the open sea by La Banya-El Trabucador spit. In spite of the low wave energy reaching the shore, the inner side of El Trabucador beach is quite dynamic. It very commonly has small transverse bars (Falqués 1989; Ribas et al. 2015) but the important features here are the shoreline undulations that occur at a larger length scale, with an alongshore spacing ranging from 0.06 to 0.25 km. In the most recent Google Earth images (from, e.g., 2013) there are about 12 undulations. In contrast, in an image taken in 2004 the coastline was relatively straight and only two large undulations appeared at the southwest tip. Most transverse bars occur at the embayments (Fig. 12.7a), so they are not linked to the shoreline undulations.

Very peculiar are the undulations on the inner side of the barrier beach between the Nador basin and the open sea in Morocco. There are undulations all along the inner shoreline and some of them are quite irregular. In particular, at the southeast tip there are three of them that are very pronounced and have a clear spacing of between 2.7 and 3.6 km

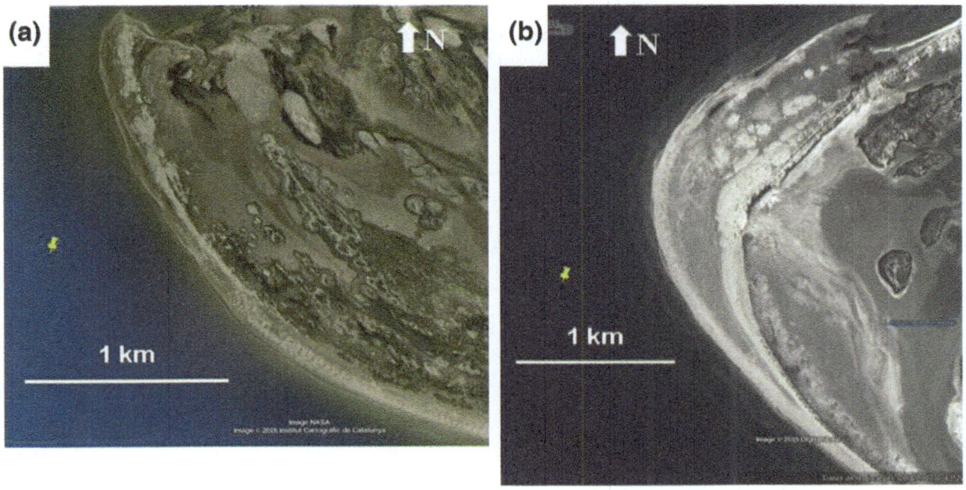

Fig. 12.6 Shoreline undulations **a** along La Banya spit (southern lobule of the Ebro Delta; Catalonia, Spain) (*Source* Google Earth, image from 2015 Institut Cartogràfic i Geològic de Catalunya and NASA, taken on 1/06/2004) and **b** along Beauduc (Rhône delta, France) (*Source* Google Earth, image from 2015 DigitalGlobe, taken on 28/04/2010)

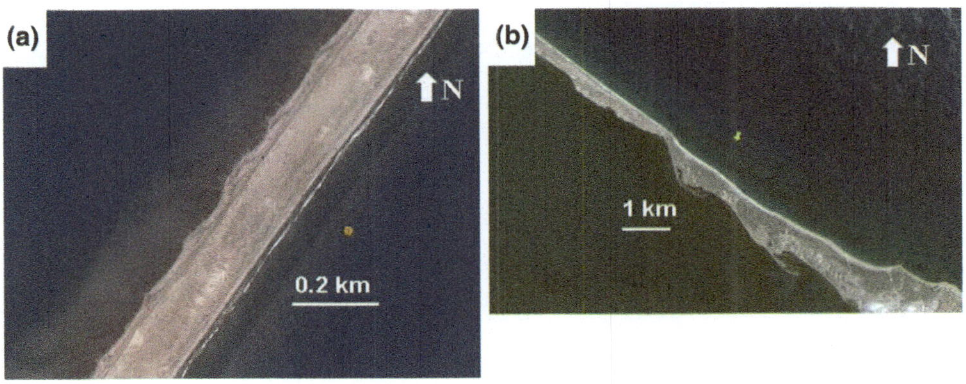

Fig. 12.7 Shoreline undulations in enclosed or semi-enclosed basins: **a** undulations on the inner side of the Trabucador barred beach (Ebro delta, Catalonia, Spain) (*Source* 2015 Institut Cartogràfic i Geològic de Catalunya, taken in 2013) and **b** sand waves developing into sand spits along Nador lagoon (Morocco) (*Source* Google Earth, image from 2015 DigitalGlobe, taken on 16/07/2015)

(Fig. 12.7b). The two south-easternmost ones are very asymmetrical and have developed a spit indicating a littoral drift to the southeast. They are probably triggered by high-angle waves from the west (Ashton et al. 2009).

Acknowledgments This research has been funded by the Spanish government through the research projects CTM2012-35398 and CTM2015-66225-C2-1-P (co-funded by FEDER, U.E.).

References

Ashton, A.D., Murray A.B., Littlewood, R., Lewis, D.A. and Hong., P. Fetch-limited self- organization of elongate water bodies (2009). Geology, 37, 187–190, doi:10.1130/G25299A.1.

Caballeria, M., A. Falqués, and N. van den Berg (2011). Potential instabilities of Catalan coastline induced by high-angle waves, in Proceeding of the 7th IAHR Symposium on River, Coastal and Estuarine Morphodynamics, pp. 2133–2143, Tsinghua Univ. Press, Beijing, China. [CD-ROM].

Falqués, A. (1989). Formación de topografía rítmica en el Delta del Ebro. Revista de Geofísica, 45 (2), 143–156.

Magaña, P., Lopez-Ruiz, A., Lira, A., Ortega-Sanchez, M. and Losada, M. A. (2014). A public, open Western Europe database of shoreline undulations based on imagery. Applied Geography, 55, 278–291.

Ortega-Sánchez, M., Losada, M.A., and Baquerizo, A. (2003). On the development of large- scale cuspate features on a semi-reflective beach: Carchuna beach, Southern Spain. Marine Geology 198, 209–223.

Ortega-Sánchez, M., Losada, M.A. and Baquerizo, A. (2004). Reply to Comment on "On the development of large scale features on a semi-reflective beach: Carchuna beach, Southern Spain" by A. Ashton and A. Brad Murray. Marine Geology 206, 285–288.

Ortega-Sánchez, M., Quevedo, E., Baquerizo, A. and Losada, M. A. (2008). Comment on "High-angle wave instability and emergent shoreline shapes: 1. Modeling of sand waves, flying spits, and capes" by Andrew D. Ashton and A. Brad Murray. J. Geophys. Res. 113, F01005, doi:10.1029/2007JF000860.

Ribas, F., Falqués, A., de Swart, H. E., Dodd., N., Garnier, R. and Calvete, D. (2015). Understanding coastal morphodynamic patterns from depth-averaged sediment concentration. Rev. Geophys., 53, doi:10.1002/2014RG000457.

Nearshore Sand Bars

13

Francesca Ribas, Albert Falqués, and Roland Garnier

Abstract

This review summarizes the morphological characteristics and dynamics of nearshore sand bars observed in the surf zone of sandy beaches worldwide, with length scales ranging from tens to hundreds of meters and time scales ranging from hours to weeks. They include shore-parallel bars (straight and crescentic) and transverse bars of different types. Furthermore, the present knowledge on the physical processes behind their formation and development is discussed.

Keywords

Shore-parallel bars • Crescentic bars • Transverse bars • Physical processes • Self-organization • Sediment transport • Coastal morphodynamics

13.1 Motivation

Subtidal nearshore bars are sand deposits (i.e. shallow areas) occurring in the surf zone of sandy beaches. They have been observed on both open and protected coasts, with fine to medium sand and surf-zone slopes smaller than 0.05, in predominantly nontidal to microtidal settings (they have been occasionally reported on beaches with a significant tidal range) (Wijnberg and Kroon 2002; Van Enckevort and Ruessink 2003). They are important because waves dissipate part of their energy on the bars and the bars can also provide sand to the beach if they migrate onshore (Ribas et al. 2015). Furthermore, alongshore rhythmic bars (Fig. 13.1b) can have a direct impact on the shoreline by creating areas of erosion and deposition, and they are coupled to spatial patterns in surf zone currents (e.g. rip currents) that affect transport and exchange of pollutants, plankton, nutrients or other floating matter. Rip currents are also dangerous for swimmers, being one of the most lethal natural hazards worldwide. Moreover, studying the sand bar dynamics allows us to identify important physical mechanisms that control coastal evolution. In particular, it increases our understanding of the effective sediment transport in areas of the coastal zone (e.g. the swash zone and the inner surf zone) where there is still a significant lack of knowledge on this important process.

This article is a short review of the morphological characteristics and dynamics of the different types of nearshore bars (Sect. 13.2) and the physical processes behind their evolution (Sect. 13.3). Furthermore, Ribas et al. (2016) describes observations of nearshore sand bars on beaches of the western Mediterranean Sea.

13.2 Classification and Morphological Characteristics and Dynamics

13.2.1 Shore-Parallel Bars (Straight or Crescentic)

One to three shore-parallel bars are frequently observed on micro- to mesotidal beaches worldwide (van Enckevort et al. 2003). They are elongated, narrow, shore-parallel ridges that

F. Ribas (✉) · A. Falqués
Department of Physics, Universitat Politècnica de Catalunya, Barcelona, Spain
e-mail: francesca.ribas@upc.edu

R. Garnier
Environmental Hydraulics Institute (IH Cantabria), Universidad de Cantabria, Santander, Spain

J. Guillén et al. (eds.), *Atlas of Bedforms in the Western Mediterranean*,
DOI 10.1007/978-3-319-33940-5_13

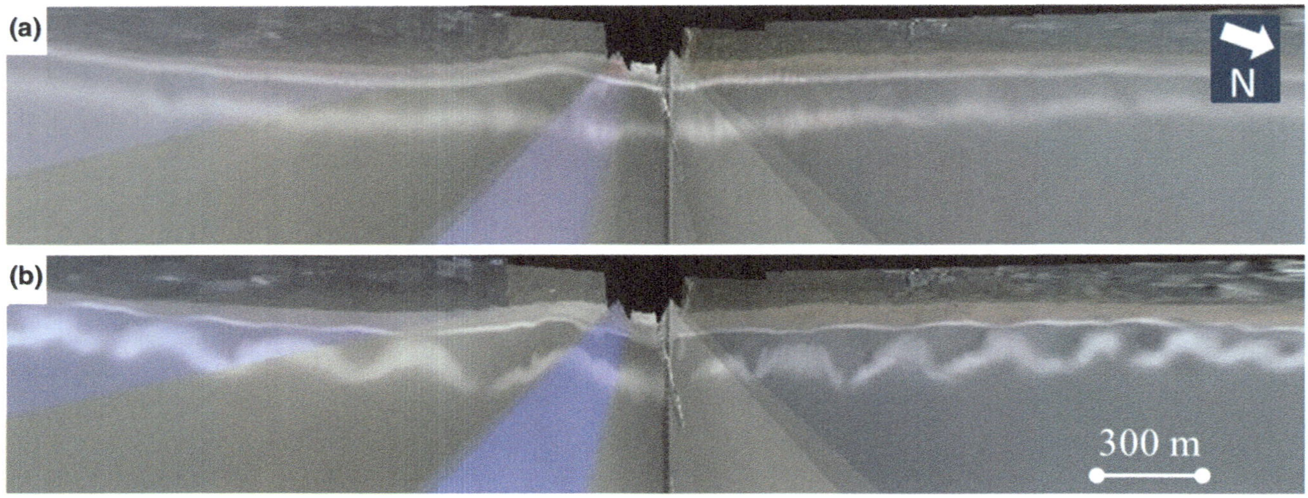

Fig. 13.1 Transition between a straight bar (**a**) and a crescentic bar (**b**), (mean wavelength of about 300 m) at Duck beach, USA, in August–September 1998. The plan views of the beach are created from time-averaged video-images, where the white zones indicate bar presence due to preferable wave breaking on the shallows. The coastline is at the top. Images courtesy of Dr Nathaniel Plant, USGS, USA

are separated from the shoreline by a deeper area or trough. Gentler nearshore slopes seem to favour a larger number of bars (Wijnberg and Kroon, 2002). They can vary in configuration from alongshore-uniform or *straight bars* (Fig. 13.1 a) to *crescentic bars* that undulate in plan view (Fig. 13.1b) (Wright and Short 1984). Sometimes the undulation is quite subtle, the bar being almost straight, but sometimes it is very pronounced, featuring crescent moons with the horns (shoals) pointing shoreward and the bays (deeps) located seaward (van Enckevort et al. 2004). Crescentic bars are sometimes called rip channel systems because strong rip currents always flow offshore in the deep areas, which are called rip channels. On the other hand, straight bars are sometimes slightly oblique instead of being alongshore-parallel (Guillén and Palanques 1993).

The wavelength or alongshore spacing between rip channels (i.e. between crescentic bar horns) tends to be relatively constant alongshore and ranges from tens of metres up to 2–3 km (van Enckevort et al. 2004). Very often, along beaches with crescentic bars the shoreline features undulations with a similar wavelength. Since this wavelength is typically larger than the spacing between ordinary beach cusps, they are called megacusps (Wright and Short 1984). The horns of the crescentic bars can connect to the shoreline and to the megacusp system during long-lasting conditions of low wave energy (down-state transition, see Wright and Short (1984) and Ranasinghe et al. (2004)). The resulting transverse bar system (Fig. 13.2 and 13.3b) is one of the four different types of transverse bar systems that will be discussed in Sect. 13.2.2.

Apart from the slow onshore migration that crescentic bars experience during the down-state transition, at rates of O(1 m/d), crescentic bars can also migrate alongshore in the downdrift direction in the case of oblique wave incidence at rates of up to O(100 m/d) (van Enckevort et al. 2004). During storms of certain characteristics (see Sect. 13.2.2), crescentic bars become straight and migrate offshore at rates of O(10 m/d) (Ruessink et al. 2009). Finally, at longer inter-annual time scales, the shore-parallel bars on open multi-barred beaches experience the following cycle: bar formation near the shoreline, net offshore migration for one or more years, and decay in deep water (van Enckevort and Ruessink 2003, among others). The duration of the cycle varies from about a year (observed at a Japanese site) to more than a decade (observed at several Dutch sites) (Ruessink et al. 2009). On embayed single-barred beaches, the net offshore migration trend is not observed (Ranasinghe et al. 2004; Ojeda et al. 2011).

13.2.2 Transverse Bars

The second kind of nearshore sand bars are generally called transverse bars because they are typically attached to the shoreline and extend offshore either in the shore-normal direction or with an oblique orientation in case of oblique wave incidence. If the crests are shifted in (against) the direction of the alongshore current, we use the term down-current (up-current) oriented bars. Several transverse bars separated by an approximately constant alongshore

Fig. 13.2 Transverse bars at Palm beach, Australia, corresponding to type 1 (TBR bars), with spacing between bars of a few hundred metres. Photograph taken by the authors

distance are often observed (Fig. 13.2). In the presence of an alongshore current, they migrate downdrift with migration rates of up to 40 m/d (Hunter et al. 1979; Konicki and Holman 2000; Ribas and Kroon 2007; Pellón et al. 2014). Four different types of transverse bars have recently been characterized (Pellón et al. 2014; Ribas et al. 2015).

Type 1: TBR bars (Figs. 13.2 and 13.3b). The most common type is that corresponding to the transverse bar and rip (TBR) state in the standard beach state classifications (Wright and Short 1984). The TBR bars are commonly observed on open microtidal beaches under medium-energy conditions. They are typically wide and short-crested and their origin is the merging of a crescentic bar into the beach (see Sect. 13.2.1), so their spacing is strongly related to that of the pre-existing crescentic bar. They can be approximately perpendicular to the shore when shore-normal waves dominate, or down-current oriented (Fig. 13.3b) when incoming waves arrive with a predominant obliquity (Holman et al. 2006; Castelle et al. 2006). As in the case of crescentic bars, TBR bars also show strong and narrow rip currents flowing seaward in the troughs and wider and weaker onshore flows over the crests.

Type 2: Medium-energy finger bars (Fig. 13.3d). These transverse bars have been observed on open microtidal beaches under medium-energy conditions (Konicki and Holman 2000; Ribas and Kroon 2007) and they always coexist with shore-parallel (or crescentic) bars. The term finger bars refers to their thin and elongated nature, and distinguishes them from the wider and shorter TBR bars. These bars are ephemeral, having a residence time ranging from 1 day to 1 month. They are attached to the low-tide shoreline or, occasionally, to the shore-parallel bar (Konicki and Holman 2000). Ribas and Kroon (2007) and Ribas et al. (2014) have shown that they are linked to the presence of obliquely incident waves that create a significant alongshore current, and that they are up-current oriented.

Type 3: Low-energy finger bars (Fig. 13.3c). These transverse bars are persistent features in fetch-limited beaches without a shore-parallel bar (Falqués 1989; Pellón et al. 2014). Bruner and Smosna (1989) and Pellón et al. (2014) gave information concerning both their orientation and the forcing direction. At the two sites, the bars were down-current oriented with respect to the alongshore current generated by the wind-waves.

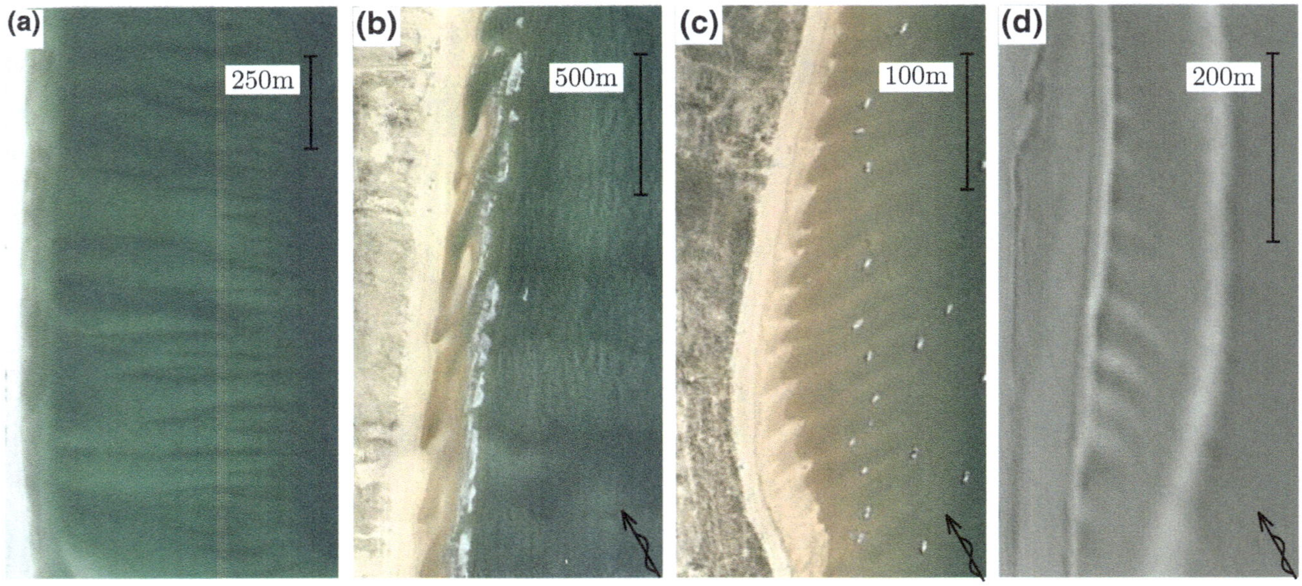

Fig. 13.3 Examples of observed transverse bars with different orientations: **a** shore-normal large-scale finger bars at Anna Maria Island, USA (*Source* Google Earth, US Geological Survey and USDA Farm Service Agency), **b** down-current-oriented TBR bars on the French Atlantic coast (*Source* Google Earth), **c** down-current-oriented low-energy finger bars at El Puntal, Santander, Spain (*Source* Google Earth) and **d** up-current-oriented medium-energy finger bars at Noordwijk, the Netherlands (time-averaged video image). The coastline is on the left. Figure from Ribas et al. (2015)

Type 4: Large-scale finger bars (Fig. 13.3a). These transverse bars have long cross-shore spans of O(1 km) and develop across both the surf and the shoaling zone. They are generally observed to be persistent features in low-energy microtidal environments (Niedoroda and Tanner 1970; Gelfenbaum and Brooks 2003), and are typically oriented almost perpendicular to the shore. Although their dynamics is less well understood, the wave focusing caused by refraction of normal incident waves by the bars seems to be essential (Niedoroda and Tanner 1970).

13.3 Physical Processes for Their Development and Dynamics

13.3.1 Shore-Parallel Straight Bars

According to the widely accepted beach state classification (Wright and Short 1984), shore-parallel bars occur under moderate-energy conditions. In the situations when sand bars are alongshore uniform, it is commonly assumed that cross-shore rather than alongshore processes control the formation and migration of bars (Fernández-Mora et al. 2015). During storms, intense breaking of strong waves drives near-bottom offshore-directed flow (undertow) that transports sand offshore, causing the offshore migration of the bars. Onshore bar migration occurs under intermediate wave conditions, when the undertow is less intense and the cross-shore sediment transport is mainly due to wave non-linearity (velocity skewness and acceleration skewness). Recent process-based models have been able to reproduce these processes separately (Fernández-Mora et al. 2015; Dubarbier et al. 2015) but such models are still struggling to reproduce both processes accurately without changing the parameter values. Finally, the inter-annual net offshore migration of straight bars is accepted to be the result of gradual onshore movement during calm periods combined with episodic strong offshore movement during storms (van Enckevort and Ruessink 2003; Walstra et al. 2014).

Although the processes behind the cross-shore migration of shore-parallel bars are relatively well-known, the causes of initial bar formation are not yet clear (Wijnberg and Kroon 2002; Aagaard et al. 2008) and it is unknown whether bars detach from the shoreline (as argued in, e.g., Walstra et al. 2014) and then subsequently migrate offshore, or accrete in situ at some distance from the shoreline, in the commonly accepted framework of the breakpoint hypothesis (in which the bar forms near the break point of the incident waves, e.g. Mariño-Tapia et al. (2007)). The main reason for this uncertainty is that detailed experimental data during bar formation (including current and sediment transport measurements) are scarce. Aagaard et al. (2008) documented in detail the formation of a straight bar in the inner surf zone of a Danish beach, in which an erosion-accretion cross-shore pattern (generated by cross-shore variations in the speed and direction of the cross-shore transport processes discussed in the paragraph above) created the simultaneous development of a new trough near the shoreline and a bar further seaward. The cross-shore sediment transport gradients are due to a delicate balance between the offshore transport driven by undertow and the onshore transport driven by wave asymmetry, processes that are not yet well understood. Consequently, numerical sediment transport models often have difficulties simulating the initiation of a new bar-trough development (e.g. Mariño-Tapia et al. 2007).

13.3.2 Crescentic Bars and Transverse TBR Bars (Type 1)

Crescentic bars and TBR bars also occur on medium sand beaches under moderate-energy conditions (Wright and Short 1984). Within that framework, crescentic bars develop out of a straight bar for decreasing wave energy and TBR bars develop when the crescentic bar welds to the shore. With increasing wave energy when a storm arrives, the system experiences a reverse (up-state) transition, developing again into a straight, shore-parallel bar. The latter process is called bar straightening or morphological reset, and for decades it was simply associated with high-energy waves. However, recent studies have stressed the important role of wave obliquity in the transitions between a straight and a crescentic bar, revisiting the traditional classification. They found that crescentic bars develop mainly for normal wave incidence and bar straightening occurs for high oblique waves (Price and Ruessink 2011).

Results of recent morphodynamic models indicate the same tendency (Calvete et al. 2005; Garnier et al. 2008, 2013). In the framework of these models, the physical mechanism for crescentic bar formation is the coupling between the morphology and the induced rip-current circulation (self-organization). First, an incipient crescentic bar creates a pattern of wave breaking (waves break stronger on the shoals), which creates extra shoreward-directed forces and an over-elevation occurs on the shoreward part of the shoals. Small feeder currents flow from there to the shoreward part of the rip channels and flow offshore through these channels as rip currents. The circulation cells close due to wide onshore currents over the shoals. Second, since waves break somewhere on the seaward flank of the shore-parallel bar, there is a maximum in the depth-averaged sediment concentration (DASC) there. The onshore currents over the shoals flow to regions with less sediment load and thereby deposit sediment to the shoals, and the opposite occurs in the channels, where sediment is eroded. This is called positive feedback and the crescentic bar continues growing. Alongshore migration of crescentic bars is produced by obliquely incident waves, due to a downdrift shift of the rip-current circulation (Garnier et al. 2008). Bar straightening can also be understood with the same mechanisms: in cases of significant wave obliquity, the generated alongshore current weakens the rip-current circulation and shifts it downdrift, so alongshore migration is induced and the positive feedback weakens so much that the bar eventually becomes straight. A more detailed explanation can be found in Ribas et al. (2015).

13.3.3 Transverse Medium-Energy Finger Bars (Type 2)

Shore-oblique finger bars like the up-current-oriented, transverse medium-energy finger bars always coexist with a significant alongshore current. In such conditions, hydrodynamic processes that induce a meandering of the alongshore current can be more important than the breaking-induced currents of the previous section. Measurements of the hydrodynamics occurring over these bars are not available, but models have proved that, due to frictional forces and mass conservation, the alongshore current experiences a seaward deflection over up-current-oriented bars (and a shoreward deflection over the up-current troughs) (Garnier et al. 2006; Ribas et al. 2012). Moreover, these medium-energy finger bars occur in the steep inner surf zone of beaches with shore-parallel bars. In this situation, incident waves shoal before the crest of the shore-parallel bar, break over the bar, then reform over the trough, and finally break again in the inner surf zone. Somewhere in the inner surf zone, a local maximum in the DASC occurs, related to the second breaker zone, and it turns out to be located close to the shoreline because (i) strong breaking-induced turbulent vortices occur near the shore, (ii) waves dissipate their remaining energy in a relatively narrow area and (iii) the local maximum in the alongshore current profile is quite close to the shoreline. Thereby, up-current-oriented bars grow because the shoreward increasing DASC enhances the convergence of sediment transport in the seaward-directed current that occurs over the up-current crests (Garnier et al. 2006; Ribas et al. 2012).

13.3.4 Transverse Low-Energy and Large-Scale Finger Bars (Types 3 and 4)

For the case of low energy and large scale finger bars, there have been few observations of the induced currents but they indicate the same type of circulation as for the TBR bars. An interesting experiment in a laboratory wave basin was performed by Niederoda and Tanner (1970). On a shore-normal (short-crested) finger bar, an onshore current was measured over the bar crest and it diverged close to the beach to flow in the seaward direction through the troughs. An onshore-directed current over the crest of a low-energy finger bar (with a shore-oblique orientation) was also observed in the field by Falqués (1989). The physical processes driving this hydrodynamic circulation over approximately shore-normal transverse bars can be qualitatively explained by the focusing of wave energy (due to wave refraction and breaking) that occurs over transverse bars, which creates the onshore-directed currents. In the case of down-current-oriented bars, the alongshore current veers toward the shore over the crests and toward the sea over the troughs, so the corresponding current perturbations are reinforced by those created by wave-induced forces. These low-energy and large-scale finger bars typically emerge in terraced profiles with gentle slopes under normal and oblique waves. Waves dissipate their energy slowly across a wide-saturated surf zone, with the wave orbital velocity amplitude decreasing onshore across the surf zone. In the case of oblique wave incidence, an alongshore current is also generated, and typically has a maximum somewhere in the middle of the surf zone. Under such conditions, the combined action of the wave orbital velocities and the depth-averaged current produces a DASC profile that has a maximum in the outer part of the surf zone. Thereby, down-current or shore-normal transverse bars, with their onshore current perturbations on the crests, grow due to positive feedback created by the seaward increasing DASC (Garnier et al. 2006).

Acknowledgments This research has been funded by the Spanish government through the research projects CTM2012-35398 (cofunded by FEDER, U.E.) and CTM2015-66225-C2-1-P (MINECO/FEDER).

References

Aagaard, T., A. Kroon, M. G. Hughes and B. Greenwood (2008). Field observations of nearshore bar formation, Earth Surf. Process. Landforms 33, 1021–1032, doi:10.1002/esp.1599.

Bruner, K. R., and R. A. Smosna (1989). The movement and stabilization of beach sand on transverse bars, Assateague Island, Virginia, J. Coastal Res., 5(3), 593–601.

Dubarbier, B., B. Castelle, V. Marieu and G. Ruessink (2015). Process-based modeling of cross-shore sandbar behavior, Coastal. Eng., 95, 35–50, doi:10.1016/j.coastaleng.2014.09.004.

Calvete, D., N. Dodd, A. Falqués, and S. M. van Leeuwen (2005). Morphological development of rip channel systems: Normal and near normal wave incidence, J. Geophys. Res., 110(C10006), doi:10.1029/2004JC002803.

Castelle, B., P. Bonneton, N. Senechal, H. Dupuis, R. Butel, and D. Michel (2006). Dynamics of wave-induced currents over an alongshore non-uniform multiple-barred sandy beach on the Aquitanian Coast, France, Cont. Shelf Res., 26(1), 113–131, doi:10.1016/j.csr.2005.08.027.

Falqués, A. (1989). Formación de topografía rítmica en el Delta del Ebro, Revista de Geofísica, 45(2), 143–156.

Fernández-Mora, A., D. Calvete, A. Falqués, and H. E. de Swart (2015). Onshore sandbar migration in the surf zone: New insights into the wave-induced sediment transport mechanisms, Geophys. Res. Lett., 42, doi:10.1002/2014GL063004.

Garnier, R., D. Calvete, A. Falqués, and M. Caballeria (2006). Generation and nonlinear evolution of shore-oblique/transverse sand bars, J. Fluid Mech., 567, 327–360, doi:10.1017/S0022112006002126.

Garnier, R., D. Calvete, A. Falqués, and N. Dodd (2008). Modelling the formation and the long-term behavior of rip channel systems from the deformation of a longshore bar, J. Geophys. Res., 113(C07053), doi:10.1029/2007JC004632.

Garnier, R., A. Falqués, D. Calvete, J. Thiébot, and F. Ribas (2013). A mechanism for sandbar straightening by oblique wave incidence, Geophys. Res. Lett., 40, 2726–2730, doi:10.1002/grl.50464.

Gelfenbaum, G., and G. R. Brooks (2003). The morphology and migration of transverse bars off the west-central Florida coast, Mar. Geol., 200, 273–289, doi:10.1016/S0025-3227(03)00187-7.

Guillén, J., and A. Palanques (1993). Longshore bar and trough systems in a microtidal, storm-wave dominated coast: The Ebro Delta (Northwestern Mediterranean), Mar. Geol., 115, 239–252, doi:10.1016/0025-3227(93)90053-X.

Holman, R. A., G. Symonds, E. B. Thornton, and R. Ranasinghe (2006). Rip spacing and persistence on an embayed beach, J. Geophys. Res., 111, C01006, doi:10.1029/2005JC002965.

Hunter, R. E., H. E. Clifton, and R. L. Phillips (1979). Depositional processes, sedimentary structures, and predicted vertical sequences in barred nearshore systems, Southern Oregon coast, J. Sediment. Petrol., 49(3), 711–726, doi:10.1306/212F7824-2B24-11D7-8648000102C1865D.

Konicki, K. M., and R. A. Holman (2000). The statistics and kinematics of transverse bars on an open coast, Mar. Geol., 169, 69–101, doi:10.1016/S0025-3227(00)00057-8.

Mariño-Tapia, I. J., T. J. O'Hare, P. E. Russell, M. A. Davidson and D. A. Huntley (2007). Cross-shore sediment transport on natural beaches and its relation to sandbar migration patterns: 2. Application of the field transport parameterization. J. Geophys. Res. 112, C03002, doi:10.1029/2005JC002893

Niederoda, A. W., and W. F. Tanner (1970). Preliminary study on transverse bars, Mar. Geol., 9, 41–62, doi:10.1016/0025-3227(70)90079-4.

Ojeda, E., J. Guillén and F. Ribas (2011). Dynamics of single-barred embayed beaches, Mar. Geol., 280, 76–90, doi:10.1016/j.margeo.2010.12.002.

Pellón, E., R. Garnier, and R. Medina (2014). Intertidal finger bars at El Puntal, Bay of Santander, Spain: Observation and forcing analysis, Earth Surf. Dyn., 2, 349–361, doi:10.5194/esurf-2-349-2014.

Price, T. D., and B. G. Ruessink (2011). State dynamics of a double sandbar system, Cont. Shelf Res., 31, 659–674, doi:10.1016/j.csr.2010.12.018.

Ranasinghe, R., G. Symonds, K. Black, and R. Holman (2004). Morphodynamics of intermediate beaches: A video imaging and numerical modelling study, Coastal Eng., 51, 629–655, doi:10.1016/j.coastaleng.2004.07.018.

Ribas, F., and A. Kroon (2007). Characteristics and dynamics of surfzone transverse finger bars, J. Geophys. Res., 112, F03028, doi:10.1029/2006JF000685.

Ribas, F., A. ten Doeschate, H. E. de Swart, G. Ruessink, and D. Calvete (2014). Observations and modelling of surf-zone transverse finger bars, Ocean Dyn., 64, 1193–1207, doi:10.1007/s10236-014-0719-4.

Ribas, F., A. Falqués, H. E. de Swart, N. Dodd, R. Garnier, and D. Calvete (2015). Understanding coastal morphodynamic patterns from depth-averaged sediment concentration, Rev. Geophys., 53, doi:10.1002/2014RG000457.

Ribas, F., A. Falqués and R. Garnier (2016). Nearshore sand bars on western Mediterranean beaches, in *Atlas of Bedforms in the Western Mediterranean*, ch. 14, Springer, ISBN: 978-3-319-33940-5.

Ribas, F., H. E. de Swart, D. Calvete, and A. Falqués (2012). Modeling and analyzing observed transverse sand bars in the surf zone, J. Geophys. Res., 117, F02013, doi:10.1029/2011JF002158.

Ruessink, B. G., L. Pape, I. L. Turner (2009). Daily to interannual cross-shore sandbar migration: Observations from a multiple sandbar system, Cont. Shelf Res., 29, 1663–1677, doi:10.1016/j.csr.2009.05.011.

Van Enckevort, I. M. J., and B. G. Ruessink (2003). Video observations of nearshore bar behaviour. Part 1: alongshore uniform variability, Cont. Shelf Res., 23, 501–512, doi:10.1016/S0278-4343(02)00234-0.

Van Enckevort, I. M. J., B. G. Ruessink, G. Coco, K. Suzuki, I. L. Turner, N. G. Plant, and R. A. Holman (2004). Observations of nearshore crescentic sandbars, J. Geophys. Res., 109 (C06028), doi:10.1029/2003JC002214.

Walstra, D. J. R., B. G. Ruessink, A. J. H. M. Reniers and R. Ranasinghe (2014). Process-based modeling of kilometer-scale alongshore sandbar variability, Earth Surf. Process. Landforms, doi:10.1002/esp.3676.

Wijnberg, K.M., and A. Kroon (2002). Barred beaches, Geomorphology, 48, 103–120.

Wright, L. D., and A. D. Short (1984). Morphodynamic variability of surf zones and beaches: A synthesis, Mar. Geol., 56, 93–118, doi:10.1016/0025-3227(84)90008-2.

14 Nearshore Sand Bars on Western Mediterranean Beaches

Francesca Ribas, Albert Falqués, and Roland Garnier

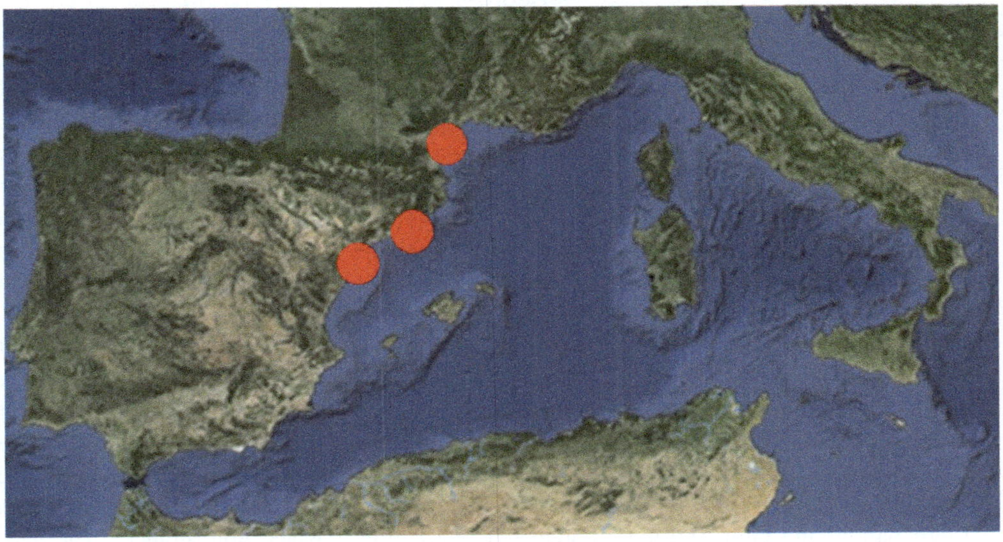

Abstract

This contribution reviews the existing observations of nearshore sand bars (including shore-parallel straight and crescentic bars, and transverse bars) on western Mediterranean beaches. In general, their morphological characteristics and the dynamic processes they experience do not differ qualitatively from those measured on other coasts. A peculiarity of the bars in western Mediterranean is their slower dynamics due to the milder climate, characterized by short-period waves and long episodes of small waves. During these calm periods, the morphology is arrested because there is no significant sediment transport.

Keywords

Shore-parallel bars • Crescentic bars • Transverse bars • Catalan coast • Gulf of Lions • Sediment transport • Coastal morphodynamics

F. Ribas (✉) · A. Falqués
Department of Physics, Universitat Politècnica de Catalunya, Barcelona, Spain
e-mail: francesca.ribas@upc.edu

R. Garnier
Environmental Hydraulics Institute (IH Cantabria), Universidad de Cantabria, Santander, Spain

J. Guillén et al. (eds.), *Atlas of Bedforms in the Western Mediterranean*, DOI 10.1007/978-3-319-33940-5_14

14.1 Introduction

Nearshore bars are sand deposits with length scales of tens to hundreds of metres and time scales of hours to weeks occurring in the surf zone of sandy beaches (see a detailed description in Ribas et al. 2016). These morphological features have been intensively studied in different parts of the world, but they have only been sporadically surveyed on beaches of the western Mediterranean Sea (Table 14.1).

King and Williams (1949) were pioneers in describing (rather qualitatively) the *shore-parallel sand bars* (both *straight* and *crescentic*) occurring in several open and embayed beaches of the west of Italy (e.g. to the south of Rome and in the Gulf of Salerno), the south of France (e.g. in Les Karantes, in the eastern part of the Gulf of Lions) and Algeria. (The italicized terms throughout the text are the names of the morphological patterns related to nearshore bars described in Ribas et al. 2016.) Other Algerian beaches showing crescentic bars were studied in the 1960s (see reviews by Sonu (1973) and Barusseau and Saint-Guily (1981)). Also, multiple offshore sand bars are a common feature in north-eastern part of Jerba Island, Tunisia (Boczar-Karakiewicz et al. 2001).

Nowadays, most of the research studies focus on the western part of the Gulf of Lions, Languedoc-Roussillon, France (see Sect. 14.2) and the coast of Catalonia, Spain (see Sect. 14.3). Nearshore bars have also been monitored on other beaches of the Spanish coast. Backstrom et al. (2008) studied a double shore-parallel bar system in the southern part of the Velez delta, Andalucía, Spain. They quantified the dynamics during a major eastern ('Levante') storm. The single shore-parallel bar present in Cala Millor, in Mallorca, was monitored monthly by Gómez-Pujol et al. (2011) for eight months. This sandy beach is embayed, with a length of 1700 m. During the monitoring period, characterized by mild wave conditions, a crescentic bar migrated onshore, welded to the upper beach and then flattened under energetic wave conditions.

14.2 Observations of Nearshore Sand Bars in the Western Gulf of Lions, France

The western part of the Gulf of Lions (Languedoc-Roussillon, France) has been studied in detail by several authors (Barusseau and Saint-Guily 1981; Certain and Barusseau 2005; Gervais et al. 2011; Aleman et al. 2011, 2013, 2015). In particular, Aleman et al. (2011, 2015) used a large topo-bathymetric LIDAR survey conducted in summer 2009 along the whole Languedoc-Roussillon coast, from Argelès up to Saintes-Maries-de-la-Mer. This 200-km-long coast consists of quasi-continuous sandy beaches with a well-developed double sandbar system. In half of the surveyed area, both shore-parallel bars were shown to switch between the straight and the crescentic configurations, the latter occurring most often in the inner bar. In some parts, the crescentic inner bar was attached to the shore, becoming a system of *transverse TBR bars*. A limitation of that LIDAR survey is that the bar dynamics could not be described because only one survey was available. Examples of the double bar system near Port-la Nouvelle (southern part of the Gulf of Lions) in 2003 and 2008 are displayed in Fig. 14.1, clearly showing the spatial and temporal variability of the bar configurations. In the northern system, both bars were straight in 2008 (b) but the inner bar was crescentic in 2003 (a). In the southern system, the inner bar was in a transverse TBR bar configuration and the outer one was crescentic in both years, but in 2008 the 3D patterns were less pronounced (c, d). Aleman et al. (2013) evaluated the interannual cross-shore migration of the shore-parallel bars on the Languedoc-Roussillon coast, finding *net offshore migration* with different cycle periods on the different parts of the coast and an absence of net offshore migration on other parts more sheltered.

Certain and Barusseau (2005) had previously described the cross-shore migration (at different time scales) of the bars in the double shore-parallel bar system in Sète (Languedoc-Roussillon, France). The bars displayed the standard behaviour of seaward migration during storms and shoreward migration when the energy conditions decreased. The net offshore migration behaviour was triggered by extreme storm events, during which the outer bar moved strongly seaward and degenerated following heavy swell. The inner bar, exposed to the swell, then moved seaward to replace the initial outer bar, and a new inner bar was created at the coast. Finally, Gervais et al. (2011) described the short-term 3D evolution of the crescentic inner bar in Sète during two major storms in winter 2008–2009. Their results evidence the important role of the pre-storm morphology in the final configuration.

14.3 Observations of Nearshore Sand Bars on Catalan Beaches, Spain

14.3.1 Ebro Delta

A system of *low-energy transverse finger bars* was surveyed by Falqués (1989) in the inner part of El Trabucador, Ebro delta. The bars are located inside the Alfacs bay, a fetch-limited site affected by waves of low height. The alongshore spacing between bars was measured to be 60 m on average and the bars were obliquely oriented some 10°–40° (with respect to the shore normal) towards the SW. It is

Table 14.1 Description of published observations of nearshore sand bars on western Mediterranean beaches (in journals indexed in the Journal Citation Reports)

References	Sites	Type of bars	Measurements[a]	Complementary data
King and Williams (1949)	Beaches of western Italy, Algeria and southern France	Shore-parallel straight and crescentic bars	Sporadic '2D surveys' & aerial photographs	Dynamics qualitative Forcing qualitative
Barusseau and Saint-Guily (1981)	Several beaches of Languedoc-Roussillon, France	Double shore-parallel bar system with straight/crescentic bars and TBR bars	Aerial photographs	Dynamics not described Forcing qualitative
Aleman et al. (2011, 2015)			1 LIDAR '3D survey' & aerial photographs	Dynamics not described Forcing quantitative
Aleman et al. (2013)			'2D surveys' every 1–2 y for 24 y	Dynamics quantitative Forcing quantitative
Certain and Barusseau (2005)	Sète beach, Languedoc-Roussillon, France	Double shore-parallel bar system with straight and crescentic bars	'2D surveys' for several years	Dynamics quantitative Forcing quantitative
Gervais et al. (2011)			4 '3D surveys', before and after 2 major storms	Dynamics quantitative Forcing quantitative
Backstrom et al. (2008)	Velez delta, Andalucía, Spain	Shore-parallel straight bar	2 '2D surveys', before and after a major storm	Dynamics quantitative Forcing quantitative
Gómez-Pujol et al. (2011)	Cala Millor, Mallorca Balearic Island, Spain	Shore-parallel straight/crescentic bar and TBR bars	Monthly '3D surveys' for 8 months	Dynamics quantitative Forcing quantitative
Falqués (1989)	Alfacs Bay, Ebro delta, Catalonia, Spain	Low-energy transverse finger bars	1 '3D survey' & aerial photographs and maps	Dynamics qualitative Forcing qualitative
Guillén and Palanques (1993)	Ebro Delta plain, Catalonia, Spain	Shore-parallel straight bars	'2D surveys' every 4 months for 1 y and 2 months	Dynamics quantitative Forcing not described
Ojeda et al. (2011)	Barcelona city beaches, Catalonia, Spain	Shore-parallel straight/crescentic bar and TBR bars	4.3 y: hourly video observations & annual '3D surveys'	Dynamics quantitative Forcing quantitative
de Swart et al. (2016)	Castelldefels beach, Catalonia, Spain	Double shore-parallel bar system with straight/crescentic bars and TBR bars	4.2 y: hourly video observations & '3D surveys' every 6 months	Dynamics quantitative Forcing quantitative

[a]Time period and morphologic measurements ('3D surveys' refers to topo-bathymetric surveys when successive profiles are close enough (<100 m) to show the 3D morphology, and '2D surveys' otherwise)

a quite persistent bar system that can be observed in many of the available aerial photographs of the "Institut Cartogràfic i Geològic de Catalunya" since 1973, as shown in Fig. 14.2. Interestingly, it is nevertheless dynamic since bars are not present in some other photographs.

The whole outer beaches of the Ebro delta were monitored for 1 year and 2 months by Guillén et al. (1993), with 27 cross-shore profiles measured every 4 months. This two-directional drift, open coast displayed a stable system of two shore-parallel bars and a more dynamic inner bar/terrace (an inner terrace and the first outer bar are visible in Fig. 14.2a, b). The bars were not exactly shore-parallel but they displayed a certain obliquity: the distance from the bar crest to the shoreline increases northward and southward from Cape Tortosa, according with the littoral drift direction. The authors suggested this could be due to an alongshore growth of the outer bar during storms. The seasonal migration rate of the bars was also quantified.

14.3.2 Llobregat Delta (Barcelona and Castelldefels)

With the advent of video monitoring in the 1990s (see a review in Holman and Haller 2013), the study of nearshore sand bar dynamics made a step forward because data sets with good spatial and temporal resolution became available for long time periods. The first video monitoring station installed on a western Mediterranean beach was that of Barcelona city (northern part of the Llobregat delta). Five cameras located at 142 m height offer a 180° view of four of the city's beaches (protected by shore-perpendicular groins) and, since October 2001, the images obtained have been used to study them (Ojeda and Guillén 2006; Ribas et al. 2010; Ojeda et al. 2011). Figure 14.3 shows plan views of two of the beaches, obtained after rectifying and merging the 10-minute exposure images of the five cameras. Ojeda et al. (2011) described the characteristics and dynamics of the

Fig. 14.1 Double sand bar system near Port-la Nouvelle (**a**, **b**, 7 km north; **c**, **d**, 5 km south), Languedoc-Roussillon, France, in 2003 (**a**, **c**) and in 2008 (**b**, **d**). The coordinates of the centre of the shorelines are 43° 04′ 27″N–03° 05′ 38″E (**a**, **b**) and 42° 58′ 38″N–03° 02′ 54″E (**c**, **d**). Google Earth images in December 2003 and 2008 (**a**, **c**, **d**, image from NASA; **b**, image from DigitalGlobe). The morphological configurations of the bars vary in time and space: the inner bar is most often crescentic and the outer bar is most often straight

shore-parallel bars on these beaches for 4.3 years. Bogatell (600 m length) displayed a terrace and La Barceloneta (1100 m length) displayed a shore-parallel bar (data obtained from 3D topo-bathymetric surveys, Ribas et al. 2010). The terrace edge of Bogatell and the bar of La Barceloneta switched between a straight configuration during storms (Fig. 14.3a, f), and a crescentic shape configuration (Fig. 14.3b, g) and a system of transverse TBR bars (Fig. 14.3c–e, h–j) during milder conditions, when the system also migrated slowly onshore (Ojeda et al. 2011). The time scales of the variability of the cross-shore bar position and the 3D morphology were different on the two Barcelona beaches: the smaller terraced-bar at Bogatell beach underwent a larger number of changes. An arrest of the beach

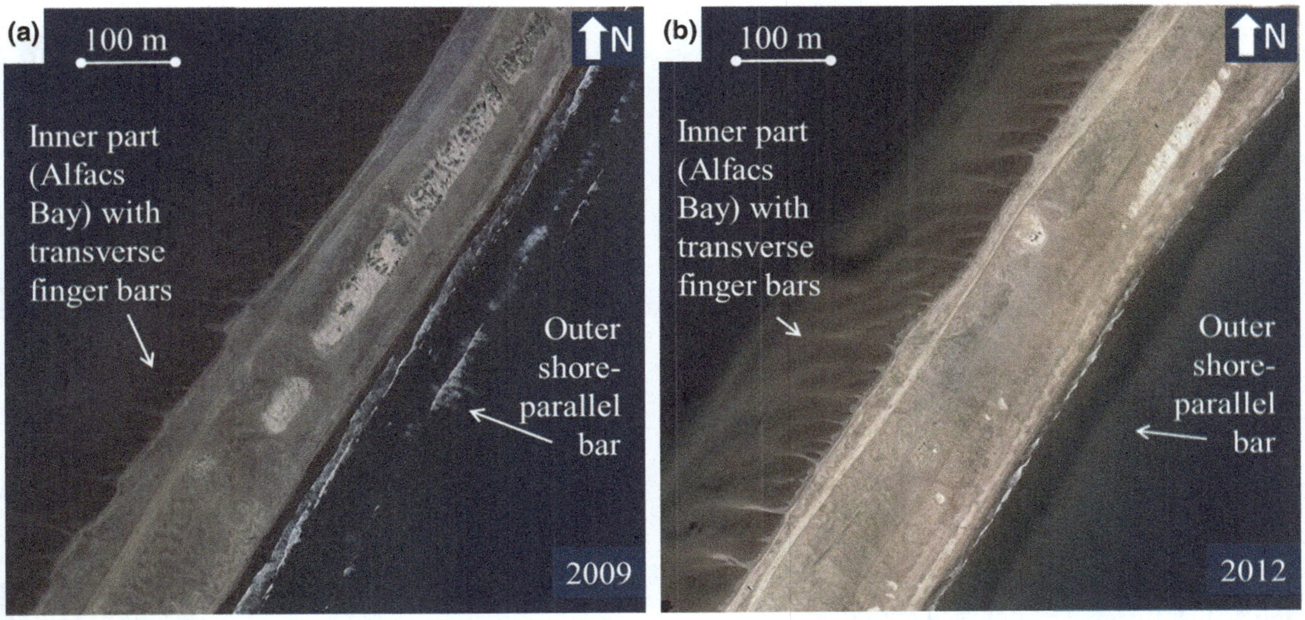

Fig. 14.2 Low-energy transverse finger bars in the inner part of El Trabucador, Ebro Delta, and shore-parallel straight bars in the outer part in 2009 (**a**) and 2012 (**b**). The coordinates of the centre of the image are 40° 36′ 21″N–0° 43′ 21″E. Orthophotos from the "Institut Cartogràfic i Geològic de Catalunya"

configuration at these beaches typically occurs during long periods, mostly associated with the long summer season typical of Mediterranean wave conditions. On the interannual time scale, Barcelona bars showed no net offshore migration trend during the study period.

A video monitoring station was more recently installed atop a 30-m-high structure on Castelldefels open beach, in the southern part of the Llobregat delta, 20 km south of Barcelona, and has been providing images since October 2010. These images, together with 3D topo-bathymetric surveys, are being used to study the dynamics of the double shore-parallel sand bar system over four years and three months (de Swart et al. 2016). The outer bar was always straight and the inner bar was also approximately straight in 2011–2012 but it behaved dynamically in 2013–2014, varying between a straight and a crescentic configuration, and often welding to the beach as transverse TBR bars. The straightening of crescentic bars was shown to be related to oblique waves. Figure 14.4 shows the typical stages during a crescentic bar event: initial straight bar (Fig. 14.4a), crescentic bar of small amplitude (Fig. 14.4b), crescentic bar of large amplitude with *megacusps* (Fig. 14.4c), and bar straightening with a complex configuration in the inner surf zone after a second storm with oblique wave incidence (Fig. 14.4d). This beach also displays sporadically *medium-energy transverse finger bars* (Fig. 14.4e), which are attached to the shoreline within the horns of the crescentic inner bar.

14.4 Summary and Comparison with Nearshore Bars on Other Coasts

Nearshore bars appear on sandy western Mediterranean beaches, both open and embayed, as often as on other micro- to mesotidal coasts but they have been monitored much less intensively. As occurs in other parts of the world (see Ribas et al. 2016), on coasts with fine to medium sediment the profile is mild and two or three shore-parallel bars can occur (southern part of the Gulf of Lions, the Velez delta, the Ebro Delta plain and Castelldefels). On more sheltered open beaches with medium to coarse sediment, the profile is steeper and only one shore-parallel bar appears (Barcelona and Cala Millor, Spain). In a fetch-limited environment with a very mild profile, low-energy transverse finger bars are quite persistent features (Alfacs bay, Spain). The morphological characteristics of the bars on western Mediterranean beaches and their dynamic processes do not differ qualitatively from those measured at other sites. The fact that Mediterranean coasts have a minimum tidal range (of a few

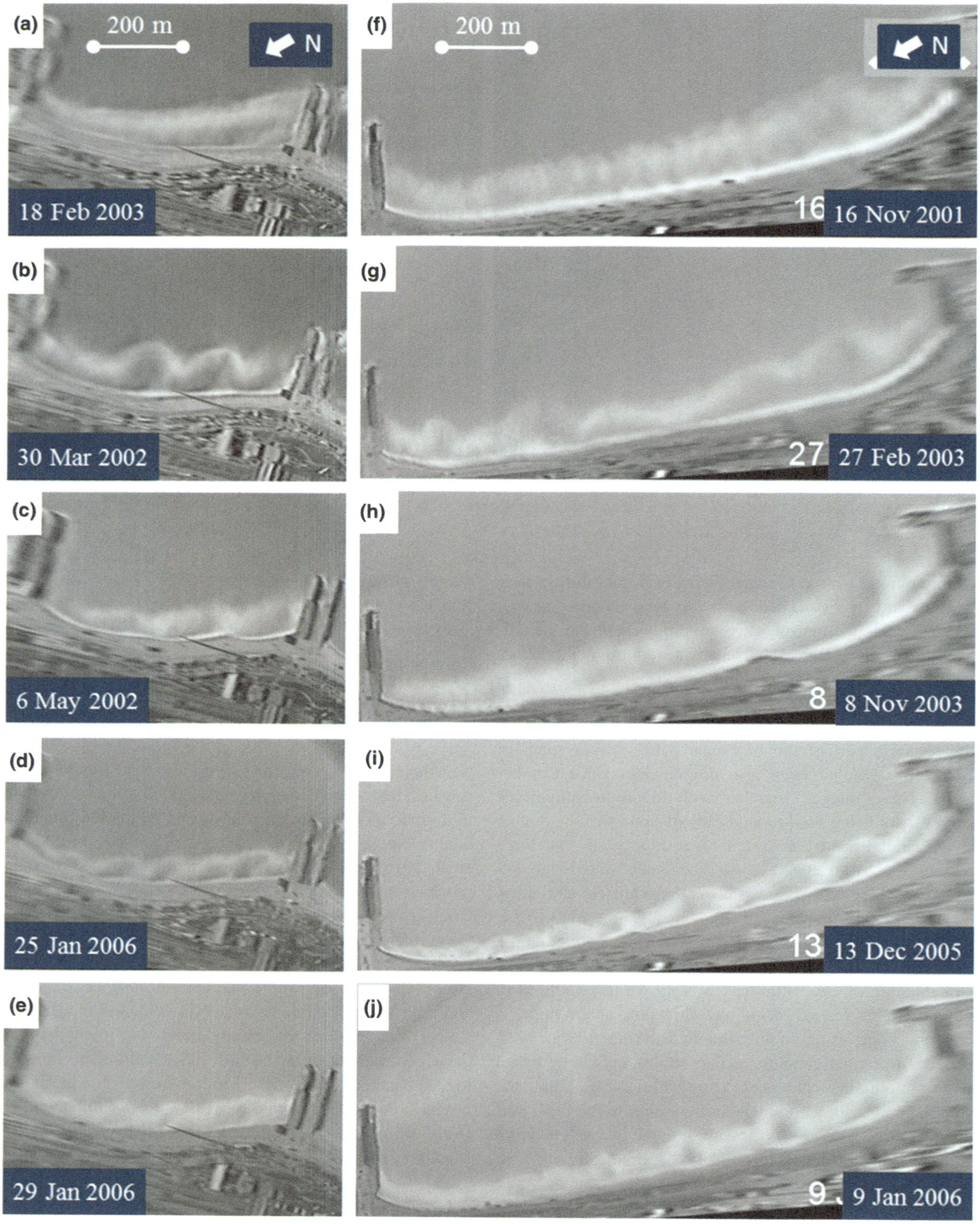

Fig. 14.3 Shore-parallel bars of Barcelona city beaches studied by Ojeda et al. (2011): Bogatell (**a–e**) and La Barceloneta (**f–j**). The plan views of the beach are created from time-averaged video-images, where the white zones indicate bar presence due to preferential wave breaking on the shallows. The terrace edge in Bogatell and the bar in La Barceloneta switch between different configurations. The coastline is at the bottom. The coordinates of the centre of the shoreline are 41° 23′ 39″N–2° 12′ 24″E (**a–e**) and 41° 22′ 32.72″N–02° 11′ 28.11″E (**f–j**). Reprinted from Ojeda et al. (2011), with permission from Elsevier

tens of centimetres) has not so far been shown to produce any specific morphologic characteristics.

As occurs on other coasts (Ribas et al. 2016), shore-parallel bars experience a fast offshore migration and become straight during storms, to subsequently migrate onshore more slowly during post-storm conditions. During the latter process, they gradually change to a crescentic shape that eventually welds to the beach, becoming a TBR transverse bar system. However, the bars in the western Mediterranean experience slower dynamic processes due to the milder climate, with short-period waves and long episodes of small waves, during which the morphology is arrested because the wave energy is too low to cause significant sediment transport (Certain and Barusseau 2005; Ojeda et al. 2011; Gómez-Pujol et al. 2011; de Swart et al. 2016). Recent data from bars on western Mediterranean beaches confirm the important role of oblique wave incidence in straightening crescentic bars (Álvarez-Ellacuría et al. 2012; de Swart et al. 2016). Interannual net offshore migration has been reported on open beaches (Certain et al. 2005; Aleman et al. 2013) but not on more protected, embayed beaches (Ojeda et al. 2011; Aleman et al. 2013), in agreement with the available literature on other coasts (Ribas et al. 2016). Low-energy transverse finger bars observed in the inner part of the Ebro delta (Falqués 1989) and medium-energy transverse finger bars observed in Castelldefels (Fig. 14.4e) have not been analysed quantitatively but their morphological characteristics are similar to those monitored on other coasts.

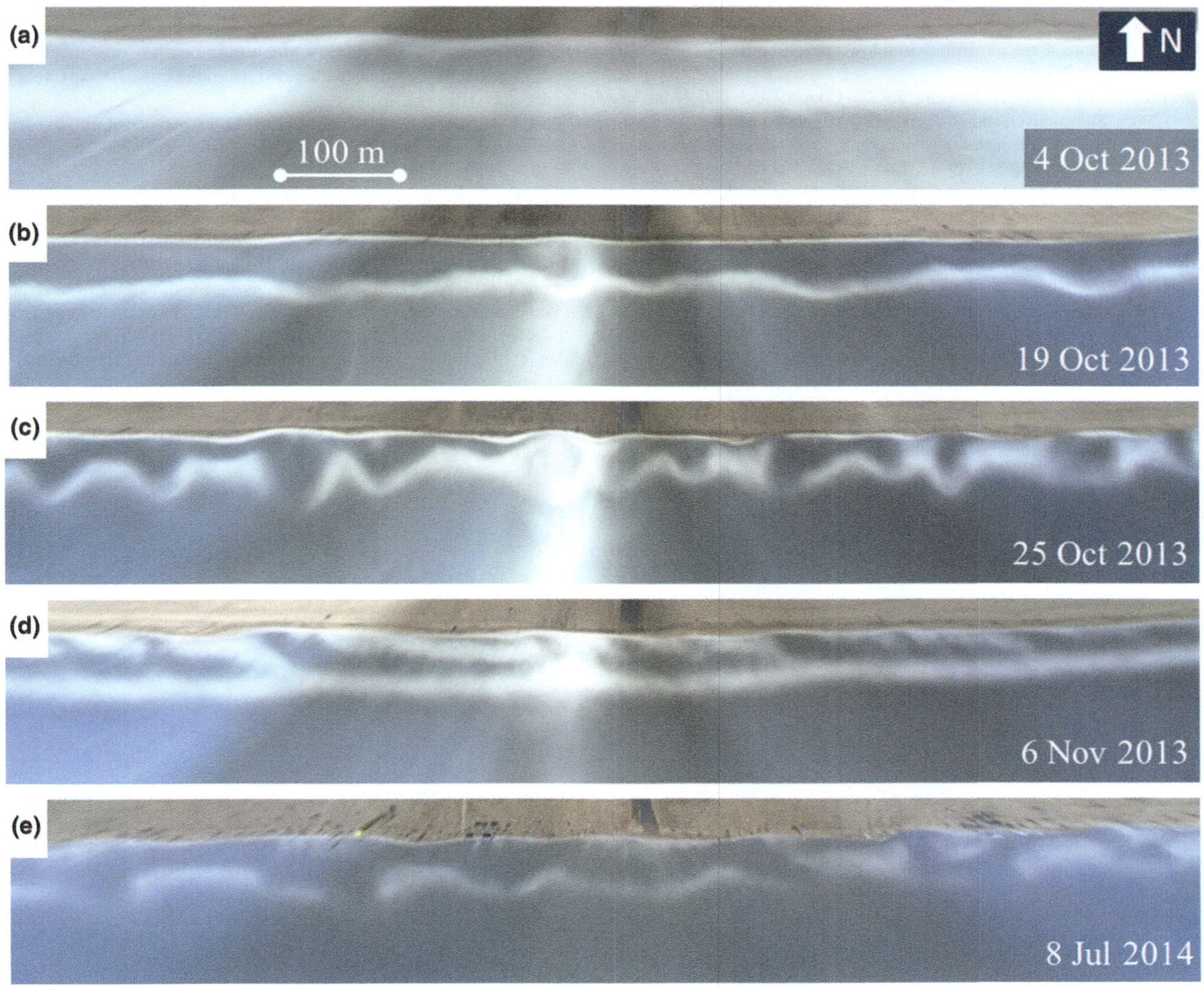

Fig. 14.4 Shore-parallel inner bar of Castelldefels, Catalonia, Spain, switching between different configurations during a crescentic bar event (**a**–**d**). This beach also displays sporadically medium-energy transverse finger bars attached to the shoreline (**e**). The coastline is at the top. The coordinates of the centre of the shoreline are 41° 15′ 50″N–1° 59′ 30″E

Acknowledgments This research has been funded by the Spanish government through the research projects CTM2012-35398 (cofunded by FEDER, U.E.) and CTM2015-66225-C2-1-P (MINECO/FEDER).

References

Aleman, N., N. Robin, R. Certain, E.J. Anthony and J.-P. Barusseau (2015). Longshore variability of beach states and bar types in a microtidal, storm-influenced, low-energy environment, Geomorphology, 241, 175–191, doi:10.1016/j.geomorph.2015.03.029.

Aleman, N., N. Robin, R. Certain, J.-P. Barusseau and M. Gervais (2013). Net offshore bar migration variability at a regional scale: inter-site comparison (Languedoc-Roussillon, France), J. Coast. Res. (SI65), 1715–1720, doi:10.2112/SI65-290.1.

Aleman, N., N. Robin, R. Certain, C. Vanroye, J.-P. Barusseau and F. Bouchette (2011). Typology of nearshore bars in the Gulf of Lions (France) using LIDAR technology, J. Coastal Res., SI64, 721–725.

Álvarez-Ellacuría, A., A. Orfila, L. Gómez-Pujol, and J. Tintoré (2012). Intermediate states transitions in a low microtidal beach, in Coastal Eng. 2012, Poster.

Backstrom, J.T., D.W.T. Jackson, J.A.G. Cooper, and G.C. Malvárez (2008). Storm-driven shoreface morphodynamics on a low-wave energy delta: the role of nearshore topography and shoreline orientation, J. Coastal Res., 24(6), 1379–1387, doi:10.2112/07-0926.1.

Barusseau, J.-P., and B. Saint-Guily (1981). Disposition, caractères et formation des barres d'avant-côte festonnées du Golfe du Lion. Oceanologica acta, 4(3), 297–304.

Boczar-Karakiewicz, B., W. Romanczyk, B. Long and J. L. Bona (2001). Nourishment of a barred nearshore: Jerba, Tunisia, AGU Fall Meeting 2001, abstract #OS31B-0419.

Certain, R., and J.-P. Barusseau (2005). Conceptual modeling of sand bars morpho-dynamics for a microtidal beach (Sète, France). Bull. Soc. Géol. Fr., 176 (4), 343–354, doi:10.2113/176.4.343.

de Swart, R.L., F. Ribas, G. Ruessink, R. Garnier, G. Simarro and J. Guillén (2016). Characteristics and dynamics of crescentic bar events in an open tideless double-barred beach, in preparation.

Falqués, A. (1989). Formación de topografía rítmica en el Delta del Ebro, Revista de Geofísica, 45(2), 143–156.

Gervais, M., Y. Balouin, J. Thiebot, R. Certain, R. Bélon, R. Pedreros, N. Robin and S. Berne (2011). Morphodynamic evolution of nearshore bars in response to winter storms (Lido de Sète, NW Mediterranean), J. Coastal Res., SI64, 1855–1860.

Gómez-Pujol, L., A. Orfila, A. Álvarez-Ellacuría and J. Tintoré (2011). Controls on sediment dynamics and medium-term morphological change in a barred microtidal beach (Cala Millor, Mallorca, Western Mediterranean), Geomorphology, 132, 87–98, doi:10.1016/j.geomorph.2011.04.026.

Guillén, J., and A. Palanques (1993). Longshore bar and trough systems in a microtidal, storm-wave dominated coast: The Ebro Delta (Northwestern Mediterranean), Mar. Geol., 115, 239–252, doi: 10.1016/0025-3227(93)90053-X.

Holman, R.A., and Merrick C. Haller (2013). Remote Sensing of the Nearshore. Annu. Rev. Mar. Sci., 5, 95–113, doi:10.1146/annurev-marine-121211-172408.

King, C. A. M., and W. W. Williams (1949). The formation and movement of sand bars by wave action, The Geographical J., 113, 70–85, doi:10.2307/1788907.

Ojeda, E., and J. Guillén (2006). Monitoring beach nourishment based on detailed observations with video measurements. J. Coast. Res. SI 48, 100–106.

Ojeda, E., J. Guillén and F. Ribas (2011). Dynamics of single-barred embayed beaches, Mar. Geol., 280, 76–90, doi:10.1016/j.margeo.2010.12.002.

Ribas, F., A. Falqués and R. Garnier (2016). Nearshore sand bars, in *Atlas of Bedforms in the Western Mediterranean*, ch. 13, Springer, ISBN: 978-3-319-33940-5.

Ribas, F., E. Ojeda, T. Price and J. Guillén (2010). Assessing the suitability of video imaging for studying the dynamics of nearshore sandbars in tideless beaches. IEEE Trans. Geosci. Remote Sens. 48 (6), 2482–2497, doi:10.1109/TGRS.2009.2039576.

Sonu, C. J. (1973). Three-dimensional beach changes. The Journal of Geology, 42–64.

15 Contemporary Subaqueous Dune Field Development Over an Abandoned River Mouth (Ebro Delta)

Q. Guerrero, J. Guillén, R. Durán, and R. Urgeles

Abstract

High-resolution multibeam bathymetry, bottom sediment samples and time series of current intensities and directions were collected over a dune field in the Ebro Delta coastal area. Aerial photographs were used to analyse the morphological changes of the river mouth since 1946. The bathymetry show a $\sim$6.4 km^2 dune field located at a water depth of 6–15 m, with dunes of 1.3 m median height and 250 m median wavelength. The location of the dune field coincides partially with a former river mouth, an area that emerged in 1946. Onset of dune field development is believed to have started in the 1940s, when the Cape Tortosa river mouth was abandoned and the shoreline underwent a severe retreat. Currently, the recorded high-energy current events induced by northwesterly winds result in a dynamic seabed over the dunes, including ripple development, suggesting that the dune field is active mostly during these high-energy periods.

Keywords

Subaqueous dune field development • Coastal retreat • Deltaic area • Present-day dynamics

Q. Guerrero (✉) · J. Guillén · R. Durán · R. Urgeles
Institut de Ciències del Mar (ICM), CSIC, Passeig Marítim 37-49, Barcelona, Spain
e-mail: queralt@icm.csic.es

J. Guillén et al. (eds.), *Atlas of Bedforms in the Western Mediterranean*,
DOI 10.1007/978-3-319-33940-5_15

15.1 Introduction

Subaqueous dunes are highly dependent on the local hydrodynamic conditions. Sediment transport, sediment grain size, water depth, local morphology, flow intensity and direction are the main factors controlling their development and result in dunes of different dimensions, orientation, shape and migration rates (Ikehara and Kinoshita 1994; Lobo et al. 2000; Kubicki 2008). In the micro-tidal context of the western Mediterranean, several dune fields have been described on the middle and outer continental shelf and have been interpreted to form in coastal areas during low sea-level periods (Durán et al. 2013; Simarro et al. 2015). However, current examples of dune field formation in shallow areas are scarce, probably because sea waves tend to wash out these features.

This work analyses a dune field located on the shoreface of the Ebro Delta. We aim to determine the present-day sediment dynamics and the morphological evolution of the delta plain during the last century, suggesting a formation mechanism in an erosional deltaic context.

15.2 General Setting

The Ebro Delta, covering an area of about 325 km^2, is one of the most important wet coastal ecosystems in the western Mediterranean (Fig. 15.1a). It is located in a micro-tidal wave-dominated coast (maximum range of astronomical tide, 0.25 m). A seasonal wave regime predominates, with the high-energy period between October and March, when the most intense storms occur, and the low-energy season between June and September, when the weather is mainly mild. The most common storm directions come from the E, NW, and S, with the most intense swell-dominated storms coming from the E (locally known as *Llevantades*). These intense storms have an annual return period and significant wave height of 3.5 m (Bolaños et al. 2009). The wind regime is characterized by northwestern winds (Mistral),

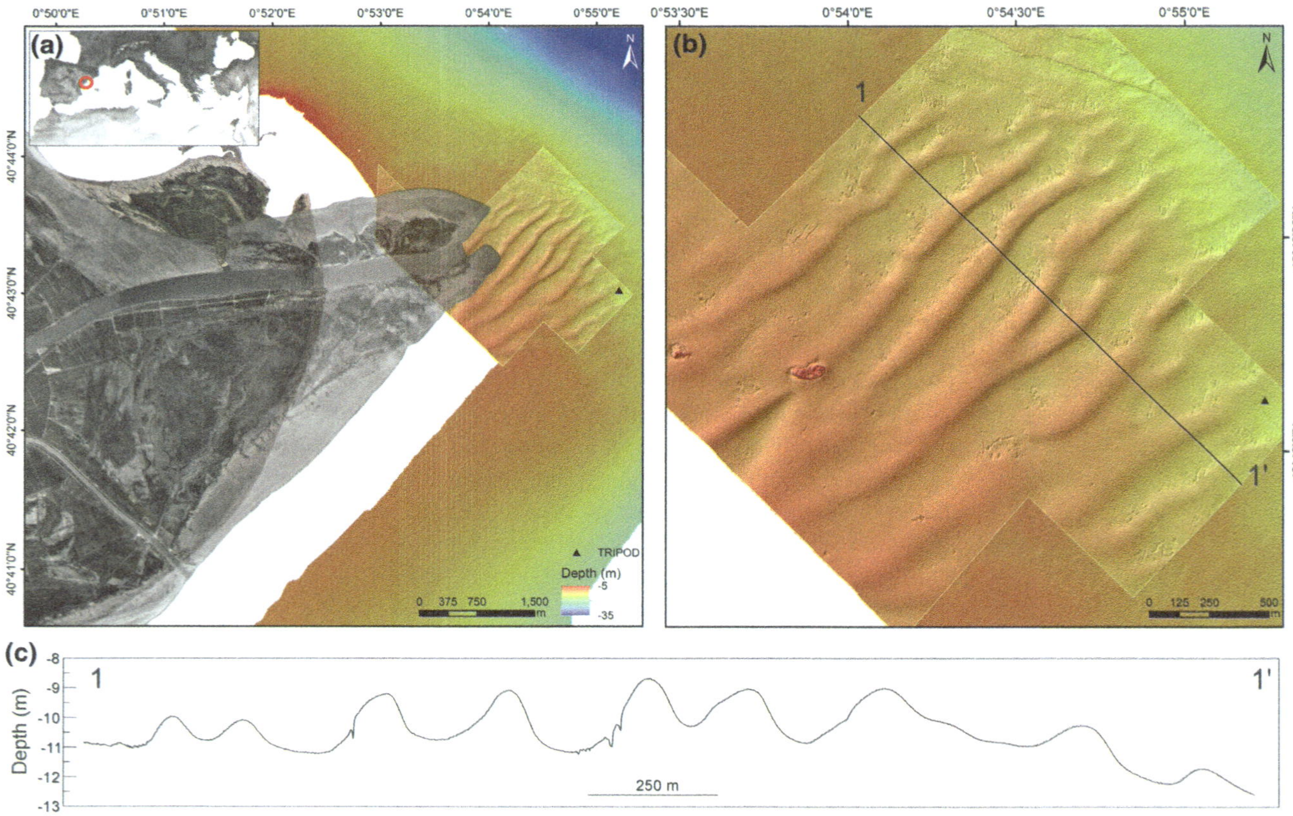

Fig. 15.1 **a** Location of the study area and bathymetry of the dune field. Comparison between the 1946 (*black and white photograph*) and 2014 aerial photographs. **b** Detail of the dune field with the very high-resolution bathymetry acquired in 2015. The *black triangle* indicates the location of the tripod while the *yellow line* indicates the extent of the very high-resolution bathymetry (0.5 m). **c** An example of a bathymetric profile showing the geometry of the dunes

which are extremely intense and persistent in this area because of the orography, leading to sea-dominated storms (García et al. 1993).

Historically, changes in the position of the Ebro River mouth were the main factor responsible for the morphological evolution of the delta plain (Guillén and Palanques 1997). During the last 75 years, the evolution of the delta plain has been mainly influenced by opening of the new river mouth in 1937 and the progressive abandonment of the former river mouth of Cape Tortosa. The new river mouth did not become the main river channel until the mid-twentieth century, when the abandoned Cape Tortosa mouth underwent severe coastal retreat (2300 m from 1946 to 2014) (Fig. 15.1a).

15.3 Data Collection

High-resolution swath-bathymetry data were acquired in summer 2004 with a Simrad EM3002d multibeam echosounder on board the *Arraix* survey boat in the framework of the PRODELTA project. Two additional cruises were carried out in the framework of the FORMED project: in October 2013 using an ELAC NautikSeaBeam 1050D multibeam echosounder system on the *B/O García del Cid* and in November 2015 with an R2SONIC 2024 broadband multibeam echosounder system on the *Mirafons Nova* boat. The latter provided very high-resolution bathymetry (gridded at 0.5 m).

The FORMED project also included the deployment of an instrumented benthic tripod over the dune field at 13.3 m water depth from 15 October 2013 to 10 January 2014 off Cape Tortosa (Fig. 15.1). The tripod was equipped with an Aanderaa current meter (RCM9) at 1 m above the bottom (mab) measuring time series of current intensities and directions every 30 min and a GOPRO camera at 1.63 mab recording sequences of 10 s every 4 h. A sediment sample was recovered by a HAPS corer at the tripod location.

Wave-field time series data from the Tarragona offshore buoy located at 688 m depth were provided by the Spanish Ports Authority (Puertos del Estado). This buoy provides hourly measurements and statistics data. The wave-field data were propagated from the buoy position to the tripod location considering only the shoaling effect.

Aerial photographs were obtained from the Cartographic and Geologic Institute of Catalonia (ICGC). The images showed changes in the river mouth configuration and the morphological evolution.

15.4 Results and Discussion

The dune field develops in fine sand sediment between the 6 and 15 m isobaths and extends over 6.4 km^2 of the swath-mapped deltaic lobe off the abandoned Cape Tortosa mouth (Fig. 15.1a). The dune height ranges from 0.5 to almost 3 m and the wavelength from 100 to 375 m, with median values of 1.3 and 250 m, respectively (Fig. 15.1b). The dune crests are fairly rectilinear, oriented ~45°N, and occasionally display bifurcation. A slight change in the crest-alignment orientation is observed because of their arrangement perpendicular to the bathymetric contours (deltaic lobe shape) (Fig. 15.1b). Smaller-scale superimposed bedforms, ripples, were observed over the dunes during the deployment of the tripod (Fig. 15.2a, b). Their evolution and morphology were governed by changes in hydrodynamic conditions (Fig. 15.2c).

The development of the dune field on the shoreface seems to be closely related to the change in the river mouth location, which led to an erosional context and the progressive abandonment of the Cape Tortosa mouth. Under these conditions severe shoreline retreat (ca. 2300 m from 1946 to 2014) occurred (Fig. 15.1a). The eroded sediment was available for transport by waves and currents and was redistributed along the coast. Indeed, the location of the dune field is very close to the shoreline position of Cape Tortosa during the 1940s (Fig. 15.1a). Therefore, the onset of the dune field probably occurred with the progressive abandonment of the former river mouth, whose erosion provided the main sediment supply.

The hydrodynamic measurements showed maximum near-bottom current intensities of ~0.6 m/s flowing towards the SSE (Fig. 15.2d). These currents were driven by NW winds and were arranged roughly perpendicular to the crestline of the dunes at the tripod location. Formation of ripples was associated with these high-energy events, with their morphologies varying depending on the hydrodynamic conditions: either wave-dominated or wave- and current-dominated (Fig. 15.2a, b, c). Therefore, the high-energy hydrodynamic periods were able to mobilize bottom sediments in the area, suggesting that the dune field might be currently active. Considering the orientation of the dunes crestline, the dune field formation and evolution can be confidently related to the SE-flowing currents induced by NW winds.

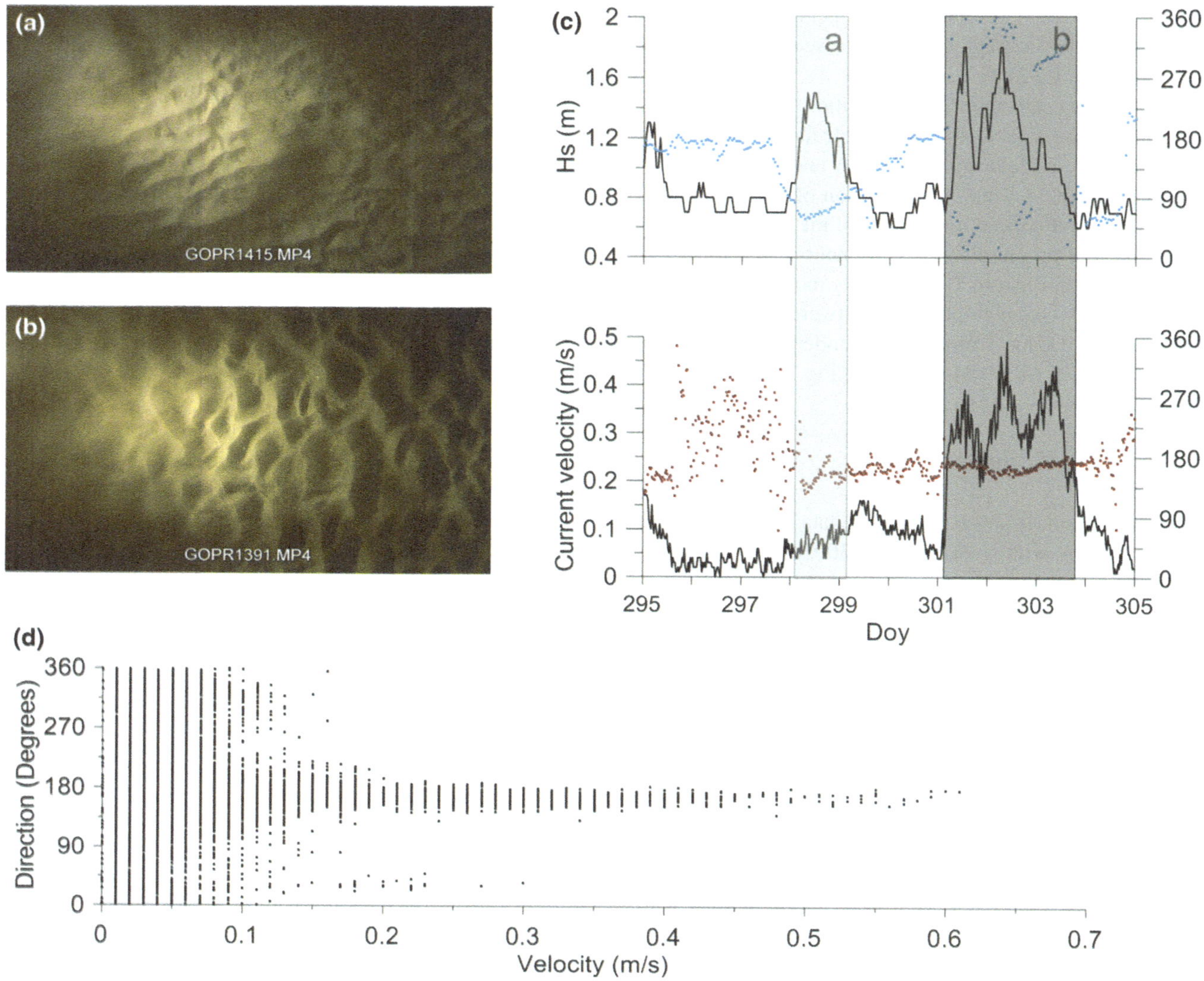

Fig. 15.2 Hydrodynamics over the dune field at the location of the tripod. **a** Example of rippled seabed resulting from wave-dominated hydrodynamic conditions. **b** Example of rippled morphology resulting from wave and current-dominated hydrodynamic conditions. **c** Significant wave height (*black line*) and wave direction (*blue dots*) at the top and current velocities (*black line*) and current directions (*red dots*) *below*; The *light grey band* highlights formation of wave-dominated ripples and the *dark grey band* highlights wave- and current-dominated ripples. **d** Relation between current speed and current direction during the tripod deployment

15.5 Conclusions

High-resolution seafloor mapping allowed a dune field to be identified over the former Cape Tortosa river mouth in the Ebro Delta. The dune field occupies 6.4 km^2 of the swath-mapped area. The dunes display maximum heights of almost 3 m and wavelengths of 375 m. They are composed of fine sand and arranged perpendicular to the bathymetric contours.

The genesis of the dune field is closely related to the contemporary evolution of the Ebro River mouth, probably favoured by the progressive abandonment of the former river mouth of Cape Tortosa since the 1940s. The measured currents and waves time series data suggest that the dune field is currently active. Moreover, development of ripples during high-energy periods evidences the dynamism and sediment remobilization in this area.

Recently developed subaqueous dune fields such as the one analysed in this study can provide new insights for better understanding the genetic mechanisms of bedforms during transgressive episodes in the geological past. In fact, although most of the western Mediterranean continental shelves are widely occupied by relict, waning or low-dynamic bedforms presumably formed in coastal environments (see examples in this book), there is little evidence of contemporary formation. Finally, the time-scale over which the dune field has evolved (several decades) may provide a useful example for predicting the morphodynamic shoreface evolution in a future sea-level rise scenario.

Acknowledgments Q. Guerrero was supported by the Spanish Government through a FPI grant (Ref. BES-2013-066261). This research was supported by the project FORMED (CGL2012-33989). R. Durán is supported by a CSIC JAE-Doc contract co-funded by the FSE. We thank the captains and crews of the R/Vs *García del Cid*, *Arraix* and *Mirafons Nova* for their kind assistance. Puertos del Estado (Ministerio de Fomento) provided wave data information.

References

Bolaños, R., Jorda, G., Cateura, J., Lopez, J., Puigdefabregas, J., Gomez, J., Espino, M. (2009). The XIOM: 20 years of a regional coastal observatory in the Spanish Catalan coast. J. Mar. Syst. 77, 237–260.

Durán, R., Rivera, J., Guillen, J., Cárdenas, E. De, Muñoz, A., Acosta, J. (2013). Sandy Subaqueous Dunes on the Murcia Continental Shelf (Western Mediterranean Sea). Rapp. Comm. Int. Mer Médit. 40.

García, M. A., Sánchez-Arcilla, A., Sierra, J.P., Sospedra, J., Gómez, J. (1993). Wind waves off the Ebro Delta, NW Mediterranean. J. Mar. Syst. 4, 235–262.

Guillén, J., Palanques, A. (1997). A historical perspective of the morphological evolution in the lower Ebro river. Environ. Geol. 30, 174–180.

Ikehara, K., Kinoshita, Y. (1994). Distribution and origin of subaqueous dunes on the shelf of Japan. Mar. Geol. 120, 75–87.

Kubicki, A. (2008). Large and very large subaqueous dunes on the continental shelf off southern Vietnam, South China Sea. Geo-Marine Lett. 28, 229–238.

Lobo, F.J., Hernández-Molina, F.J., Somoza, L., Rodero, J., Maldonado, A., Barnolas, A. (2000). Patterns of bottom current flow deduced from dune asymmetries over the Gulf of Cadiz shelf (southwest Spain). Mar. Geol. 164, 91–117.

Simarro, G., Guillén, J., Puig, P., Ribó, M., Lo Iacono, C., Palanques, A., Muñoz, A., Durán, R., Acosta, J. (2015). Sediment dynamics over sand ridges on a tideless mid-outer continental shelf. Mar. Geol. 361, 25–40.

16 Bedforms on the Lowermost Reach of the Tiber River (Rome, Italy): Preliminary Results from Integrated Geophysical Surveys and Samplings

Alessandro Bosman and Luciana Orlando

Abstract

An ultra high-resolution multibeam survey of the lowermost reach of the Tiber River (Rome, Italy) was carried out in February 2011 on a total length of 56 km from the northernmost part of Rome city to the river mouth. The morpho-bathymetric data reveal that the main morphological features of the river bed are widespread dune fields, covering about 45 % of the surveyed area. Three main types of bedform were identified, with wavelengths ranging from a few metres up to hundreds of metres. The seismic profiles show a transparent layer(s) of 1–2 m, corresponding to a traction carpet of coarse-grained sediment, where the bedforms migrate. By integrating bathymetric data with high-resolution seismic profiling and sampling, one can better define the genesis of these bedforms and provide insights into the present-day sedimentary processes acting on the river bed.

Keywords

Dune fields • River floor mapping • Multibeam bathymetry • High-resolution seismic profiling

---

A. Bosman (✉)
Istituto di Geologia Ambientale e Geoingegneria, Consiglio Nazionale delle Ricerche, Rome, Italy
e-mail: alessandro.bosman@uniroma1.it

L. Orlando
Dipartimento di Ingegneria Civile, Edile e Ambientale, University of Rome, Sapienza, Italy

J. Guillén et al. (eds.), *Atlas of Bedforms in the Western Mediterranean*, DOI 10.1007/978-3-319-33940-5_16

16.1 Introduction

In the last few years, seafloor mapping systems have undergone impressive technological advances, especially through the improvement of the new high-resolution multibeam echo-sounders and global navigation satellite positioning systems. Multibeam bathymetry is used to extensively map continental margins, greatly enhancing the knowledge of the geological processes that are acting on the seafloor. Recently, this mapping has been extended to inland waters such as the Po River (Bosman et al. 2014), and has shown complex morphologies as well as active erosional and depositional processes due to the dynamics of hydraulic flows. The lowermost reach of the Tiber River is located in the Tiber Valley, where the city of Rome was built (Fig. 16.1a). The substrate of the river is composed of an almost continuous superimposition of Pliocene–Quaternary marine and continental sedimentary deposits (Amorosi and Milli 2001; Raspa et al. 2008; Funiciello and Giordano 2005). Currently the channel bed is composed of a discontinuous layer of alluvial sediment and a relict substratum that is exposed on the channel bed and is known as the Tiber depositional sequence (Milli et al. 2013).

16.2 Methods

The multibeam surveys were performed using the Teledyne Reson SeaBat 7125 SV1 echo-sounder. This system works at a frequency of 400 kHz, emitting up to 512 soundings across 140°. The vessel positioning was supplied in real-time by an Applanix Position and Attitude System (POS/MV 320 V5) using RTK corrections received by a GPS master base-station belonging to the GNSS National Dynamic Network (http://www.igmi.org/rdn/). Very highly accurate positioning of the navigation lines was obtained by post-processing kinematic (PPK) techniques using POSPac MMS software in order to reduce horizontal and vertical positioning uncertainties (Bosman et al. 2015). Depth elevation data were referenced to an IGM datum (http://www.igmi.org/) corrected by means of GPS PPK techniques. Multibeam data were processed using Caris Hips & Sips 8.1 in order to generate a digital elevation model (DEM) with a 0.2 m cell size. The sub-bottom profilers were collected using Benthos Chirp III 2.7 kHz source and processed through Geo Suite All Works software. River bed samples were collected by a 5-L Van Veen grab.

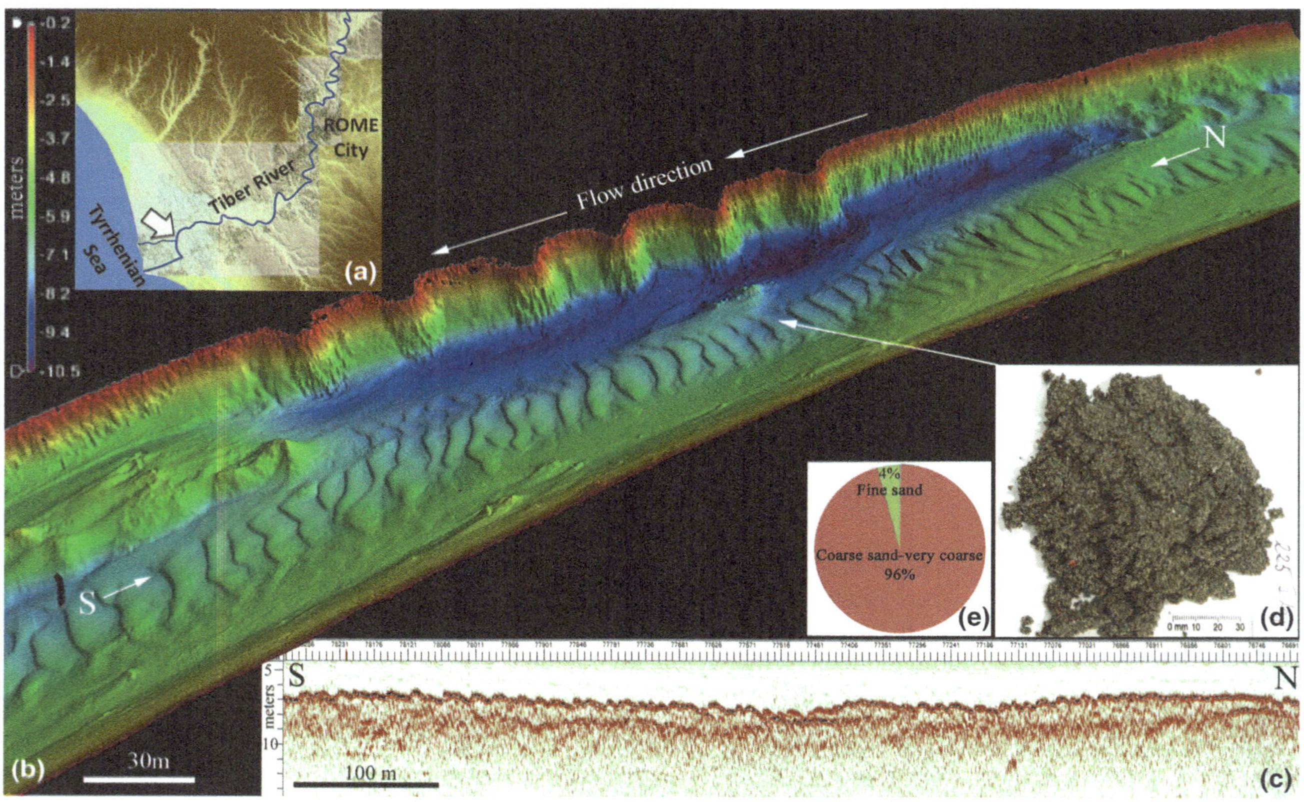

Fig. 16.1 Dune fields on the lowermost reach of the Tiber River close to the mouth. **a** Key map of the area surveyed in **b**. **b** High-resolution digital elevation model showing widespread dunes (*colour scale in metres*). **c** Chirp profile across the dunes showing the base of the traction carpet. **d** Grab sample recovered from the dune field. **e** Grain size analysis from the sample, showing coarse and very coarse sand

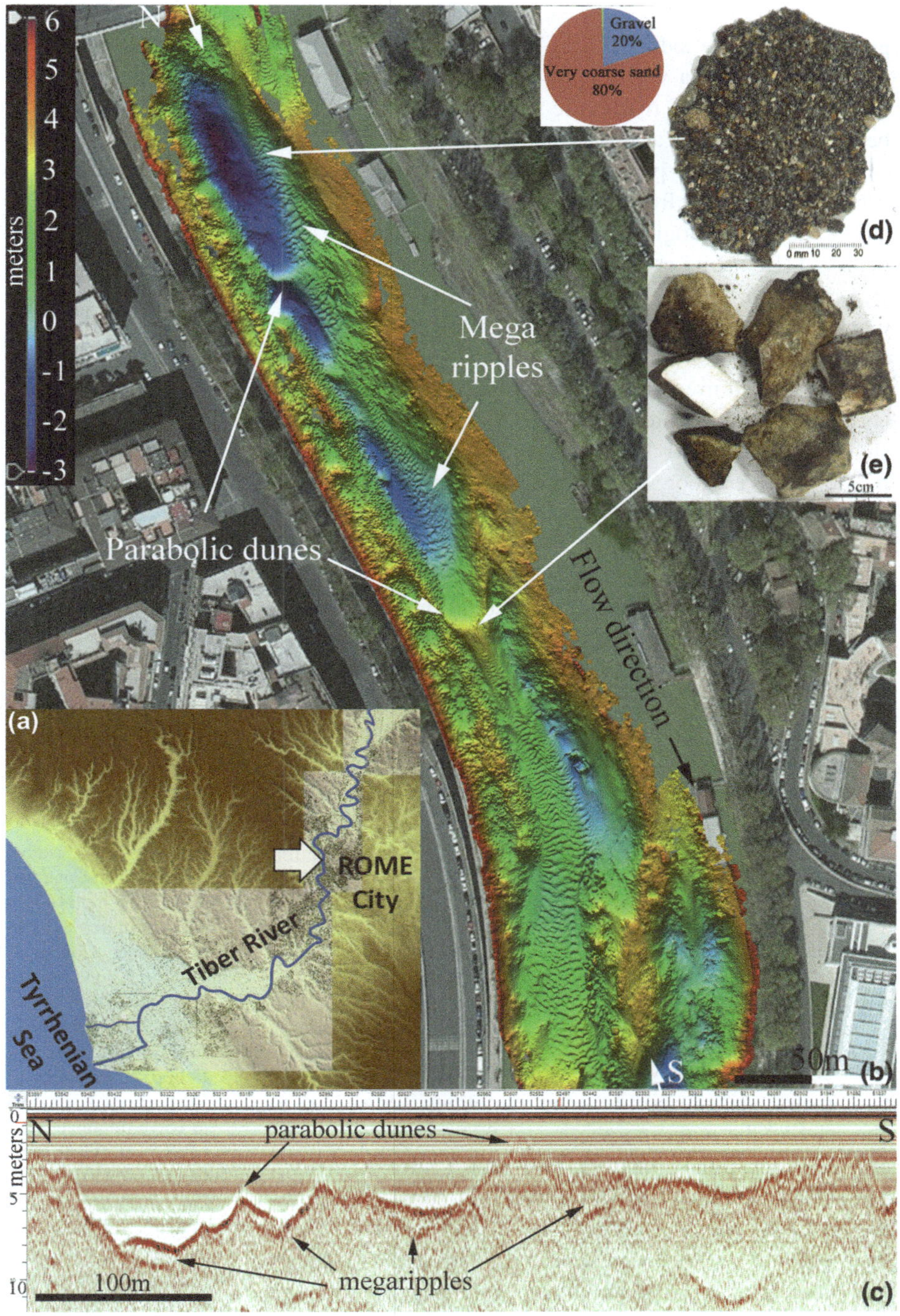

Fig. 16.2 Parabolic dunes and mega ripple fields on the lowermost reach of the Tiber River in the centre of Rome. **a** Key map of the area surveyed in **b**. **b** High-resolution digital elevation model showing megaripples superimposed on the parabolic dunes (*colour scale in metres*). **c** Chirp profile shows the lack of internal reflectors on the parabolic dunes, whereas a transparent layer corresponds to the megaripples. **d** Very coarse sand and gravel sampled on the mega ripples. **e** Anthropogenic cobbles and boulders sampled on the parabolic dunes

16.3 Results

Multibeam bathymetry The high-resolution DEM shows several examples of morphological features related to erosive-depositional fluvial processes, such as erosive furrows, enclosed depressions, slide scars and bedforms. The river bed morphology is mainly dominated by widespread waveforms (Fig. 16.1b) characterized by different wavelengths and shapes, confined within the river banks. Three main types of bedform field were identified. The most common were dunes with a height of 0.3–0.4 m and a wavelength of 10 m. The dunes are characterized by a gently sloping stoss side (2°–4°) and a markedly steeper lee side (10°–20°). The second type consists of parabolic dunes with a height of about 2 m high and a length of up to 150 m. They were identified only in the stretch of river located in the historical centre of Rome City (Fig. 16.2). These features show a typical parabolic shape elongated towards the current, and they are very wide (10–20 m), with a steep stoss side (up to 15°) and a gently sloping lee side (3°–5°). The third type consists of mega ripples with a wavelength of 1–3 m and a height of 0.1–0.2 m. The mega ripples are widespread on the channel bed, often occurring as isolated features or in other cases superimposed on the parabolic dunes. Occasionally, a relict substratum crops out from the channel bed and from sidewalls and some archaeological findings of Roman age are also observed (Bernabini et al. 2006).

Seismic stratigraphy Chirp profiles recorded along the main dune field show a shallow high-amplitude reflector located between 0.5 and 2.5 m below the river bed and some deep minor and irregular reflectors located at variable depth (Fig. 16.1c). Chirp profilers along the mega ripples also show a thin reflector located between 0.5 and 2 m depth below the river floor. In contrast, chirp profiles along the parabolic dune do not show seismic penetration of the signal but rather an irregular morphology, apart from the area where megaripples are present (Fig. 16.2c).

Sampling The grab samples carried out on the bedforms show different grain sizes and sediment compositions: gravel and very coarse sand are located mainly in the middle portion of the channel (Figs. 16.1d and 16.2d) and correspond to the dune fields and megaripples (Figs. 16.1 and 16.2). Pebbles and cobbles, often represented by anthropogenic sediments mixed with the natural debris, are recovered in the areas where parabolic dunes are present. This arrangement is almost constant in the investigated areas, but consolidated clays have been found on the relict substratum and along the banks, where mud and silt can also be easily found.

16.4 Final Remarks

The integrated analysis of high-resolution multibeam bathymetry, seismic profiles, and samplings allowed us to characterize the main morphological features related to the erosive-depositional processes acting in the lowermost reach of the Tiber River. In particular, the most common features are bedforms affecting about 45 % of the river bed. Three main types of bedform were identified, with different shapes and sizes: dunes, parabolic dunes and megaripples. The genesis of these bedforms can be related to the interaction between the river discharge and the sediment flooring the river bed. The monitoring of the Tiber River bed evolution through repeated multibeam bathymetry highlighted a fast evolution of the bedforms (Orlando et al. 2012), indicating a significant and very active sediment transport towards the mouth. The most significant morphological changes occur during the floods that repeatedly affect the river, and these changes mainly involve dunes and megaripples.

Acknowledgments This study was funded by the Centro Funzionale Regionale di Protezione Civile belonging to the Regione Lazio and by the National Protection Department and was conducted in collaboration with Earth Sciences Department of the University of Rome. We gratefully thank Ing. Francesco Mele and Ing. Domenico Spina for their insightful discussion on fluvial themes. We especially thank Dr. André Fascetti, Dr. Erika Lai and Dr. Alfredo Pazzini for their valuable support during the surveys. We also thank Alan Lounds for the English revision. Finally, we want to thank the RITMARE (Ricerca ITaliana per il MARE) Project.

References

Amorosi, A., Milli, S. (2001). Late Quaternary depositional architecture of Po and Tevere river deltas (Italy) and worldwide comparison with coaeval deltaic successions. Sedimentary Geology 144, 357–375.

Bernabini, M., Bosman A., Chiocci, F., Macelloni, L., Orlando, L. (2006). First multibeam and high resolution reflection seismics survey on the Tiber River lower course. EAGE Special Publication Geological Society. ISSN: 1369–4081.

Bosman, A., Casalbore, D., Anzidei, M., Muccini, F., Carmisciano, C., Chiocci, F. (2015). The first ultra-high resolution Digital Terrain Model of the shallow-water sector around Lipari Island (Aeolian Islands, Italy). Annals of Geophysics Vol 58. 2. In press.

Bosman, A., Madricardo, F., Remia, A., Correggiari, A., Romagnoli, C., Kruss, A., Casalbore, D., Morelli, E., Moscon, G. (2014). First morphological mapping of the Po delta (North Adriatic Sea, Italy) from ultra high-resolution multibeam bathymetry and backscatter data. Conference Paper. Ocean Sciences Meeting Hawaii Convention Center. Honolulu, Hawaii USA, February 2014.

Funiciello, R., Giordano, G. (2005). Geologic map of Rome, Carta Geologica del Comune di Roma, Vol. 1. CDROM. Università di Roma TRE-Comune di Roma-DDS Apat.

Milli, S., D'Ambrogi, C., Bellotti, P., Calderoni, G., Carboni, M.G., Celant, A., Di Bella, L., Di Rita, F., Frezza, V., Magri, D., Pichezzi, R.M., Ricci, V. (2013). The transition from wave-dominated estuary to wave-dominated delta: The Late Quaternary stratigraphic architecture of Tiber River deltaic succession (Italy). Sedimentary Geology 284–285, 159–180.

Orlando, L., Bosman, A., Lai, E., Mele, F., Spina, D. (2012). Analysis of active sedimentary process in the lower coarse of the Tiber river (Rome, Italy) through high-resolution geophysical data and samples. 34th International Geological Congress, 5–10 August 2012, Brisbane, Australia.

Raspa, G., Moscatelli, M., Stigliano, F., Patera, A., Marconi, F., Folle, D., Vallone, R., Mancini, M., Cavinato, G.P., Milli, S., Coimbre, Lite, Costra, J.F. (2008). Geotechnical characterization of the upper Pleistocene–Holocene alluvial deposits of Roma (Italy) by means of multivariate geostatistics: Cross-validation results. Engineering Geology 101 (2008) 251–268.

Part III

Bedforms in Prodeltas and Sorted Bedforms

Holocene Muddy Bedforms on the Llobregat River Prodelta Wedge

17

Roger Urgeles, Ben De Mol, Marc De Batist, and John E. Hughes-Clarke

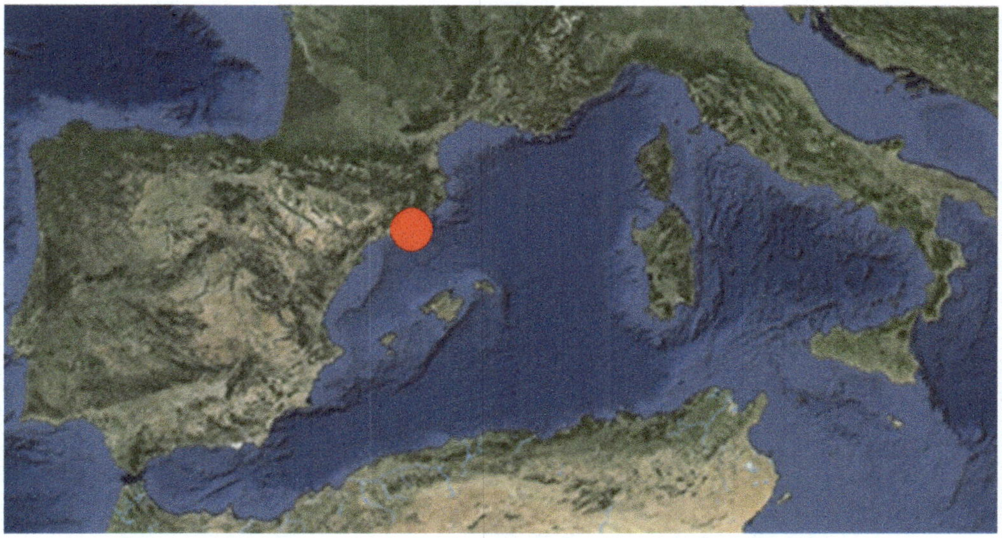

Abstract

A field of sediment waves on the Llobregat River prodelta, Catalonia, Spain, is examined using high-resolution multibeam bathymetry and seismic reflection profiles. The sediment waves develop on the prodelta front on slope gradients between 3° and 0.2°, within the Late-Holocene highstand mud wedge. Their characteristics and the most likely process at the origin of these structures are evaluated.

Keywords

Prodelta • Sediment waves • Late-Holocene mud wedge • Llobregat river • Barcelona

R. Urgeles (✉)
Dept. Geociències Marines, Institut de Ciències del Mar (CSIC), Pg. Marítim de la Barceloneta, 37-49, 08003 Barcelona, Catalonia, Spain
e-mail: urgeles@icm.csic.es

M. De Batist
Renard Centre for Marine Geology, University of Ghent, 9000 Ghent, Belgium

J.E. Hughes-Clarke
Ocean Mapping Group, Dept. Geodesy and Geomatics Engineering, University of New Brunswick, Fredericton, NB, Canada

B. De Mol
VNG Norge As, Oslo, Norway

J. Guillén et al. (eds.), *Atlas of Bedforms in the Western Mediterranean*,
DOI 10.1007/978-3-319-33940-5_17

17.1 Introduction-Study Area

Sediment waves are commonly observed on the seafloor off river outlets, deep below the wave base (e.g., Bornhold and Prior 1990; Cattaneo et al. 2004; Chiocci et al. 1996; Correggiari et al. 2001; Mosher and Thomson 2002; Trincardi and Normark 1988). It has been found that these sedimentary features are most often shaped by diverse sedimentary processes under different current regimes (Urgeles et al. 2011). One such set of sedimentary bedforms occurs off the Llobregat River delta, Barcelona, Spain (Fig. 17.1). The Llobregat River drains an area of 5045 km^2 (the submerged prodelta has an extension of about 165 km^2) and discharges ~527 hm^3 $year^{-1}$ (16.7 m^3s^{-1}) of water at the river mouth.

The thickness of Holocene deposits in the Llobregat delta ranges from 20 m close to the delta apex, 8 km landward from the present coastline, to about 60 m at the present coastline (Manzano 1986; Marquès 1975). Sedimentation rates vary between 5 and 34 mm $year^{-1}$ throughout the Holocene period (Manzano 1986; Marquès 1975). The delta has little tidal influence and the mean wave climate in the area is of low energy with a significant wave height of 0.74 m, a wave period of 4 s and major storms produced by NE winds (Barrera 2004). Sea water circulation along the Catalan shelf is dominated by the Liguro-Provençal Current, which flows southwestward along the shelf break (Arnau et al. 2004). The grain size of surface sediments in the Llobregat River prodelta has a high temporal variability (Puig et al. 1999). It may range from well-sorted, coarse sediments (sands) to fine sediments (mud).

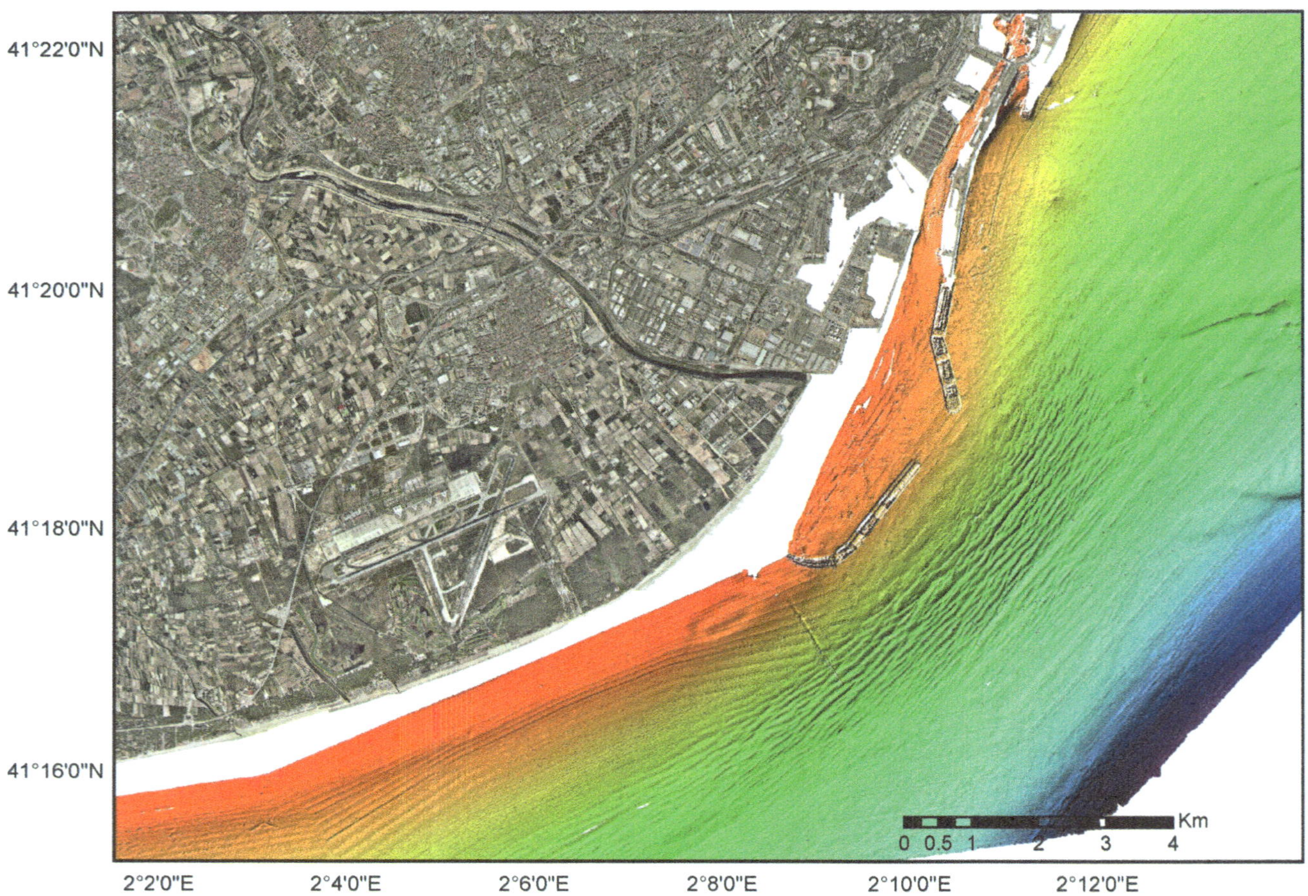

Fig. 17.1 Orthophoto map (onshore; courtesy of Institut Cartogràfic de Catalunya 2005) and shaded relief multibeam map (offshore) displaying sediment waves on the Llobregat River prodelta, offshore of Barcelona, Spain. Data were acquired in 2004 prior to diversion of the Llobregat River 2.5 km south of the mouth on the orthophoto to accommodate the enlargement of the Port of Barcelona. The dredging operations and port construction works are already visible

17.2 Methods

This work is based on Simrad's 300 kHz EM3000 dual multibeam echosounder data and seismic reflection profiles collected with a 300 J SIG sparker system and an ultra-high resolution Simrad TOPAS PS-040 5 kHz parametric echosounder. These systems respectively provide a vertical resolution of ~2 and 0.1 m and a horizontal resolution of ~9 and 0.25 m. The final bathymetric DTMs were produced at 2 m resolution (for further details see Urgeles et al. 2011, 2007).

17.3 Results-Discussion

The sediment waves on the Llobregat prodelta develop between 35 and 90 m water depth, on the steepest part of the prodelta (Fig. 17.2). They are present in the concave upward prodelta front on slopes of 3°–0.3°. The area occupied by these sediment waves is about 25 km^2 (2.2 km wide and 12 km long). In plan view the sediment waves appear as a series of ridges and swales with an intricate pattern of branching and truncated ridges (Fig. 17.1). Most of them can be traced for 300–400 m along isobaths but they can be as

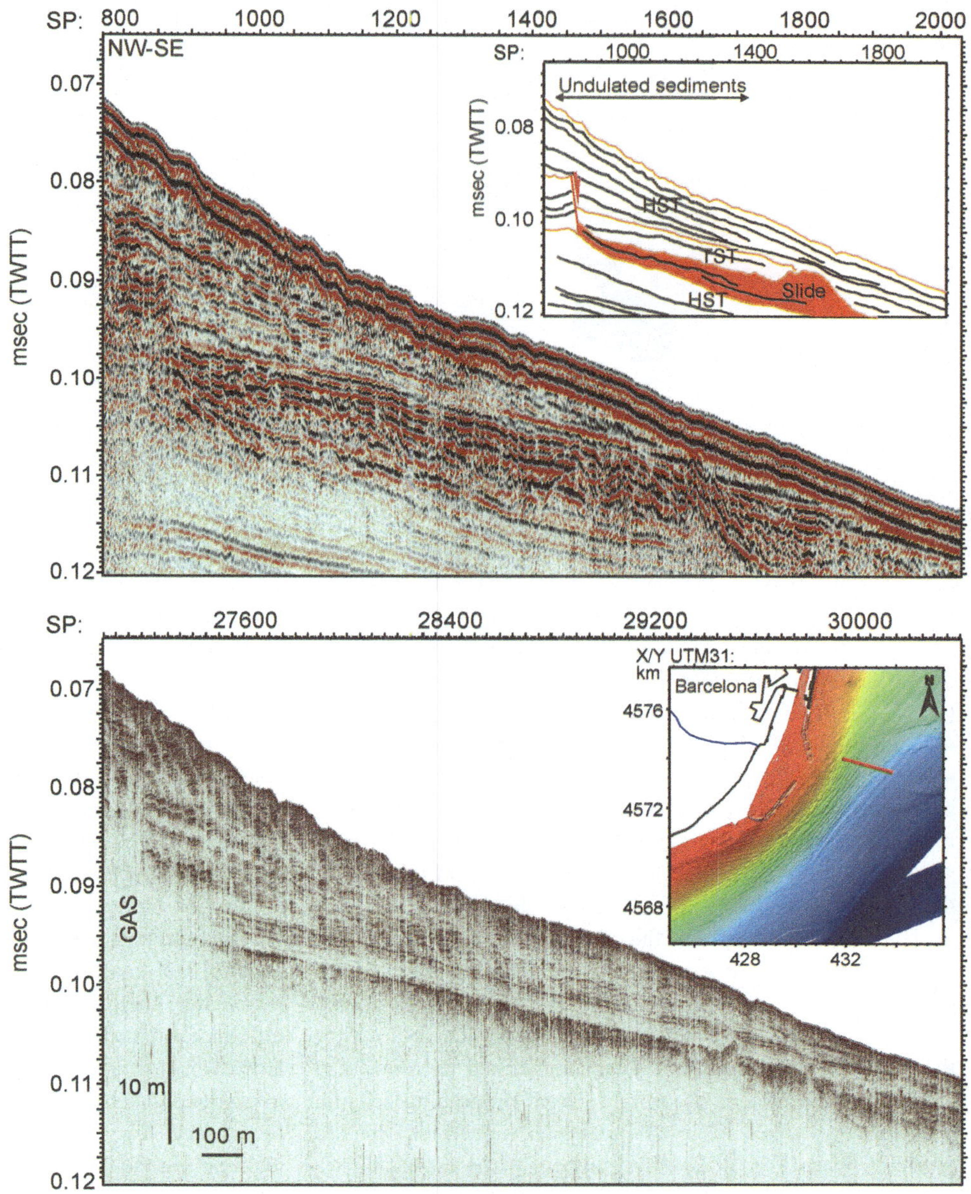

Fig. 17.2 Sparker seismic reflection profile (*top*) with line drawing (*top inset*) showing the location of the sediment waves with respect to the overall stratigraphy of the Llobregat River prodelta and 5 kHz sub-bottom profile over the same *line* at the same *horizontal* and *vertical scale* displaying details of the sediment waves. Inset map on *bottom figure* shows the location of the profiles at the northern edge of the sediment wave field

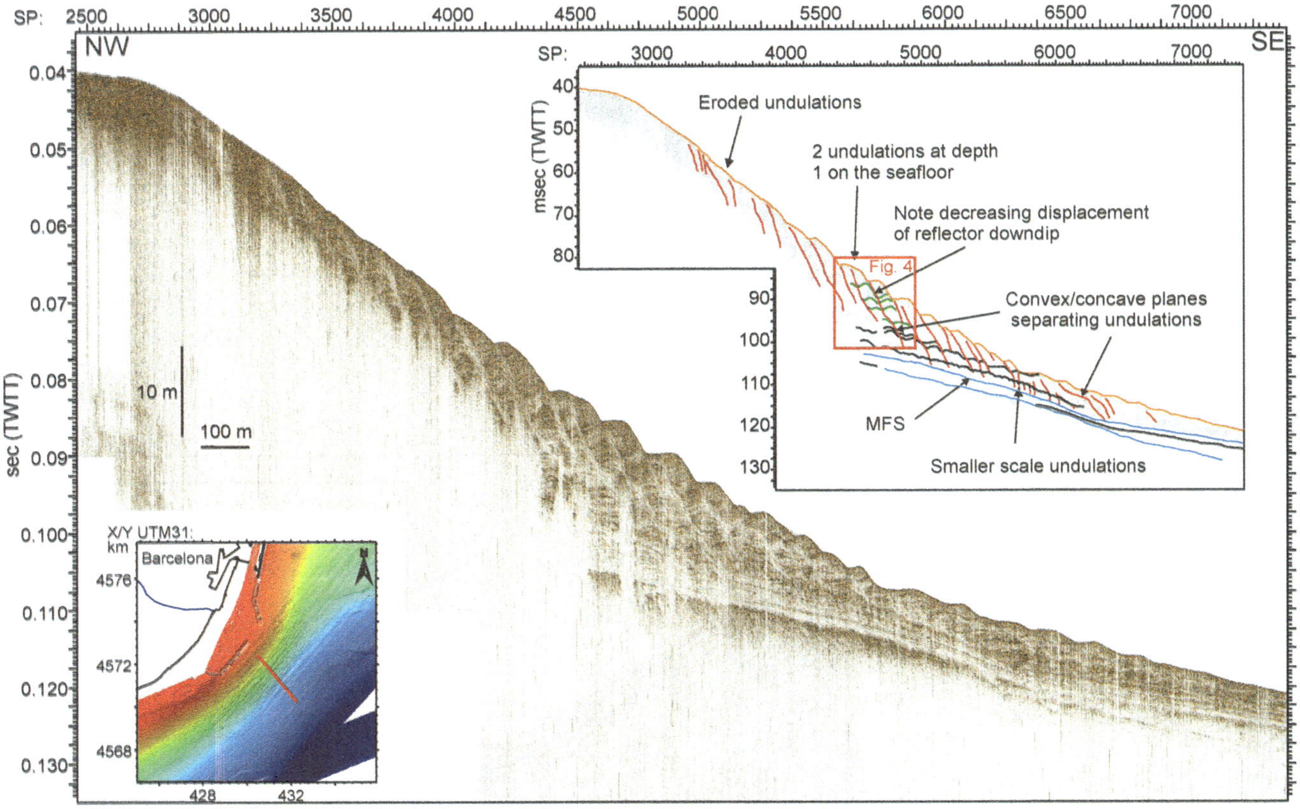

Fig. 17.3 5 kHz *sub-bottom* profile showing overall aspect of sediment waves and internal structure on the Llobregat prodelta foresets. *Top right inset* shows a line drawing with the main characteristics of the sediment waves. *Lower left inset* map shows the location of the profile in the central part of the sediment wave field

long as 2 km. From crest to trough the sediment waves range on the land-facing side from as high as 1.3 m to a few centimetres. The wavelength (L) ranges from about 200 m on the shallower parts to about 40 m on the deepest parts of the prodelta, but no clear trend is observed in wave height (H). The H/L ratio of the sediment waves is between 100 and 400. In shallow waters the sediment waves show relatively short upslope limbs and long downslope limbs (near 0 asymmetry values), while at deeper water depths the sediment waves appear more symmetric. The upslope limb of the sediment waves may face landward or seaward, while the downslope limb always faces seaward (Fig. 17.3).

The seismic data show that the sediment waves on the Llobregat prodelta develop downslope from an acoustically chaotic area, interpreted as gas-enriched sediments (Fig. 17.2). The sediment waves develop exclusively within the Holocene highstand and are characterized by a uniform wavy stratified pattern of strong to faint prograding seismic reflectors (up to 20 ms two-way travel time thick) on the prodelta front (Fig. 17.2). They appear to be rooted at the top of the maximum flooding surface. Both wavelength and amplitude tend to decrease with increasing stratigraphic depth (Fig. 17.3). Areas that do not presently show sediment waves at the seafloor may show evidence of seafloor sediment waves down-section, with the sediment waves being truncated at the seafloor (Fig. 17.3). This pattern occurs at the shallowest water depths. The crest and trough angle of climb of most sediment waves are not homogeneous down-section (Figs. 17.3 and 17.4). Variations are also not consistent, i.e. the angle may increase or decrease or change in trend down-section, i.e. it may display alternations between convex and concave shapes (Fig. 17.3).

The sediment waves can be explained by processes in the bottom boundary layer. Evidence of erosion in the shallowest sediment waves is present on the Llobregat prodelta in relatively shallow waters up to at least 30 m below sea level (mbsl) (Fig. 17.3), which suggests that storms have an influence on recent reshaping of the sediment wave field. However, storms with a recurrence period of 5 years are able to produce water motion only up to 63 mbsl, and it therefore seems necessary to invoke an additional mechanism for maintaining the sediment waves in the observed depth range (30–90 mbsl). Puig et al. (2007) have shown that internal waves can play a role in resuspending and transporting

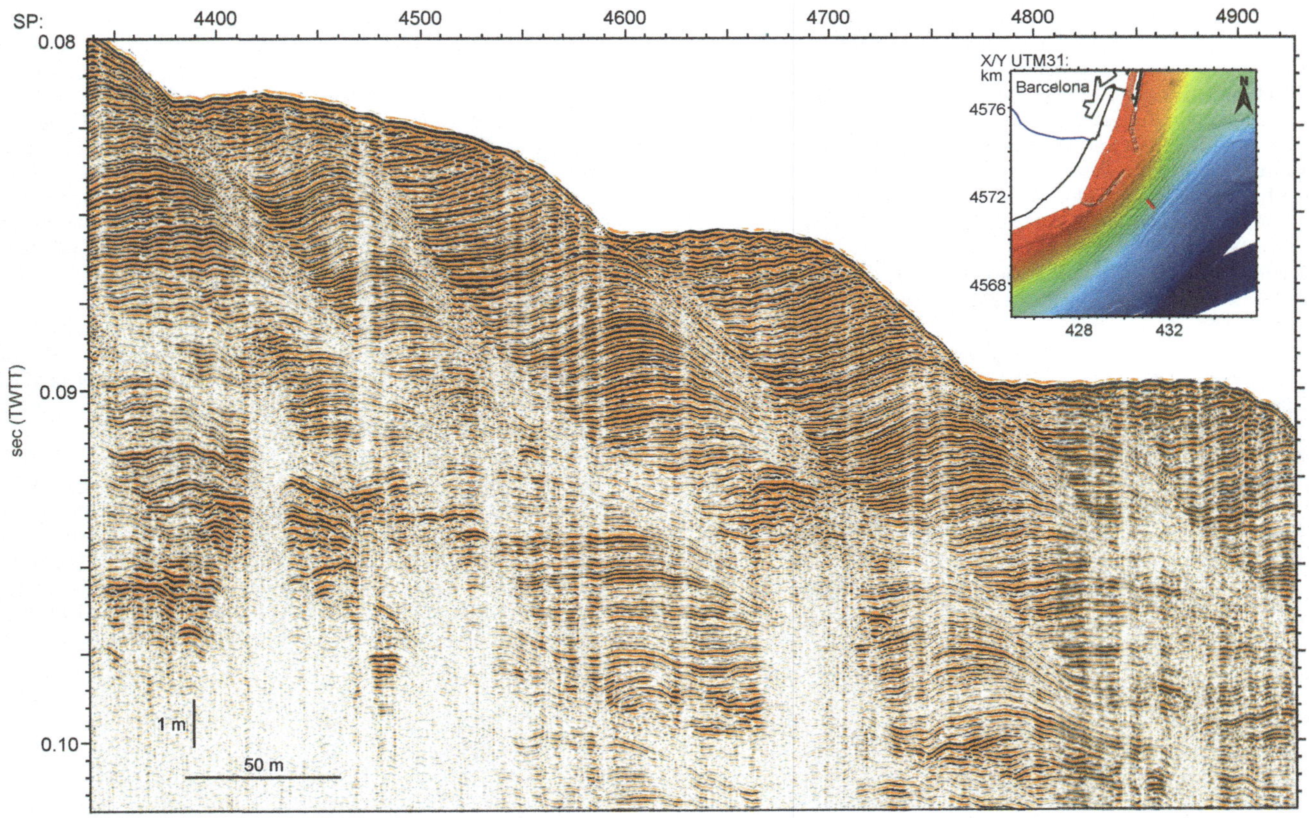

Fig. 17.4 Detail of 5 kHz *sub-bottom* profile of Fig. 17.3 showing detailed internal geometry of the reflections within the sediment wave field. *Top right inset* shows the location of the profile. See also inset with line drawing on Fig. 17.3 for location

sediment in prodeltaic areas. Near-inertial internal waves induced by local wind pulses tend to propagate across the water column through isopycnals and concentrate their energy at the shelf regions where the seasonal thermocline intersects with the seabed, which turns out to be the depth range characterized by sediment waves (see examples from the Llobregat prodelta observed using high-resolution temperature sensors in van Haren (2016)). The sediment waves would mainly develop along the prodelta front because here is where sediment transported by the Llobregat River is available for resuspension by internal waves. Also, in this area the slope gradient is such that the incidence angle of internal waves on the slope may meet critical conditions and the internal wave energy become trapped along the slope, thus enhancing sediment reworking (Lamb 2014; Ribó et al. 2016).

17.4 Conclusions

The prodelta in front of the Llobregat River is dominated by a field of sediment waves at water depths of 35–90 m on slopes of between 3° and 0.3°. Sediment waves are rooted on the maximum flooding surface, develop within the highstand systems tract and occur beyond the clinoform rollover point. Detailed morphologic and seismostratigraphic analysis of the sediment waves indicates they are most likely generated from sediment resuspension by internal waves.

Acknowledgments Data acquisition was made possible thanks to the project Prodelta (REN2002-02323) funded by the Spanish Ministry for Science and Technology. We are grateful to IHS for an educational Kingdom Suite licence. Constructive reviews by Alan Lounds and an anonymous reviewer greatly improved the manuscript

References

Arnau, P., Liquete, C., Canals, M. (2004). River mouth plume events and their dispersal in the Northwestern Mediterranean Sea. Oceanography 17, 22–31.

Barrera, N. (2004). Influencia del rebase del oleaje en playas sobre lagunas costeras: el caso de la laguna de la Magarola.

Bornhold, B.D., Prior, D.B. (1990). Morphology and Sedimentary Processes on the Subaqueous Noeick River Delta, British Columbia, Canada, in: Colella, A., Prior, D.B. (Eds.), Coarse-Grained Deltas. Blackwell Publishing Ltd., pp. 169–181.

Cattaneo, A., Correggiari, A., Marsset, T., Thomas, Y., Marsset, B., Trincardi, F. (2004). Seafloor undulation pattern on the Adriatic shelf and comparison to deep-water sediment waves. Mar. Geol. 213, 121–148. doi:10.1016/j.margeo.2004.10.004

Chiocci, F.L., Esu, F., Tommasi, P., Chiappa, V., (1996). Stability of the submarine slope of the Tiber River delta, in: Senneset, K. (Ed.), Landslides: Proceedings of the 7th International Symposium on Landslides. Balkema, the Netherlands, pp. 521–526.

Correggiari, A., Trincardi, F., Langone, L., Roveri, M. (2001). Styles of Failure in Late Holocene Highstand Prodelta Wedges on the Adriatic Shelf. J. Sediment. Res. 71, 218–236. doi:10.1306/042800710218

Lamb, K.G. (2014). Internal wave breaking and dissipation mechanisms on the continental slope/shelf. Annu. Rev. Fluid Mech. 46, 231–254.

Manzano, M. (1986). Estudio sedimentológico del prodelta holoceno del Llobregat.

Marquès, M.A. (1975). Las formaciones cuaternarias del delta del Llobregat. Acta Geol. Hispanica 10, 21–28.

Mosher, D.C., Thomson, R.E. (2002.) The Foreslope Hills: large-scale, fine-grained sediment waves in the Strait of Georgia, British Columbia. Mar. Geol. 192, 275–295.

Puig, P., Ogston, A.S., Guillén, J., Fain, A.M.V., Palanques, A. (2007). Sediment transport processes from the topset to the foreset of a crenulated clinoform (Adriatic Sea). Cont. Shelf Res. 27, 452–474. doi:10.1016/j.csr.2006.11.005

Puig, P., Palanques, A., Sanchez-Cabeza, J.A., Masqué, P. (1999). Heavy metals in particulate matter and sediments in the southern Barcelona sedimentation system (North-western Mediterranean). Mar. Chem. 63, 311–329. doi:10.1016/S0304-4203(98)00069-3

Ribó, M., Puig, P., Muñoz, A., Lo Iacono, C., Masqué, P., Palanques, A., Acosta, J., Guillén, J., Gómez Ballesteros, M. (2016). Morphobathymetric analysis of the large fine-grained sediment waves over the Gulf of Valencia continental slope (NW Mediterranean). Geomorphology 253, 22–37. doi:10.1016/j.geomorph.2015.09.027

Trincardi, F., Normark, W.R. (1988). Sediment waves on the Tiber prodelta slope: interaction of deltaic sedimentation and currents along the shelf. Geo-Mar. Lett. 8, 149–157.

Urgeles, R., Cattaneo, A., Puig, P., Liquete, C., Mol, B.D., Amblàs, D., Sultan, N., Trincardi, F. (2011). A review of undulated sediment features on Mediterranean prodeltas: distinguishing sediment transport structures from sediment deformation. Mar. Geophys. Res. 32, 49–69. doi:10.1007/s11001-011-9125-1

Urgeles, R., De Mol, B., Liquete, C., Canals, M., De Batist, M., Hughes-Clarke, J.E., Amblas, D., Arnau, P.A., Calafat, A.M., Casamor, J.L., Centella, V., De Rycker, K., Fabres, J., Frigola, J., Lafuerza, S., Lastras, G., Sanchez, A., Zuniga, D., Versteeg, W., Willmott, V. (2007). Sediment undulations on the Llobregat prodelta: Signs of early slope instability or sedimentary bedforms? J. Geophys. Res.-SOLID EARTH 112. doi:10.1029/2005JB003929

van Haren, V. (2016). Internal waves and bedforms. This book.

18 Prodeltaic Undulations and Hyperpycnal Flows (I): Morphological Observations

Patricia Bárcenas, Francisco José Lobo, Luis Miguel Fernández-Salas, Miguel Ortega-Sánchez, Isabel Mendes, and Jorge Macías

P. Bárcenas · J. Macías
Dpto. Análisis Matemático, Facultad de Ciencias, 29080 Málaga, Spain
e-mail: pbarcenas@uma.es

J. Macías
e-mail: jmacias@uma.es

F.J. Lobo (✉)
Instituto Andaluz de Ciencias de la Tierra, CSIC-Universidad de Granada, 18100 Armilla, Granada, Spain
e-mail: pacolobo@iact.ugr-csic.es

L.M. Fernández-Salas
Instituto Español de Oceanografía, Centro Oceanográfico de Cádiz, 11006 Cádiz, Spain
e-mail: luismi.fernandez@cd.ieo.es

M. Ortega-Sánchez
Andalusian Institute for Earth System Research, University of Granada, Edificio CEAMA, Avda. del Mediterraneo s/n, 18006 Granada, Spain
e-mail: miguelos@ugr.es

I. Mendes
CIMA, Universidade do Algarve, Campus de Gambelas, 8005-139 Faro, Portugal
e-mail: imendes@ualg.pt

J. Guillén et al. (eds.), *Atlas of Bedforms in the Western Mediterranean*,
DOI 10.1007/978-3-319-33940-5_18

Abstract

Fields of submarine undulations occurring over prodeltaic slopes may be caused in some circumstances by the imprint of high-density sediment flows (i.e. hyperpycnal flows). This is the case of numerous Mediterranean settings, where marked seasonal climates and abrupt inshore physiography are conducive to the generation of these flows. In this contribution we examine the morphological patterns of undulations that are compatible with such an origin. Prodeltaic undulations occur off the offlap breaks on relatively low slopes (1°–2°) and show dominance of landward-directed forms. Morphological parameters of undulations such as heights, widths and lengths exhibit distinctive dimensions and tend to follow consistent patterns around nearby river mouths. In particular, height distributions tend to follow symmetric patterns, and vertical form indexes are lower than those of other sediment waves with a different genesis. These observations are compatible with geological processes under the influence of high-density sediment flows, such as different deposition rates due to enhanced bedload transport. Lateral changes of river mouths involving modifications of sediment flows are also imprinted in the geomorphological parameters of this kind of undulation.

Keywords

Multibeam bathymetry • Prodeltaic undulations • Hyperpycnal flows • Alborán sea

18.1 Introduction

The coverage of multibeam bathymetric data over extensive shallow-water areas highlights the occurrence of surface undulations (or crenulations) in numerous prodeltaic environments. These undulations have been particularly evidenced in the Mediterranean Sea (e.g. see Urgeles et al. 2011 for a comprehensive summary), but also in other shallow-water submarine deltaic bodies (Lee et al. 2002; Carle and Hill 2009) and in coastal embayments such as fjords (Bøe et al. 2004). Diverse hypotheses have been formulated to explain the generation of these shallow-water submarine morphologies, such as deformational features (i.e. creep) (e.g. Correggiari et al. 2001; Lee and Chough 2001), bedforms generated by oceanographic phenomena such as along-shelf currents at different water depths (Trincardi and Normark 1988) and internal waves (Puig et al. 2007; Urgeles et al. 2011), and bedforms generated by high-density (e.g. hyperpycnal, turbidite current) sediment flows (e.g. Lee et al. 2002; Urgeles et al. 2007). In the most complex cases, a composite origin involving sediment deformation and sediment wave forming processes has been postulated (Cattaneo et al. 2004). Recent studies would indicate that, at least for the Mediterranean area, most of these seafloor features could be the result of boundary layer processes (Urgeles et al. 2011).

In this contribution we summarize the morphological characteristics of submarine undulations occurring on prodeltaic slopes that seem to be compatible with origins linked to the episodic activity of high-density river flows. We base this deduction on observations reported in diverse settings, but with a focus on undulation fields of southern Iberia (Fig. 18.1), where physiographic and climatic constrains are conducive to the generation of hyperpycnal flows.

18.2 Sediment Supply of Short, Mountainous Rivers

Hyperpycnal flows in the marine environment may be generated by volume concentrations of negatively buoyant plumes exceeding the ambient density of water bodies (Mulder and Syvitski 1995). These processes are important because they may provide a link between inland sediment sources and marine sinks (Lamb and Mohrig 2009).

It is believed that small- to medium-sized mountainous rivers are more important than large rivers in transporting sediments between land and sea through hyperpycnal flows. Small- to medium-sized rivers in arid climates can attain high concentrations during exceptional floods, enabling the plunging of sediment plumes (Mulder et al. 2003). In contrast, the discharges from large river systems usually exhibit lower sediment concentrations (Mulder and Syvitski 1995). The frequency and volume of gravity (hyperpycnal) flows also seem to be controlled by the physiographic setting of the respective drainage basins, as sediment yields are directly related to high hinterland elevations and decreasing size of catchments (Milliman and Syvitski 1992; Mulder and Syvitski 1996). Soft sediment availability, seawater dilution

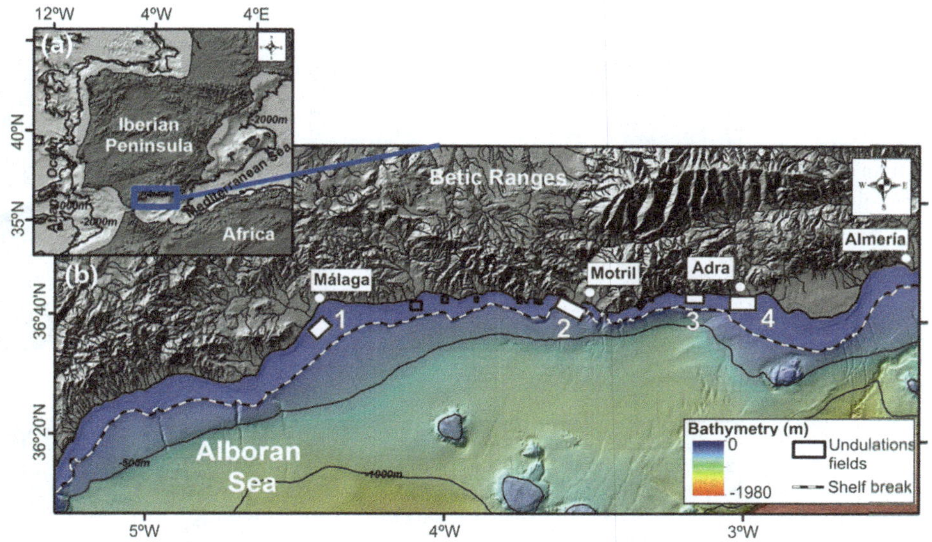

Fig. 18.1 Geographical location of seafloor undulation fields observed in the northern Alborán Sea. **a** Emplacement of the Alborán Sea, in the westernmost sector of the western Mediterranean Basin, between the Iberian Peninsula and Africa (image generated from the ETOPO database); **b** the study area is located on the northern margin of the Alborán Sea, off the Betic Ranges (the inland physiography is extracted from the Digital Terrain Model of the "Junta de Andalucía" and the submarine bathymetry is extracted from the Alborán Sea submarine database of the "Secretaría General del Mar"). Location of the Guadalhorce River *a*, Guadalfeo River *b*, Albuñol Stream *c* and Adra River *d* seafloor undulation fields is shown

and erosion of coastal features are other factors that enable the triggering of hyperpycnal flows (Mulder et al. 2003).

Many fluvial catchments in the Mediterranean region are characterized by small- to medium-sized rivers and ephemeral streams with abrupt topographies, resulting from proximity between mountain ranges and adjacent coasts (Liquete et al. 2009). In addition, the discharge regime of Mediterranean systems shows marked seasonality (Anthony et al. 2014), inducing a torrential character of the fluvial systems (Sacchi et al. 2009) and high sediment yields (Anthony et al. 2014). These general conditions are exacerbated in the fluvio-deltaic systems of southern Iberia draining into the Alborán Sea (Fig. 18.1), where climatic conditions are particularly arid and vegetation cover very sparse, resulting in highly erodible soils that feed large sediment loads during peak fluvial discharges. As a consequence, sediment yields are very high and transport from land to sea to construct submarine deltas is very effective (Liquete et al. 2005).

18.3 Morphological Observations

Undulations regarded as migrating sediment waves commonly develop at intermediate shelf water depths (i.e. 30–90 m) over concave-upward surfaces of prodeltaic deposits (Lee et al. 2002) with average slopes of 1°–2° (Bøe et al. 2004; Urgeles et al. 2007). These sediment waves have widths (L_n, or extension in the normal-to-strike direction) of tens to hundreds of metres and heights (H) of metric scale (Bøe et al. 2004; Urgeles et al. 2007). Characteristic lengths (L_s, or extension in the strike direction) are of hundreds of metres, although some larger features may occur. Wavelengths and wave heights usually decrease with distance from the source (Lee et al. 2002). Vertical form indexes (L_n/H) range between 100 and 400 (Urgeles et al. 2007). Plan view patterns of sediment waves are variable, with a prevalence of sinuous patterns (Bøe et al. 2004; Urgeles et al. 2007).

Seafloor undulation fields observed on the northern Alborán Sea margin occur beyond the offlap break, covering areas of 1.4–11 km^2 at water depths ranging from 16 to 90 m (Figs. 18.2 and 18.3), where average slopes show progressive seaward-declining trends (Fernández-Salas et al. 2007; Bárcenas et al. 2009; Bárcenas 2013; Lobo et al. 2015).

The Alborán Sea undulations show H values lower than 1.3 m, L_n values of hundreds of metres and L_s values of up to 2000 m. L_n/H ratios are lower than 300 (Fernández-Salas et al. 2007; Bárcenas et al. 2009; Lobo et al., 2015). Most of the geometric parameters are arranged in consistent patterns around river mouths, particularly H and L_n/H (Figs. 18.2 and 18.3). H values are largest off the river mouths, decreasing sidewards, whereas L_n/H values follow opposite trends (Fig. 18.4). The highest H values tend to occur at mid-depths, whereas L_n/H values increase downslope.

The undulations show either landward-directed asymmetries or a symmetric character. Undulation crestlines are nearly straight or sinuous, and tend to follow the direction of

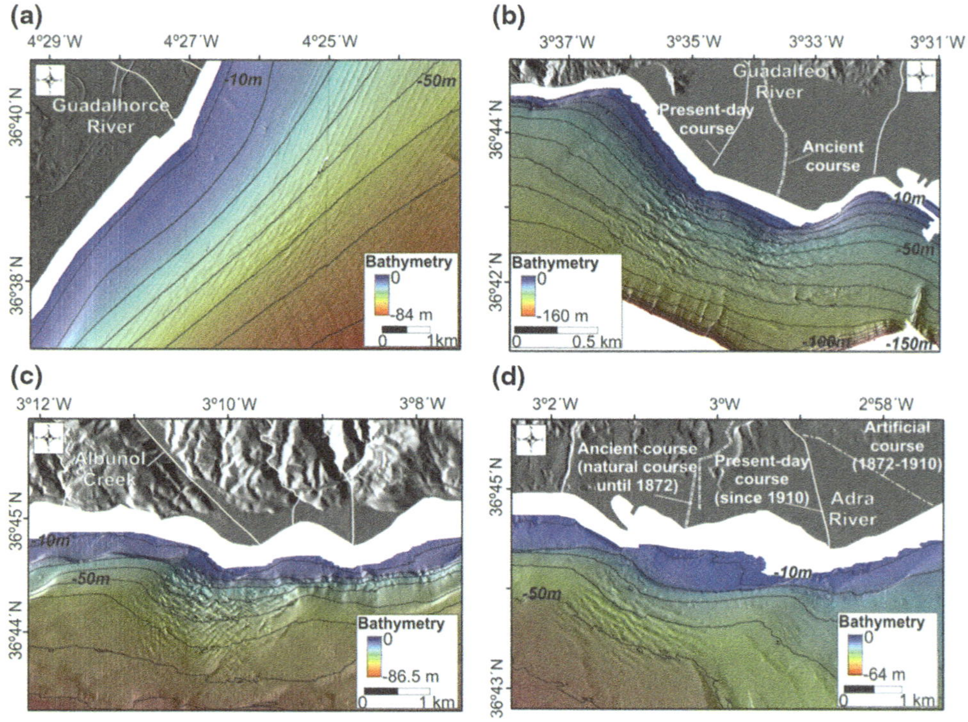

Fig. 18.2 Hill-shaded reliefs of several seafloor undulation fields located in southern Iberia. **a** Guadalhorce River, **b** Guadalfeo River, **c** Albuñol Stream and **d** Adra River. Lateral morphological changes from river mouths are clearly displayed in *b* and *d*. See location in Fig. 18.1

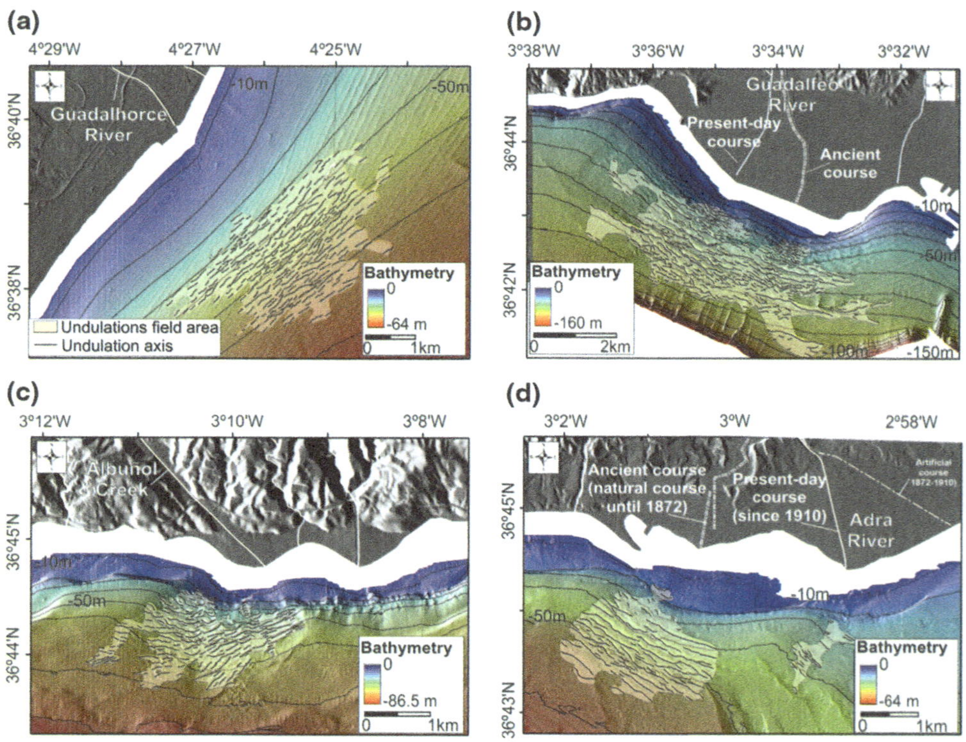

Fig. 18.3 Mapping of several undulation fields and individual undulation axis in the northern Alborán Sea. **a** Guadalhorce River, **b** Guadalfeo River, **c** Albuñol Stream and **d** Adra River. Lateral changes of undulation morphology away from river mouths are displayed in *b* and *d*. See location in Fig. 18.1

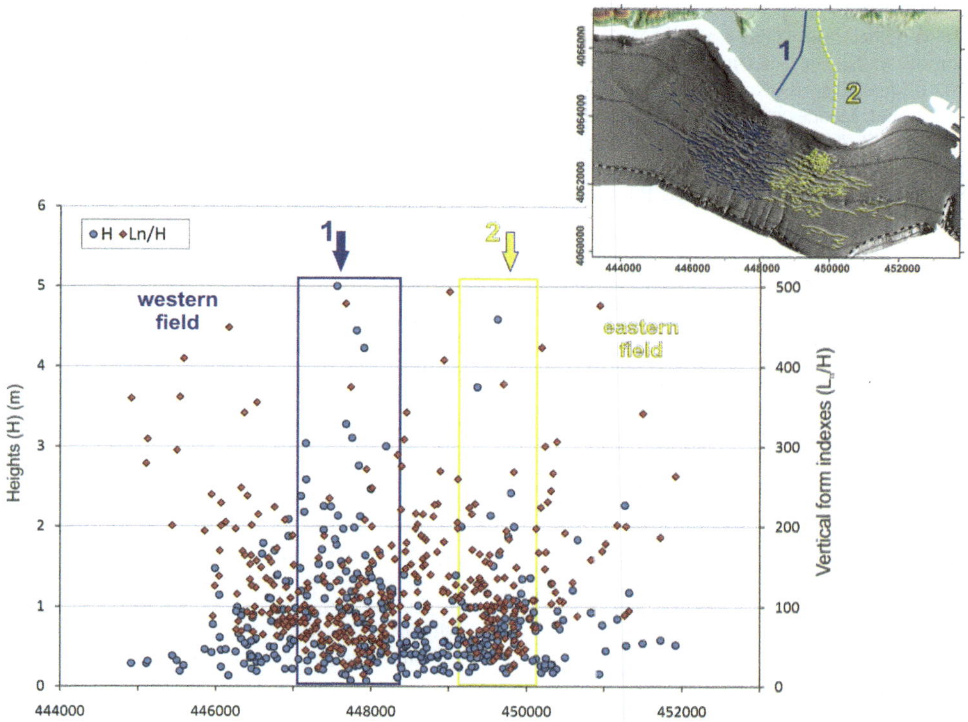

Fig. 18.4 Grapth showing the alongshelf variation of significant morphometric parameter such as H(m) and L_n/H of submarine undulations composing the western and eastern undulation fields on the Guadalfeo River prodeltaic wedge. The present-day river mouth is indicated by (*1*), whereas the ancient river mouth is indicated by (*2*). Areas off the river mouths are characterized by the largest H values, particularly the present-day river mouth, where values higher than 2 m are relatively frequent, and most L_n/H values are lower than 100 m. In contrast, the external areas of the undulation fields and the transitional area between both river mouths tend to exhibit H values lower than 2 m and L_n/H values higher than 100

bathymetric contours (Fernández-Salas et al. 2007; Bárcenas et al. 2009; Lobo et al. 2015).

18.4 Diagnostic Morphological Criteria Compatible with Hyperpycnal Flow Genesis

Several geomorphic patterns such as undulation occurrence over concave surfaces with gentle slopes and dominance of landward-directed asymmetries and of linear crestlines are considered as indicators of sediment wave occurrence (Lee et al. 2002; Levchenko and Roslyakov 2010; Ponce and Carmona 2011). The coast-directed asymmetry is attributed to differential deposition rates (Lee et al. 2002). In contrast, sediment deformation features tend to be linked to higher slopes, and creep folds tend to exhibit arcuate patterns (Wynn and Stow 2002).

Geomorphic parameters of undulations show organized patterns around nearby river mouths (Fig. 18.3). These distinctive trends in dimensions are typical of sediment waves (Lee et al. 2002; Urgeles et al. 2007), as dune dimensions are most likely controlled by the relative rates of bedload versus suspended load transport (Carle and Hill 2009). In particular, the overall low values and the seaward decrease of vertical form indexes are indicative of sediment wave generation by high-density sediment fluxes (Bøe et al. 2004; Urgeles et al. 2011; Lobo et al. 2015).

In the undulation fields of the Alborán Sea, undulation dimensions appear to be related to the regional significance and dimensions of effluents, also suggesting development controlled by fluvial sediment flows (Bárcenas et al. 2009). In cases where diverse undulation fields occur within a single deltaic system, geomorphic parameters with contrasting differences have been linked to temporal changes in the hyperpycnal flows associated with the modification of the river outlets. Thus, lower energy sedimentary regimes are characterized by lower H values and higher L_n and L_n/H values (Fig. 18.4), and H distributions only show moderately symmetric patterns. In contrast, intensification and focusing of riverine inflows is inferred by higher H values and lower Ln/H values (Fig. 18.4), and by increased symmetric patterns around river mouths (Lobo et al. 2015).

Acknowledgments This work is a contribution to the projects MOSAICO (P06-RNM-01594) and TESELA (P11-787 RNM7069) funded by the "Junta de Andalucía", and it is also related to project CGL2011-30302-C02-02 funded by the Spanish Ministry for Economy and Competitiveness. Isabel Mendes thanks the Portuguese Science Foundation (FCT) for grant SFRH/BPD/72869/2010. The manuscript was improved thanks to useful comments and suggestions made by Roger Urgeles (CSIC-Institut de Ciències del Mar, Barcelona, Spain) and by an anonymous reviewer.

References

Anthony EJ, Marriner N, Morhange C (2014). Human influence and the changing geomorphology of Mediterranean deltas and coasts over the last 6000 years: From progradation to destruction phase? Earth-Sci Rev 139 (0):336–361. doi:10.1016/j.earscirev.2014.10.003.

Bárcenas P (2013). Procesos Morfogenéticos y Evolución Reciente de los Depósitos Prodeltaicos del Sureste de la Península Ibérica: Aplicaciones de los Modelos Matemáticos. PhD, Málaga.

Bárcenas P, Fernández-Salas LM, Macías J, Lobo FJ, Díaz del Río V (2009). Estudio morfométrico comparativo entre las ondulaciones de los prodeltas de los ríos de Andalucía Oriental. Rev Soc Geol España 22 (1–2):43–56.

Bøe R, Bugge T, Rise L, Eidnes G, Eide A, Mauring E (2004). Erosional channel incision and the origin of large sediment waves in Trondheimsfjorden, central Norway. Geo-Mar Lett 24 (4):225–240. doi:10.1007/s00367-004-0180-3.

Carle L, Hill PR (2009). Subaqueous Dunes of the Upper Slope of the Fraser River Delta (British Columbia, Canada). J Coast Res 25 (2):448–458. doi:10.2112/06-0796.1.

Cattaneo A, Correggiari A, Marsset T, Thomas Y, Marsset B, Trincardi F (2004). Seafloor undulation pattern on the Adriatic shelf and comparison to deep-water sediment waves. Mar Geol 213 (1–4):121-148. doi:10.1016/j.margeo.2004.10.004.

Correggiari A, Trincardi F, Langone L, Roveri M (2001). Styles of Failure in Late Holocene Highstand Prodelta Wedges on the Adriatic Shelf. Jour Sediment Res 71 (2):218–236. doi:10.1306/042800710218.

Fernández-Salas LM, Lobo FJ, Sanz JL, Díaz-del-Río V, García MC, Moreno I (2007). Morphometric analysis and genetic implications of pro-deltaic sea-floor undulations in the northern Alboran Sea margin, western Mediterranean Basin. Mar Geol 243 (1–4):31–56. doi:10.1016/j.margeo.2007.04.013.

Lamb MP, Mohrig D (2009). Do hyperpycnal-flow deposits record river-flood dynamics? Geology 37 (12):1067–1070. doi:10.1130/g30286a.1.

Lee SH, Chough SK (2001). High-resolution (2–7 kHz) acoustic and geometric characters of submarine creep deposits in the South Korea Plateau, East Sea. Sedimentology 48 (3):629–644. doi:10.1046/j.1365-3091.2001.00383.x.

Lee HJ, Syvitski JPM, Parker G, Orange D, Locat J, Hutton EWH, Imran J (2002). Distinguishing sediment waves from slope failure deposits: field examples, including the 'Humboldt slide', and modelling results. Mar Geol 192 (1–3):79-104. doi:10.1016/S0025-3227(02)00550-9.

Levchenko OV, Roslyakov AG (2010). Cyclic sediment waves on western slope of the Caspian Sea as possible indicators of main transgressive/regressive events. Quat Inter 225 (2):210–220. doi:10.1016/j.quaint.2009.12.001.

Liquete C, Arnau P, Canals M, Colas S (2005). Mediterranean river systems of Andalusia, southern Spain, and associated deltas: A source to sink approach. Mar Geol 222–223:471–495. doi:10.1016/j.margeo.2005.06.033.

Liquete C, Canals M, Ludwig W, Arnau P (2009). Sediment discharge of the rivers of Catalonia, NE Spain, and the influence of human impacts. Jour Hydrol 366 (1–4):76–88. doi:10.1016/j.csr.2006.11.005.

Lobo FJ, Goff JA, Mendes I, Bárcenas P, Fernández-Salas LM, Martín-Rosales W, Macías J, Díaz del Río V (2015). Spatial variability of prodeltaic undulations on the Guadalfeo River prodelta: support to the genetic interpretation as hyperpycnal flow deposits. Mar Geophys Res 36 (4): 309–333. doi:10.1007/s11001-014-9233-9.

Milliman JD, Syvitski JPM (1992). Geomorphic/Tectonic Control of Sediment Discharge to the Ocean: The Importance of Small Mountainous Rivers. Jour Geol 100 (5):525–544. doi:10.1086/629606

Mulder T, Syvitski JPM (1995). Turbidity Currents Generated at River Mouths during Exceptional Discharges to the World Oceans. Jour Geol 103 (3):285–299. doi:10.1086/629747

Mulder T, Syvitski JPM (1996). Climatic and Morphologic Relationships of Rivers: Implications of Sea-Level Fluctuations on River Loads. Jour Geol 104 (5):509–523. doi:10.1086/629849

Mulder T, Syvitski JPM, Migeon S, Faugères J-C, Savoye B (2003). Marine hyperpycnal flows: initiation, behavior and related deposits. A review. Mar Petrol Geol 20 (6–8):861–882. doi:10.1016/j.marpetgeo.2003.01.003.

Ponce JJ, Carmona N (2011). Coarse-grained sediment waves in hyperpycnalclinoform systems, Miocene of the Austral foreland basin, Argentina. Geology 39 (8):763–766. doi:10.1130/g31939.1.

Puig P, Ogston AS, Guillén J, Fain AMV, Palanques A (2007). Sediment transport processes from the topset to the foreset of a crenulated clinoform (Adriatic Sea). Cont Shelf Res 27 (3–4):452–474. doi:10.1016/j.csr.2006.11.005.

Sacchi M, Molisso F, Violante C, Esposito E, Insinga D, Lubritto C, Porfido S, Toth T (2009). Insights into flood-dominated fan-deltas: very high-resolution seismic examples off the Amalfi cliffed coasts, eastern Tyrrhenian Sea. In: Violante C (ed) Geohazard in Rocky Coastal Areas. Geol Soc, London, Spec Publ, vol 322. Geological Society, London, pp 33–71. doi:10.1144/sp322.2.

Trincardi F, Normark W (1988). Sediment waves on the tiberprodelta slope: Interaction of deltaic sedimentation and currents along the shelf. Geo-Mar Lett 8 (3):149–157. doi:10.1007/bf02326091.

Urgeles R, De Mol B, Liquete C, Canals M, De Batist M, Hughes-Clarke JE, Amblàs D, Arnau PA, Calafat AM, Casamor JL, Centella V, De Rycker K, Fabrés J, Frigola J, Lafuerza S, Lastras G, Sànchez A, Zuñiga D, Versteeg W, Willmott V (2007). Sediment undulations on the Llobregat prodelta: Signs of early slope instability or sedimentary bedforms? Jour Geophys Res: Solid Earth 112 (B5):B05102. doi:10.1029/2005JB003929.

Urgeles R, Cattaneo A, Puig P, Liquete C, De Mol B, Amblàs D, Sultan N, Trincardi F (2011). A review of undulated sediment features on Mediterranean prodeltas: distinguishing sediment transport structures from sediment deformation. Mar Geophys Res 32 (1):49–69. doi:10.1007/s11001-011-9125-1.

Wynn RB, Stow DAV (2002). Classification and characterisation of deep-water sediment waves. Mar Geol 192 (1–3):7-22. doi:10.1016/S0025-3227(02)00547-9.

Prodeltaic Undulations and Hyperpycnal Flows (II): Evolutionary Trends

19

Francisco José Lobo, Patricia Bárcenas, Isabel Mendes, Miguel Ortega-Sánchez, Jorge Macías, and Luis Miguel Fernández-Salas

F.J. Lobo (✉)
Instituto Andaluz de Ciencias de la Tierra, CSIC-Universidad de Granada, 18100 Armilla, Granada, Spain
e-mail: pacolobo@iact.ugr-csic.es

P. Bárcenas · J. Macías
Dpto. Análisis Matemático, Facultad de Ciencias, 29080 Málaga, Spain
e-mail: pbarcenas@uma.es

J. Macías
e-mail: jmacias@uma.es

I. Mendes
CIMA, Universidade do Algarve, Campus de Gambelas, 8005-139, Faro, Portugal
e-mail: imendes@ualg.pt

M. Ortega-Sánchez
Andalusian Institute for Earth System Research, University of Granada, Edificio CEAMA, Avda. del Mediterraneo s/n, 18006 Granada, Spain
e-mail: miguelos@ugr.es

L.M. Fernández-Salas
Instituto Español de Oceanografía, Centro Oceanográfico de Cádiz, 11006 Cádiz, Spain
e-mail: luismi.fernandez@cd.ieo.es

J. Guillén et al. (eds.), *Atlas of Bedforms in the Western Mediterranean*,
DOI 10.1007/978-3-319-33940-5_19

Abstract

Undulation fields found on Mediterranean prodeltaic slopes usually display sub-surface precursory features recognized through seismic profiling and sediment core data. In this contribution we examine the stratigraphic patterns of sub-surface undulations occurring in wedge-shaped Holocene deposits that are in agreement with the episodic activity of hyperpycnal flows. Shallow sub-surface undulation facies show two fundamental features: lateral continuity of reflections and upward migration trends. Upward evolutionary trends may also be observed at the scale of the entire prodeltaic wedges. The most simple cases document repeated wavy stratified patterns or upward transition from sub-parallel to undulate. In some cases, however, the upward evolution can be very complex, with several stages ranging from non-development of undulations to pronounced wave migration, in addition to intermediate aggradational stages. Sedimentologically, sets of undulations exhibit variations between coarse and fine fractions. The sub-surface stratigraphic patterns of prodeltaic undulations are suggestive of enhanced development during evolving environmental conditions, mediated by the alternation of wet and dry climates characteristic of Mediterranean regions. Those conditions would have eventually favoured the activity of episodic hyperpycnal flows. Lateral changes of the internal undulation patterns also reveal the modification of fluvial flows and their imprint in the submarine environment.

Keywords

Prodeltaic undulations • Wave migration • Hyperpycnal flows • Holocene • Mediterranean sea

19.1 Introduction

Prodeltaic undulations in the Mediterranean region are characterized by a conspicuous morphological fingerprint documented through extensive multibeam imagery (Bárcenas et al. This Volume). The analysis of seismic data of various resolutions reveals that in many settings the surface undulations also have precursory features in the sub-surface (e.g. Lee et al. 2002; Cattaneo et al. 2004; Berndt et al. 2006; Urgeles et al. 2007, 2011; Levchenko and Roslyakov 2010; Lobo et al. 2015). The sub-surface undulations also display distinctive sedimentological (e.g. Milia et al. 2008; Lobo et al. 2015) and geotechnical signatures (Sultan et al. 2008). The precursory undulations usually occur along the stratigraphic column of wedge-shaped deposits constituting the submarine portions of deltaic systems, which are believed to have formed preferentially during the Holocene sea-level stabilization (Anthony et al. 2014).

In most cases evidence suggests different evolutionary stages of the precursory undulations that could be compatible with a sediment wave forming hypothesis, such as oceanographic phenomena or high-density hyperpycnal flows (Bárcenas et al. This Volume). In other cases, however, the stratigraphy of Holocene highstand prodeltaic wedges is complicated by the existence of several geological phenomena, such as gas-charged sediments, slide surfaces or mud diapirs, which have been attributed to the prevalence of instability processes (e.g., Díaz and Ercilla 1993; Correggiari et al. 2001).

In this chapter we summarize the available evidence indicating the existence of distinct evolutionary stages of undulation development that could be linked to the preferential activity of high-density hyperpycnal flows, as opposed to the leading influence of sediment deformation.

19.2 Holocene Prodeltaic Wedges in the Mediterranean Sea

Wedge-shaped prodeltaic deposits found off major fluvial sources are one of the most distinctive depositional features found in the Mediterranean region. They occur off major rivers, such as the Po, Nile, Ebro and Rhône, but also off the minor rivers and streams that are very abundant in the region, particularly on the Mediterranean coasts of Iberia (Anthony et al. 2014). The onset of most of these deltaic systems was ignited when the sea-level stabilized in the mid-Holocene (Stanley and Warne 1994); the subsequent growth occurred through progradational pulses as a consequence of increasing interactions between natural processes and human perturbations, which were constructive during the Roman Period and the Little Ice Age. However, the most recent interval (i.e. the last 50 years) is mainly characterized by a substantial decrease in sediment flows and the overall erosion of most deltaic systems (Anthony et al. 2014).

Deltaic Mediterranean systems linked to major rivers and some small seasonal rivers (e.g. Fluvià-Muga, Ombrone) are characterized by external sigmoid shapes or by an offshore change from convex-upward to concave-upward (Díaz and Ercilla 1993; Tortora 1999). Some of these deltaic deposits may extend up to hundreds of kilometres along the coast and exhibit a moderate maximum thickness of a few tens of metres (Díaz et al. 1996; Sacchi et al. 2009). In contrast, the depositional geometries of deltaic deposits with superimposed sediment undulations that are genetically related hyperpycnal flows exhibit poorly developed deltaic plains, very high sediment thickness (in the range of 50–80 m) with lobate depocentres close to the present-day coastlines (Urgeles et al. 2007; Bárcenas et al. 2015; Lobo et al. 2015), and dominance of high-angle planar or concave-upward cross-sectional profiles (Lobo et al. 2006; Bárcenas et al. 2015).

The distinction of depositional units formed during the Holocene sea-level stabilization on Mediterranean prodeltas is attributed to the imprint of climatic changes affecting river basins and their supply to the adjacent shores. In general, wet-dry changes led to lateral depocentre redistribution through successive river avulsion and delta lobe switching processes (Correggiari et al. 2005; Fanget et al. 2014). In particular, the Little Ice Age is considered as the most significant climatic event affecting deltaic architecture because it triggered significant changes of sediment supply (Cattaneo et al. 2003; Correggiari et al. 2005; Fanget et al. 2014).

19.3 Evidence of Progressive Undulation Development

A number of sub-surface features on submarine deltaic slopes are assumed to be related to a sediment wave origin. The most unequivocal indications are (a) lateral continuity between sediment wave internal reflections and lateral parallel bedding (Lee et al. 2002; Urgeles et al. 2007; Milia et al. 2008), and (b) upslope migrating or climbing sediment waves. Internal reflections tend to mimic the external morphology, and sediment accumulation is enhanced on the upstream flanks (Lee et al. 2002), although this pattern may be spatially variable (Milia et al. 2008). These lateral changes are accompanied by changes in acoustic impedance on the seafloor, suggesting differential deposition (Berndt et al. 2006).

From a longer-term perspective, prodeltaic sediment waves tend to be persistent features, as they may occur along the stratigraphic column through the vertical stacking of internal reflections (Lee et al. 2002). In fact, most studies indicate increased development of sediment waves over time (Fig. 19.1). The Noeick Delta is a pioneering example of prodeltaic sediment waves rooted in deep stratigraphic horizons (Bornhold and Prior 1990). In the Mediterranean region, the most significant cases have been documented in deltaic deposits off short fluvial systems in the Iberian Peninsula (Fernández-Salas et al. 2007; Urgeles et al. 2007; Lobo et al. 2015). Prodeltaic sediment waves tend to show upward evolutionary trends. Sediment undulations usually show a wavy stratified pattern (aggrading or upward migrating) that is repeated from the base of the highstand progradational wedges towards the surface (Urgeles et al. 2007), or instead there is an upward transition from sub-parallel to undulated (Fig. 19.1a). This upward evolution may be accompanied by increased spatial distributions of undulatory patterns (Lobo et al. 2015) or by lateral changes of facies. Areas with well-marked crests and troughs may evolve seaward to landward dipping undulations with fewer channels, and seaward to sub-parallel configurations where undulation patterns become subdued (Fig. 19.1a). Both amplitudes and wavelengths tend to increase upward (Bøe et al. 2004; Urgeles et al. 2007).

In some settings, sediment waves show more complex upward trends, as several evolutionary stages may be identified (Levchenko and Roslyakov 2010) (Fig. 19.1b). In the lower part of the highstand wedge sediment waves are very scarce (pre-sediment wave stage); upward, sediment waves show a complete evolutionary sequence from embryonic development (initiation stage) to well-marked migration (upslope migration stage); and in between, sediment wave development is mainly vertical (aggradational stage). These stages may be separated by intervals with lower sediment wave development (quiescence stages) (Fig. 19.1b).

Sedimentologically, fields of prodeltaic undulations exhibit variations between coarse fractions (sands) and fine fractions (mixtures of silts and clays) (Milia et al. 2008; Lobo et al. 2015) (Fig. 19.2).

Numerical modelling also predicts the generation and evolution of undulatory patterns. Figure 19.3 shows how a simple 1D numerical model can be useful to analyze on a conceptual basis the genetic and evolutionary mechanisms of seabed undulations. Nevertheless, this kind of simple models cannot reproduce the actual rate of growing and evolution of the bedforms, due to the underlying simplifying hypothesis. The most critical one concerns the fact that the sediment transported by the river and deposited on the seabed will not be affected by the movement of the ambient fluid. Besides, the temporal scale (8 days) of the results presented here (Fig. 19.3) represents a further simplification to real processes and need to be properly interpreted.

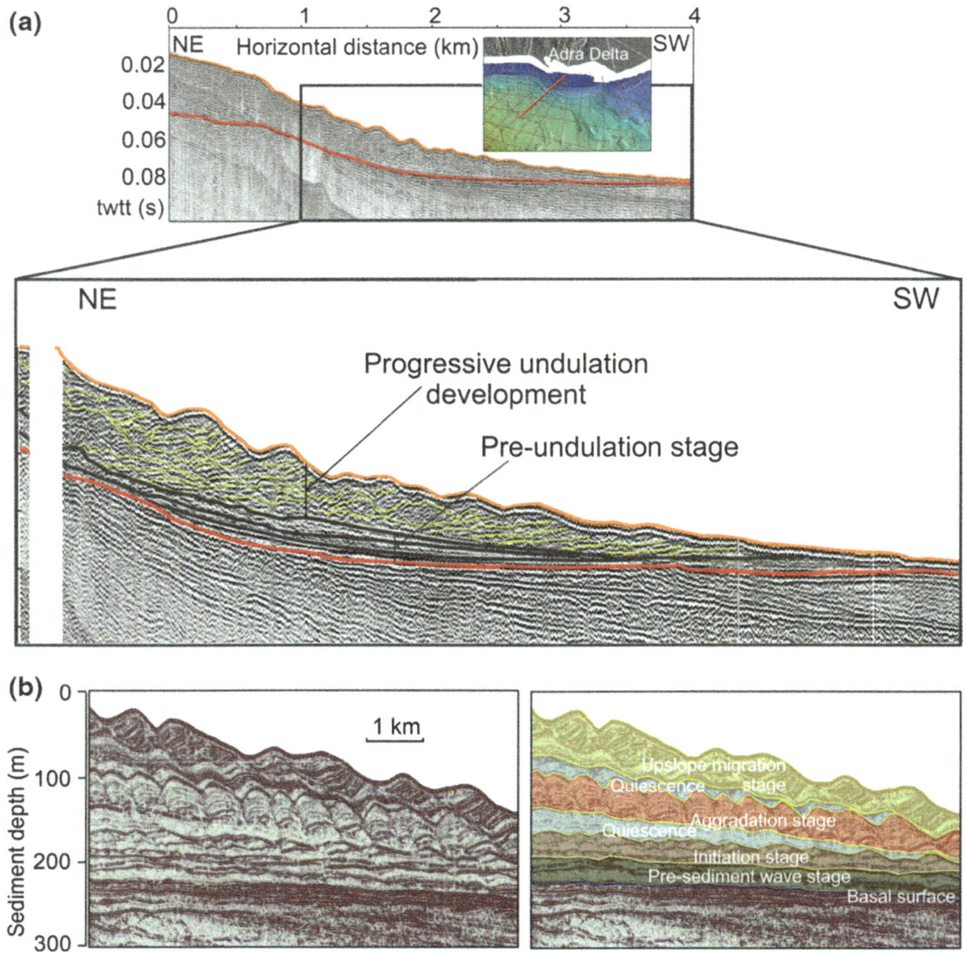

Fig. 19.1 Examples of undulation development throughout the stratigraphic column. **a** Undulation development in the prodeltaic wedge off the Adra River, southeast Iberian Peninsula. The lower part of the Holocene sediment wedge exhibits sub-parallel reflections; the upper part shows abundant undulation features whose landward migration trend increases upward. **b** Undulation development in the Caspian Sea (modified after Levchenko and Roslyakov 2010). A full sequence of undulation development is observed above a basal surface, from a precursory, incipient stage to a full development stage, with active landward migration

19.4 Evolutionary Trends Compatible with Sediment Flow Activity

The lack of sediment disruptions between undulation fields and adjacent sub-parallel layering does not favour the leading activity of sediment deformation processes. Instead, the sub-bottom evolutionary patterns observed in most of the above examples show enhanced development of undulations over time and, in particular, enhanced shoreward migrations in the upper parts of the stratigraphic columns. In some exceptional cases, several generations of sediment waves have even been documented. These stratigraphic patterns are explained by considering progressive sediment wave generation in response to changing environmental conditions (Lee et al. 2002; Levchenko and Roslyakov 2010; Lobo et al. 2015). These conditions seem to be mostly influenced in many Mediterranean deltas by changing (seasonal) climates with alternation of wet and dry periods leading to torrential regimes, and subsequent modifications of fluvial loads. Many of those fluvial systems are able to produce sediment hyperpycnal flows (Urgeles et al. 2011). It is therefore logical to assume that the evolutionary patterns of sub-surface undulations are linked to the dynamics of mountainous rivers, whose intense seasonal runoffs may produce high-density fluvial flows.

The fact that the undulations occur on stratigraphic intervals that decrease their thickness seaward indicates that the undulation fields are sourced upslope (Urgeles et al. 2007). In addition, in the few examples where changes of river courses have been documented, the sub-surface pattern of the undulations also reveals the changes of fluvial activity. For example, in the Guadalfeo River delta an older undulation field shows few intra-undulation reflections; in contrast, a younger undulation field exhibits more frequent reflections actively migrating, so a connection with the activity of river flows was made (Lobo et al. 2015). In the case of the Adra River delta, the evolutionary pattern is just the opposite, with widespread undulation occurrence over an older prodeltaic lobe and scarce occurrence over the most recent sediment wedge; this change is probably related to a very significant decrease of fluvial supply which caused the transformation of the submarine depositional system, which

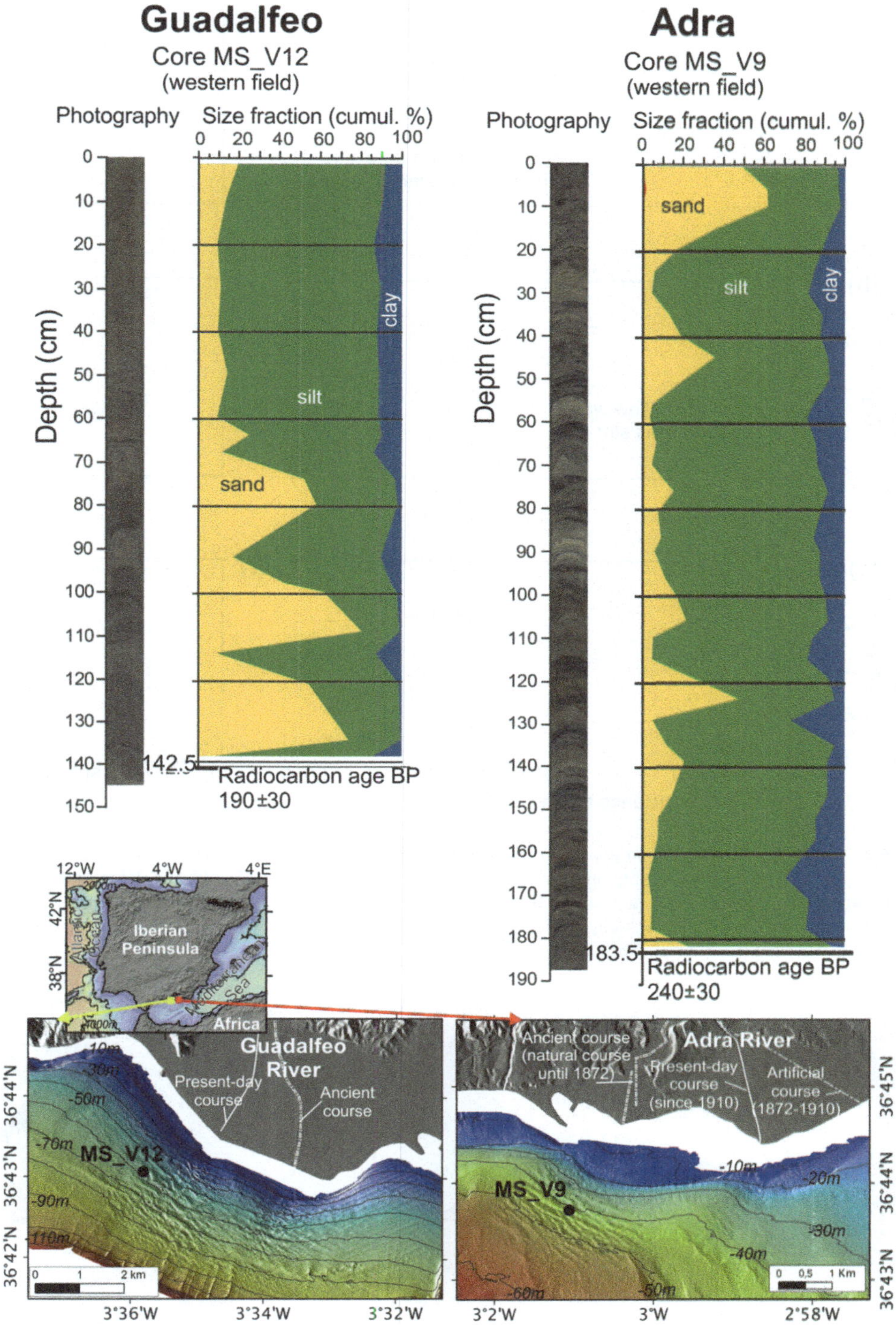

Fig. 19.2 Sedimentological characterization of prodeltaic undulations in sediment wedges off the Guadalfeo and Adra rivers located in the southeast of the Iberian Peninsula, on the northern margin of the Alborán Sea. Undulation occurrence is evidence from the existence of sandy layers within an overall muddy deposit. The sedimentary column off the Guadalfeo River was collected in the most recent prodeltaic lobe; the uppermost interval shows a decrease in sand content indicative of upstream sand retention due to drainage basin regulation. In contrast, the sediment core of the Adra delta was collected in an older prodeltaic lobe, where intercalated sandy layers occur up to the core top, suggestive of continuous undulation development

became controlled by hydrodynamic conditions (Jabaloy et al. 2010).

River influence in the changing sediment fluxes is also supported by the occurrence of coarse-grained peaks related to undulation development (Fig. 19.2), as these coarse-grained layers have been related to flood-derived rapid deposition (Budillon et al. 2005; Milia et al. 2008). The observed upward changes to fine-grained deposition may reflect the increased river regulation taking place in recent times, causing the drastic reduction in coarse-grained sediment deposition and sediment draping with partial smoothing of the undulation topography. As a consequence, fields of sediment undulations may become “frozen” or inactive (Fig. 19.4).

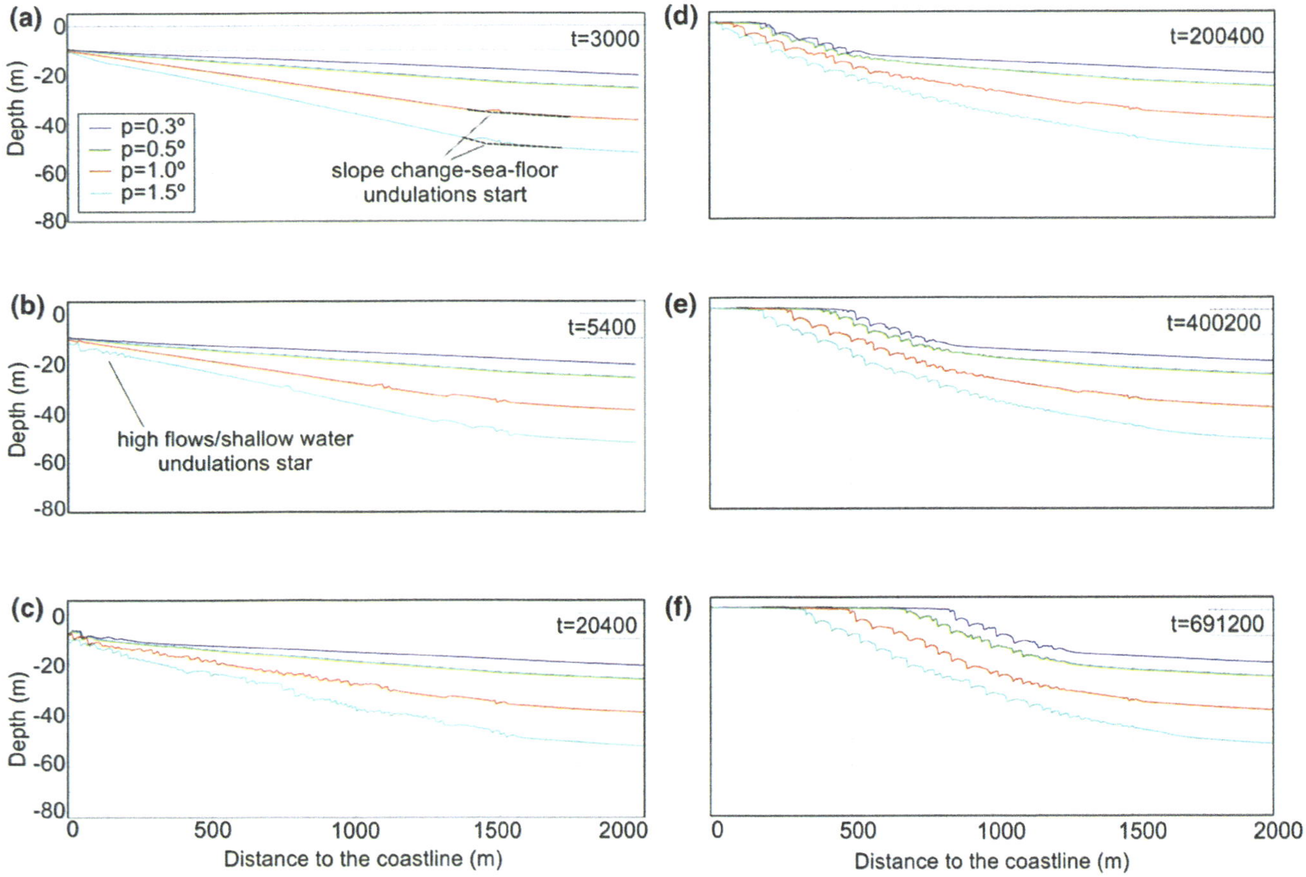

Fig. 19.3 Seafloor undulation evolution every 8 days of simulations, over the piecewise linear profile and for constant input flow (1 m^3/s). Captures at different times: **a** t = 50′, **b** t = 1 h 30′, **c** t = 5 h 40′, **d** t = 2 days, 7 h 40’, **e** t = 4 days, 15 h 10’ and **f** t = 8 days. The piecewise linear geometry profile has an initial variable slope (p = 0.3°, 0.5°, 1° and 1.5°) to 1500 m distance and constant slope of 0.3° to the end

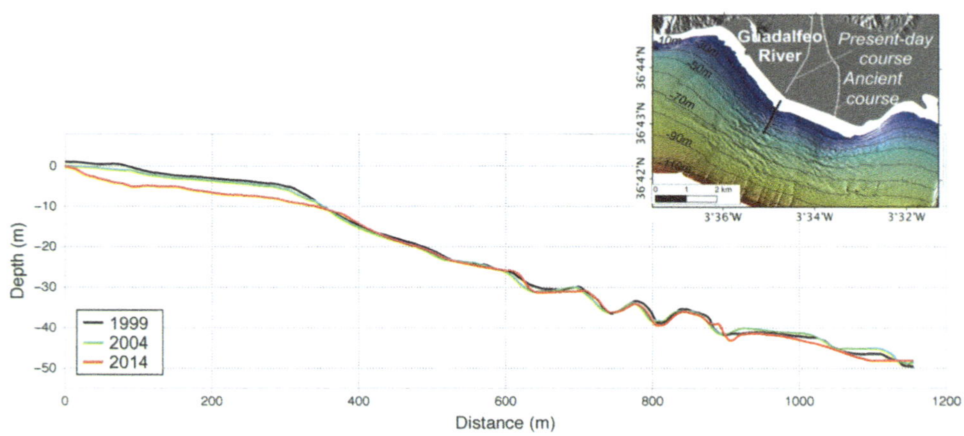

Fig. 19.4 Bathymetric profiles collected off the present-day mouth of the Guadalfeo River, southeast Iberian Peninsula, during the last 15 years. The upper part of the submarine deltaic feature has been eroded, but in contrast most of the undulations occurring in deeper water have remained stationary, indicating inactivity of the undulation field

Acknowledgments This chapter is a contribution to the projects MOSAICO (P06-RNM-01594) and TESELA (P11-787 RNM7069) funded by the "Junta de Andalucía", and it is also related to project CGL2011-30302-C02-02 funded by the Spanish Ministry for Economy and Competitiveness. Isabel Mendes thanks the Portuguese Science Foundation (FCT) for grant SFRH/BPD/72869/2010. The manuscript was improved thanks to useful comments and suggestions made by Roger Urgeles (CSIC-Institut de Ciències del Mar, Barcelona, Spain) and by an anonymous reviewer.

References

Anthony EJ, Marriner N, Morhange C (2014). Human influence and the changing geomorphology of Mediterranean deltas and coasts over the last 6000 years: From progradation to destruction phase? Earth-Sci Rev 139: 336–361. doi:10.1016/j.earscirev.2014.10.003.

Bárcenas P, Lobo FJ, Macías J, Fernández-Salas LM, López-González N, Díaz del Río V (2015). Submarine deltaic geometries linked to steep, mountainous drainage basins in the northern shelf of the Alboran Sea: Filling the gaps in the spectrum of deltaic deposition. Geomorphology 232: 125–144. doi:10.1016/j.geomorph.2014.11.028.

Bárcenas P, Lobo FJ, Mendes I, Ortega-Sánchez M, Macías J, Fernández-salas, LM (2017). Prodeltaic undulations and hyperpycnal flows (i): morphological observations. In: J. Guillén et al. (eds) Atlas of Bedforms in the Western Mediterranean. Springer, Heidelberg, pp 107–112

Berndt C, Cattaneo A, Szuman M, Trincardi F, Masson D (2006). Sedimentary structures offshore Ortona, Adriatic Sea – Deformation or sediment waves? Marine Geology EUROSTRATAFORM VOL 1: Source to Sink Sedimentation on the European Margin 234 (1–4): 261–270. doi:10.1016/j.margeo.2006.09.016.

Bøe R, Bugge T, Rise L, Eidnes G, Eide A, Mauring E (2004). Erosional channel incision and the origin of large sediment waves in Trondheimsfjorden, central Norway. Geo-Mar Lett 24 (4): 225–240. doi:10.1007/s00367-004-0180-3.

Bornhold BD, Prior DB (1990). Morphology and sedimentary porcesses on the subaqueous Noeick River delta, British Columbia, Canada. In: Spec. Publs int. Ass. Sediment., vol 10. IAS, pp 169–181. doi:10.1002/9781444303858.ch9.

Budillon F, Violante C, Conforti A, Esposito E, Insinga D, Iorio M, Porfido S (2005). Event beds in the recent prodelta stratigraphic record of the small flood-prone Bonea Stream (Amalfi Coast, Southern Italy). Mar Geol 222–223: 419–441. doi:10.1016/j.margeo.2005.06.013.

Cattaneo A, Correggiari A, Langone L, Trincardi F (2003). The late-Holocene Gargano subaqueous delta, Adriatic shelf: Sediment pathways and supply fluctuations. Mar Geol 193 (1–2): 61–91. doi:10.1016/S0025-3227(02)00614-X.

Cattaneo A, Correggiari A, Marsset T, Thomas Y, Marsset B, Trincardi F (2004). Seafloor undulation pattern on the Adriatic shelf and comparison to deep-water sediment waves. Mar Geol 213 (1–4): 121–148. doi:10.1016/j.margeo.2004.10.004.

Correggiari A, Trincardi F, Langone L, Roveri M (2001). Styles of Failure in Late Holocene Highstand Prodelta Wedges on the Adriatic Shelf. Journal of Sedimentary Research 71 (2): 218–236. doi:10.1306/042800710218.

Correggiari A, Cattaneo A, Trincardi F (2005). The modern Po Delta system: Lobe switching and asymmetric prodelta growth. Mar Geol 222–223: 49–74. doi:10.1016/j.margeo.2005.06.039.

Díaz JI, Ercilla G (1993). Holocene depositional history of the Fluviá—Muga prodelta, northwestern Mediterranean Sea. Mar Geol 111 (1–2): 83–92. doi:10.1016/0025-3227(93)90189-3.

Díaz JI, Palanques A, Nelson CH, Guillén J (1996). Morpho-structure and sedimentology of the Holocene Ebro prodelta mud belt (northwestern Mediterranean Sea). Cont Shelf Res 16 (4): 435–456. doi:10.1016/0278-4343(95)00019-4.

Fanget A-S, Berné S, Jouet G, Bassetti M-A, Dennielou B, Maillet GM, Tondut M (2014). Impact of relative sea level and rapid climate changes on the architecture and lithofacies of the Holocene Rhone subaqueous delta (Western Mediterranean Sea). Sediment Geol 305: 35–53. doi:10.1016/j.sedgeo.2014.02.004.

Fernández-Salas LM, Lobo FJ, Sanz JL, Díaz-del-Río V, García MC, Moreno I (2007). Morphometric analysis and genetic implications of pro-deltaic sea-floor undulations in the northern Alboran Sea margin, western Mediterranean Basin. Mar Geol 243 (1–4): 31–56. doi:10.1016/j.margeo.2007.04.013.

Jabajoy-sánchez A, Lobo FJ, Azor A, Bárcenas P, Fernández-Salas LM, Díaz del Río V, Pérez Peña JV (2010). Human-driven coastline changes in the Adra River deltaic system, southeast spain. Geomorphology 119 (1–2): 9–22. doi:10.1016/j.geomorph.2010.02.004.

Lee HJ, Syvitski JPM, Parker G, Orange D, Locat J, Hutton EWH, Imran J (2002). Distinguishing sediment waves from slope failure deposits: field examples, including the 'Humboldt slide', and modelling results. Mar Geol 192 (1–3): 79–104. doi:10.1016/S0025-3227(02)00550-9.

Levchenko OV, Roslyakov AG (2010). Cyclic sediment waves on western slope of the Caspian Sea as possible indicators of main transgressive/regressive events. Quaternary International 225 (2): 210–220. doi:10.1016/j.quaint.2009.12.001.

Lobo FJ, Fernández-Salas LM, Moreno I, Sanz JL, Maldonado A (2006). The sea-floor morphology of a Mediterranean shelf fed by small rivers, northern Alboran Sea margin. Cont Shelf Res 26 (20): 2607–2628. doi:10.1016/j.csr.2006.08.006.

Lobo FJ, Goff JA, Mendes I, Bárcenas P, Fernández-Salas LM, Martín-Rosales W, Macías J, Díaz del Río V (2015). Spatial variability of prodeltaic undulations on the Guadalfeo River prodelta: support to the genetic interpretation as hyperpycnal flow deposits. Mar Geophys Res 36 (4): 309–333. doi:10.1007/s11001-014-9233-9.

Milia A, Molisso F, Raspini A, Sacchi M, Torrente MM (2008). Syneruptive features and sedimentary processes associated with pyroclastic currents entering the sea: the AD 79 eruption of Vesuvius, Bay of Naples, Italy. Journal of the Geological Society 165 (4): 839–848. doi:10.1144/0016-76492007-110.

Sacchi M, Molisso F, Violante C, Esposito E, Insinga D, Lubritto C, Porfido S, Toth T (2009). Insights into flood-dominated fan-deltas: very high-resolution seismic examples off the Amalfi cliffed coasts, eastern Tyrrhenian Sea. In: Violante C (ed) Geohazard in Rocky Coastal Areas. Geological Society, London, Special Publications, vol 322. Geological Society, London, pp 33–71. doi:10.1144/sp322.2.

Stanley DJ, Warne AG (1994). Worldwide Initiation of Holocene Marine Deltas by Deceleration of Sea-Level Rise. Science 265 (5169):228–231. doi:10.1126/science.265.5169.228.

Sultan N, Cattaneo A, Urgeles R, Lee H, Locat J, Trincardi F, Berné S, Canals M, Lafuerza S (2008). A geomechanical approach for the genesis of sediment undulations on the Adriatic shelf. Geochemistry Geophysics Geosystems 9 (Q04R03). doi:10.1029/2007GC001822

Tortora P (1999). Sediment distribution on the Ombrone river delta seafloor and related dispersal processes. Geologica Romana 35: 211–218.

Urgeles R, De Mol B, Liquete C, Canals M, De Batist M, Hughes-Clarke JE, Amblàs D, Arnau PA, Calafat AM, Casamor JL, Centella V, De Rycker K, Fabrés J, Frigola J, Lafuerza S, Lastras G, Sànchez A, Zuñiga D, Versteeg W, Willmott V (2007). Sediment undulations on the Llobregat prodelta: Signs of early slope instability or sedimentary bedforms? Journal of Geophysical Research: Solid Earth 112 (B5): B05102. doi:10.1029/2005JB003929.

Urgeles R, Cattaneo A, Puig P, Liquete C, DE Mol B, Amblàs D, Sultan N, Trincardi F (2011). A review of undulated sediment features on mediterranean prodeltas: distinguishing sediment transport structures from sediment deformation. Mar Geophys Res 32 (1): 49–69. doi:10.1007/s11001-011-9125-1.

Sorted Bedforms Along the Egadi Islands Continental Shelf (Southern Tyrrhenian)

20

Claudio Lo Iacono and Jorge Guillén

Abstract

The analysis of sidescan sonar data and sediment samples around the Egadi Islands, on the NW Sicilian margin (southern Tyrrhenian), revealed the existence of two main sorted bedform fields. Sorted bedforms were mapped along the shelf sectors of the Marettimo and Favignana Islands, organized on elongated patches almost perpendicular to the coastline for a depth range of 15–50 m. Sediment sampling revealed that the bedforms are composed of sandy sediments lying over coarse sandy, gravelly and pebbly substrates. We assessed the long-term stability of the sorted bedforms, which could be related to along-shore transverse bottom currents associated with exceptional storms coming from the NE.

Keywords

Sorted bedforms • Inner shelf • Sidescan sonar • Egadi insular margin • Tyrrhenian sea

C.L. Iacono (✉)
Marine Geosciences, National Oceanography Centre, European Way, SO14 3ZH Southampton, UK
e-mail: cllo@noc.ac.uk

J. Guillén
Institut de Ciències del Mar, CSIC, Passeig Marítim de la Barceloneta 37-49, 08830 Barcelona, Spain

J. Guillén et al. (eds.), *Atlas of Bedforms in the Western Mediterranean*,
DOI 10.1007/978-3-319-33940-5_20

20.1 Introduction-Study Area

Sorted bedforms are defined as elongate sedimentary features having a patchy, sinuous or linear geometry and consisting of medium to coarse sands ("transversal sand patches" in Kenyon 1970; Cirac et al. 2000) adjacent to gravelly and pebbly deposits (Coco et al. 2007a; Murray and Thieler 2004). Sorted bedforms have generally been referred to as "rippled scoured depressions", due to their scoured morphology of a few metres and the occurrence of ripples above them (Cacchione et al. 1984; Goff et al. 2005; Davis et al. 2013). They mainly develop perpendicular to the coast and are generally observed along inner shelf regions at depths ranging from 15 to 70 m, with a length of several kilometres and a width of hundreds of metres (Ferrini and Flood 2005; Diesing et al. 2006; Lo Iacono and Guillén 2008).

Despite these bedforms are barely evident in the multibeam bathymetric images due to their low or absent relief, they are easily recognizable in the backscatter imagery, owing to their strong and sharp contrast of sediment texture. The improvement of methods for acquiring and processing backscatter data in the last few years has allowed sorted bedforms to be recognized as ubiquitous features along several continental margins around the world, including the Mediterranean Sea, having in common moderate to strong hydrodynamic regimes and reduced inputs of fine-grained sediments (Murray and Thieler 2004; Lo Iacono and Guillén 2008; Galparsoro et al. 2011; Goldstein et al. 2011, Durán et al. 2013; De Falco et al. 2015).

Mechanisms of formation and maintenance of sorted bedforms invoke self-organizing sedimentary processes (Coco et al. 2007b; Goldstein et al. 2011). The differences in roughness between sandy and gravelly sediments induce differential transport rates and shear stresses, favouring feedback mechanisms between transversal flows, seabed composition and evolving morphology (Green et al. 2004; Murray and Thieler 2004; Coco et al. 2007a; Van Hoyen et al. 2010). Sorted bedforms have already been recognized on the inner shelf of Marettimo Island (Egadi Archipelago, western Tyrrhenian) as gravelly and sandy features showing a long-term stability (Lo Iacono and Guillén 2008).

In this chapter, Marettimo sorted bedforms are described and compared with similar features recognized on the inner shelf of Favignana Island, in the same Archipelago. The Egadi Archipelago is located in the westernmost sector of the Sicilian margin (Catalano et al. 1996; Agate et al. 1996). Holocene marine deposits of this area show a general reduced thickness (1–2 m) compared with the usual values observed on the Sicilian shelf, due to the moderate to strong hydrodynamics and reduced sediment supply from river input (Agate et al. 1998). The dominant sediment textures along the inner shelf are coarse carbonatic sands in Favignana and heterogeneous mixtures composed of pebbles and gravels with medium to coarse sands in Marettimo (Lo Iacono 2004; Lo Iacono and Guillén 2008).

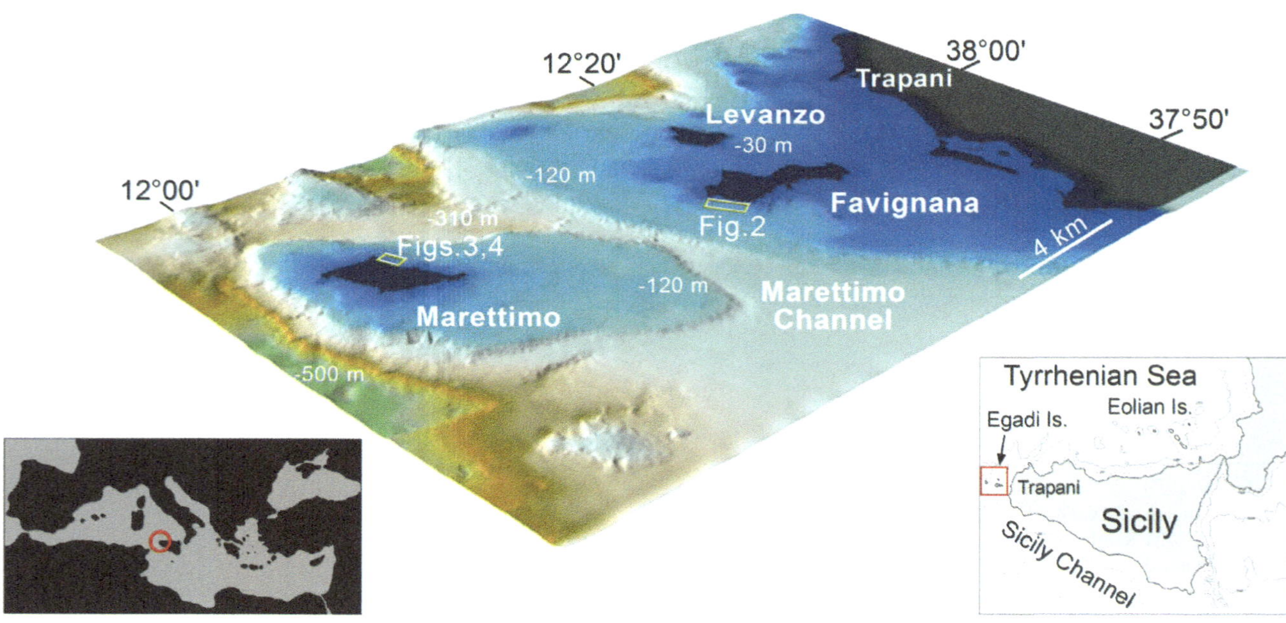

Fig. 20.1 3D bathymetric model of the Egadi Archipelago

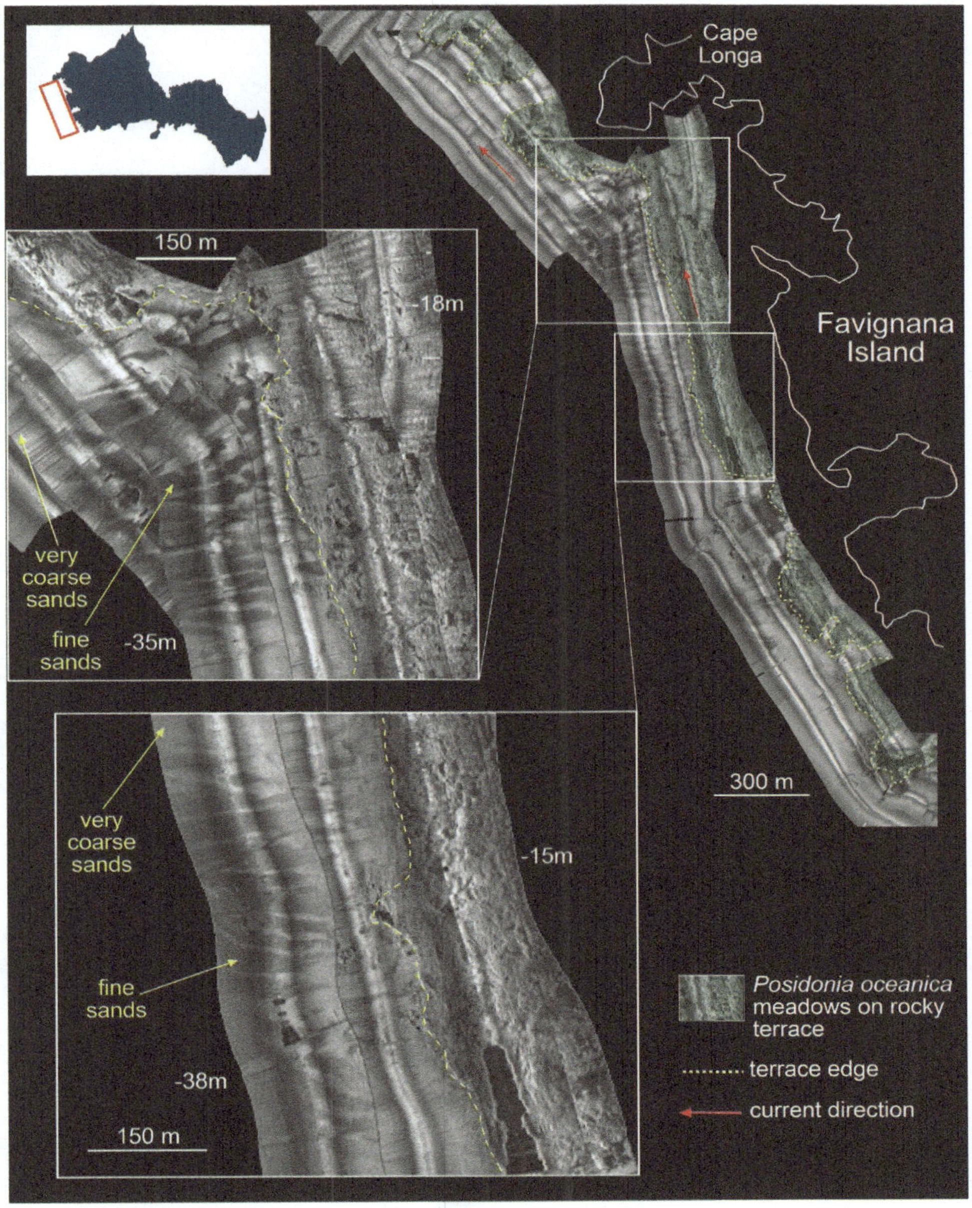

Fig. 20.2 Sidescan sonar mosaic of the favignana sorted bedforms. (Pixel resolution: 20 cm)

20.2 Methods

Side scan sonar (SSS) data were collected along the continental shelf of the Egadi Islands in 2002 using the SIS 1500 Benthos-Datasonics model operating at 100 kHz and towed at a depth of about 15 m above the sea bed. The SSS acquisition range was 75 m per channel for a total swath coverage of 150 m and a spatial resolution, perpendicular to the track line, of 0.3 m. By comparing the 2002 SSS mosaic with a 1989 SSS mosaic carried out by the Italian Geologic Service (analog EGG 272 SSS system), we estimated the stability of the Marettimo sorted bedforms. Sediment samples for the characterization of bottom sediments and calibration of acoustic facies were collected by scuba divers or through a van Veen grab (Fig. 20.1).

20.3 Results-Discussion

Favignana sorted bedforms are perpendicular to the coast of the western inner shelf of Favignana, and are closely interspaced over a depth range of 28–40 m, although the backscatter mosaic does not cover their deep terminations (Fig. 20.2). They display straight and linear boundaries and are 10–35 m wide and up to 220 m long (Fig. 20.2). Favignana sorted bedforms consist of fine sand bands laying over

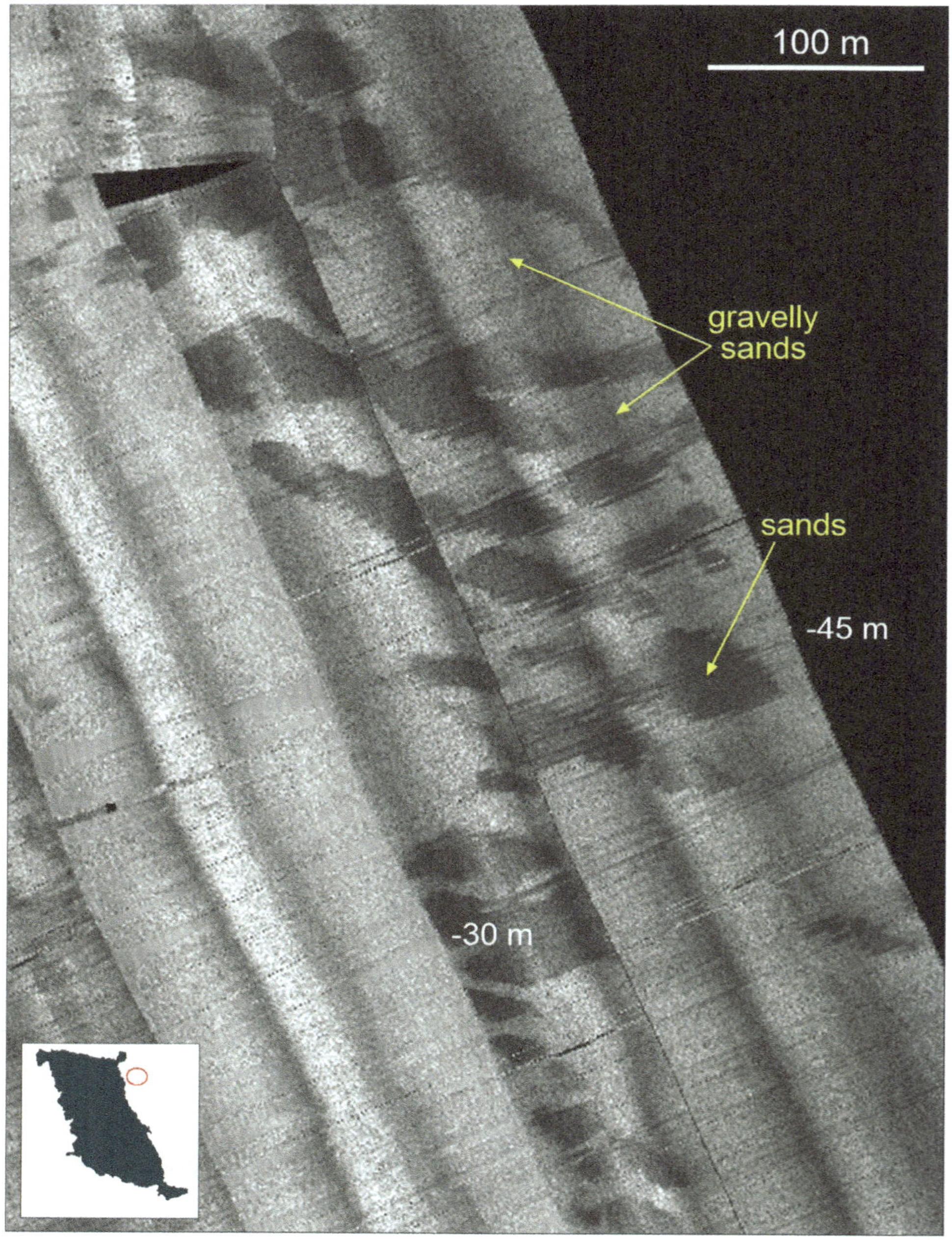

Fig. 20.3 Sidescan sonar mosaic of the marettimo sorted bedforms. (Pixel resolution: 20 cm). Modified from Lo Iacono and Guillén 2008

very coarse sand deposits (Fig. 20.2) and do not have any morphological expression on the seafloor.

Marettimo sorted bedforms develop perpendicular to the coast along the northeastern sector of the Marettimo shelf, on a particularly steep shelf portion (2.5° gradient) (Fig. 20.3). They show a width ranging from 15 to 50 m and a longitudinal extension ranging from 50 to almost 300 m. They display linear bands, changing to bifurcations and irregular patchy shapes in the deepest part of the area (50–60 m) (Fig. 20.3). Sediment samples and in situ observations reveal that the Marettimo sorted bedforms correspond to medium sands over strongly bioturbated, poorly sorted sandy gravel and pebbly sediments, having a thickness of around 10 cm. As in the case of the Favignana shelf, Marettimo sorted bedforms display a smooth and sub-horizontal morphology, though most of these bedforms generally present slight bathymetric depressions along the gravelly bands and ripples on the sandy sediments (Murray and Thieler 2004; Goff et al. 2005; Diesing et al. 2006, De Falco et al. 2015). These morphological and sedimentary differences may be the consequences of the reduced sediment inputs along the Egadi insular margin. Available data do not allow us to fully understand the hydrodynamic forcing which favoured the creation of the Marettimo and Favignana sorted bedforms. However, based on morphological, textural and hydrodynamic

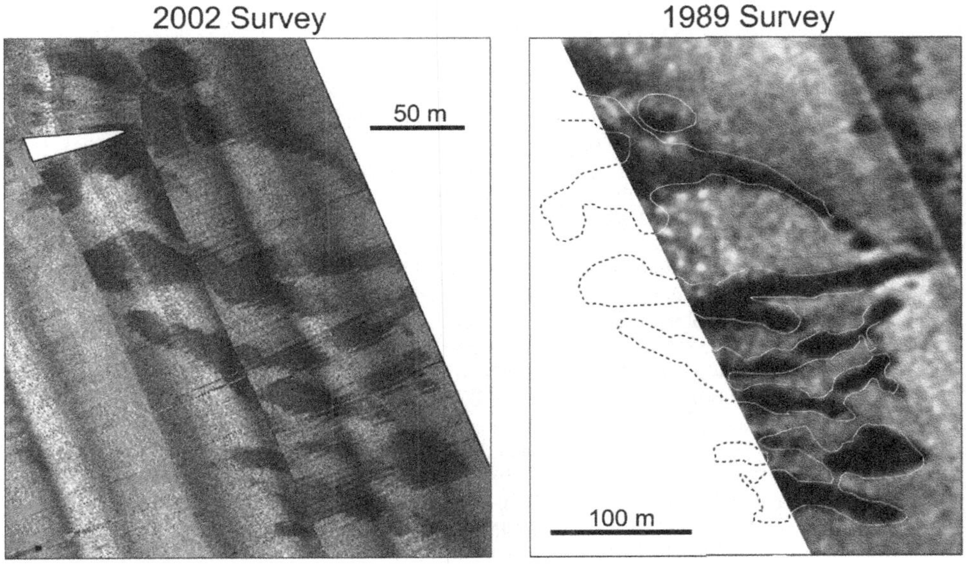

Fig. 20.4 Sidescan sonar mosaics of the sorted bedforms from data acquired in 1989 and 2002. *Dotted lines* in the 1989 mosaic correspond to the drawing of the 2002 bedforms

similarities, it is assumed that their genetic mechanisms are equivalent to those described in other areas (Murray and Thieler 2004; Coco et al. 2007a, b; Durán et al. 2013), where these bedforms are interpreted as transverse sedimentary features generated by along-shelf currents. Supporting this hypothesis, moderate to strong along-shelf currents were observed at the sea surface and close to the seafloor along the Favignana sorted bedforms and along the Marettimo Channel, flowing almost parallel to the coastline (Lo Iacono et al. 2004; Polizzi 2013). However, we believe that strong bottom currents generated during exceptional storm events are required to transport the coarse-grained sediments of the studied bedforms (Lo Iacono and Guillén 2008). The location and orientation of the mapped bedforms (Figs. 20.2 and 20.3) could reflect the genetic role of the storms coming from the SE, which are associated with the "Scirocco" wind and are among the strongest storms affecting the area. The comparison between two SSS mosaics taken in the Marettimo area in 1989 and 2002 shows a morphological stability of sorted bedforms for at least 13 years (Lo Iacono and Guillén 2008) (Fig. 20.4). These field observations support the long-term stability of these coarse bedforms, as also reported in previous works (Murray and Thieler 2004; Goff et al. 2005; Markert et al. 2013).

20.4 Conclusions

We described two sorted bedform fields in the inner shelf sectors of the Egadi Archipelago (southern Tyrrhenian). Sorted bedforms, consisting of coarse-grained sediments, are up to 40 m wide and 300 m long, and develop in a depth range of 15–50 m. Our study confirms the persistence of sorted bedforms in the Mediterranean Sea, along inner shelf settings dominated by moderate hydrodynamic regimes. In comparison with similar bedforms described in oceanic settings, the mobility of the Mediterranean sorted bedforms is probably related to events with a longer recurrence period. Though their genetic mechanisms still need to be highlighted, the Egadi sorted bedforms could be related to along-shelf transverse bottom currents associated with exceptionally strong NE "Scirocco" storms.

Acknowledgments Data collection was carried out in the framework of the GebecSud Project (P.I. Prof. R Catalano) supported by the "Ministero Istruzione Università e Ricerca" (MIUR, 488/92, Cluster 10) and directed by the Geology and Geodesy Department of Palermo University.

References

Agate M., Buscemi N., Catalano R., D'Angelo S., Di Maio D., Di Stefano P., Lucido M., Macaluso T., Marsella E., Pantaleone N. A., Pepe F., Sacchi L., Sulli A. (1996). Il Foglio geologico N.0 604 "Isole Egadi". Un prototipo delle nuova cartografia marina. Atti del convegno del Gruppo Informale di Sedimentologia. 30–34.

Agate M., D'Argenio A., Di Maio D., Lo Iacono C., Lucido M., Mancuso M., Scannavino M. (1998). La dinamica sedimentaria dell'Offshore della Sicilia Nord-Occidentale durante Il Tardo Quaternario. Atti del 791 Congresso Nazionale Societa` Geologica Italiana, Guida alle escursioni, 1, Palermo, Italy, 157–167.

Cacchione D.A., Drake D.E., Grant W.E., Tate G.B. (1984). Rippled scour depressions of the inner continental shelf off central California. Journal of Sedimentary Petrology 54, 1280–1291.

Catalano R., Di Stefano P., Sulli A., Vitale F. (1996). Paleogeography and structure of the Central Mediterranean. Sicily and its offshore. Tectonophysics 260, 291–323.

Cirac, P., Berne, S., Castaing, P., Weber, O. (2000). Processus de mise en place et d'e´volution de la couverture se´dimentaire superficielle de la plate-forme nord-aquitaine. Oceanologica Acta 23, 663–686.

Coco, G., Murray, A.B., Green, M.O. (2007a). Sorted bedforms as self-organized patterns: 1. Model development. Journal of Geophysical Research 112, F03015.

Coco, G., Murray, A.B., Green, M.O., Thieler, E.R., Hume, T.M. (2007b). Sorted bedforms as self-organized patterns: 2. Complex forcing scenarios. Journal of Geophysical Research 112, F03016.

Davis A.C.D., Kvitek R.G., Mueller C.B.A., Young M.A., Storlazzi C. D., Phillips E.L. (2013). Distribution and abundance of rippled scour depressions along the California coast. Continental Shelf Research 69, 88–100.

De Falco G., Budillon F., Conforti A., Di Bitetto M., Di Martino G., Innangi S., Simeone S., Tonielli R. (2015). Sorted bedforms over transgressive deposits along the continental shelf of western Sardinia (Mediterranean Sea). Marine Geology 359, 75–88.

Diesing M., Kubicki A., Winter C., Schwarzer K. (2006). Decadal scale stability of sorted bedforms, German Bight, southeastern North Sea. Continental Shelf Research 26, 902–916.

Durán R., Canals M., Lastras G., Micallef A., Amblas D., Pedrosa-Pàmies R., Sanz J.L. (2013). Sediment dynamics and post-glacial evolution of the continental shelf around the Blanes submarine canyon head (NW Mediterranean). Progress in Oceanography, doi:http://dx.doi.org/10.1016/j.pocean.2013.07.031.

Ferrini V.L., Flood R.D. (2005). A comparison of rippled scour depression identified with multibeam sonar: evidence for sediment transport in inner shelf environments. Continental Shelf Research 25, 1979–1995.

Galparsoro I., Borja A., Legorburu I., Hernández C., Chust G., Liria P., Uriarte A. (2011). Morphological characteristics of the Basque continental shelf (Bay of Biscay, northern Spain); their implications for Integrated Coastal Zone Management. Geomorphology 118, 314–329.

Goff J.A., Mayer L.A., Traykovski P., Buynevich I., Wilkens R., Raymond R., Glang G., Evans R.L., Olson H., Jenkins C. (2005). Detailed investigation of sorted bedforms, or "rippled scour depressions", within the Martha's Vineyard Coastal Observatory, Massachusetts. Continental Shelf Research 25, 461–484.

Goldstein E. B., Murray A. B., Coco G. (2011). Sorted bedform pattern evolution: Persistence, destruction and self-organized intermittency. Geophys. Res. Lett., 38, L24402, doi:10.1029/2011GL049732.

Green M.O., Vincent C.E., Trembanis A.C. (2004). Suspension of coarse and fine sand on a wave-dominated shoreface, with implications for the development of rippled scour depressions. Continental Shelf Research 24, 317–335.

Kenyon N.H. (1970). The origin of some transverse sand patches in the Celtic Sea. Geological Magazine 107, 389–394.

Lo Iacono C. (2004). Aspetti geomorfologici, sedimentologici ed ecologici dell'offshore delle Isole Egadi. PhD Thesis in Sedimentary Geology. University "Federico II" of Naples, Italy, pp. 158.

Lo Iacono, C., Guillén, J. (2008). Environmental conditions for gravelly and pebbly dunes and sorted bedforms on a moderate-energy inner shelf (Marettimo Island, Italy, western Mediterranean). Continental Shelf Research 28, 245–256.

Markert E., Holler P., Kröncke I., Bartholomä A. (2013). Benthic habitat mapping of sorted bedforms using hydroacoustic and ground-truthing methods in a coastal area of the German Bight/North Sea. Estuarine, Coastal and Shelf Science 129, 94–104.

Murray A.B., Thieler E.R. (2004). A new hypothesis and exploratory model for the formation of large-scale inner-shelf sediment sorting and "rippled scour depressions". Continental Shelf Research 24, 295–315.

Polizzi S. (2013). Caratterizzazione geomorfologica e sedimentaria di un margine continentale sottoalimentato: analisi sismo-acustica del sistema piattaforma-scarpata delle Isole Egadi (Sicilia nord-occidentale). PhD Thesis in Sedimentary Geology, University of Palermo, pp. 158.

Van Oyen T., de Swart H. E., Blondeaux P. (2010). Bottom topography and roughness variations as triggering mechanisms to the formation of sorted bedforms, Geophys. Res. Lett., 37, L18401, doi:10.1029/2010GL043793.

Sorted Bedforms Developed on Sandy Deposits Derived from Small Ephemeral Streams (Catalan Continental Shelf)

21

Ruth Durán, Jorge Guillén, and Araceli Muñoz

Abstract

Multibeam echosounder data and sediment samples were used to characterize sorted bedforms on the inner shelf off Lloret and Tossa de Mar (NW Mediterranean Sea). The sorted bedforms are formed by a sequence of coarse-grained (coarse sand) and fine-grained (fine to medium sand) domains, with little topographic relief (up to 1 m). They exhibit elongated shapes and are oriented nearly perpendicular to the shoreline, at water depths ranging from 10 to 40 m. The sorted bedforms display lateral symmetry in backscatter and bathymetric relief with high backscatter centred on the bathymetric depression. They appear associated with elongated sand deposits fed by short, ephemeral streams that extend across-shelf over the infralittoral prograding wedge down to 40 m water depth. Sorted bedforms are better developed in deeper waters (20–40 m), probably due to stronger

R. Durán (✉) · J. Guillén
Institut de Ciències del Mar (ICM-CSIC), Barcelona, Spain
e-mail: rduran@icm.csic.es

A. Muñoz
Tragsa-SGP, C/Julián Camarillo 6B, 28037 Madrid, Spain

J. Guillén et al. (eds.), *Atlas of Bedforms in the Western Mediterranean*,
DOI 10.1007/978-3-319-33940-5_21

hydrodynamic conditions in the shallower sector of the shelf, which prevent the development or maintenance of these morphological features. The morphological evolution of these bedforms indicates that they are persistent features, showing small changes in their boundaries over a decadal timescale.

Keywords
Persistent sorted bedforms • Bedform evolution • Inner shelf • Ephemeral streams • NW mediterranean

21.1 Introduction

Sorted bedforms are sedimentary features characterized by a sharply edged sequence of coarse-grained and fine-grained domains, with little topographic relief (of the order of a metre) relative to the bedform spacings, which vary from metres to kilometres (Murray and Thieler 2004). They have been observed on a variety of continental shelves at water depths ranging from 10 to 90 m (e.g. Hume et al. 2003; Murray and Thieler 2004; Ferrini and Flood 2005; Goff et al. 2005; Diesing et al. 2006; Coco et al. 2007b). In the Mediterranean Sea, sorted bedforms have been described in the eastern sector of the Marettimo shelf (NW Sicily) at a water depth of 45–50 m (Lo Iacono and Guillén 2008) and on the Sardinia shelf at a water depth of 43–70 m (De Falco et al. 2015). The formation of sorted bedforms has been related to a feedback between bed composition and sediment flux in a self-organized pattern (Murray and Thieler 2004; Coco et al. 2007a). Several studies have shown that sorted bedforms tend to be ephemeral in shallow waters (down to 15–20 m) (Hume et al. 2003; Ferrini and Flood 2005), while in deeper waters they are more persistent (Eittreim et al. 2002; Lo Iacono and Guillén 2008; De Falco et al. 2015).

In this work, we present the morphology, sedimentological characteristics and decadal stability of sorted bedforms identified on the Catalan continental shelf (NW Mediterranean Sea) off the towns of Lloret and Tossa de Mar (Fig. 21.1). The inner continental shelf off Lloret and Tossa de Mar extends down to 40–55 m water depth (Durán et al. 2013). The seafloor is markedly irregular, with numerous relieves of up to 10 m corresponding to a discontinuous infralittoral prograding wedge and rocky outcrops. The continental shelf is deeply incised by the Blanes submarine canyon with its head only 5 km from the shoreline. The Lloret de Mar and Tossa de Mar streams cross the coastal mountain ranges formed by igneous rocks (granite) and feed this sector of the continental shelf. The amount of sediment input depends directly on the rainfall regime, which is characterized by long dry periods interrupted by short, strong events that can result in floods within a few hours (Martin-Vide et al. 2008). Waves and currents can remobilize this sediment on the nearshore and continental shelf. The Catalan continental shelf is a wave-dominated, microtidal (<0.2 m) environment that has a seasonal wave climate with high-energy events occurring mostly in late autumn, winter and early spring, with the highest waves (Hs up to 5 m) coming from the NE and E.

21.2 Methods

The characterization of sorted bedforms is based on swath bathymetry and backscatter data gathered in 2004 using the SIMRAD EM3000D multibeam echosounder. Data were compared with recently acquired bathymetric data collected in July 2013 using the Elac SeaBeam 1050D 50 kHz system to assess the decadal evolution of these bedforms. Multibeam data were processed (including correction for heading, depth, pitch, heave and roll) and interpolated to a 4 × 4 m regular grid using the CARIS HIPS and SIPS Hydrographic Data Processing System. Morphological analysis of seabed and bedform features and bedform mobility were performed using ArcGIS software. Detailed geo-referenced shaded-relief images have sun illumination with an azimuth of 355° and an elevation of 35°. In addition, 14 bottom samples were collected in the area of sorted bedforms using a box corer dredge. A Horiba LA 950 laser diffraction particle size analyser was used to determine the grain size distribution of the sediment.

21.3 Results

Sorted bedforms were observed on the inner shelf off Lloret and Tossa de Mar at water depths of 10–40 m (Fig. 21.1). The inner shelf is characterized by a relatively steep seafloor (mean slope of 1.5°–2°) of high backscatter that extends down to 40 m water depth, where it becomes gentler (0.2°–0.5°). The backscatter imagery shows several elongated patches of low backscatter that extend across-shelf from the shoreface down to 15–40 m or appear detached from the

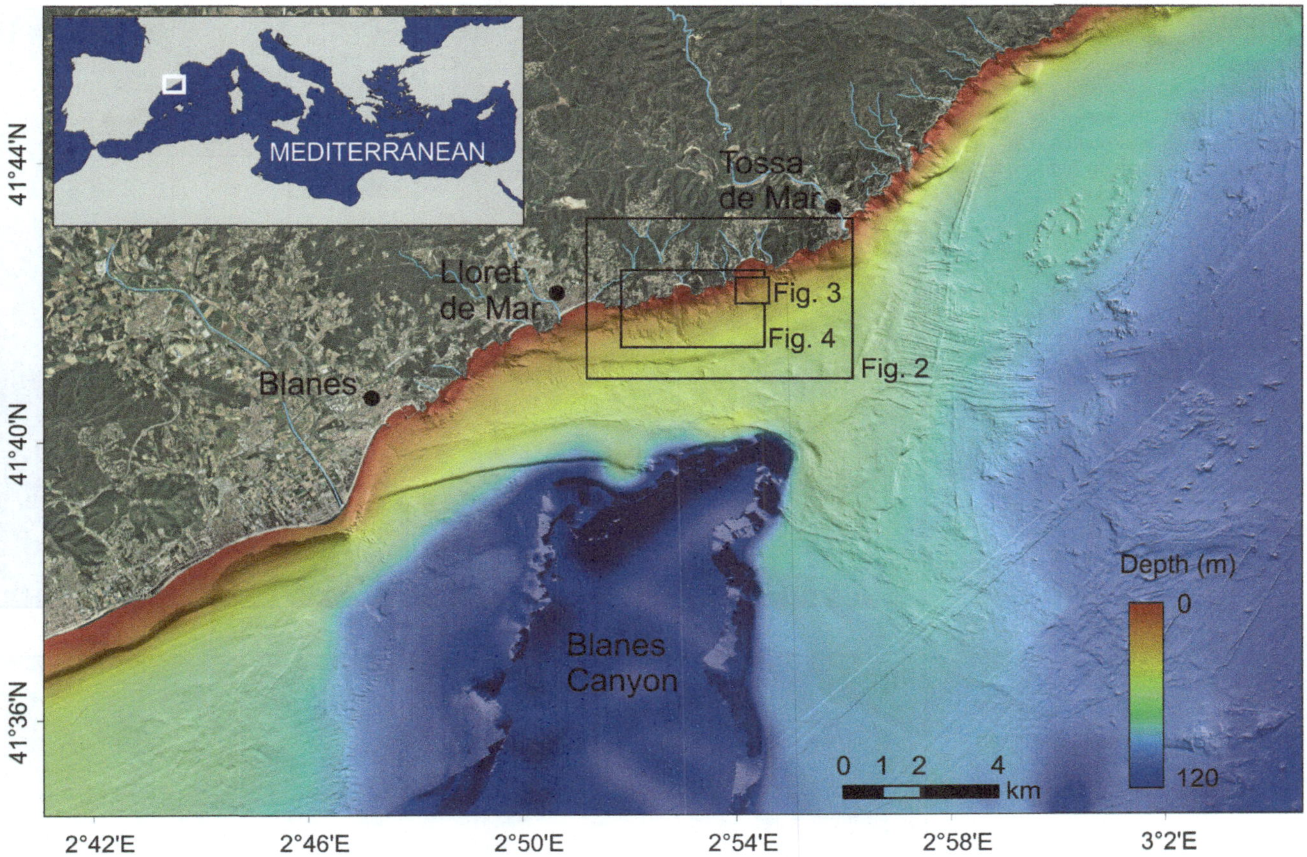

Fig. 21.1 *Location map and shaded-relief colour* multibeam bathymetry of the continental shelf off Lloret and Tossa de Mar. Illumination from the NNW. Locations of Figs. 21.2, 21.3 and 21.4 are shown. Orthophotos and topographic data from the Institut Cartogràfic i Geològic de Catalunya (http://www.icgc.cat/)

shoreface at water depths of 20–40 m. These patches are located off small bays where short streams discharge coarse sediment (Fig. 21.2). Elongated sorted bedforms have developed on some of these patches, showing a wide range of dimensions: from small bedforms that are 75 m long and 20 m wide to large features that are 600 m long and 100 m wide. The sorted bedforms are displayed in the backscatter imagery as elongated patches of low and high backscatter with subtle relief (0.4–1 m in depth), which lie nearly perpendicular to the isobaths (Fig. 21.2). They are better developed in deeper areas (20–40 m), whereas they are poorly developed or even absent near the shoreline.

Within individual sorted bedforms, backscatter data display an abrupt transition between high intensities in the shallow depressions and low values in the highs (Fig. 21.3). High backscatter values correspond to poorly to moderately sorted coarse sand (median size of 0.55–0.96 mm) with a gravel and mud content of 0.6–13.6 % and 0–0.7 %, respectively. Lower values of backscatter correspond to well-sorted fine to medium sand (median sized of 0.22–0.35 mm) with a mud content of 0.3–2.7 % (Fig. 21.3). The morphological evolution of the sorted bedforms revealed high stability of these bedforms over a 10-year time span, with small changes in the bedform morphology, as well as the formation of new sorted bedforms (Fig. 21.4).

21.4 Discussion and Conclusions

The sorted bedforms observed off Lloret and Tossa de Mar are associated with elongated sand deposits composed of fine sand that extend down to 40 m water depth and are superimposed on the infralittoral prograding wedge, which is characterized by high backscatter corresponding to coarse sand (Figs. 21.2 and 21.3). The finer-grained sand deposits are generated by sedimentary inputs from ephemeral streams that have their headwaters in the coastal mountain ranges. The sorted bedforms are characterized by a sharply edged sequence of coarse and fine sandy grained domains, with little topographic relief (up to 1 m), as described on other continental shelves (Murray and Thieler 2004; Goff et al. 2005). The fine domain is composed of well-sorted fine to

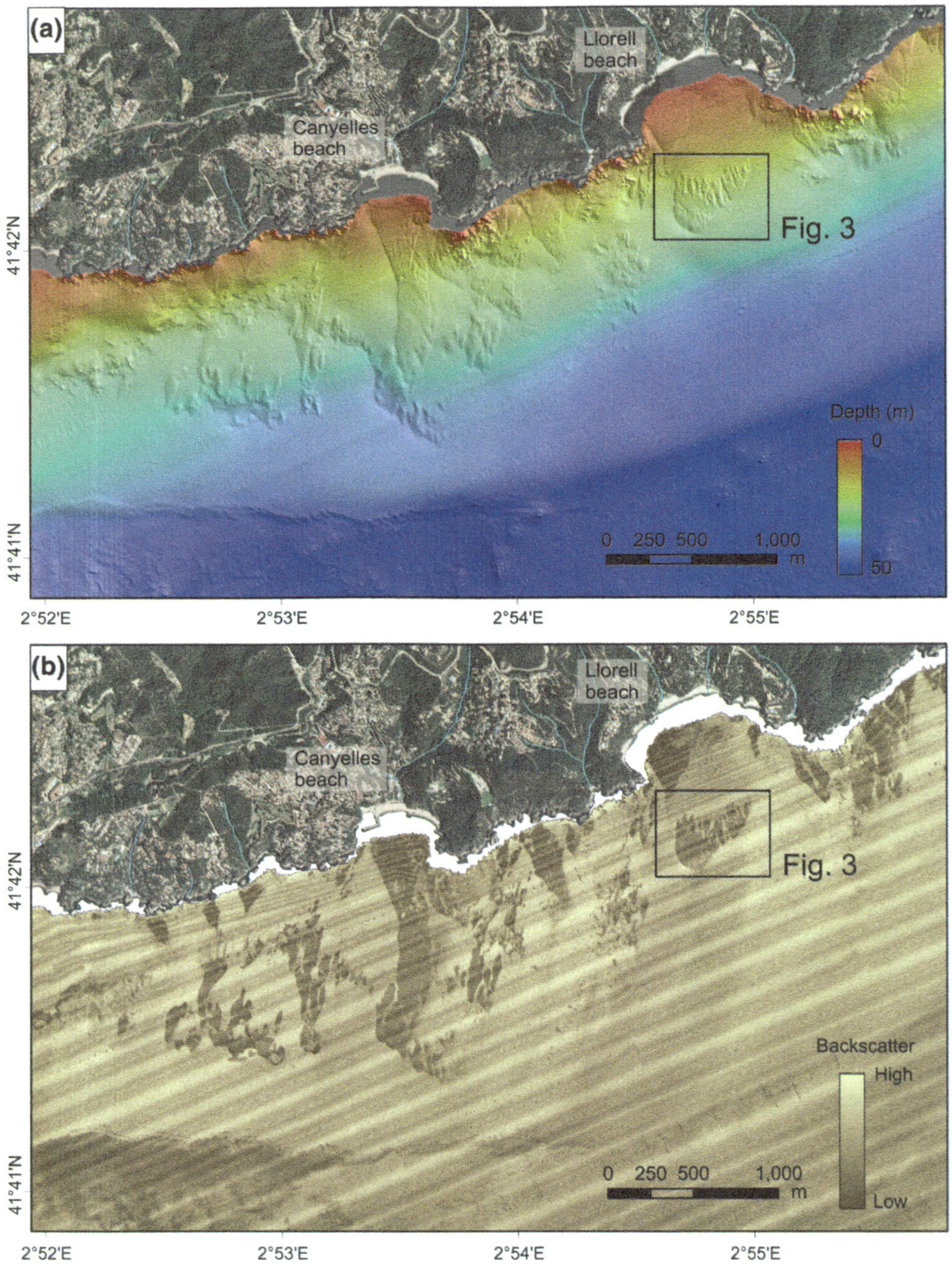

Fig. 21.2 a *Shaded-relief colour* multibeam bathymetry (illumination from the NNW) and **b** backscatter imagery of the inner shelf illustrating the sorted bedforms. High backscatter intensity corresponds to *lighter tones* while lower backscatter corresponds to *darker tones*

medium sands and forms the bathymetric highs, whereas poorly sorted coarse sand dominates in the bathymetric lows and on the adjacent continental shelf (Fig. 21.3). The sorted bedforms display lateral symmetry in both bathymetric expression and grain size, with the coarse domain centred on the bathymetric lows, as observed on the continental shelves of Marenttimo (Lo Iacono and Guillén 2008), New Zealand (Trembanis and Hume 2011) and California (Davis et al. 2013). However, the sorted bedforms observed off Lloret and Tossa de Mar have some peculiarities in relation to other sorted bedforms, as they are associated with finer-grained mounds that provide the required substratum for sorted bedform development. The local contribution of well-sorted fine sand by ephemeral streams over the coarse sand domain contributes to the bed sediment heterogeneity (mixture of sediment), which is further reorganized to form sorted bedforms.

The sorted bedforms are better developed in deeper waters (20–40 m) than near the shoreline, probably due to local hydrodynamic conditions. Stronger wave forcing in the shallower sector of the shelf may prevent the development or maintenance of these morphological features. A similar

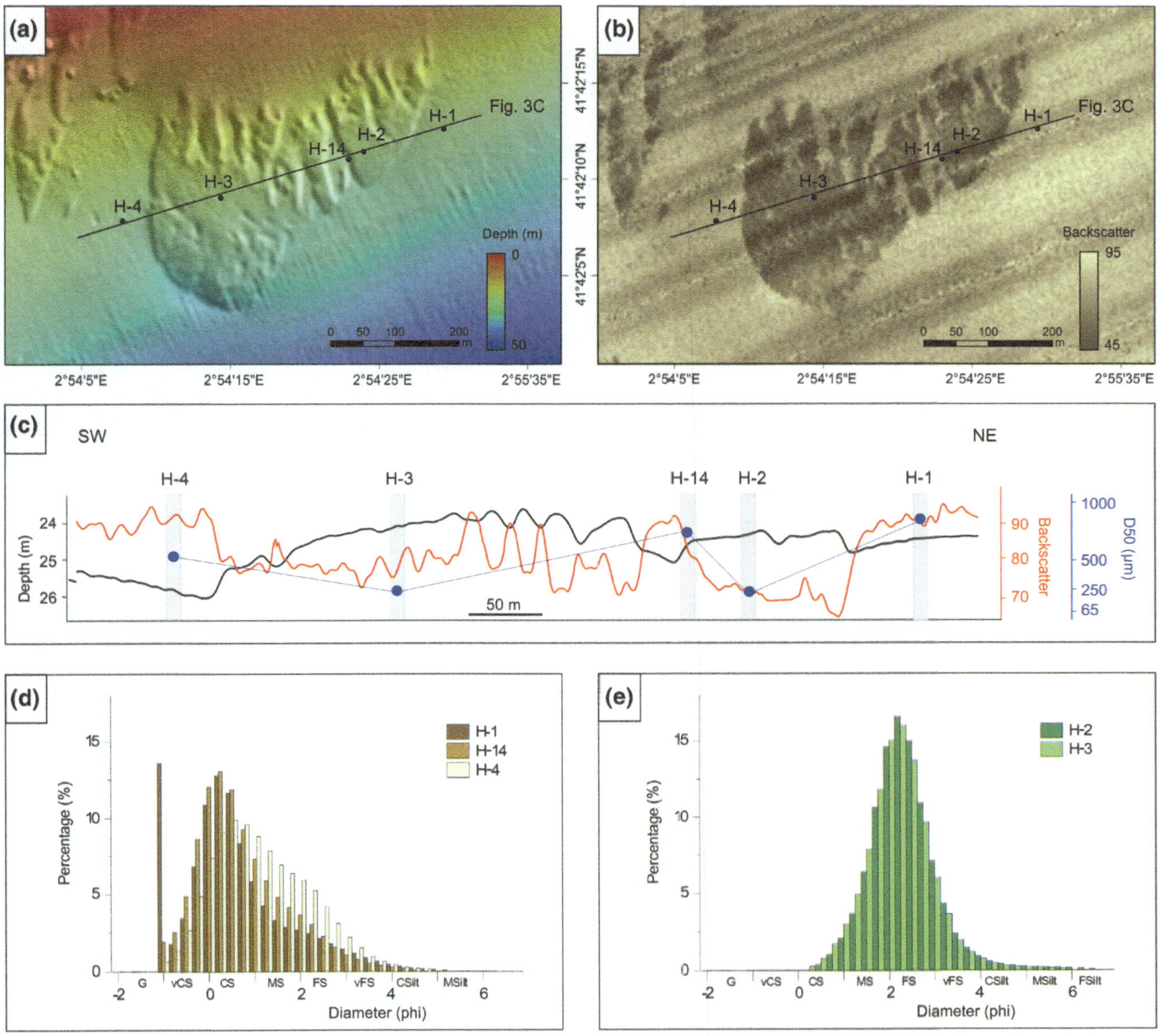

Fig. 21.3 **a** *Shaded-relief colour* multibeam bathymetry (illumination from the NNW) and **b** backscatter imagery of the sorted bedforms. High backscatter (*light tones*) corresponds to coarse sand, whereas low backscatter (*dark tones*) corresponds to fine sand. **c** Across-shelf bathymetric and backscatter profiles, including the median grain size of sediment samples collected on the highs (H-2 and H-3) and depressions (H-1, H-4 and H-14) of the sorted bedforms. **d** and **e** *grain* size distribution of sediment composing the highs and depressions

behaviour has been observed on other inner shelves, where shallower sorted bedforms are described as ephemeral features resulting from more complicated hydrodynamics (Hume et al. 2003; Ferrini and Flood 2005; Coco et al. 2007b). This hypothesis is also supported by numerical simulations that show variations in the bedform pattern or even the disappearance of sorted bedforms with variations in the wave height (Coco et al. 2007b; Goldstein et al. 2011). At a decadal time scale, sorted bedforms appear as persistent features with small changes in their boundaries, in agreement with previous observations and numerical simulations that highlighted the persistence and long-term stability of sorted bedforms at water depths greater than 15–20 m over annual or even decadal timescales (Eittreim et al. 2002; Diesing et al. 2006; Lo Iacono and Guillén 2008; De Falco et al. 2015).

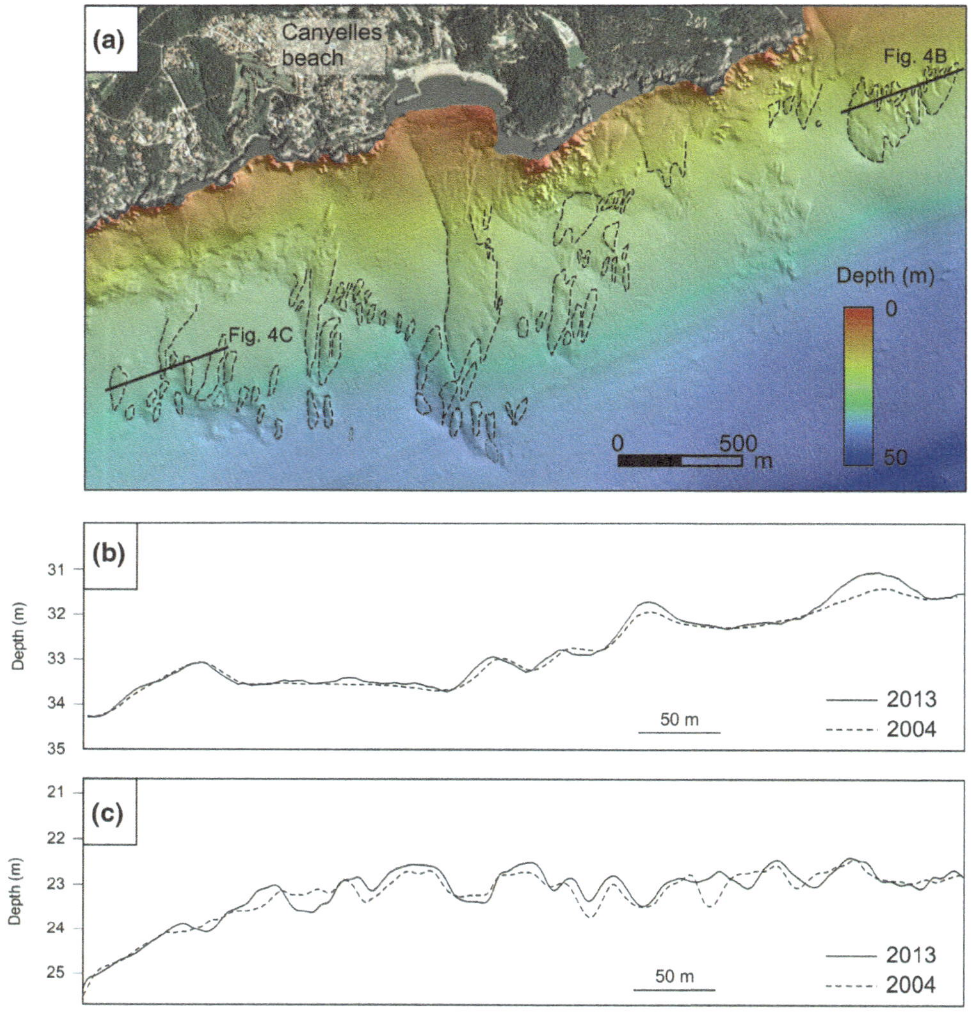

Fig. 21.4 a *shaded-relief colour* bathymetry (illumination from the NNW) of sorted bedforms derived from a bathymetric survey in 2013. The location of the bedform boundaries in 2004 is displayed by a *dotted line*. **b** and **c** Bathymetric profiles across the sorted bedforms

Acknowledgments This work was funded by the FORMED project (CGL2012-33989) funded by the Spanish Ministry of Economy and Competitiveness. The authors wish to thank the Secretaría General de Pesca and Tragsa for the 2004 Espace Project dataset, the captain and the crew of the R/V *García del Cid* and the UTM technicians for their assistance. R. Durán is supported by a CSIC JAE-Doc contract co-funded by the FSE.

References

Coco, G., Murray, a. B., Green, M.O. (2007a). Sorted bed forms as self-organized patterns: 1. Model development. J. Geophys. Res. 112, F03015. doi:10.1029/2006JF000665

Coco, G., Murray, a. B., Green, M.O., Thieler, E.R., Hume, T.M. (2007b). Sorted bed forms as self-organized patterns: 2. Complex forcing scenarios. J. Geophys. Res. 112, F03016. doi:10.1029/2006JF000666

Davis, A.C.D., Kvitek, R.G., Mueller, C.B. A., Young, M. A., Storlazzi, C.D., Phillips, E.L. (2013). Distribution and abundance of rippled scour depressions along the California coast. Cont. Shelf Res. 69, 88–100. doi:10.1016/j.csr.2013.09.010

De Falco, G., Budillon, F., Conforti, A., Di Bitetto, M., Di Martino, G., Innangi, S., Simeone, S., Tonielli, R. (2015). Sorted bedforms over transgressive deposits along the continental shelf of western Sardinia (Mediterranean Sea). Mar. Geol. 359, 75–88. doi:10.1016/j.margeo.2014.11.008

Diesing, M., Kubicki, A., Winter, C., Schwarzer, K. (2006). Decadal scale stability of sorted bedforms, German Bight, southeastern North Sea. Cont. Shelf Res. 26, 902–916. doi:10.1016/j.csr.2006.02.009

Durán, R., Canals, M., Lastras, G., Micallef, A., Amblas, D., Pedrosa-Pàmies, R., Sanz, J.L. (2013). Sediment dynamics and post-glacial evolution of the continental shelf around the Blanes submarine canyon head (NW Mediterranean). Prog. Oceanogr. 118, 28–46. doi:10.1016/j.pocean.2013.07.031

Eittreim, S. L., Anima, R. J., Stevenson A. J. (2002). Seafloor geology of the Monterey Bay area continental shelf, Mar. Geol., 181, 3–34. doi:10.1016/S0025-3227(01)00259-6

Ferrini, V.L., Flood, R.D. (2005). A comparison of rippled scour depression identified with multibeam sonar: evidence for sediment transport in inner shelf environments. Cont. Shelf Res. 25, 1979–1995. doi:10.1016/j.csr.2005.07.002

Goff, J. A., L. A. Mayer, P. Traykovski, I. Buynevich, R. Wilkens, R. Raymond, G. Glang, R. L. Evans, H. Olson, and C. Jenkins. (2005). Detailed investigation of sorted bedforms, or "rippled scour depressions," within the Martha's Vineyard Coastal Observatory, Massachusetts, Cont. Shelf Res. 25, 461–484. doi:10.1016/j.csr.2004.09.019

Goldstein, E.B., Murray, a. B., Coco, G. (2011). Sorted bedform pattern evolution: Persistence, destruction and self-organized intermittency. Geophys. Res. Lett. 38, L24402. doi:10.1029/2011GL049732

Hume, T. M., Trembanis, A. C., Hill, A., Liefting R., Stephens S. (2003). Spatially variable, temporally stable, sedimentary facies on an energetic inner shelf, in Coastal Sediments'03 edited by A. Press, Am. Soc. Civ. Eng., Reston, Va.

Lo Iacono, C., Guillén, J. (2008). Environmental conditions for gravelly and pebbly dunes and sorted bedforms on a moderate-energy inner shelf (Marettimo Island, Italy, western Mediterranean). Cont. Shelf Res. 28, 245–256. doi:10.1016/j.csr.2007.08.005

Martin-Vide, J., Sanchez-Lorenzo, A., Lopez-Bustins, J.A., Cordobilla, M.J., Garcia-Manuel, A., Raso, J.M. (2008). Torrential rainfall in northeast of the Iberian Peninsula: synoptic patterns and WeMO influence. Advances in Science and Research 2, 99–105.

Murray, A.B., Thieler, E.R. (2004). A new hypothesis and exploratory model for the formation of large-scale inner-shelf sediment sorting and "rippled scour depressions." Cont. Shelf Res. 24, 295–315. doi:10.1016/j.csr.2003.11.001

Trembanis, A.C., Hume, T.M. (2011). Sorted bedforms on the inner shelf off northeastern New Zealand: spatiotemporal relationships and potential paleo-environmental implications. Geo-Marine Lett. 31, 203–214. doi:10.1007/s00367-010-0225-8

22 Dynamics of Sorted Bedforms on a Shallow Infralittoral Prograding Wedge Influenced by Dredging (El Masnou, NW Mediterranean)

Ruth Durán, Belén Alonso, Gemma Ercilla, Ferran Estrada, David Casas, and Araceli Muñoz

Abstract

The dynamics of sorted bedforms on the inner shelf off El Masnou coast (NW Mediterranean) was investigated based on time-series of swath bathymetry and backscatter and grain-size analysis of grab samples spanning 4 years (2006–2009). The sorted bedforms are superimposed on the edge of a shallow infralittoral wedge with an orientation nearly perpendicular to the slope and oblique to the shoreline. They commonly display lateral asymmetry in backscatter, with the high backscatter domain on the depression and on the eastern side of the bedforms. The short-term evolution of these bedforms evidenced changes in the location of the bedform boundaries, widening and narrowing, predominant migration towards the southwest in the direction of the net alongshore sediment transport, and formation of new sorted bedforms. The analysis of repeated bathymetries also revealed the impact of dredging activities on the sorted bedforms, resulting in the destruction of part of the bedform pattern, which started to recover over the subsequent months. This behaviour suggests that the sorted bedforms off El Masnou are persistent and dynamic over a 4 year time span.

R. Durán (✉) · B. Alonso · G. Ercilla · F. Estrada · D. Casas
Institut de Ciències del Mar (ICM-CSIC), Pg. Marítim de la Barceloneta, 37–49, 08003 Barcelona, Spain
e-mail: rduran@icm.csic.es

D. Casas
Instituto Geológico y Minero de España (IGME), C/Rios Rosas 23, 08003 Madrid, Spain

A. Muñoz
Tragsa-SGP, C/Julián Camarillo 6B, 28037 Madrid, Spain

J. Guillén et al. (eds.), *Atlas of Bedforms in the Western Mediterranean*,
DOI 10.1007/978-3-319-33940-5_22

Keywords
Sorted bedforms • Bedform dynamics • Anthropogenic impact • Inner shelf • NW mediterranean

22.1 Introduction

Sorted bedforms are spatially-grain-size-sorted features that are widespread on continental shelves worldwide (e.g. Murray and Thieler 2004; Lo Iacono and Guillén 2008; Trembanis and Hume 2011; Durán et al. 2013; De Falco et al. 2015). They are constituted by coarse sediment on the bathymetric lows and fine sediment on the highs. In terms of temporal evolution, variable migration patterns have been described in the literature, ranging from no net movement with small back-and-forth migration of the bedform domain (Hume et al. 2003; Goff et al. 2005) to migration rates as high as 50–150 m in a 2 year time span (Ferrini and Flood 2005). Regarding their persistence, available observations report ephemeral features in shallow waters (shallower than 15–20 m) that appear and disappear at timescales of the order of a few months (Hume et al. 2003; Ferrini and Flood 2005). In water depths greater than 15–20 m, however, sorted bedforms appear as persistent and stable features with limited changes over annual (Hume et al. 2003; Goff et al. 2005;Trembanis and Hume 2011) and decadal timescales (Diesing et al. 2006).

On the inner Barcelona continental shelf (northwestern Mediterranean), sorted bedforms were observed off El Masnou coast (Fig. 22.1). This work presents the morphology and sedimentological characteristics of these bedforms and provides new evidences about their recent evolution and present-day activity, based on the analysis of repeated multibeam bathymetries and sediment samples.

The inner shelf off El Masnou coast is morphologically defined by large modern and relict infralittoral prograding wedges (Fig. 22.1). The older infralittoral wedge extends down to 26–30 m, whereas the recent one reaches water depths ranging from 7 to 15 m (Ercilla et al. 2010). The continental shelf is a wave-dominated, microtidal shelf (tidal

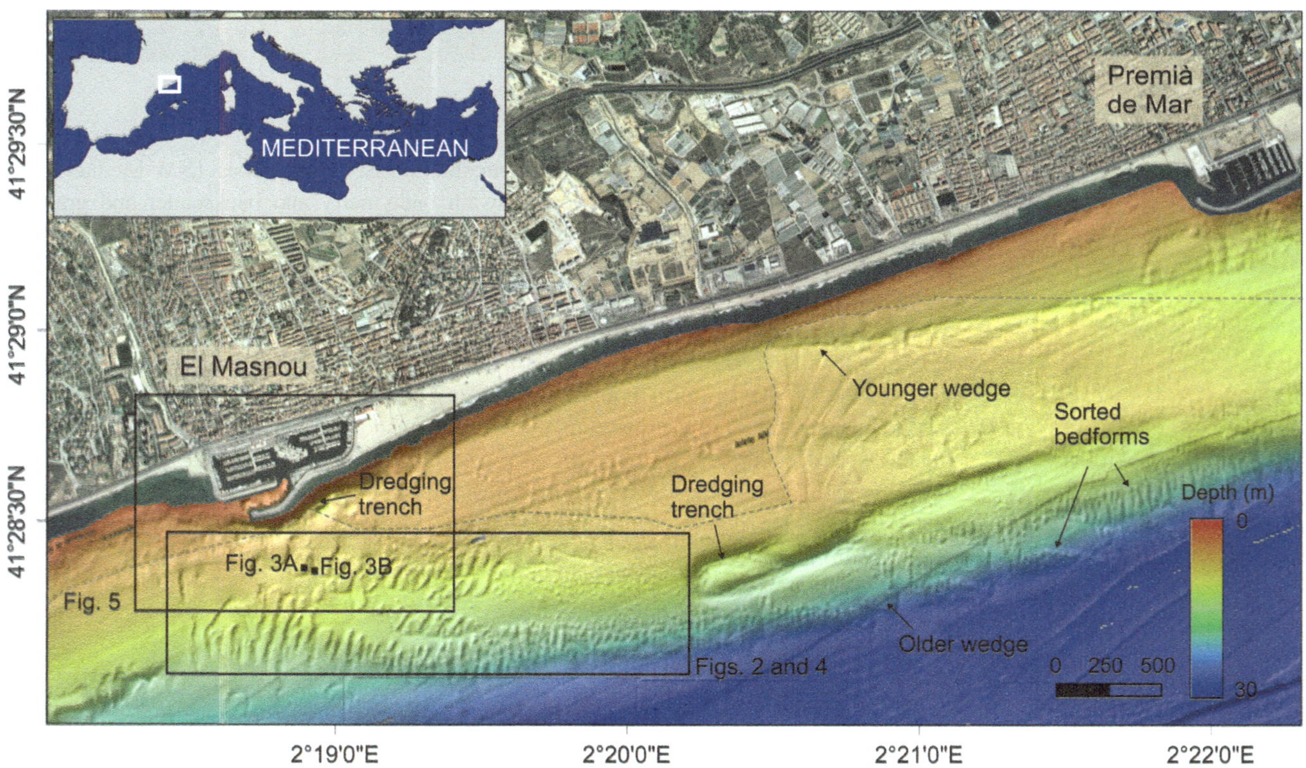

Fig. 22.1 Location map and bathymetry of the continental shelf off El Masnou coast showing the sorted bedform pattern superimposed on large prograding sediment wedges. A *dashed grey line* indicates the area of high-resolution multibeam data (4 × 4 m grid) obtained in 2009. The location of Figs. 22.2 to 22.5 is also shown. Orthophoto from the *Institut Cartogràfic i Geològic de Catalunya* (ICGC)

range 0.4 m). Prevailing NE-E waves generate an intense southwestward alongshore transport of sediment that can reach up to 45,000 m^3 y^{-1} (DGCP 1986). The continental shelf has been impacted by human activities during the last few decades, including dredging activities, trawling, sewage pipes and artificial reefs (Liquete et al. 2007). In particular, submarine dredging operations have been carried out in the El Masnou area since the 1980s, with a local impact on the modern infralittoral sedimentary system (Durán et al. 2008; Ercilla et al. 2010).

22.2 Methods

This study was based on repeated bathymetric surveys carried out between 2006 and 2009 using Elak-Nautik SeaBeam 1185 and SIMRAD EM3000D multibeam echosounders. A total of five surveys were conducted in March, June and November 2006, May 2007 and 2009. The multibeam data were post-processed using the CARIS HIPS and SIPS Hydrographic Data Processing System, including correction for heading, depth, pitch, heave and roll. Finally, digital elevation models were produced at 4 m resolution over the inner shelf and at 0.5 m resolution in the area adjacent to El Masnou harbour. Bedform mobility was analysed by comparing the multibeam data from the different surveys using ArcGIS 10.3 software. In addition, 20 surficial sediment samples were collected using a van Veen grab within the area of sorted bedforms between March and October 2007.

22.3 Results

22.3.1 Sorted Bedform Characteristics

Sorted bedforms on the inner shelf off the El Masnou coast appear at a depth range of 8–23 m (Fig. 22.1). They show a nearly N-S orientation oblique to the shoreline, which changes slightly to the NNE-SSE off El Masnou harbour (Fig. 22.1). Sorted bedforms exhibit a linear morphology and form sequences of parallel aligned highs along several belts that appear superimposed on the edge of several prograding wedges (Fig. 22.1). The seabed where the bedforms occur corresponds to areas of higher seabed gradient (0.8°–2°) in comparison with the surroundings (0.2°–0.5°). The bathymetric expression of these bedforms is very subtle, varying from 0.2 to 0.5 m high, 50 to 70 m wide and 80 to 200 long (Fig. 22.2). In cross-section, they display both symmetric and asymmetric profiles. Backscatter displays abrupt transitions from high backscatter on the depressions and the eastern side of the bedforms and low backscatter on the topographic highs (Fig. 22.2c). In general, coarse sand with a median size of 0.62–0.96 mm dominates on the topographic troughs and the eastern side, whereas the sediment on the crest is characterized by finer sand with a median size of 0.17–0.19 mm (Fig. 22.3).

22.3.2 Short-Term Bedform Evolution

The short-term evolution of the sorted bedforms was investigated by comparing the sorted bedform patterns observed in 2009 with their morphology in 2006 and 2007. The results evidenced a variety of responses, including changes in the bedform boundaries that result in widening and narrowing of the sorted bedforms, migration and formation of new bedforms (Fig. 22.4). In general, the sorted bedforms migrated towards the southwest, with migration rates of up to 30 m over a time span of 2 years. Moreover, the sorted bedforms off El Masnou harbour also showed changes in orientation caused by higher displacements of the seaward end of the bedform boundaries (Fig. 22.3). Finally, new sorted bedforms were observed in the most recent surveys at water depths of 7–15 m (Figs. 22.2 and 22.3). The evolution of the sorted bedforms was also examined over 1 year in the area adjacent to El Masnou harbour (Fig. 22.5). In this area, the shallowest sorted bedforms (down to 8 m deep) were disrupted by dredging activities in March and May 2006. As illustrated in Figs. 22.5a, b, dredging activities resulted in the destruction or alteration of the shallowest part of the bedform pattern. Over the six subsequent months, from June to November 2006, the sorted bedforms remained stationary (Fig. 22.5c). However, in May 2007, new small sorted bedforms emerged at the shallowest part of the bedform pattern (Fig. 22.5d).

22.4 Discussion and Conclusions

The sorted bedforms of the inner continental shelf off El Masnou show the main characteristics of linearly shaped sorted bedforms described on other inner shelves (Murray and Thieler 2004; Coco et al. 2007; Lo Iacono and Guillén 2008; De Falco et al. 2015). They appear as alternating linear bands of coarse and fine sands with subtle relief and are located in areas of higher seafloor gradient (Fig. 22.1). The lateral grain size asymmetry, with the coarse domain extending from the bathymetric low to the crest along the eastern side, corresponds to the updrift side relative to the dominant alongshore sediment transport towards the southwest (DGCP 1986). This grain size asymmetry was also described on other inner shelves as a response to a transverse, alongshore flow (Murray and Thieler 2004; Goff et al. 2005). The sorted bedforms off El Masnou are shore-oblique, differing from previous observations in which bedforms are nearly normal to the shoreline (e.g. Murray and Thieler 2004; Diesing et al. 2006). The

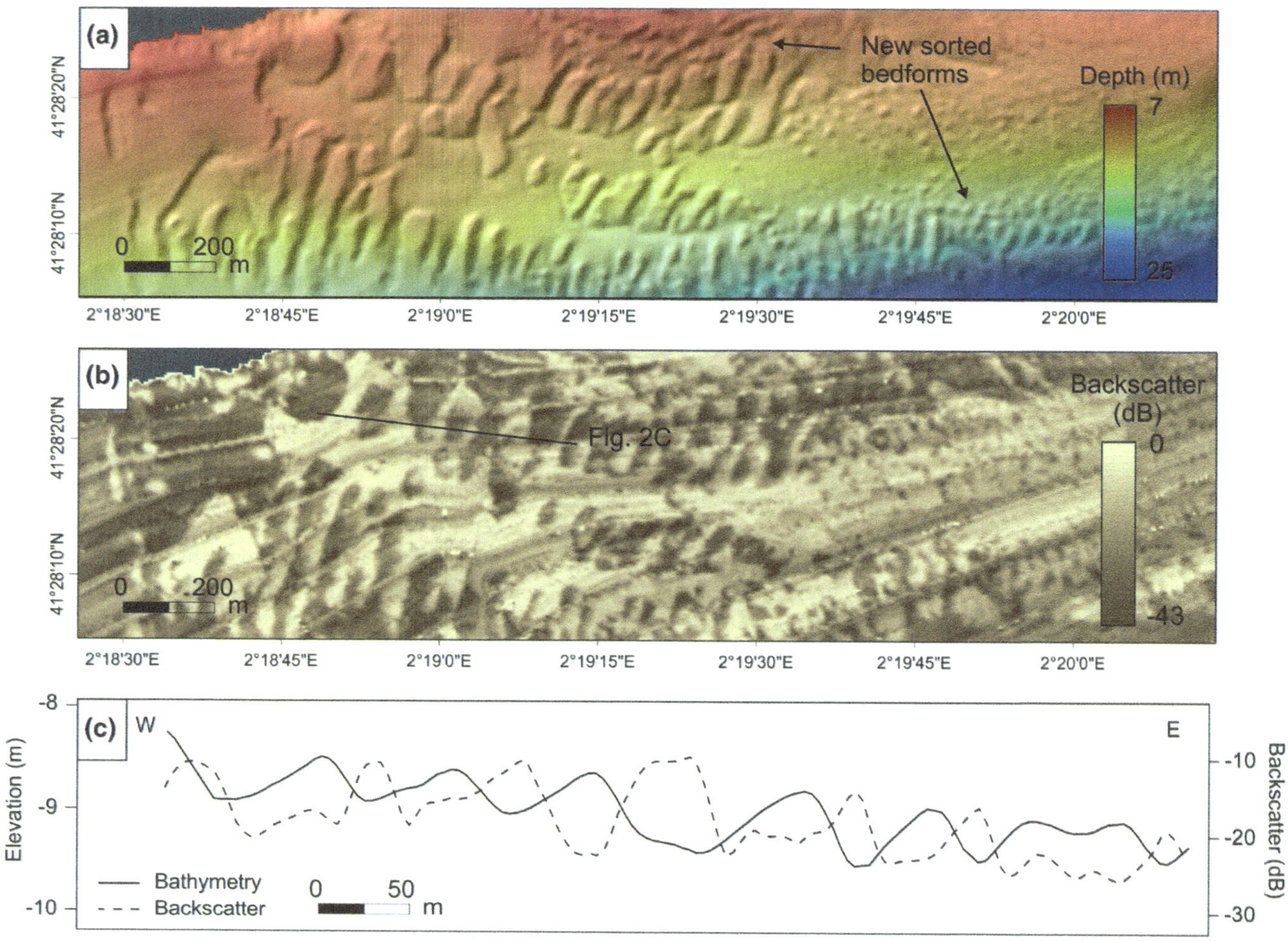

Fig. 22.2 a Detailed shaded-relief colour multibeam bathymetry and b backscatter imagery illustrating sorted bedforms. High backscatter intensity is shown in *light tones*, whereas lower backscatter is shown in *dark tones*. c Bathymetric and backscatter profile across the sorted bedforms. See Fig. 22.1 for location

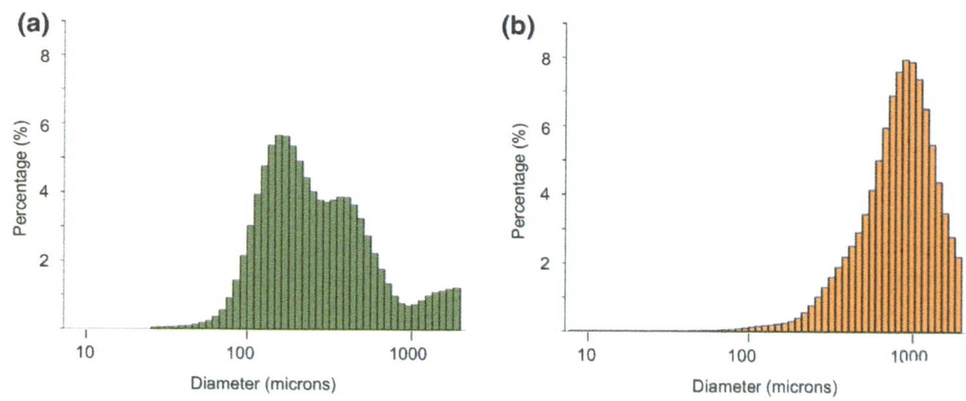

Fig. 22.3 Grain-size distribution of sediment composing the highs (a) and depressions (b). See Fig. 22.1 for location of sediment sampling stations

oblique angle of the sorted bedforms on the El Masnou inner shelf can be related to the preexisting large-scale seabed morphology, as they lie perpendicular to the isobaths along the edges of large-scale prograding sediment bodies, which are oblique to the shoreline (Fig. 22.1).

The short-term evolution of sorted bedforms off El Masnou indicates that these bedforms are persistent over a 4 year time span, with some modifications in their morphology. The main changes include bedform migration, variations in the position of the bedform boundaries and

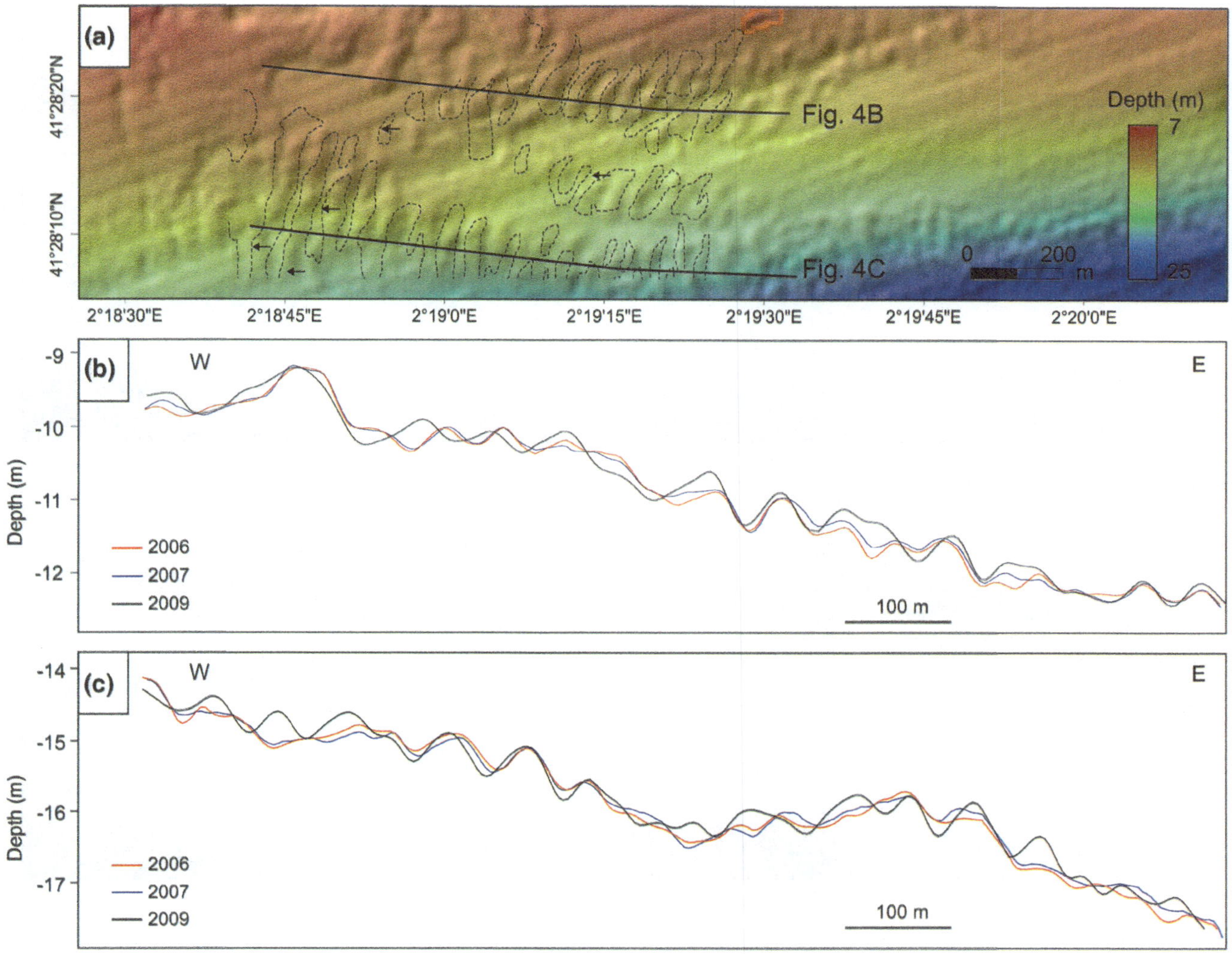

Fig. 22.4 **a** Detailed shaded-relief colour multibeam bathymetry illustrating the sorted bedforms off El Masnou in 2007. *The dotted line* represents the location of the bedform boundaries in 2009. Note the predominant bedform migration towards the southwest, indicated by *black arrows*. **b** and **c** bathymetric profiles across the sorted bedforms off El Masnou at different water depths: 9–12 m and 14–17 m, respectively

rotation of some bedforms. The predominant migration of the sorted bedforms is towards the southwest, in the direction suggested by their backscatter asymmetry and the predominant alongshore currents (DGCP 1986; Durán et al. 2008; Ercilla et al. 2010). The migration rate (up to 30 m in 2 years) is consistent with those observed on other inner shelves (Ferrini and Flood 2005). The differential migration along individual bedforms, particularly off El Masnou harbour, results in an overall rotation in orientation from oblique to nearly perpendicular to the direction of the maximum seafloor gradient (Figs. 22.2a and 22.3b). Finally, the displacements of the bedforms boundaries can also occur in opposite directions, resulting in the widening and narrowing of some bedforms.

The analysis of repeated bathymetries revealed the formation of small bedforms at water depths of 7–15 m over time scales of months to years (Figs. 22.2 and 22.3). Particularly, off El Masnou harbour, the formation of new sorted bedforms occurs over a span of 6 months after their destruction by dredge activities (Fig. 22.5). The occurrence of new sorted bedforms in the last surveys can be explained by higher wave conditions between 2007 and 2009 than in the previous period. Moderate to severe southern and eastern storms occurred between December 2007 and December 2008, with maximum Hs of up to 4.3 m, whereas the period from March 2006 to November 2007 was characterized by a low-energy wave climate (mean Hs of 0.55 m) with no significant storms. The formation of new bedforms in shal-

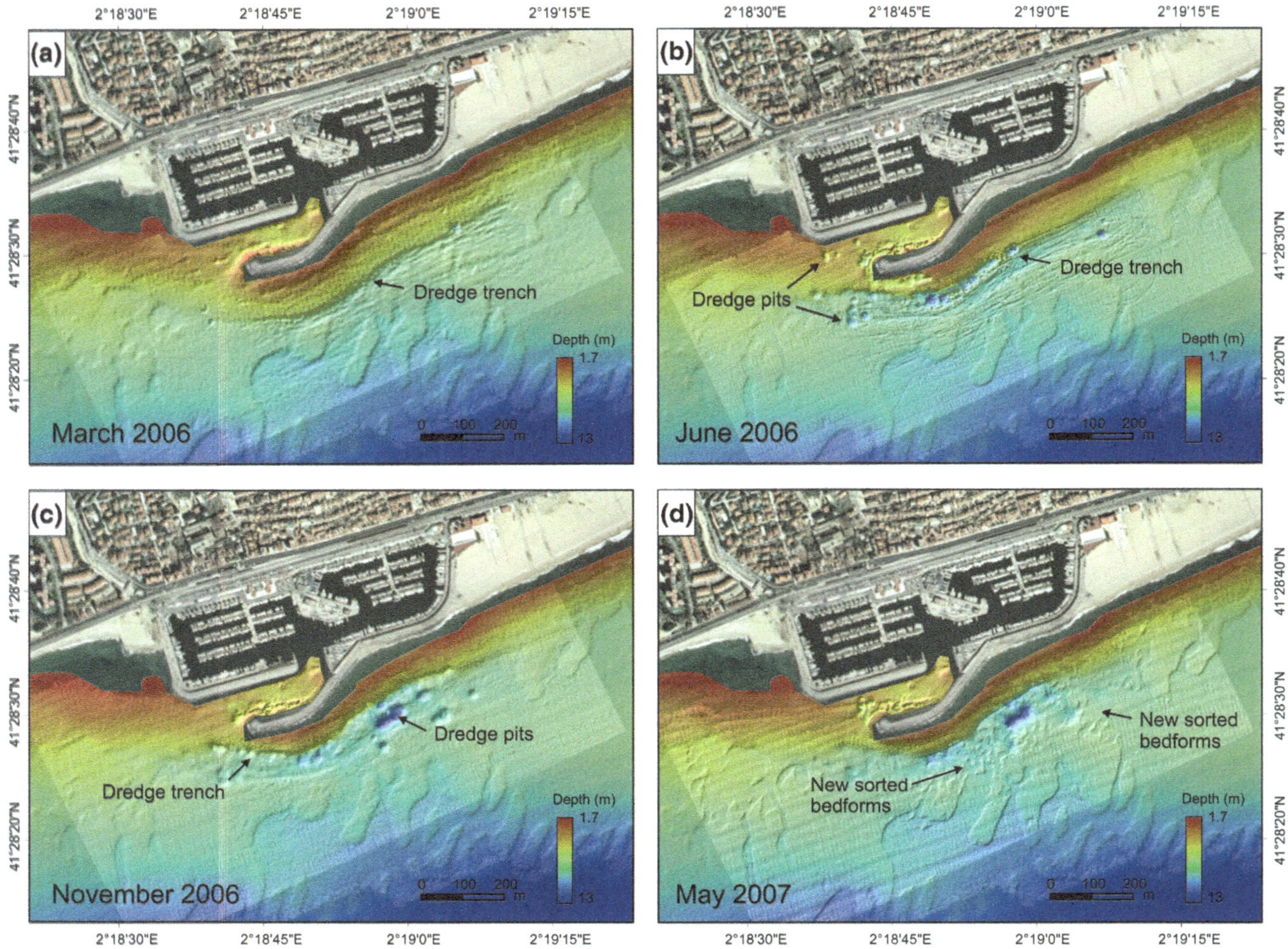

Fig. 22.5 Repeated multibeam bathymetries (0.5 × 0.5 m grid) over 1 year in the sorted bedform area adjacent to El Masnou harbour: **a** March 2006, **b** June 2006, **c** November 2006, and **d** May 2007. Note the impact of the sand extraction in the shallowest sorted bedforms and the formation of new features a few months after cessation of dredging activities

low water is in agreement with previous works that observed temporally intermittent sorted bedforms that appear and disappear at water depths shallower than 10–15 m as a result of changes in the forcing conditions (Hume et al. 2003; Ferrini and Flood 2005; Coco et al. 2007; Trembanis and Hume 2011). Numerical modelling supports these observations and proposes that bedforms subject to high wave climates may be temporally intermittent features as a result of increased wave orbital velocity during storms (Goldstein et al. 2011).

Acknowledgments This work was funded by the European BEACHMED-e project, an INTERREG IIIC Regional Framework Operation, and the FORMED project (CGL2012-33989) funded by the Spanish Ministry of Economy and Competitiveness. The authors wish to thank the Secretaría General de Pesca and Tragsa for the 2009 Espace Project dataset. R. Durán is supported by a CSIC JAE-Doc contract co-funded by the FSE.

References

Coco, G., Murray, A. B., Green, M.O., Thieler, E.R., Hume, T.M. (2007). Sorted bed forms as self-organized patterns: 2. Complex forcing scenarios. J. Geophys. Res. 112, F03016. doi:10.1029/2006JF000666.

De Falco, G., Budillon, F., Conforti, A., Di Bitetto, M., Di Martino, G., Innangi, S., Simeone, S., Tonielli, R. (2015). Sorted bedforms over transgressive deposits along the continental shelf of western Sardinia (Mediterranean Sea). Mar. Geol. 359, 75–88. doi:10.1016/j.margeo.2014.11.008.

DGPC (Direcció General de Protecció Civil). (1986). Investigación tecnológica de las acciones a tomar para la estabilidad de las playas del Maresme. Generalitat de Catalunya, Barcelona, Spain.

Diesing, M., Kubicki, A., Winter, C., Schwarzer, K. (2006). Decadal scale stability of sorted bedforms, German Bight, southeastern North Sea. Cont. Shelf Res. 26, 902–916. doi:10.1016/j.csr.2006.02.009.

Durán, R., Nuez, M., Alonso, B., Ercilla, G., Estrada F., CasasD., Farrán M. (2008). Assessment of sand trapped by coastal structures towards better management. El Masnou (Maresme, Catalunya). Geotemas 10, 511–514.

Durán, R., Canals, M., Lastras, G., Micallef, A., Amblas, D., Pedrosa-Pàmies, R., Sanz, J.L. (2013). Sediment dynamics and post-glacial evolution of the continental shelf around the Blanes submarine canyon head (NW Mediterranean). Prog. Oceanogr. 118, 28–46. doi:10.1016/j.pocean.2013.07.031.

Ercilla, G., Estrada, F., Casas, D., Durán, R., Nuez, M., Alonso, B., Farrán, M. (2010). The El Masnou infralittoral sedimentary environment (Barcelona province, NW Mediterranean Sea): morphology and Holocene seismic stratigraphy. Sci. Mar. 74, 179–196. doi:10.3989/scimar.2010.74n1179.

Ferrini, V.L., Flood, R.D. (2005). A comparison of rippled scour depressions identified with multibeam sonar: evidence of sediment transport in inner shelf environments. Cont. Shelf Res. 25, 1979–1995. doi:10.1016/j.csr.2005.07.002.

Goff, J. A., L. A. Mayer, P. Traykovski, I. Buynevich, R. Wilkens, R. Raymond, G. Glang, R. Evans, L., Olson, H., Jenkins, C. (2005). Detailed investigation of sorted bedforms, or "rippled scour depressions", within the Martha's Vineyard Coastal Observatory, Massachusetts, Cont. Shelf Res. 25, 461–484. doi:10.1016/j.csr.2004.09.019.

Goldstein, E.B., Murray, A. B., Coco, G. (2011). Sorted bedform pattern evolution: persistence, destruction and self-organized intermittency. Geophys. Res. Lett. 38, L24402. doi:10.1029/2011GL049732.

Hume, T. M., Trembanis, A. C., Hill, A., Liefting, R., Stephens, S. (2003). Spatially variable, temporally stable, sedimentary facies on an energetic inner shelf, in Coastal Sediments'03 edited by A. Press, Am. Soc. Civ. Eng., Reston, Va. 14 pp.

Liquete, C., Canals, M., Lastras, G., Amblas, D., Urgeles, R., De Mol, B., De Batist, M., Hughes-Clarke, J.E. (2007). Long-term development and current status of the Barcelona continental shelf: a source-to-sink approach. Cont. Shelf Res. 27, 1779–1800. doi:10.1016/j.csr.2007.02.007.

Lo Iacono, C., Guillén, J. (2008). Environmental conditions for gravelly and pebbly dunes and sorted bedforms on a moderate-energy inner shelf (Marettimo Island, Italy, western Mediterranean). Cont. Shelf Res. 28, 245–256. doi:10.1016/j.csr.2007.08.005.

Murray, A.B., Thieler, E.R. (2004). A new hypothesis and exploratory model for the formation of large-scale inner-shelf sediment sorting and "rippled scour depressions." Cont. Shelf Res. 24, 295–315. doi:10.1016/j.csr.2003.11.001.

Trembanis, A.C., Hume, T.M. (2011). Sorted bedforms on the inner shelf off northeastern New Zealand: spatiotemporal relationships and potential paleo-environmental implications. Geo-Marine Lett. 31, 203–214. doi:10.1007/s00367-010-0225-8.

23 Sorted Bedforms Along the Continental Shelf of Western Sardinia

Giovanni De Falco, Francesca Budillon, Alessandro Conforti, Gabriella Di Martino, Sara Innangi, Simone Simeone, and Renato Tonielli

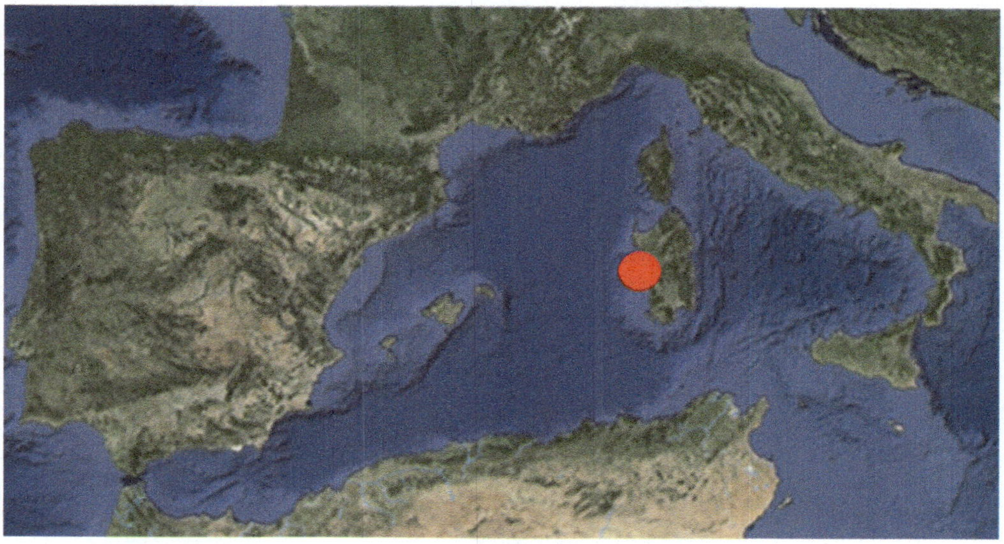

Abstract

Two sorted bedforms fields located at the western Sardinian margin are described. Bedforms are developed at the edge of prograding sedimentary wedges related to the last sea-level rise and in a small basin of the inner shelf, surrounded by rocky outcrops. The mechanisms of formation and evolution are discussed in relation with sea-level oscillations, hydrodynamic forcing and complex seabed morphology.

Keywords

Sorted bedforms • Prograding wedges • Multibeam • Backscatter • Wave hydrodynamics

23.1 Introduction

Sorted bedforms are a sequence of low ridges and depressions of alternating finer and coarser-grained sediments with the latter located in the depressions and on the sides of the bedforms facing into the mean flow direction (Murray and Thieler 2004). The formation of sorted bedforms was related to the presence of large wave-generated ripples that form in the coarse sediments. The interaction of wave motion with the rough seafloor generates a near-bed turbulence, with a winnowing of fine material and the formation of scoured depressions. The roughness reduction in the sandy patches favors finer-grained sediment deposition with a feedback between sediment grain size at the seabed and the fast changing morphology which allows for the formation of

G. De Falco (✉) · A. Conforti · S. Simeone
Istituto per l'Ambiente Marino Costiero del CNR, Località Sa Mardini, 09170 Torregrande Oristano, Italy
e-mail: giovanni.defalco@cnr.it

F. Budillon · G. Di Martino · S. Innangi · R. Tonielli
Istituto per l'Ambiente Marino Costiero del CNR, Calata Porta di Massa, 80, 80133 Naples, Italy

J. Guillén et al. (eds.), *Atlas of Bedforms in the Western Mediterranean*,
DOI 10.1007/978-3-319-33940-5_23

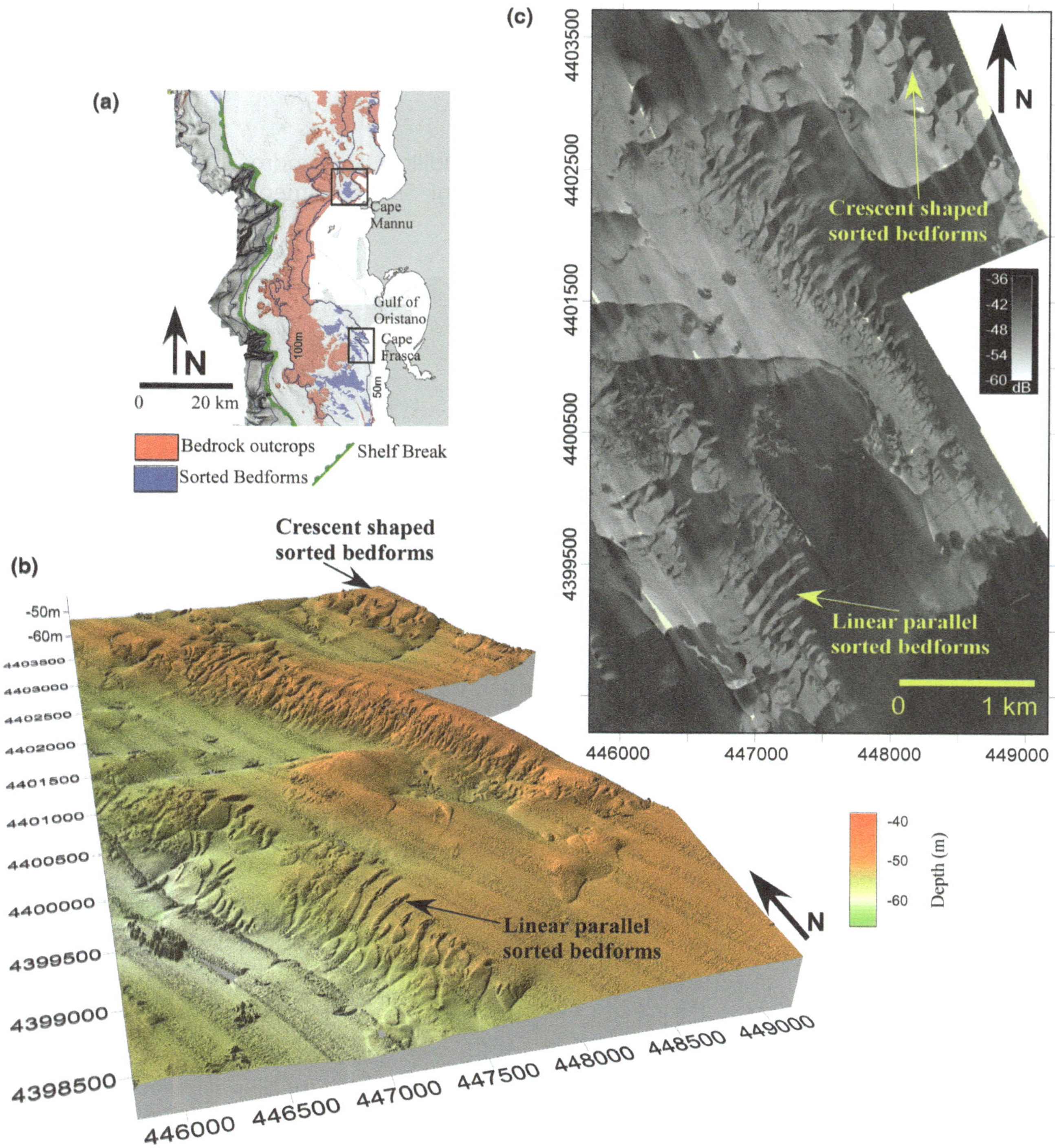

Fig. 23.1 **a** Location of the sorted bedforms along the shelf of central-western margin of Sardinia. **b** Oblique three-dimensional view of the digital seafloor terrain model off the Gulf of Oristano-Cape Frasca area (see **a** for location). It shows the terraces and associated linear and crescent-shaped sorted bedforms. **c** Backscatter intensity map of the same sector

sorted bedforms in a self-organized pattern (Murray and Thieler 2004; Coco et al. 2007).

Sorted bedforms have been found along the shelf of the central sector of the western margin of Sardinia (39°34′ N to 40°16′ N; 008°14′ E to 008°26′ E, western Mediterranean). The morphology of the region is mainly controlled by tectonic features. Along the mid-shelf, Paleozoic-Early Miocene basement rocks outcrop. Along the inner shelf small

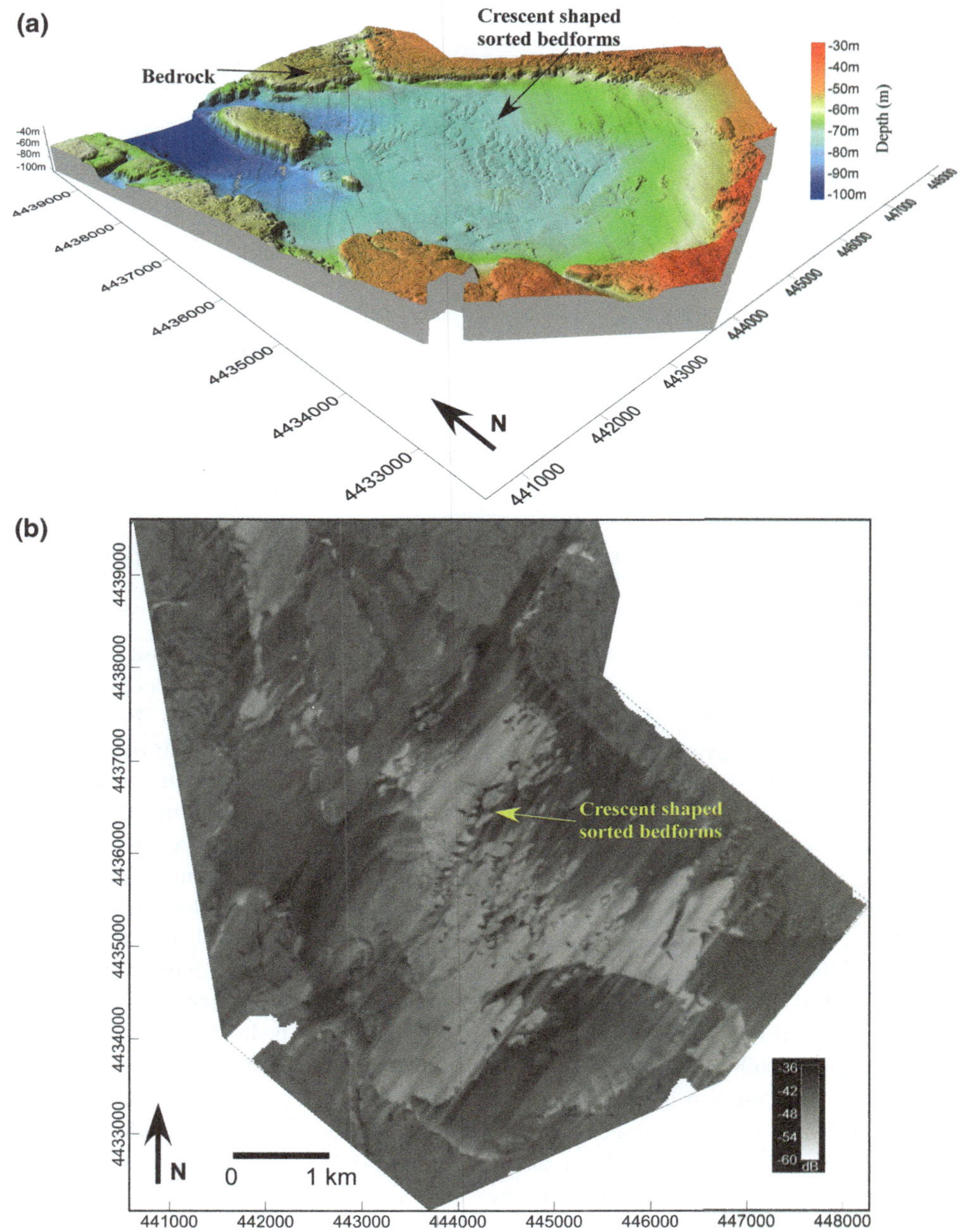

Fig. 23.2 **a** Oblique three-dimensional view of the digital seafloor terrain model off Cape Mannu (see Fig. 23.1a for location) showing the basin surrounded by rock outcrops where crescent-shaped sorted bedforms occur. **b** Backscatter intensity map of the same sector

basins filled by Pliocene sequences occur, whereas in the outer shelf the basement is overlain by Pliocene and Quaternary sequences (Conforti et al. 2015). Transgressive deposits associated with the last sea level rise are characterized by a thin layer of coarse bioclastic sediments over the bedrock in the depth range of 35–90 m, while the outer shelf is covered by clays and calcareous mud. In addition, sedimentary wedges composed of gravelly siliciclastic sands and calcareous bioclastic sands line the inner shelf (De Falco et al. 2015).

The dominant wind of the area is the Mistral, which blows from the northwestern sector (between 300° and 315°, nautical convention) (Cucco et al. 2006). Typical Mistral storm winds have a return period of 1 year and produce a JONSWAP-type wave spectrum characterized by significant wave heights (WHS) of ~8 m, peak periods of 10 s and a predominant wave approach from 308° (northwest). By comparison, storms with a return period of 10 years are characterized by a WHS of 10 m and a peak period 11.5 s.

Table 23.1 Morphological description and depth of bedform occurrence

Bedforms				Terraces	
Morphology		Depth (m)	Crest orientation	Depth (edge/base) (m)	Slope (°)
Sector 1—Cape Frasca–Gulf of Oristano					
Few irregular shaped dunes		83–87		83 ÷ 100	1.7
Sorted bedforms	Crescent shaped, partly oriented	73–80		73 ÷ 83	2.7
	Crescent shaped, parallel, oriented	67–70	0°–30°	67 ÷ 77	0.5
	Linear, parallel, oriented	52–60	50°–70°	52 ÷ 63	1.3
	Linear, parallel, oriented	48–55	20°–50°	48 ÷ 55	1.3
	Linear, parallel, oriented	43–45	50°–60°	43 ÷ 47	1.5
Isolated dunes		35–40			
Sector 2—Cape Mannu					
Sorted bedforms	Crescent shaped, partly oriented	65–75	50°–110°	No terraces	

Listed are the depth ranges (edge/base) and the slope gradients at the border of the terraces where the bedforms occur

23.2 Methods

Multibeam echosounder (MBES) data were collected throughout the western margin of Sardinia during two oceanographic cruises in 2009 and 2011 using the R/V Maria Grazia and Urania belonging to the Italian National Research Council (CNR). Data were acquired using a Kongsberg EM 3002D (293–307 kHz) MBES over a depth range of 50–70 m, providing a 200° swath coverage with 1.5° beam widths having a vertical resolution of 1 cm. The deeper sectors (70 to 700–800 m) were mapped using a Kongsberg EM 710 MBES, which operates at a frequency of 100 kHz and provides a 150° swath coverage with 0.5° beam widths. Both MBES devices also provided backscatter data which were used to interpret the characteristics of the seabed. All data were processed using the software Caris Hips and Sips. Very high-resolution seismic data and sediment samples were also available from the study area (De Falco et al. 2015).

23.3 Results

Two sectors (Gulf of Oristano-Cape Frasca and Cape Mannu, Fig. 23.1a), where sorted bedforms were recorded, are described in Figs. 23.1 and 23.2. In the sector located off the Gulf of Oristano-Cape Frasca (Fig. 23.1b), sorted bedforms occur at the edge of depositional terraces located at different depths between 43 and 83 m (Table 23.1) (De Falco et al. 2015). The sorted bedforms are either linear or crescent-shaped and occur in the depth range of 43–70 m. They comprise a sequence of slight depressions alternating with flat-topped, <1 m high elongated bodies spaced 70–200 m apart (Fig. 23.1b).

The analysis of the acoustic backscatter shows that the bedforms are located at the landward limits and on top of large sedimentary bodies that are characterized by high backscatter intensity (∼−36 dB, dark areas in Fig. 23.1c). The slightly elevated bedforms, by contrast, are characterized by low backscatter intensity (∼−48 dB, light areas in Fig. 23.1c). The depressions between the elevated bedforms, in turn, have backscatter intensities lying between those of the larger sedimentary bodies and the superimposed elevated bedforms (∼−39 dB). At shallow depth bedforms progressively disappear. The shallower linear and oriented bedforms are found at 43–45 m depth (Table 23.1), the latter consisting of smaller crescent-shaped morphologies about 0.5 m high and spaced about 60 m apart.

In the Capo Mannu sector (Fig. 23.2, see Fig. 23.1a for the sector location), sorted bedforms were observed on a small basin at a depth of 65–75 m. The basin is surrounded by tabular rock outcrops, the edges of which are located at depths of 30–50 m. The sorted bedforms are crescent-shaped, <1 m high and spaced 100–150 m apart. They exhibit a wide variation in crest orientation. The acoustic backscatter corresponds to that described for the Gulf of Oristano-Cape Frasca area (Figs. 23.1c and 23.2b).

The sediment grain-sizes range from gravels to muddy sands, the sediment being composed of mixed siliciclastic/carbonate material (De Falco et al. 2015). Gravelly sands are associated with areas of higher backscatter intensity, whereas the superimposed bedforms consisting of fine-medium sands are characterized by lower backscatter intensity. The depressions between the superimposed bedforms are also composed of gravelly sands.

A comparison of an MBES backscatter map acquired in 2007 with a side-scan sonar mosaic recorded in 2000 confirms that the sorted bedforms are active under the present hydrodynamic conditions. Distinct differences between the two maps document that the coarse-grained sediment body at a depth of 45 m is overlain by a mobile of sand carpet with a rate of migration of bedforms up to 100 m in 7 years (De Falco et al. 2015).

23.4 Discussion and Conclusions

The results demonstrate that near-bottom shear stresses related to storm waves forced by northwesterly winds exceed the threshold values for fine and medium sands motion. Storm waves propagating from the NW can thus be considered as the main hydrodynamic forcing factor of sorted bedforms of the west Sardinian shelf.

In the Gulf of Oristano-Cape Frasca sector, sorted bedforms are persistent features mainly subjected to regular hydrodynamic forcing from NW waves at depths ranging from 43 to 70 m. As sorted bedforms were not found in shallower water, it must be assumed that these are here probably ephemeral features that are rapidly destroyed by multi-directional waves and currents.

Sorted bedforms were found at the edge of prograding sedimentary wedges related to the last sea-level rise. The mechanisms of formation and evolution include the following phases: (i) wedge formation with winnowing of coarse-grained and fine-grained sediments between the top and the edge of the terrace; (ii) drowning of deeper wedges and subsequent deposition of shallower wedges associated with the landward migration of the shoreline in response to pulses of sea-level rise; (iii) formation of persistent sorted bedforms over the edges of the terraces after sea level rose by at least 40 m above the seafloor (De Falco et al. 2015). At a lower sea level the hydrodynamic conditions were probably more variable, and a regular pattern of sorted bedforms was unlikely to occur. With the increase of depth (in the range 40–70 m), the storm waves from the North West constituted a more regular forcing which can be related to the development of regular sorted bedforms morphologies.

In the sector of Cape Mannu the hydrodynamics at the seabed is conditioned be the morphology of the surrounding rock outcrops, the irregular pattern of the sorted bedforms reflecting the complexity of the hydrodynamic forcing in this environment.

In conclusion, the morphology of the seabed, characterized by the presence of depositional terraces and rocky outcrops, affects the hydrodynamic forcing in the sectors of the Gulf of Oristano-Capo Frasca and Capo Mannu, thus conditioning the morphology and distribution of sorted bedforms.

Acknowledgments This study was funded by the RITMARE Flagship Project funded by the Italian Ministry of University and Research, and MAGIC project (Marine Geo-Hazard along the Italian coast) funded by the Italian Civil Protection Department.

References

Coco, G., Murray, A.B., Green, M.O. (2007). Sorted bedforms as self-organized patterns: 1. Model development. Journal of Geophysical Research 112, F03015.

Conforti, A., Budillon, F., Tonielli, R., De Falco, G. (2015). A newly discovered Pliocene volcanic field on the western Sardinia continental margin (western Mediterranean). Geo-Marine Letters DOI 10.1007/s00367-015-0428-0.

Cucco, A., Perilli, A., De Falco, G., Ghezzo, M., Umgiesser, G. (2006). Water Circulation and Transport Time Scales in the Gulf of Oristano. Chemistry and Ecology 22 (supplement 1), 307–331.

De Falco, G., Budillon, F., Conforti, A., Di Martino, G., Di Bitetto, M., Innangi, S., Simeone, S., Tonielli, R. (2015). Sorted bedforms over transgressive deposits along the continental shelf of western Sardinia (Mediterranean Sea). Marine Geology 359, 75–88.

Murray, A.B., Thieler, E.R. (2004). A new hypothesis and exploratory model for the formation of large-scale inner-shelf sediment sorting and "rippled scour depressions". Continental Shelf Research 24, 295–315.

Part IV

Bedforms in the Continental Shelf

Sedimentary Bedforms in the Menorca Channel Region, Balearic Islands (Western Mediterranean)

24

María Druet, Juan Acosta, Araceli Muñoz, Carmen Barberá, Joan Moranta, José Manuel Jódar, and Natalia Martínez-Carreño

Abstract

A variety of sedimentary bedforms have been identified on the continental shelf between Mallorca and Menorca (Menorca Channel). The bedforms consist of different grain sizes, ranging from very fine to coarse, predominantly bioclastic sands, and include rippled scour depressions in association with fine sand patches, sand waves, sand ridges, ribbons and furrows. The morphology and acoustic appearance of these bedforms are documented in a number of maps generated on the basis of detailed multibeam bathymetric datasets, side-scan sonar images and high-resolution reflection seismic profiles (TOPAS).

M. Druet (✉)
Instituto Geológico y Minero de España, C/La Calera 1, 28760 Tres Cantos, Madrid, Spain
e-mail: m.druet@igme.es

M. Druet · J. Acosta
Instituto Español de Oceanografía, C/Corazón de María nº8, 28002 Madrid, Spain

A. Muñoz
TRAGSA, C/Valentín Beato nº6, 28037 Madrid, Spain

C. Barberá · J. Moranta
Centre Oceanogràfic de les Balears, Instituto Español de Oceanografía, Moll de Ponent, s/n., 07015 Palma de Mallorca, Spain

J.M. Jódar · N. Martínez-Carreño
TECMARIN 1990, C/Laredo 500. Urb. Calipo-Fado, 45950 Casarribios del Monte, Toledo, Spain

J. Guillén et al. (eds.), *Atlas of Bedforms in the Western Mediterranean*,
DOI 10.1007/978-3-319-33940-5_24

Keywords
Menorca channel • Sand patches (rippled scour depressions) • Sand waves • Sand ridges • Sand ribbons • Furrows

24.1 Study Area

The Menorca Channel is the passage separating the islands of Mallorca and Menorca on the Balearic Promontory (Fig. 24.1) (Canals and Ballesteros 1997; Acosta et al. 2002). This area displays a regular relief and gentle slope, and is affected by an intense hydrodynamic regime characterized by strong currents induced by northerly winds (Pinot et al. 2002). Although there are no detailed hydrodynamic data available from this area, it is well known that the hydrodynamics of the region is quite variable, being affected by wind-driven circulation induced by the Tramontane (N) and Mistral (NW) winds, the islands orography, and the presence of the channel itself (Ordines et al. 2011).

As sediment input to the Menorca Channel from the Spanish main land is negligible, the sediments on the seafloor of this area are mainly derived from local coastal erosion and island runoff, as well as variable carbonate production by different benthic communities (Barberá et al. 2012). The sediments are mainly composed of variably sized sands with high carbonate contents (>60 %) (Alonso et al. 1988).

24.2 Methods

A multibeam bathymetric dataset was acquired on board the R/V Emma Bardán within the framework of the CALMEN project (2007–2008), using a Kongsberg EM3002

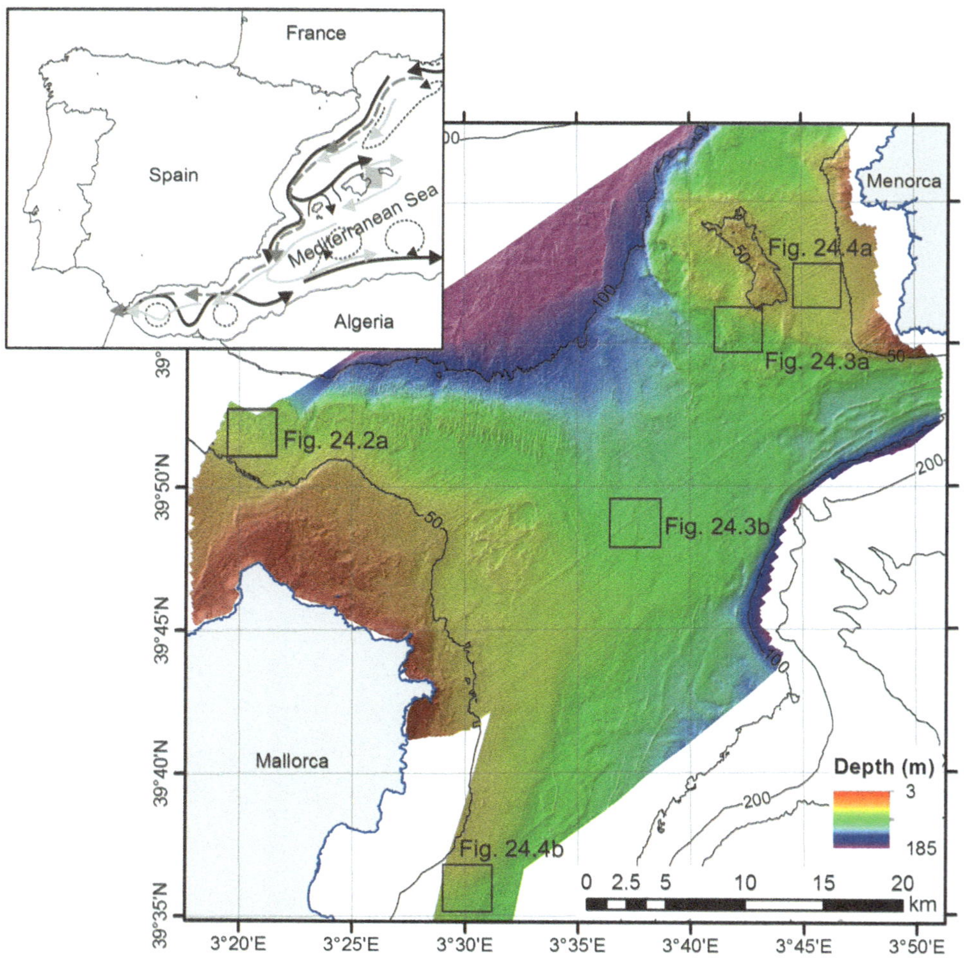

Fig. 24.1 Digital terrain model (DTM) of the Menorca Channel (illumination from the NW), including some bathymetric contours. Note the locations of Figs. 24.2, 24.3 and 24.4. *Inset* Map showing the geographic location of the study area (*grey rectangle*) and the general circulation pattern in this region of the Western Mediterranean (Vargas et al. 2010). *Black arrows* indicate the shallow Atlantic water inflow; *dashed grey arrows* indicate the intermediate Levantine waters; *light grey arrows* show the deep-water circulation; the *thin dashed black arrows* indicate possible eddies detached from the Algerian current

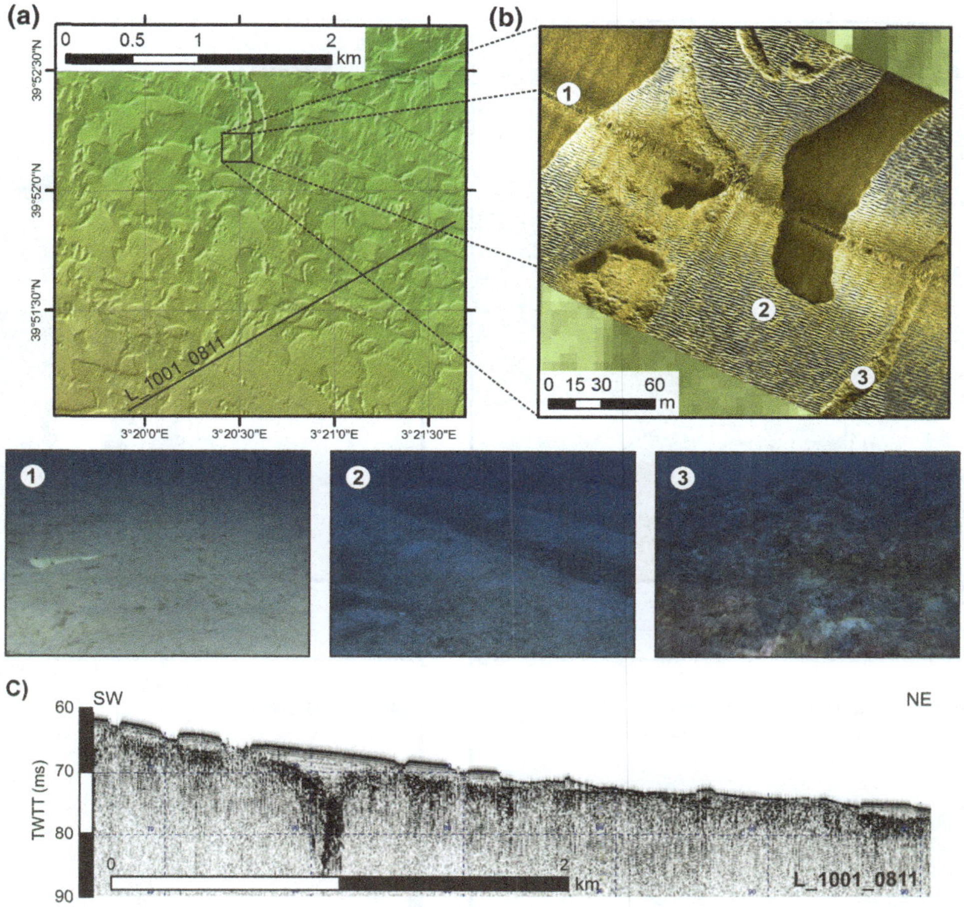

Fig. 24.2 **a** DTM detail of the sand patch area located to the NE of Mallorca (for location see Fig. 24.1). **b** Blow-up of a side-scan sonar image showing details of rippled scour depressions. ROV image *1* illustrates the existence of smaller bedforms. ROV image *2* shows sediment waves of 1–5 m in wavelength and a few decimeters in height. ROV image *3* shows bioconstructions of coralline algae. **c** TOPAS profile crossing some sediment patches (location in **a**)

multibeam echosounder. The data were processed with the NEPTUNE software to obtain a 5*5 m regular grid. Bottom sediment samples were collected during the same cruises by means of a Shipek grab sampler.

Sonographs were obtained in 2009 during the CANAL 0209 survey on board the R/V MarViva Med using a Klein Hydroscan 3000 side-scan sonar system. Data were processed with the TEI Suite to generate mosaics at 10*10 cm resolution. During this cruise sediment samples were collected with a Van Veen grab.

Finally, high-resolution reflection seismic profiles were acquired in 2011 during the INDEMARES-CANAL DE MENORCA 0811 cruise on board the R/V Miguel Oliver using a TOPAS PS 018 subbottom profiling system. In addition, seabed photographs were acquired by Remotely Operated Vehicle (ROV) deployment.

24.3 Results and Discussion

Five different bedforms types were identified in the Menorca Channel region: rippled scour depressions (in association with fine sand patches), sand waves, sand ridges, sand ribbons and furrows.

Rippled scour depressions (Cacchione et al. 1984) associated with fine sand patches were identified north of Mallorca in depths ranging from 34 to 67 m (Fig. 24.2, for location see Fig. 24.1). The fine sand patches

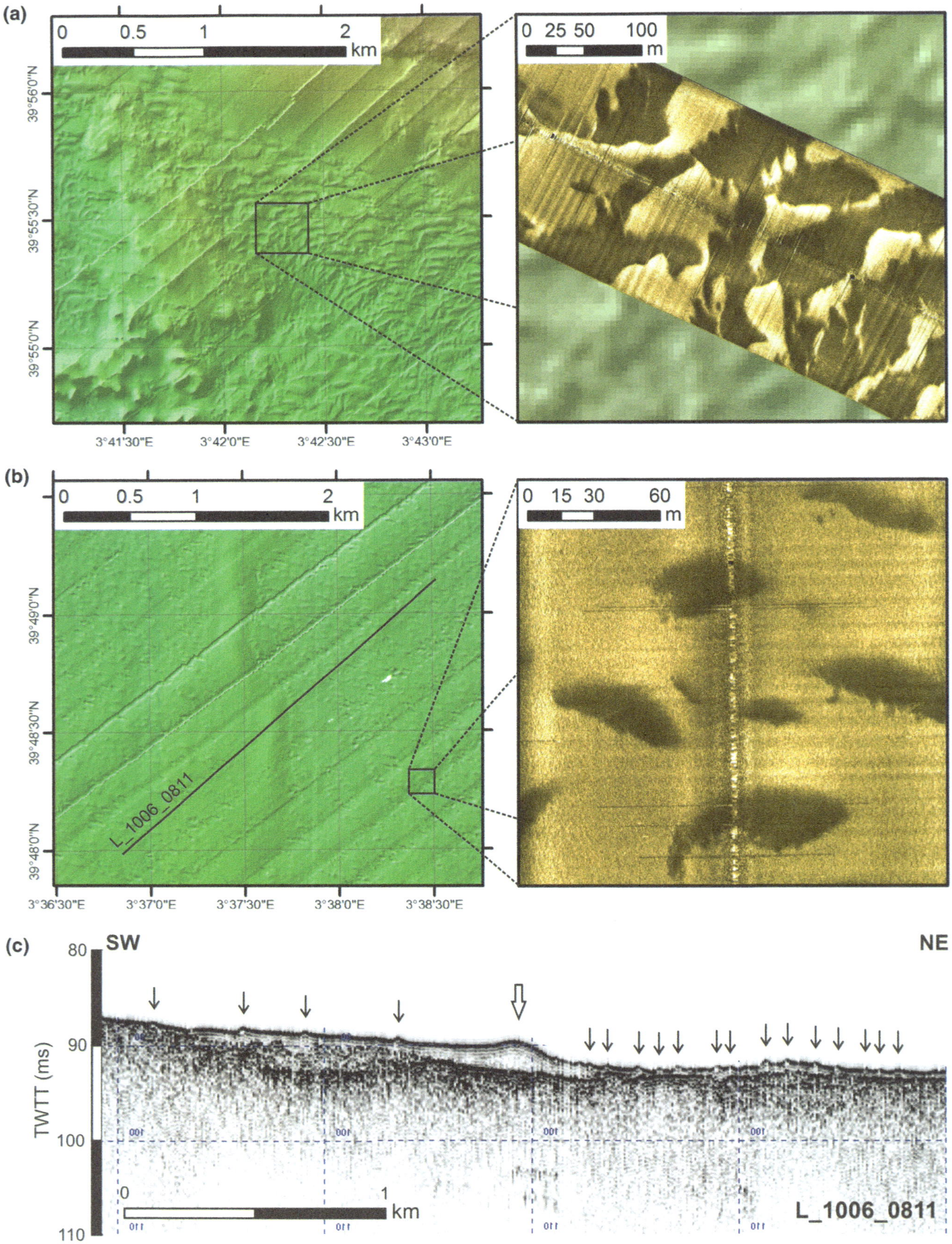
(a)
0 0.5 1 2 km
0 25 50 100 m
39°56'0"N
39°55'30"N
39°55'0"N
3°41'30"E
3°42'0"E
3°42'30"E
3°43'0"E
(b)
0 0.5 1 2 km
0 15 30 60 m
39°49'0"N
39°48'30"N
39°48'0"N
L_1006_0811
3°36'30"E
3°37'0"E
3°37'30"E
3°38'0"E
3°38'30"E
(c)
SW
NE
80
90
100
110
TWTT (ms)
0 1 km
L_1006_0811

◀ **Fig. 24.3** **a** Detail of the DTM showing irregularly shaped sand waves west of Menorca. The blow-up displays a side-scan sonar image of these sand waves. **b** Detail of the DTM showing sand waves in the center of the Menorca Channel, as well as smooth sand ridges. The blow-up illustrates the shape of the elongate sand waves, which are oriented in two different directions (WNW and ENE). **c** TOPAS profile across a sand ridge (*white arrow*). Small sand waves are also shown (*black arrows*; for location see panel **b**; locations of panels **a** and **b** are shown in Fig. 24.1)

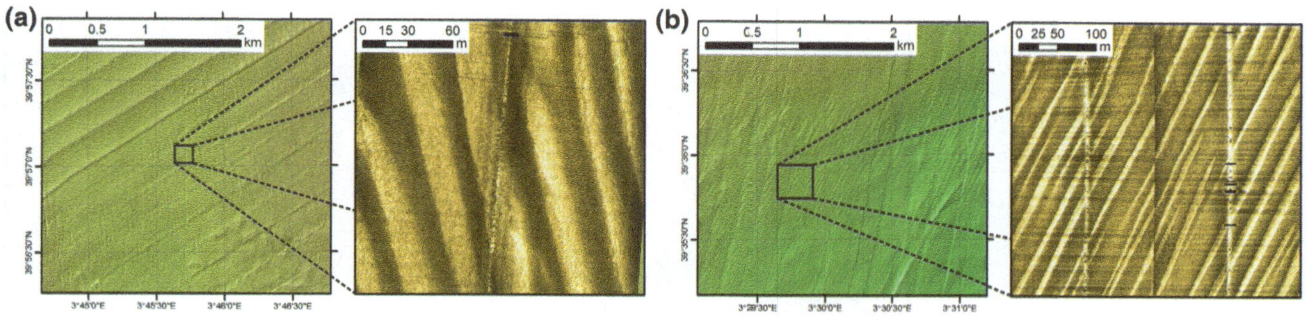

Fig. 24.4 Detail of the DTM showing elongated bedforms of both depositional (sand ribbons, **a**) and erosional (sand furrows, **b**) character. Blow-ups display side-scan sonar images of the two features (for location see Fig. 24.1)

characteristically have irregular shapes and almost flat tops with smooth, low backscatter surfaces. The horizontal dimensions of the sand patches and the interspersed rippled scour depressions are variable, ranging from tens to hundreds of meters in width, the depths of the depressions commonly ranging from 0.5 to 1 m. The sediment patches are composed of very fine sand, whereas the rippled scour depressions are composed of coarse bioclastic sand displaying ripple trains. The ripples have elongated crests, which persist laterally from a few meters to several tens of meters, crest spacing of ~2 m, and heights from a few centimeters to about 0.15 m (see side-scan sonar mosaic in panel (B) and ROV image 2 in Fig. 24.2). The rippled scour depressions frequently coexist with bioconstructions of coralline algae (coralligenous platforms, see ROV image 3 in Fig. 24.2).

Extensive fields of *sand waves* are mainly located in the center and to the southeast of the Menorca Channel. The most abundant type are regularly shaped symmetric sand waves composed of coarse sand located in the center of the Menorca Channel (Fig. 24.3b, see Fig. 24.1 for location). They have heights under 0.3 m and wavelengths ranging from 10 to 40 m. The small height/length ratio and the short lateral continuity of these bedforms suggest they are detached or isolated morphological features, probably reflecting sediment starvation in this remote area. The crests of the sand waves are mainly oriented WNW-ESE and WSW-ENE. In other areas such as southwest of Menorca, sand waves are irregularly shaped and composed of medium to coarse sand, with heights under 0.5 m and a tendency for crests to coalesce (Fig. 24.3a, see Fig. 24.1 for location).

Seven isolated elongate *sand ridges* with smooth surfaces were identified to the southwest of the center of the Menorca Channel. They are surrounded by sand waves of the type described above, are 1.5–3 m high and have crest lengths of 500–3200 m oriented N-S (Fig. 24.3b, c). Although no sediment samples are available from these sand ridges, the similarity of the acoustic backscatter intensity to that of the coarse-grained sand waves (also see seismic profile in Fig. 24.2c) suggests that they are composed of similar sediment.

Sand ribbons were identified close to Menorca at depths ranging from 44 to 55 m, and also in the center of the channel at depths up to 85 m (Fig. 24.4a). They are oriented NNW-SSE, are composed of medium to fine sand, are less than 0.1 m high, have a mean wavelength of 30 m, and crest lengths extending for hundreds of meters.

South of the survey area and fairly close to the eastern shore of Mallorca, *longitudinal furrows* of distinctly erosional character are developed in medium to coarse sand (Fig. 24.4b). They have U-shaped transverse profiles, are hundreds of meters in length with a NNE-SSW orientation, and have heights ranging from 0.05 to 0.1 m.

The high diversity of bedforms developed in mixed siliciclastic/bioclastic sediment on the continental shelf of

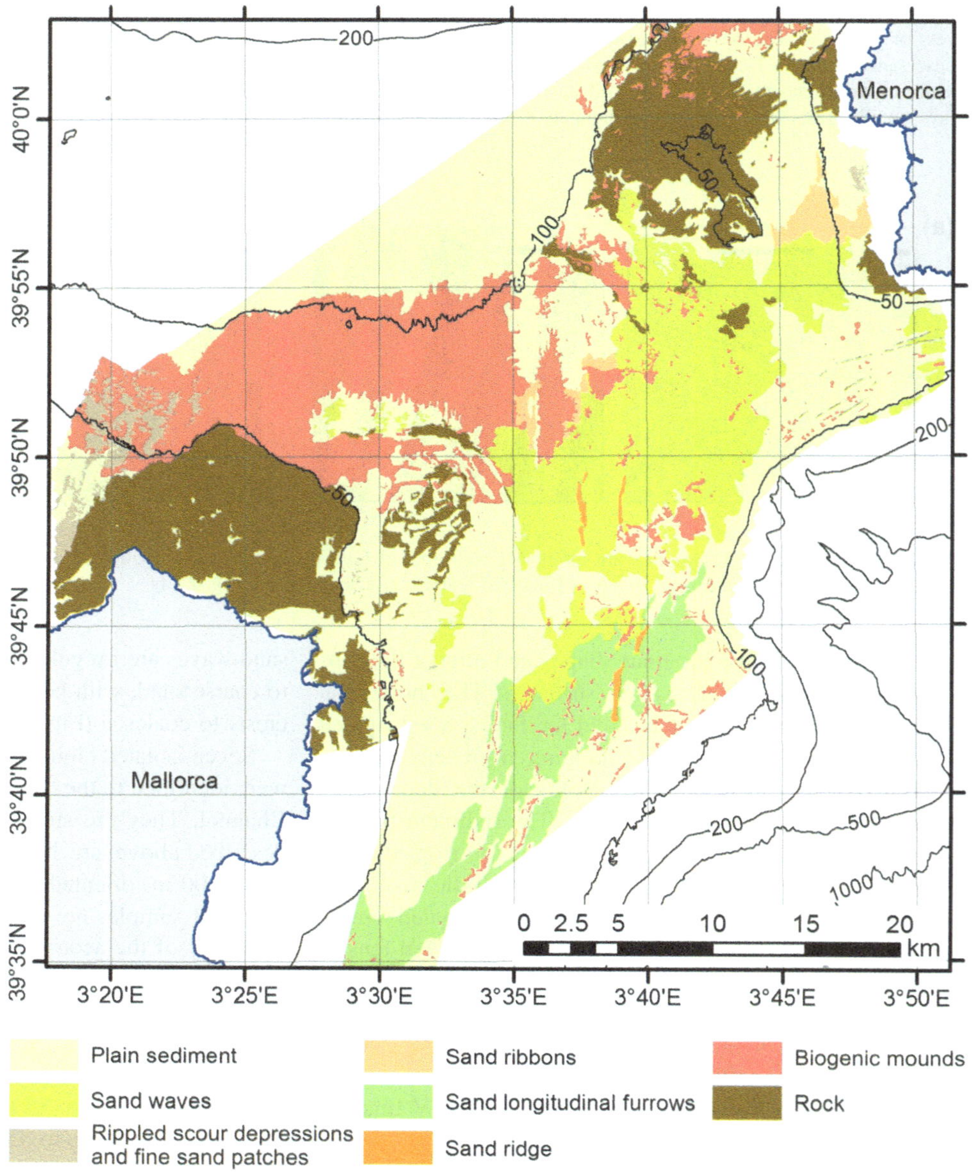

Fig. 24.5 Map showing the distribution of bedforms recorded in the Menorca Channel. Note the bathymetric contours for depth orientation

the Menorca Channel is described here in detail for the first time. The areal distribution of the various bedforms is shown in Fig. 24.5.

The low sediment input may be a determinant factor for the configuration of the observed sand waves and the development of sand ribbons. Further investigations on the influence of sediment input and the hydrodynamics at the seafloor are needed to explain the complex distribution patterns of bedforms in the Menorca Channel.

Acknowledgments We want to thank INDEMARES-CANAL DE MENORCA working group for their cooperation during the development of the project. Spanish Fishing General Secretary gave up CALMEN multibeam bathymetry datasets to the INDEMARES Project. We are thankful for the suggestions made by Dr. J. Guillén and Prof. Dr. B. W. Flemming, which improved the quality of the final manuscript.

References

Acosta J, Canals M, López-Martínez J, Muñoz A, Herranz P, Urgeles R, Palomo C, Casamor JL (2002). The Balearic Promontory geomorphology (western Mediterranean): morphostructure and active processes. Geomorph 49:177–204.

Alonso B, Guillén J, Canals M, Serra J, Acosta J, Herranz P, Sanz JL, Calafat A, Catafau E (1988). Los sedimentos de la plataforma continental balear. ActaGeolHisp 23:185–196.

Barberá C, Moranta J, Ordines F, Ramón M, de Mesa A, Díaz-Valdés M, Grau AM, Massutí E (2012). Biodiversity and habitat mapping of Menorca Channel (western Mediterranean): implications for conservation. BiodivConserv 21:701–728.

Cacchione, DA, Drake, DE, Grant, WD, Tate, GB (1984). Rippled scour depression on the inner continental shelf off central California. J Sediment Petrol 54 (4): 1280–1291.
Canals M, Ballesteros E (1997). Production of carbonate particles by phytobenthic communities on the Mallorca-Menorca shelf, northwestern Mediterranean Sea. Deep-Sea Res II 44(3):611–629.
Ordines F, Jordà G, Quetglas A, Flexas M, Moranta J, Massutí E (2011). Connections between hydrodynamics, benthic landscape and associated fauna in the Balearic Islands, western Mediterranean. Cont Shelf Res 31:1835–1844.
Pinot JM, López-Jurado JL, Riera M (2002). The CANALES experiment (1996–1998). Interannual, seasonal, and mesoscale variability of the circulation in the Balearic Channels.ProgOceanogr 55:335–370.
Vargas M, García MC, Moya F, Tel E, Parrilla G, Plaza F, Lavín A, García MJ (2010). Cambio climático en el Mediterráneo español. Temas de oceanografía 3, Instituto Español de Oceanografía. 176 pp.

Dome Dunes on the Inner to Middle Shelf Transition on a Temperate-Water Carbonate Sediment Shelf. Pitiusas Islands

25

J. Rivera, L. Pomar, N. Hermida, G. Mateu, and J. Acosta

Abstract

Ibiza and Formentera, also known as the Pitiusas Islands, are the westernmost islands of the Balearic archipelago in the western Mediterranean Sea. Around them, typical temperate-water carbonate shelves occur. Particular sedimentary processes take place on these shelves, where the lack of a significant terrigenous sediment supply and a microtidal regime reinforces the importance of seagrass meadows and wind-driven coastal currents in the sediment dynamics. The dome dune field described here is placed just downslope of a centuries-old *Posidonia oceanica* meadow close to a shallow passage between two islands where wind-driven currents are particularly important.

Keywords

Sediment waves • Inner shelf • Carbonates • Sand waves • Dome dunes • Western Mediterranean • Internal waves

25.1 Introduction

To the east of Es Freus, the local name of the strait between Ibiza and Formentera islands (Fig. 25.1), islets and shallow passages alternate. The largest passage is Freu Gros, located between Es Penjats and S'Espalmador islets and its

J. Rivera (✉) · N. Hermida · J. Acosta
Instituto Español de Oceanografia, IEO, Corazón de María 8, 28002 Madrid, Spain
e-mail: jesus.rivera@md.ieo.es

L. Pomar · G. Mateu
Càtedra "Guillem Colom Casasnovas", Edifici Guillem Colom Casasnovas - Universitat de les illes Balears, Cra. de Valldemossa Km. 7, 5 - 07122 Palma de Mallorca, Spain

J. Guillén et al. (eds.), *Atlas of Bedforms in the Western Mediterranean*,
DOI 10.1007/978-3-319-33940-5_25

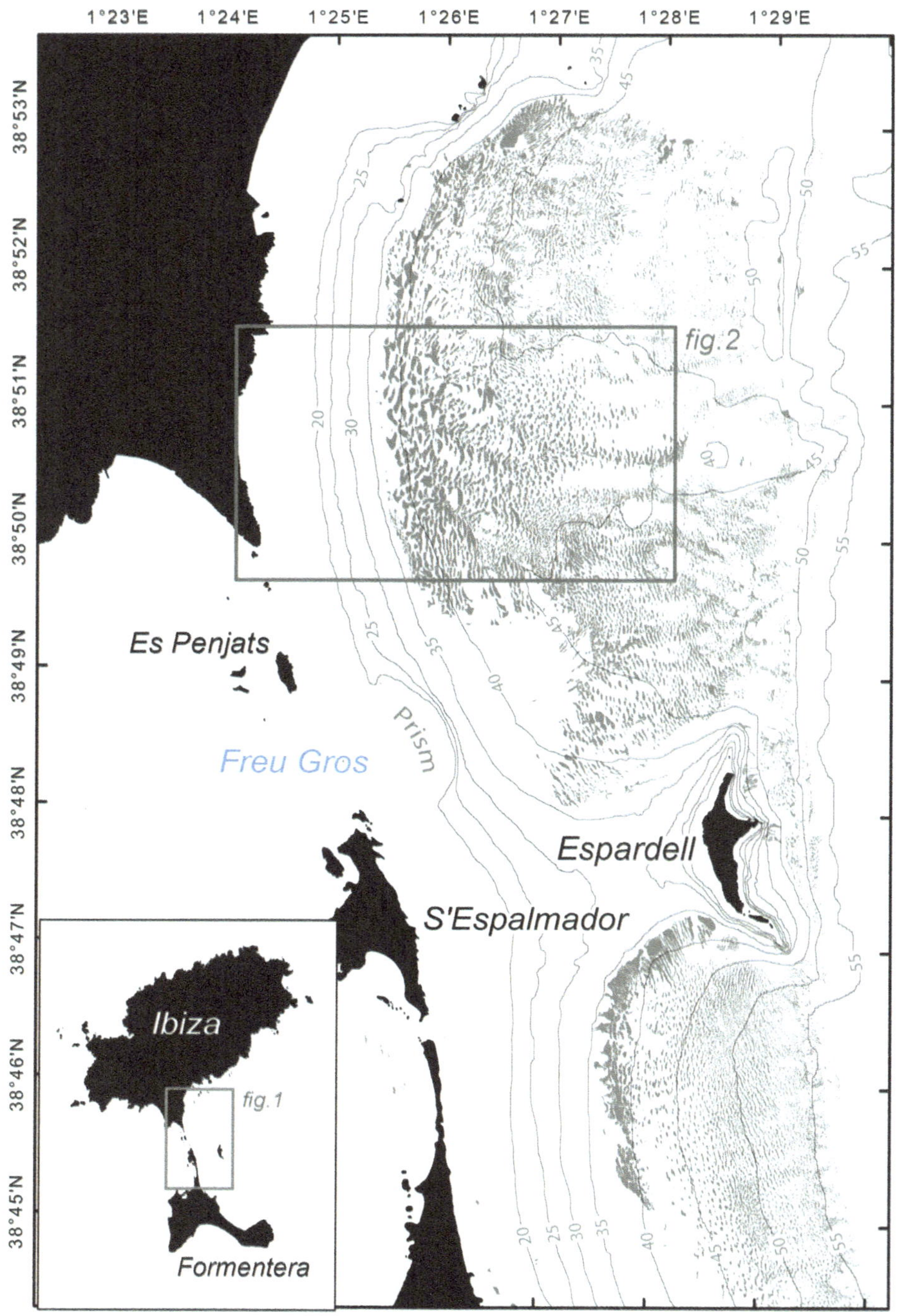

Fig. 25.1 Hummocky bedform field location. Ibiza and Formentera islands a.k.a. the Pitiusas Islands. Espardell islet divides the extension into two sectors, the north sector and the south sector. Silhouettes of each hummock have been digitized from backscatter information and displayed as *gray* polygons. Freu Gros, in *blue font*, is the main passage between the Pitiusas Islands. Figure extents are shown in *dark gray*

maximum depth is around 10 m. To the east of this passage, a small, fan-shaped sedimentary prism faces eastward, as evidenced in the bathymetric record (Fig. 25.1). To the east, below a gently dipping slope (<1°), a series of bedforms occur north and south of Espardell islet. This type of bedform is not exclusive to this site, as similar bedforms have been reported in the Menorca Channel (Druet et al. in this volume) and on the Mallorca southwestern middle shelf (Dominguez et al. 2013).

The dome dune field located to the east of Es Freus covers a total area of 88.4 km^2. It extends from 35 to 50 m water depth and is divided into two zones. The northern zone extends from Espardell islet to latitude 38°53'N, and the southern zone extends from Espardell islet to Formentera Island, reaching 38°53'N latitude (Fig. 25.1).

The main issues of the Balearic Sea regional circulation are well described (Mason and Pascual 2013; Millot 1999), but insights into coastal currents and local circulation are restricted to a few places. The fine-scale circulation studies that have been carried out show the importance of wind-driven currents in this micro tidal environment and their relation to island-trapped waves (Jordi et al. 2011 2009).

25.2 Methods

The data set presented here was gathered during two surveys conducted in March and September 2008 on board the R/V *Odon de Buen*. The aim of these surveys was to recognize and evaluate the environmental impact of a wreck sunk on 11 July 2007 one mile south of Ibiza harbour (DGMM 2009). The multibeam echosounder used in both surveys was a dual-head Kongsberg EM 3002D model. The operational frequency was 300 kHz and the maximum number of beams per ping was 508. Bathymetry and backscatter were processed using Caris H&S 9.1 software, resulting in a 2 × 2 m digital elevation model for the bathymetry and a digital terrain model for the seabed backscatter value. A digital dual frequency (100–500 kHz) Klein 3000 side scan sonar was used and data were processed using Sonar Wiz 5 software. Two digital terrain models represent the backscatter seabed value, one for each frequency. Real-time vessel position was obtained using a differential global positioning system. Differential correction was provided by Omni Star Satellites. The multibeam echosounder and side scan sonar data sets collected during the surveys represent the main body of material used in the description of these subaqueous dome dunes. In addition, published thematic cartography (MAGRAMA 2013) was also used out of the bounds of our datasets in order to improve the geographical frame. Data analysis, visualization and editing of figures was carried out using the IVS Fledermaus 7 and Esri ArcGIS 10.2 software packages.

25.3 Results

Subaqueous dome dunes (SDD) are low-relief, hummoky structures with an average height of 0.28 m and very low angle flanks (<1.7°). In plan view they have an elliptical silhouette. Size is variable, as it decreases with depth (Figs. 25.1 and 25.2c), but on average their area is about 632 m^2, corresponding to an 80-m-long major semi-axis and a 25-m-long minor semi-axis (Fig. 25.5), the major semi-axis being roughly parallel to the isobaths (Figs. 25.1 and 25.2).These bedforms are characterized by a low backscatter signal in contrast to the highly reflective surrounding seafloor, showing a sharp contour boundary between them (Figs. 25.2 and 25.3).

Characteristically, the dunes occur between 35 and 55 m on a seabed with an average slope of 0.60° and frequent 0.25° values (Fig. 25.4). The depth histogram shows that most of the bedforms are placed between 43 and 48 m water depth and values deeper than 48 m are more frequent than values shallower than 43 m (Fig. 25.4). This is because in the shallow zone the bedforms are larger, so their number is lower in relation to the area they cover, whereas in the deeper zone many small bedforms fit in the same equivalent area (Fig. 25.1). According to the orientation histogram, the hummock-like bedforms are placed on the seabed facing the east (86° average azimuth), the fact that they face east shows the general slope trend (Figs. 25.1 and 25.4).

In the deep zone of the SDD field, these bedforms are distributed in a meandering pattern (Figs. 25.1 and 25.2) that is more evident in the northern sector, where the bathymetry shows smooth elevations about one or two metres high and a few hundred metres wide. Winding corridors made up of small SSD surround these elevations (Fig. 25.1).

25.4 Discussion and Conclusions

To the west, upslope of the SDD field, a dense millenary seagrass meadow extends in the shallower areas all over Es Freus (Arnaud-Haond et al. 2012). This habitat plays a significant role in the sedimentary system as a locus for carbonate production and for accumulation due to baffling and sediment trapping (Hansen and Reidenbach 2012). The meadow retains unsorted coarse- to fine-grained sediments but only fine-grained sand to mud-grade sediments are shed off the meadows. In Balearic coastal waters, under certain wind conditions, island-trapped waves cause currents of up to 1 m/s which cause sediment resuspension (Jordi et al. 2009). Although precise knowledge of the genetic processes is still lacking, the shape and orientation of the SSD may suggest either a combination of upslope and downslope flows, or downdip bottom currents combined with oscillatory flows. The distribution patterns show a downdip decrease in the size of the bedforms and a progressive dwindling until they are

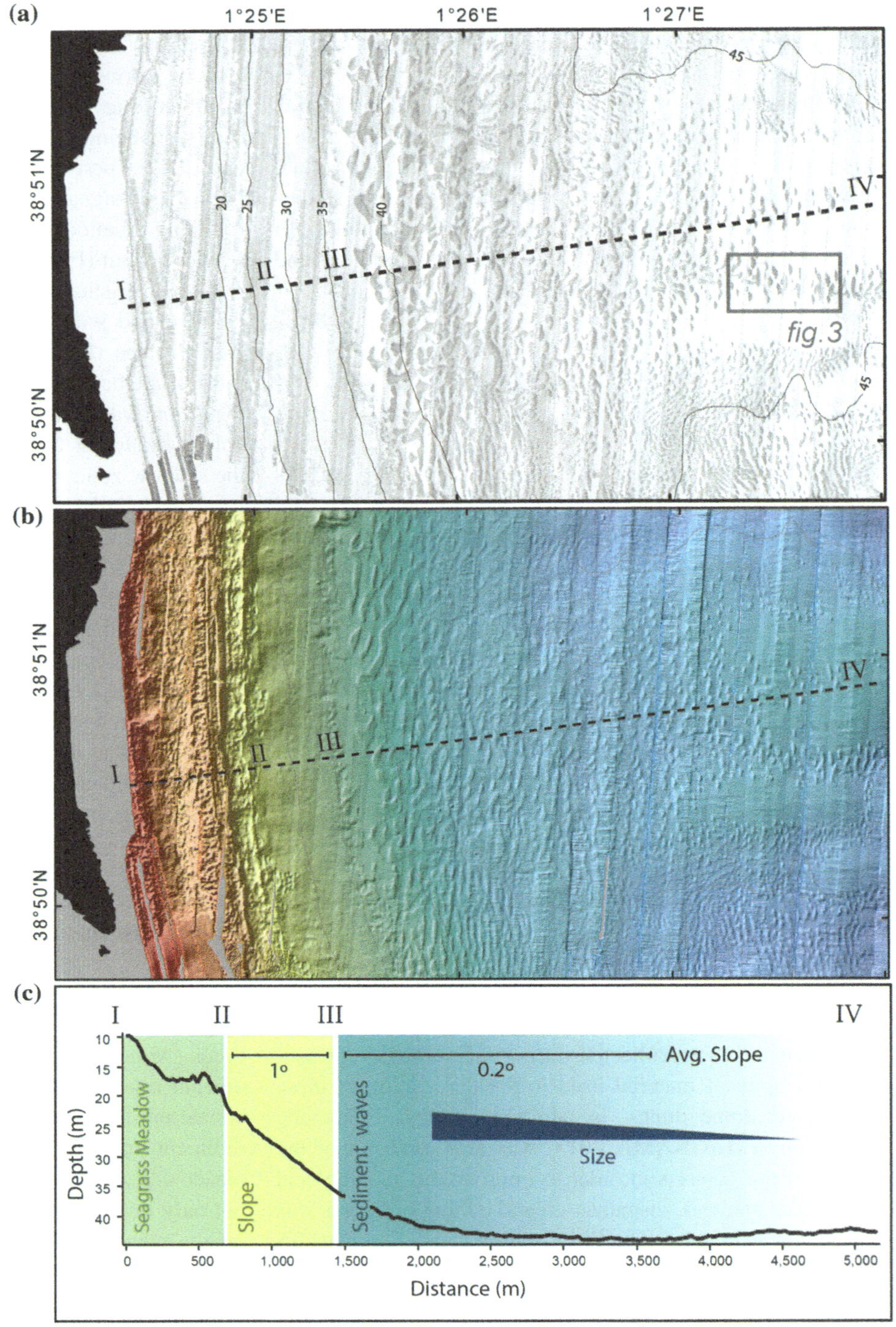

Fig. 25.2 **a** Backscatter digital terrain model; *darker areas* of the seabed are more absorbent and *lighter areas* are more reflective. **b** Digital elevation model. **c** Depth profile plotted in subfigures **a** and **b** as a *black dashed line*. *I*, *II*, *III* and *IV* indicate changes on the seabed. Seabed classes are as follows: *I–II*, Seagrass meadow; *II–III*, gentle slope; and *III–IV*, *smooth slope* where bedforms are placed. Bedform size decreases eastward

confined in slightly depressed, winding corridors (Figs. 25.1 and 25.2). This distribution can be related to a downdip waning of the flow velocity along with a reduction of the net sediment transport, in combination with the fact that the distance from the sediment source increases, gradually creating a sediment-starved environment.

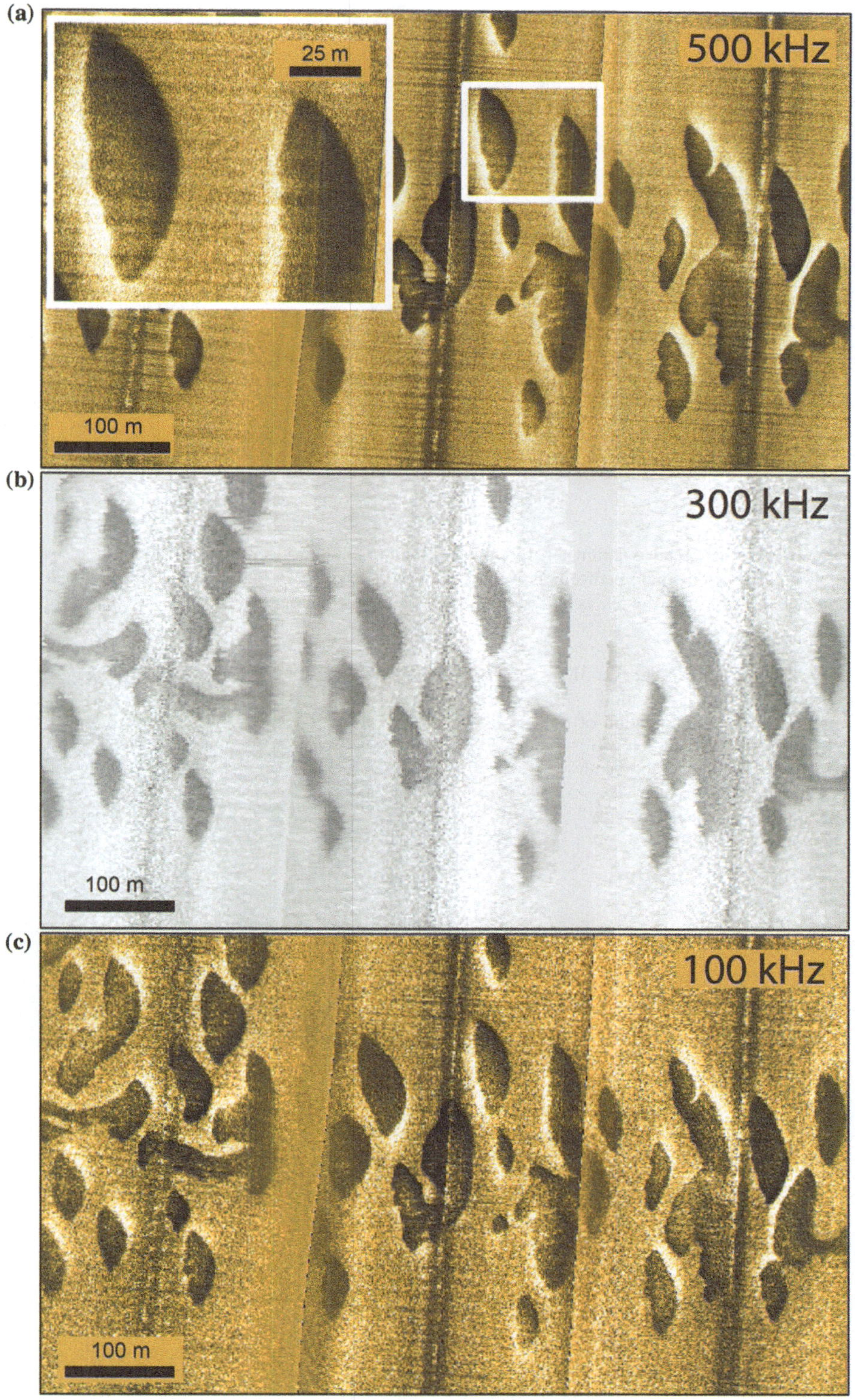

Fig. 25.3 Three backscatter digital terrain models of the same area from multibeam echosounder and side scan sonar records; *darker* areas of the seabed are more absorbent and *lighter* areas are more reflective. **a** SSS sonogram 500 kHz; high-definition frequency. **b** multibeam echosounder backscatter record 300 kHz; medium definition frequency. **c** Side scan sonogram 100 kHz; low definition/long range frequency. Location shown in Fig. 25.2

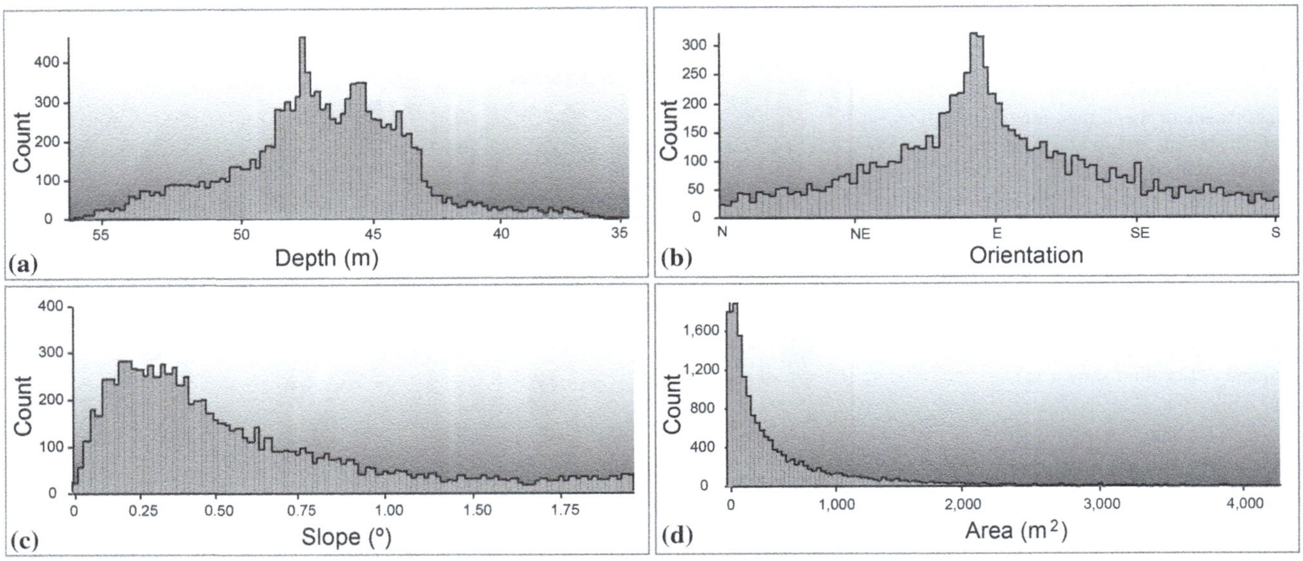

Fig. 25.4 Three seabed parameters (depth, slope and orientation) were extracted for each single bedform as well as their area in m^2. **a** Depth histogram, depth in metres below surface. **b** Orientation histogram, direction of the slope. **c** Slope histogram, seabed slope value in degrees. **d** Histogram of bedform area in m^2

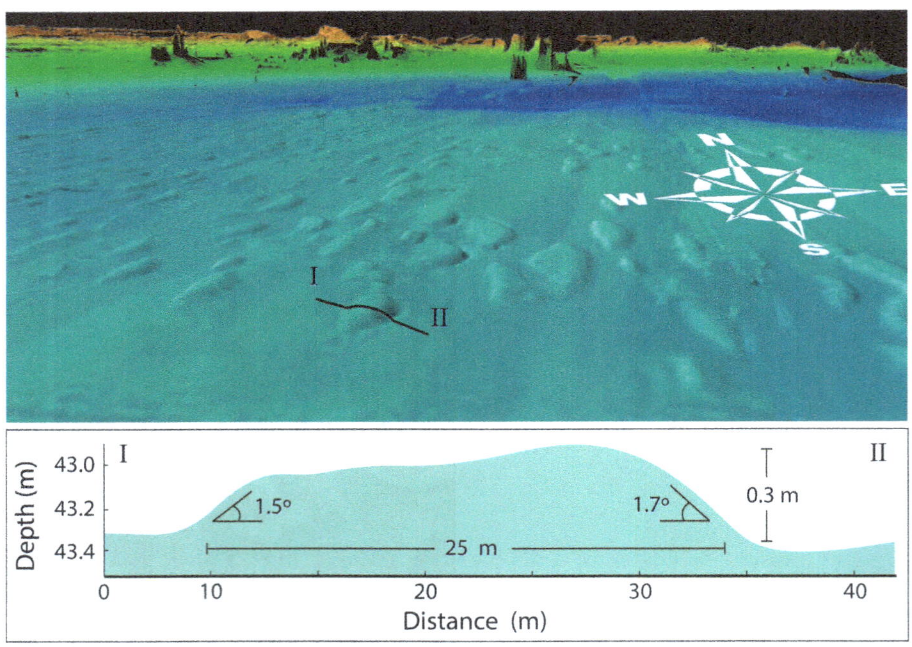

Fig. 25.5 *Upper* figure is a 3D view of the digital elevation model showing the location of a depth profile display below. *Bottom* figure shows the depth profile and morphometric parameters of a standard-size bedform

Bibliography

Arnaud-Haond, S., Duarte, C.M., Diaz-Almela, E., Marbà, N., Sintes, T., Serrão, E.A., (2012). Implications of Extreme Life Span in Clonal Organisms: Millenary Clones in Meadows of the Threatened Seagrass Posidonia oceanica. PLoS ONE 7, e30454. doi:10.1371/journal.pone.0030454.

DGMM. (2009). Informe sobre el hundimiento del buque Don Pedro en las proximidades del puerto de Ibiza el día 11 de julio de 2007, Informe de accidente marítimo. Dirección general de la marina mercante.

Dominguez, M., Fontán, A., Rivera, J., Ramón, M. (2013). Caracterización del ecosistema bentónico de la plataforma costera del área comprendida entre Sa Dragonera, Cabrera y el Cap de Ses Salines (Mallorca).

Hansen, J., Reidenbach, M. (2012). Wave and tidally driven flows in eelgrass beds and their effect on sediment suspension. Mar. Ecol. Prog. Ser. 448, 271–287. doi:10.3354/meps09225.

Jordi, A., Basterretxea, G., Wang, D.-P. (2011). Local versus remote wind effects on the coastal circulation of a microtidal bay in the Mediterranean Sea. J. Mar. Syst. 88, 312–322. doi:10.1016/j.jmarsys.2011.05.007.

Jordi, A., Basterretxea, G., Wang, D.-P. (2009). Evidence of sediment resuspension by island trapped waves. Geophys. Res. Lett. 36.

MAGRAMA. (2013). Ecocartografías - Protección de la costa - Costas y Medio Marino - magrama.es [WWW Document]. URL http://www.magrama.gob.es/es/costas/temas/proteccion-costa/ecocartografias/default.aspx (accessed 5.27.15).

Mason, E., Pascual, A. (2013). Multiscale variability in the Balearic Sea: An altimetric perspective. J. Geophys. Res. Oceans 118, 3007–3025.

Millot, C. (1999). Circulation in the Western Mediterranean Sea. J. Mar. Syst. 20, 423–442. doi:10.1016/S0924-7963(98)00078-5.

Trawl Marks and Dredge Spoils as Examples of Seabed Anthropic Alteration on Sediments (Menorca Shelf)

26

J. Rivera, G. Mateu, N. Hermida, L. Pomar, and J. Acosta

Abstract

Trawl marks and dredge spoils are two different types of anthropic seabed alterations. Both of them are footprints of the human activity on the seabed and both of them are traceable on the swath bathymetry records from certain types of seabed. High frequency multibeam echosounders allow detecting and quantifying the extent of these features so identifying their acoustic signature is a valuable issue in the field of impact assessment and environmental management. In addition, patterning and distinguishing them is something that geomorphologists need to bear in mind when interpreting geophysical records in a human influenced area in order to avoid misinterpretation of the real natural features.

Keywords

Anthropic seabed morphologies • Bedforms • Trawl fishing • Trawl marks • Dredge • Spoils • Impact assessment • Environmental management

J. Rivera (✉) · N. Hermida · J. Acosta
Instituto Español de Oceanografia, IEO, Corazón de María 8, 28002 Madrid, Spain
e-mail: jesus.rivera@md.ieo.es

G. Mateu · L. Pomar
Càtedra "Guillem Colom Casasnovas", Edifici Guillem Colom Casasnovas - Universitat de les Illes Balears, Cra. de Valldemossa Km. 7,5, 07122 Palma de Mallorca, Spain

J. Guillén et al. (eds.), *Atlas of Bedforms in the Western Mediterranean*,
DOI 10.1007/978-3-319-33940-5_26

26.1 Introduction

Trawl fishing and civil engineering works (e.g. pipe laying and dredging), may lead to seabed alteration, reshaping the submerged landscape and even locally changing sediment dynamics (Du Four and Van Lancker 2008; Palanques et al. 2006). Although the resultant entities cannot be considered as bedforms in *stricto sensu* because their origin, they actually are forms resting on the seabed that have an imprint on multibeam and SSS records.

Sediments on the continental shelf around Menorca have a variable composition. To the north, predominant siliciclastic sediments are the product of the erosion on the folded and thrusted Paleozoic rocks of the hilly Tramuntana region by the wave abrasion on the shoreline where the Tramontana wind (north wind) is frequent and intense. To the south, the Upper Miocene carbonates are preferentially dissolved, and siliciclastic terrigenous input from the Tramuntana region through the creeks, is minor. There, skeletal fragments dominate. Two main sediment types occur, and they are related to two different biotic factories. At shallower depths, down to ca. 40 mwd, *Posidonia oceanica* meadows form a particular ecosystem that sustain a diverse biota that produce sediments, mostly sand-sized skeletal grains (foraminifers, molluscs, bryozoans, etc.) (Fornós 1987; Fornos and Ahr 1997). Deeper on the shelf (down to ca. 80 mwd), in the oligophotic zone, carbonate production is dominated by calcareous red algae, including both *Peyssonnelia sp.* and corallinacean (*Lithothamnium sp., Phymatolithon calcareum and Spongites sp.*). They form coarser, gravel-sized accumulations of fine to coarse branches of red algae and rhodoliths (Canals and Ballesteros 1997; Fornos and Ahr 1997).

26.2 Methods

Data sets were gathered in the course of two successive surveys planned with the aim of assessing and monitoring the environmental impact due to the Mahón harbor refurbishment (Dredge Report 2012). First cruise took place in January 2014, before the dredging work started. The second one took place in July of the same year, after the dredging work was completed. As the same area (100 km^2) was surveyed before and after the dredging work, the resultant bathymetry and backscatter Digital Terrain Models (DTM) can be compared in order to determine the changes in the seabed due to the disposal of dredge spoils. During the first cruise 30 sediment samples (Using a Van-Veen grab) and 10 very high resolution seismic profiles (recorded by a TOPAS PS40 parametric subbottom profiler) were also acquired along with swath bathymetry.

The Multibeam Echosounder (MBES) used in both surveys was an EM3002D. This model has a dual head system and operates at a frequency of 300 kHz. Real time vessel position was obtained using a DGPS system. Differential correction was provided by Omni Star Satellites. Bathymetry and backscatter data were processed using CARIS H&S 8.0 and geomorphometric analysis was carried out using Fledermaus 7.3 and ArcGIS 9.2. Dry sieving method was performed for grain size sediment analysis.

26.3 Results

The anthropic bottom morphological signatures described below are located on the insular shelf, between 40 to 80 m water depth. In this area the sediment mean grain size is about 670 μm. Sand is the predominant fraction of the bottom sediment all over the surveyed area, mud is below 1 % of the total weight in most of the samples, and gravels are present just in the sediments collected in the dredge disposal site. In general, samples took on both types of anthropic seabed alterations show poorly shorted sediments (Fig. 26.1). At the east side of the surveyed area a sediment wave field extends (Fig. 26.1). As far as has been surveyed its area is larger than 30 km^2 and through this extension waves direction and dimensions varies from north to south. At the north end (East of La Mola) wave lengths are about 30 m and the crests axis are in the east—west direction whereas in the southern area average wave length is about 40 m and the direction of the crests axis are northeast—southwest.

26.3.1 Trawl Marks

Trawl marks found in this area are parallel grooves up to 40 m wide and about 6 cm in relief. Their dimensions are barely above MBES depth resolution, but they are very evident on the backscatter record where they can be easily traced along several thousand meters. Trawl marks are due to the contact between the lower part of the bottom trawling fishing gear and the seabed. In particular the trawl doors, also known as otter boards. These devices are two panels made of steel, placed between the net and each one of the two trawl cables to keep open the net mouth while trawling. The way the trawl doors plough the seabed depends on the nature of the seabed, as the finer the sediment is the deeper the groves are.

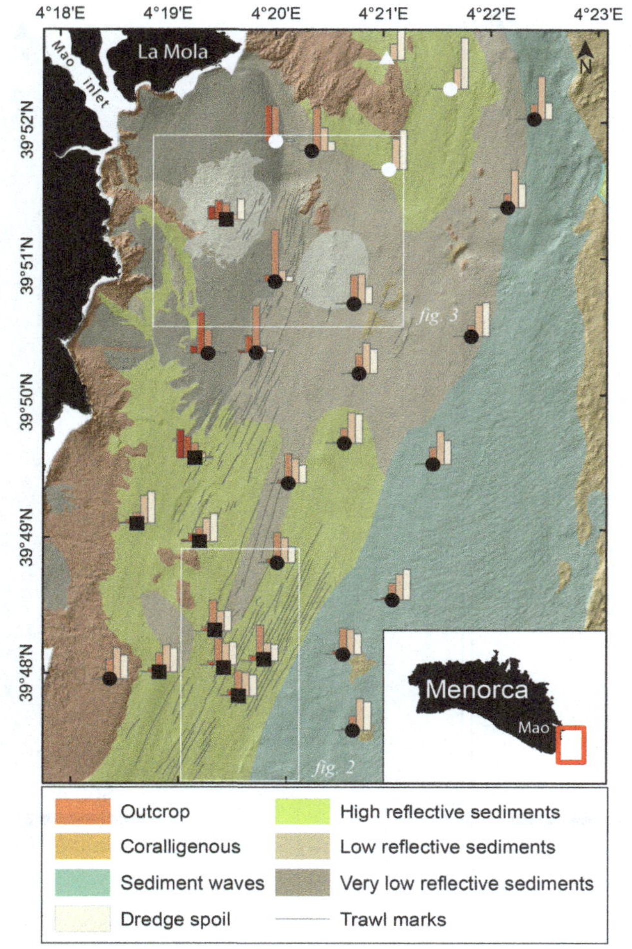

Fig. 26.1 Seabed classes from geophysical data interpretation and sediment sampling as displayed on the legend at the *bottom* of the figure. Location of sediment samples taken before dredge operations are spotted by following features: Histograms of the sand fraction; from very fine sand (*Left bar*), to very coarse sand (*Right bar*). *White triangle* means well sorted sediments. *White circle* moderately well sorted, *black circle* medium sorted and *black square* means poorly sorted sediments according to Folk and Ward sediment analysis method

26.3.2 Dumping of Dredged Sediment

Dredge spoils in the study area can be identified by a high reflective rough surface. Roughness results from the chaotic mixture of mud, sand, boulders and artifacts from the dredged material, whereas the high backscatter signal is due to the high reflective nature of spoils. Two disposal sites are located in this area. The disposal site closer to the coast extends for 1.3 km^2 and contains spoils from dredge operations carried out from 1879 to 1995. Documented information provided by the Port Authority of Balearic Islands report a total volume of spoil dumped throughout these operations of 912,871 m^3 (Fig. 26.4), with the most important dredging operation done in 1924 (720,000 m^3). The second disposal site is absent in the data set from the surveys previous to 2014 (Fig. 26.3a). Its extension is 0.5 km^2 and contains 163,890 m^3 of dredge spoils (Fig. 26.3b, c). This volume has been calculated by the depth difference between surveys (Rivera 2015), and matches the dumped volume reported by the Chief engineer. Two main differences exist between both disposal sites. First one is a low reflective band up to 100 m wide present at the southern part of the new dumping site. The low backscatter values indicate that it is made of fine sediments. The other difference is a set of high reflective strait lines from Maó inlet to the disposal site. High backscatter values indicate coarse sediments.

26.4 Discussion and Conclusions

Trawl fishing is known as an important erosive driving force able to change seabed geomorphology (Puig et al. 2012). Current Balearic laws allow fishermen to trawl between 50 m to 1,000 m depth, so trawl marks can be expected in this depth interval, where the wind driven hydrodynamic processes weakly reach the seabed. The trawl doors carve the seabed with grooves which are almost undetectable by the MBES because their small size. However the fishing gear passing through after the trawl doors generate large turbulences that increase sediment resuspension and drag suspended particles behind them (Palanques et al. 2001, 2006). This action repeated over and over leads to sediment sorting. The resultant linear footprints are 40 m wide and 0.6 m in relief, much larger than trawl door grooves, and they actually are the features displayed on the backscatter DTM (Fig. 26.2). Trawl tracks crossing the boundary between two different and well sorted sediment classes highlight their footprints on the backscatter record, as the fishing gear drag the sediments from one side to the other; i.e. trawl net running over a mud patch set on a sandy seabed will resuspend mud and drag it onto the sand resulting in low backscatter strips overlapping the higher reflective sand. Although these trawl marks are barely perceptible due to the relatively coarse sediments (670 μm) of this area, the low sedimentation rate in Balearic Islands (Zúñiga et al. 2007) and the fact that they are located under the wave base (25–30 m) mean that trawl marks can last for long time on the seabed, probably centuries, although there are no studies to confirm this point. On the other hand trawl marks interfere with the sediment wave field at the southeast (Fig. 26.1), although they are not noticeable in other sectors (North and middle) of the wave field, even though its well known that fishing activity take place in there too with the same intensity. Two circumstances could explain this. One is the resolution drop in the multibeam record due to the larger depth (greater than 80 m), but we consider that it mainly should be due to the existence of stronger bottom currents in those sectors.

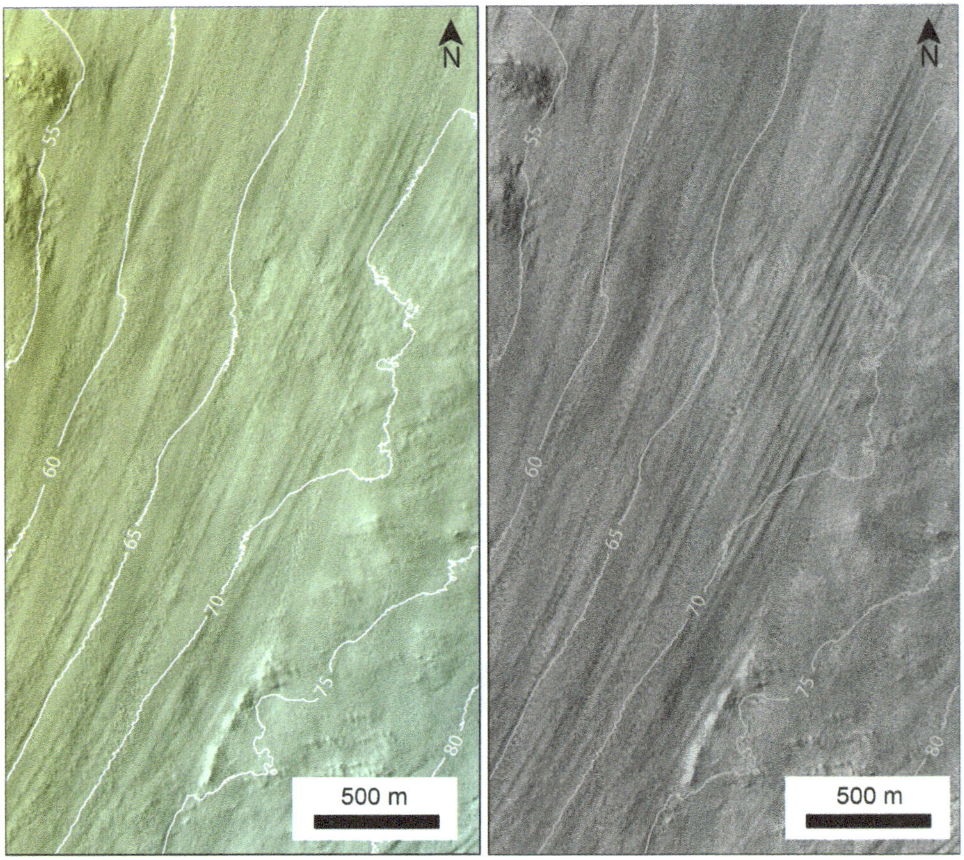

Fig. 26.2 Trawl Maks signature. *Left* Bathymetric DTM. *Right* Backscatter DTM from the same area. The extent of the represented area is shown in Fig. 26.1

Dredge spoils have a strong acoustic footprint that mask the natural seabed signature. Thanks to the multibeam records is possible to determine the extent and volume of the dumped spoils and also to have evidences about the processes occurring during disposal. Whereas coarse sediments and boulders sink straight down at the dumping location, fine sediments takes longer to reach the bottom being exposed to currents for longer. Main current direction in the area is southwards(Massutí et al. 2014), so finer fractions of sediment are carried downstream from the disposal site, reaching the bottom few meters away from dumping position. This explains the low backscatter band along the south edge of the disposal site (Fig. 26.3b, c). The traceable lines on the backscatter record (Fig. 26.3c) are due to the spoil spills from the dredge barge in its way to the disposal site. Dropped coarse spoils point the track of the barge with a high backscatter signal (Fig. 26.3c). The new spoils dumped on the new disposal site represent the 15 % of the total volume reported (Fig. 26.4), and cover 0.5 km^2. Spoils from 1982 to 1995 correspond to a similar percentage (14 %) whereas the old disposal site area is wider (1.3 km^2). This means that spoils from 1924 are still evident on backscatter record. As linear shapes and the mentioned low reflective band were absent on the spoils before the 2014, we can infer that these features fade away over time lasting for just a few years. On the contrary the acoustic signature of the spoil material core remains for decades, probably centuries, in this type of insular shelf.

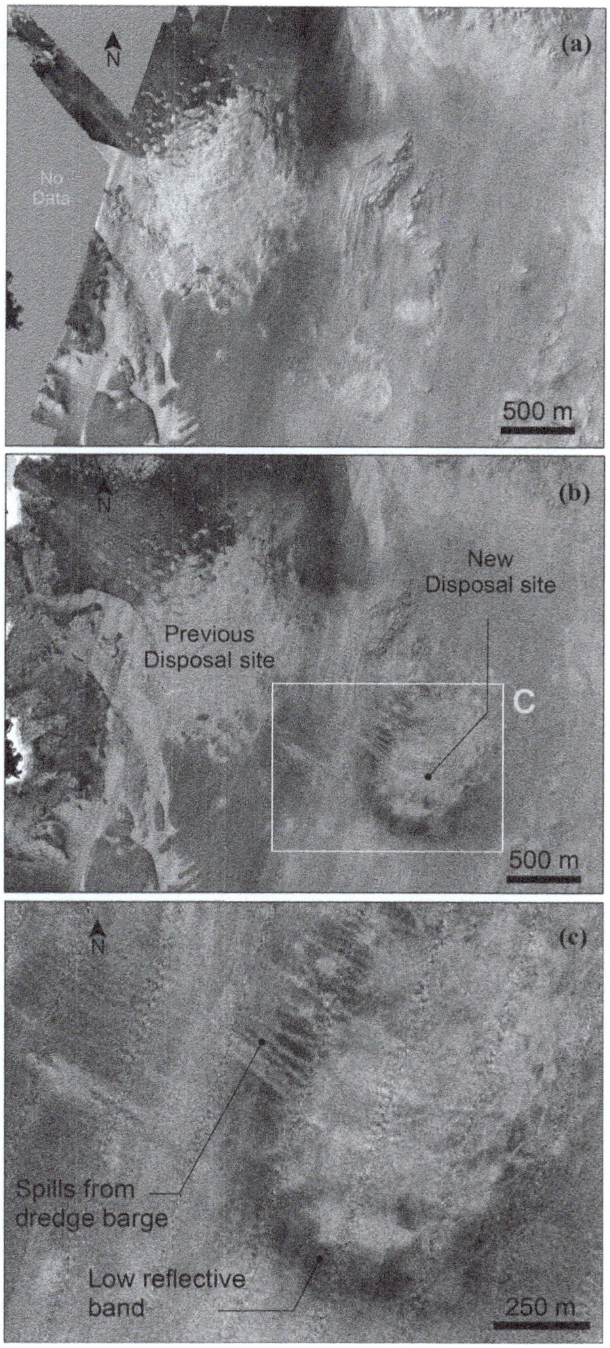

Fig. 26.3 Disposal sites. **a** Backscatter DTM before 2014 dredging work. **b** Backscatter DTM after 2014 dredging work, note linear features evidencing spills from the dredge barge in its way to the dumping location and a low reflective band in the southern half of the new disposal site due to fine sediments advection. **c** Close up of the new disposal site

Acknowledgments Data sets shown and discussed throughout text and figures were obtained in two surveys conducted by the Instituto Español de Oceanografía (IEO), on board R/V F. de Paula Navarro ("IEO Fleet," 2015); We thank the crew for their help and comradeship during both surveys. Special thanks to the Port Authority of Balearic Islands for their permission to use this material.

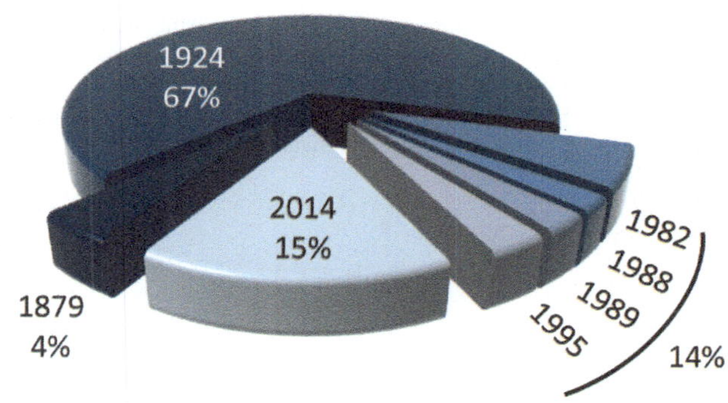

Fig. 26.4 Pie chart showing the relative volume of the dredging operations conducted in Maó. Labels indicate the year they took place and the percentage of the total dredged volume that each one represent

Bibliography

Canals, M., Ballesteros, E. (1997). Production of carbonate particles by phytobenthic communities on the Mallorca-Menorca shelf, northwestern Mediterranean Sea. Deep Sea Res. Part II Top. Stud. Oceanogr. 44, 611–629. doi:10.1016/S0967-0645(96)00095-1.

Dredge report. Dragado del Puerto de Maó: Memoria [WWW Document]. (2012). URL http://dragadopuertodemao.blogspot.com.es/p/memoria.html (accessed 7.6.15).

Du Four, I., Van Lancker, V. (2008). Changes of sedimentological patterns and morphological features due to the disposal of dredge spoil and the regeneration after cessation of the disposal activities. Mar. Geol. 255, 15–29. doi:10.1016/j.margeo.2008.04.011.

Fornós, J. (1987). Les plataformes carbonatades de les Balears, estudi sedimentològic de les plataformes miocenes de les illes Balears i la comparació amb la sedimentació actual de la seva plataforma. PhD Thesis, Universitat de Barcelona.

Fornos, J., Ahr, W. (1997). Temperate carbonates on a modern, low-energy, isolated ramp: the Balearic Platform, Spain. J. Sediment. Res. 67.

IEO Fleet [WWW Document]. (2015). URL http://www.ieo.es/web/ieo/flota (accessed 7.6.15).

Massutí, E., Santaella-Álvarez, E., López-Jurado, J.L., Balbín, R., Aparicio-González, A., Sánchez-Leal, R.F., Deudero, S., Álvarez, E., Cerdá, M., Ordines, F., Rivera, J., Abelló, P., Farriols, M.T., Orejas, C., Ramón-Herrero, M., Jhoer, S., Tovar, A., Massanet, A., Sánchez-Quiles, D., Campillo-González, J.A., León-León, V.M., Besada, V., Martínez-Gómez, C., García-Aparicio, V., Valdés-García, N.J., Navarro-Marín, C., Martínez-Franco, F., Ceruso-Juez, C. E., García-Agüera, I.M.E., Schultze, F., Reñones, O., Iglesias-Cubells, M.C. (2014). Revisión y Control del Plan de Vigilancia Ambiental de las obras de dragado del Puerto de Maó.

Palanques, A., Guillén, J., Puig, P. (2001). Impact of bottom trawling on water turbidity and muddy sediment of an unfished continental shelf. Limnol. Oceanogr. 46, 1100–1110. doi:10.4319/lo.2001.46.5.1100.

Palanques, A., Martín, J., Puig, P., Guillén, J., Company, J.B., Sardà, F. (2006). Evidence of sediment gravity flows induced by trawling in the Palamós (Fonera) submarine canyon (northwestern Mediterranean). Deep Sea Res. Part Oceanogr. Res. Pap. 53, 201–214. doi:10.1016/j.dsr.2005.10.003.

Puig, P., Canals, M., Company, J.B., Martín, J., Amblas, D., Lastras, G., Palanques, A., Calafat, A.M. (2012). Ploughing the deep sea floor. Nature 489, 286–289. doi:10.1038/nature11410.

Rivera, J. (2015). Informe de campaña y resultados MAO1407_GEOBentos. http://hdl.handle.net/10508/9045.

Zúñiga, D., Calafat, A., Sanchez-Vidal, A., Canals, M., Price, B., Heussner, S., Miserocchi, S. (2007). Particulate organic carbon budget in the open Algero-Balearic Basin (Western Mediterranean): Assessment from a one-year sediment trap experiment. Deep Sea Res. Part Oceanogr. Res. Pap. 54, 1530–1548. doi:10.1016/j.dsr.2007.06.001.

27

Sand Ridges on the Mid-Outer Valencia Continental Shelf

G. Simarro, J. Guillén, P. Puig, M. Ribó, C. Lo Iacono, A. Palanques, A. Muñoz, R. Durán, and J. Acosta

Abstract

A sand ridge field with crests oriented NE–SW (52°) located between 55 and 85 m water depth on the Valencia continental shelf (Spain) was mapped with multibeam swath bathymetry and characterized with high-resolution seismic reflection profiling and sediment sampling. Boundary layer hydrodynamic and suspended sediment concentration measurements conducted over a sand ridge at 66 m depth showed evidence of wave and current sediment resuspension during a major storm event. It is argued that these sand ridges developed in a shallow environment and there is no evidence of present-day ridge migration or morphological degradation.

Keywords

Bedforms • Sand ridge • Sediment resuspension • Continental shelf

27.1 Introduction, Study Area and Methods

The seabed of continental shelves usually shows the presence of a variety of bedforms, such as subaqueous dunes and sand ridges. Most of the bedforms located deeper than 50 m were developed in shallow environments and detached from the coast during the sea level transgression. At present, large bedforms on the outer shelf can be classified as "active" or "relict", or "moribund" if their morphology is degraded or

G. Simarro (✉) · J. Guillén · P. Puig · M. Ribó · A. Palanques · R. Durán
Institut de Ciències del Mar, CSIC, Passeig Maritim de la Barceloneta 37-49, 08830 Barcelona, Spain
e-mail: simarro@icm.csic.es

C. Lo Iacono
National Oceanography Centre–Southampton, European Way, Southampton, SO14 3ZH, UK

A. Muñoz
Grupo Tragsa, Calle Conde de Peñalver 84, Madrid, Spain

J. Acosta
Instituto Español de Oceanografía, Calle del Corazón de María 8, 28002 Madrid, Spain

J. Guillén et al. (eds.), *Atlas of Bedforms in the Western Mediterranean*,
DOI 10.1007/978-3-319-33940-5_27

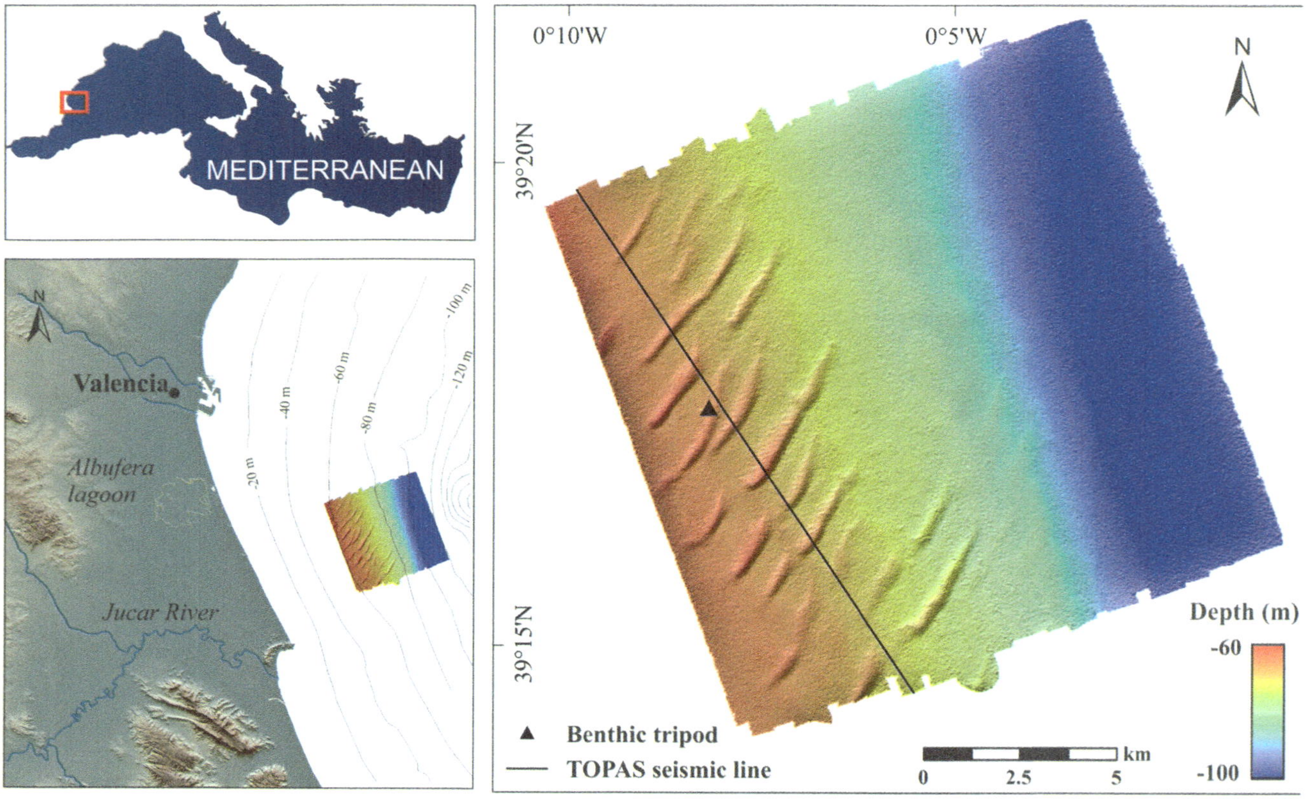

Fig. 27.1 Map of the Gulf of Valencia shelf (NW Mediterranean) including the bathymetric map of the study site, showing the sand ridges field and the location of the benthic tripod

their original characteristics are modified by sedimentary processes (Dyer and Huntley 1999). There is still an interesting debate about the possible mechanisms of maintenance or degradation of large-scale subaqueous dunes in low-energy continental shelves and about the mechanisms that can potentially remobilize the sediment (Lo Iacono et al. 2010). Specific detailed measurements of sediment dynamics in these environments are still required.

The continental shelf of the Gulf of Valencia (Fig. 27.1) has a width of up to 30 km with the shelf edge at ~150 m water depth. It is an example of a storm-dominated shelf with a source of fine sediment from riverine inputs. The circulation on the Valencia continental shelf is dominated by the wind stress and the general flow of the Catalan Sea towards the south. The wave climate is seasonal, with October–April being the most energetic period and the largest waves coming from E–NE directions. Resuspension events are frequent, occurring several times per year along the coastal zone and on the inner shelf (Guillén et al. 2006), and they have also been observed sporadically on the mid-shelf (Puig et al. 2001). Several sand ridges were identified in the central part of the Valencia shelf between the 55 and 85 m isobaths (Fig. 27.1), with heights of up to 7 m and spacing between 600 and 1100 m. Ridge transversal profiles are asymmetric with the lee face on the south side of the crests. These ridges are mainly composed of medium sands with interbedded mud layers deposited over a thin layer of very coarse sand and gravel (Durán et al. 2015). Northern and smaller sand ridges are partially blanketed by a thin layer of modern muds that can be up to 3 m thick in the troughs between ridges. The thickness of this surficial mud layer decreases southwards, as well as from the trough between the ridges towards the crest, where it is thinner than 0.5 m or even absent (Fig. 27.2). The internal structure displays medium- to high-angle dipping reflectors indicating long-term migration towards the southeast that laterally pass to aggradational facies towards the trough (Fig. 27.2).

The study of these bedforms (Simarro et al. 2015) is based on the integration of a multibeam bathymetry carried out in September 2010 and the measurements of a benthic tripod deployed from 21 September 2010 to February 2011 (local water depth ~66 m, Fig. 27.1). An acoustic backscatter system was installed on the tripod frame and was deployed approximately 110 cm above the bed to obtain the near-bed suspended sediment concentration at a high resolution in time (1 s) and space (1 cm in the vertical). A current meter to measure velocity and direction of the currents was also installed at 95 cm above the bottom. Buoy data

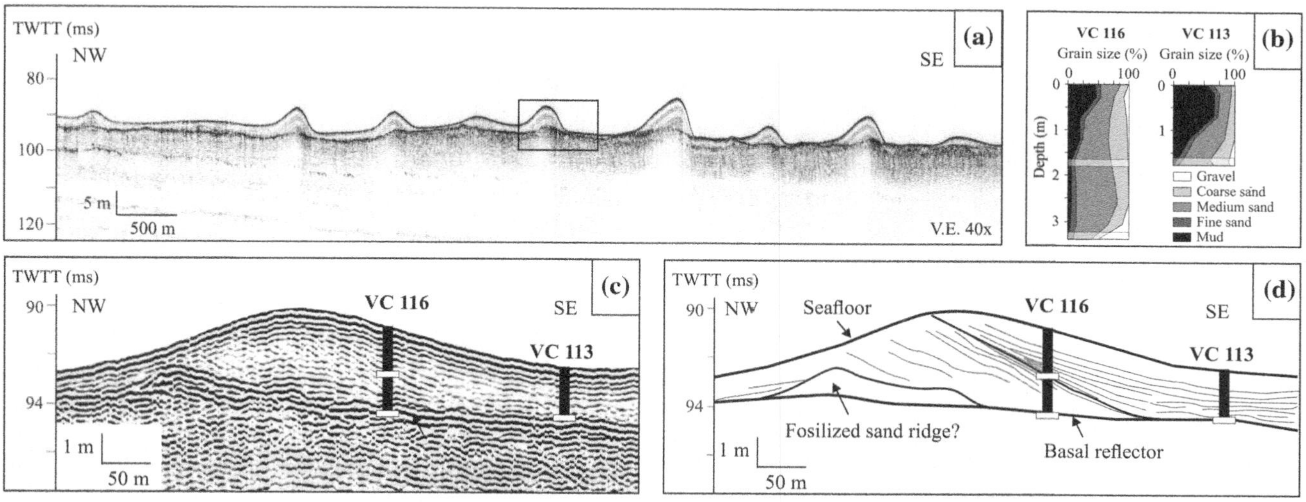

Fig. 27.2 **a** High-resolution seismic profile across the sand ridge field. **b** Description of two sediment corers located over a sand ridge. *White bars* indicate main discontinuities identified in the seismic profile. **c** Uninterpreted and **d** interpreted high-resolution seismic profile showing the internal structure of sand ridges and the location of sediment cores. A sound velocity in sediments of 1600 m/s was assumed. See Fig. 27.1 for location

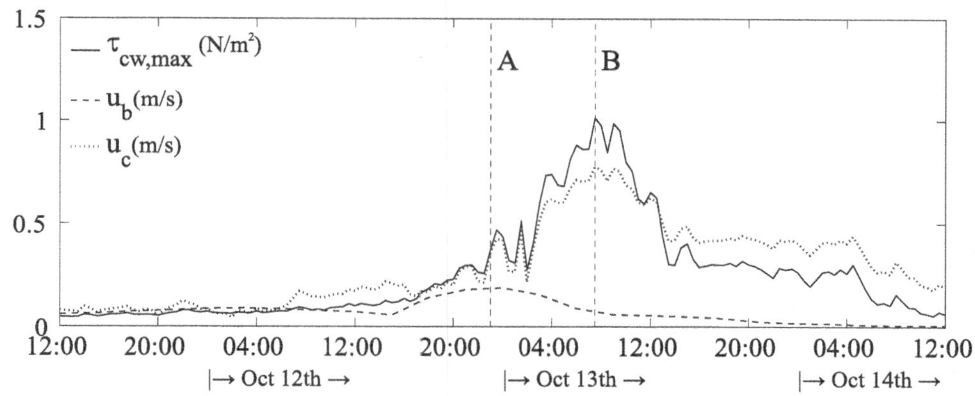

Fig. 27.3 Time series of near-bed wave and current velocities (u_b and u_c, respectively) and combined bed shear stress (τ_{cw}) during the storm event

were used to infer the statistics of the near-bed orbital velocity. Seismic profiling (April 2011) and sediment sampling data (February 2007) over the sand ridge field are also available.

27.2 Results and Discussion

A severe storm occurred in the study area from 12:00 on 11 October to 12:00 on 14 October 2010. As shown in Fig. 27.3, the near-bed orbital velocity reached values of up to 0.2 m/s during the storm (point A in Fig. 27.3). The current speed reached a maximum value of 0.8 m/s (point B in Fig. 27.3). The computed combined wave–current bed shear stress (τ_{cw}) is also shown in Fig. 27.3. The wave-current bed shear stress was dominated by the current velocity during this storm. However, waves influence the bed shear stress at the wave peak (point A), when the current velocity is far from reaching its maximum. Bottom sediment resuspension and transport is expected approximately between 20:00 on 12 October and 4:00 on 14 October, when bottom shear stresses exceeded the critical value for sediment entrainment (~ 0.21 N/m^2).

Figure 27.4 shows a detail of the intra-wave evolution of the suspended sediment concentration (SSC) at different distances from the bed and different stages of the storm (A and B in Fig. 27.3). During burst A (corresponding to the maximum orbital velocity) the current velocity u_c ~ 0.3 m/s was only slightly greater than the orbital u_b ~ 0.2 m/s, and the Fourier transform of the SSC time histories shows a

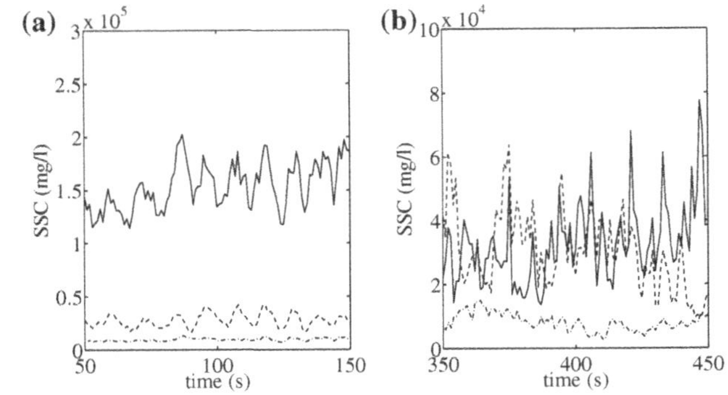

Fig. 27.4 *Left panels* Intraburst time history of SSC in the three nodes closest to the bed (extracts of) for bursts **a** and **b** (A and B in Fig. 27.3). Node closest to the bed (*full line*), node 2 cm above the closest to the bed (*dashed line*), node 4 cm above the closest to bed (*dotted line*)

dominant period ~8 s, similar to the wave period, suggesting that the wave field plays a significant role in this phase for the sediment resuspension. For burst B (when the maximum shear stress is reached) the wave influence on the bed shear stress is not distinguishable.

On a long-term perspective, the morphology and disposition of the sand ridge field on the Valencia mid-outer shelf allow us to reasonably hypothesize that its formation took place in a shallower environment and was caused by similar mechanisms to those proposed for the shoreface-connected ridges. When these ridges are gradually detached from the shoreface due to sea level rise, their growth gradually slows down, their rate of migration decreases and, finally, the ridges drown when the wave orbital velocity at the bed drops below the critical erosion of sediment. Previous studies in the study area suggest that the deepening of the former sand bodies occurred between the Younger Dryas (about 12,000 years BP) and 6250 years BP (Albarracín et al. 2014). Present-day sedimentary processes on the Valencia mid-outer shelf are characterized by sand starvation and mud delivery from rivers discharging into the area, favouring the formation of a mud layer thinning towards the SE. On the Valencia shelf the sedimentation rate is 1.5 mm/year, so it would imply a deposit of mud about 9 m thick covering the sand ridge field if we consider the last 6000 years. This means that sand ridges would be at the last stage of their geological evolution. This is not the case because muddy sediment is remobilized during storms and prevents the burial of sand ridges.

Considering that the measured hydrodynamic conditions were not exceptional (recurrence period of about 3 years), active sediment resuspension and transport episodes have been occurring frequently on these bedforms during the late Holocene highstand. This is supported by the occurrence of interbedded mud layers within the ridge sandy units. On top of that, the dipping internal reflectors suggest episodic ridge migration (Fig. 27.2), with periods of reactivation after relatively calm periods dominated by mud deposition.

The preservation of this area surrounded by muddy sediments could also be related to the action of stronger currents because of the margin configuration (Fig. 27.1). Therefore, the incorporation of sand ridges in the stratigraphic record is a very slow process.

27.3 Conclusions

In situ measurements of the hydrodynamics and SSC at 66 m depth over a sand ridge field showed the influence of wave dynamics on the resuspension process when the orbital wave velocity was similar to the current velocity. The influence of the current, though, dominates the sediment dynamics in most of the storm. The sedimentary carpet on the Valencia mid-outer shelf is active, and the influence of waves cannot be neglected for sediment mobilization in deeper areas during the most energetic storms. Though several thousand storms such as the one analysed could have impacted the detached sand ridges during the past 6000 years, no evidence of ridge migration or of morphological degradation was found. It is suggested that present-day sediment dynamics affecting the studied sand ridges during storms favour the maintenance of their morphology.

Acknowledgments This research was supported by the projects COSTEM (CTM2009-07806), BUS2 (CGL2011-22964) and FORMED (CGL2012-33989). We thank the captain and the crew of the R/V *García del Cid* and the Instrumental Service of ICM, J. Pozo and M. Lloret for their kind assistance. Puertos del Estado (Ministerio de Fomento) provided wave data information and the Subdirección General para la Protección de la Costa (Ministerio de Agricultura, Alimentación y Medio Ambiente) provided the stratigraphic results from vibrocores. G. Simarro was supported by the Spanish Government through the Ramón y Cajal programme. The studies of Dr M. Ribó were supported by an FPI doctoral grant (Ref. BES-201029949) from the Spanish Ministry of Economy and Competitiveness. R. Durán acknowledges a CSIC JAE-Doc contract co-funded by the FSE.

References

Albarracín, S., Alcántara-Carrió, J., Montoya-Montes, I., Fontán-Bouzas, Á., Somoza, L., Amos, C.L., Rey, J., (2014). Relict sand waves in the continental shelf of the Gulf of Valencia (Western Mediterranean). Journal of Sea Research 93, 33–46.

Durán, R., Guillén, J., Simarro, G., Ribó, M., Puig, P., Muñoz, A. (2015). Sand ridges in the mid-outer shelf as potential sand borrows areas (NW Mediterranean). Proceedings of the Coastal Sediments 2015.

Dyer, K.R., Huntley, D.A. (1999). The origin, classification and modelling of sand banks and ridges. Continental Shelf Research 19 (10), 1285–1330.

Guillén, J., Bourrin, F., Palanques, A., de Madron, X.D., Puig, P., Buscail, R. (2006). Sediment dynamics during wet and dry storm events on the inner shelf (SW Gulf of Lions). Marine Geology 234 (1–4), 129–142.

Lo Iacono, C., Guillén, J., Puig, P., Ribó, M., Ballesteros, M., Palanques, A., Farrán, M., Acosta, J. (2010). Large-scale bedforms along a tideless outer shelf setting in the western Mediterranean. Continental Shelf Research 30 (17), 1802–1813.

Puig, P., Palanques, A., Guillén, J. (2001). Near-bottom suspended sediment variability caused by storms and near-inertial internal waves on the Ebro mid continental shelf (NW Mediterranean). Marine Geology 178 (1–4), 81–93.

Simarro, G., Guillén, J., Puig, P., Ribó, M., Lo Iacono, C., Palanques, A., Muñoz, A., Durán, R., Acosta, J. (2015). Sediment dynamics over sand ridges on a tideless mid-outer continental shelf. Marine Geology 361, 25–40.

Subaqueous Dune Fields on the Marine Protected Area Around the Cabrera Archipelago (Balearic Islands)

28

Araceli Muñoz, César Alcalá, Laura Pascual, César León, María de la Paz Maroto, and Juan Acosta

Abstract

The continental shelf around the Cabrera Archipelago is composed of carbonate bottom sediment unaltered by human activities. Three dune field systems located to the north, east and south of the archipelago are described in this chapter. They are formed by small to medium dunes, roughly oriented N-S, and usually with symmetrical morphology. The dune morphology (orientation, size and shape) adapts to the presence of islands and outcrops on the continental shelf.

Keywords

Subaqueous dunes • Carbonate sediment • Continental shelf • Unaltered areas

A. Muñoz (✉) · C. Alcalá · L. Pascual · C. León · M. de la Paz Maroto
Tragsa, C/Valentín Beato nº6, Madrid, 28037, Spain
e-mail: amur@tragsa.es

J. Acosta
IEO, C/Corazón de María nº8, Madrid, 28006, Spain

J. Guillén et al. (eds.), *Atlas of Bedforms in the Western Mediterranean*,
DOI 10.1007/978-3-319-33940-5_28

28.1 Introduction

The National Park of the Cabrera Archipelago (south of Mallorca Island) is one of the best-preserved coastal ecosystems of the Mediterranean (Magrama 2013) (Fig. 28.1). It has a high level of protection from national and European institutions, is included in the network of Special Areas for Birds (ZEPAS), which forms part of the network of Zones of Special Protection of Mediterranean Importance (ZEPIM), and in the Natura 2000 network (Casier 2011).

The continental shelf between Mallorca Island and the Cabrera Archipelago is 35 km wide, gently sloped and occasionally disrupted by submarine terraces. The glacio-eustatic variations in sea level in the Mediterranean Sea have played an important role in fragmenting the shelf into sub-basins and forming these terraces where carbonate complex bars have developed (Acosta et al. 2002).

The circulation pattern around the archipelago is not well known (López-Jurado 1990). The dominant near-bottom current flows from NE to SW, reaching the western end of the archipelago and bordering or crossing the islands in an E-W direction (Werner et al. 1993), although this pattern is often reversed (Rio et al. 2007; Jordi and Wang 2009). The area is also affected by the sporadic presence of eddies, generally of an anticyclonic nature (Millot et al. 1997). A cyclonic eddy was also observed in the Cabrera Channel, probably associated with the interaction between the Atlantic fronts over the slope and the topography of the islands (Jordi et al. 2009). Resonance phenomena associated with trapped waves surround the islands, reinforcing each other and resulting in high-intensity currents over the continental shelf of Mallorca that produce large currents in the Cabrera Channel (>100 cm/s) (Jordi et al. 2009).

The seabed of most of the Mallorca-Cabrera shelf is composed of carbonate sediment, favoured by low continental inputs and temperate waters. The bottom sediment mainly originates from seagrass-derived biogenic components, with a percentage of carbonate content varying between 77 and 84 % (Alonso et al. 1988; Fornós and Ahr 1997).

On the continental shelf around the Cabrera Archipelago three dune fields were identified in the north, east and south (at Na Foradada Island, Na Redona Island and Es Estells islands, respectively) (Fig. 28.1). They are examples of bedforms developed on a temperate carbonate shelf that is almost unaffected by human activities.

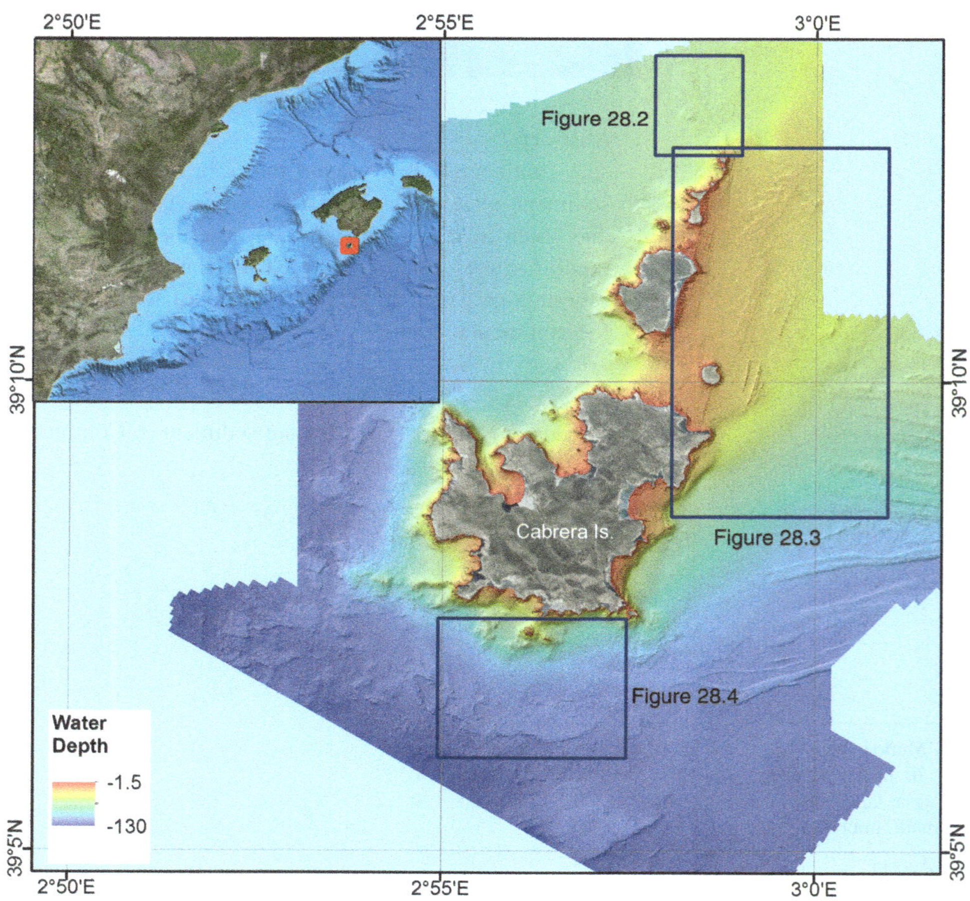

Fig. 28.1 Location map of the study area

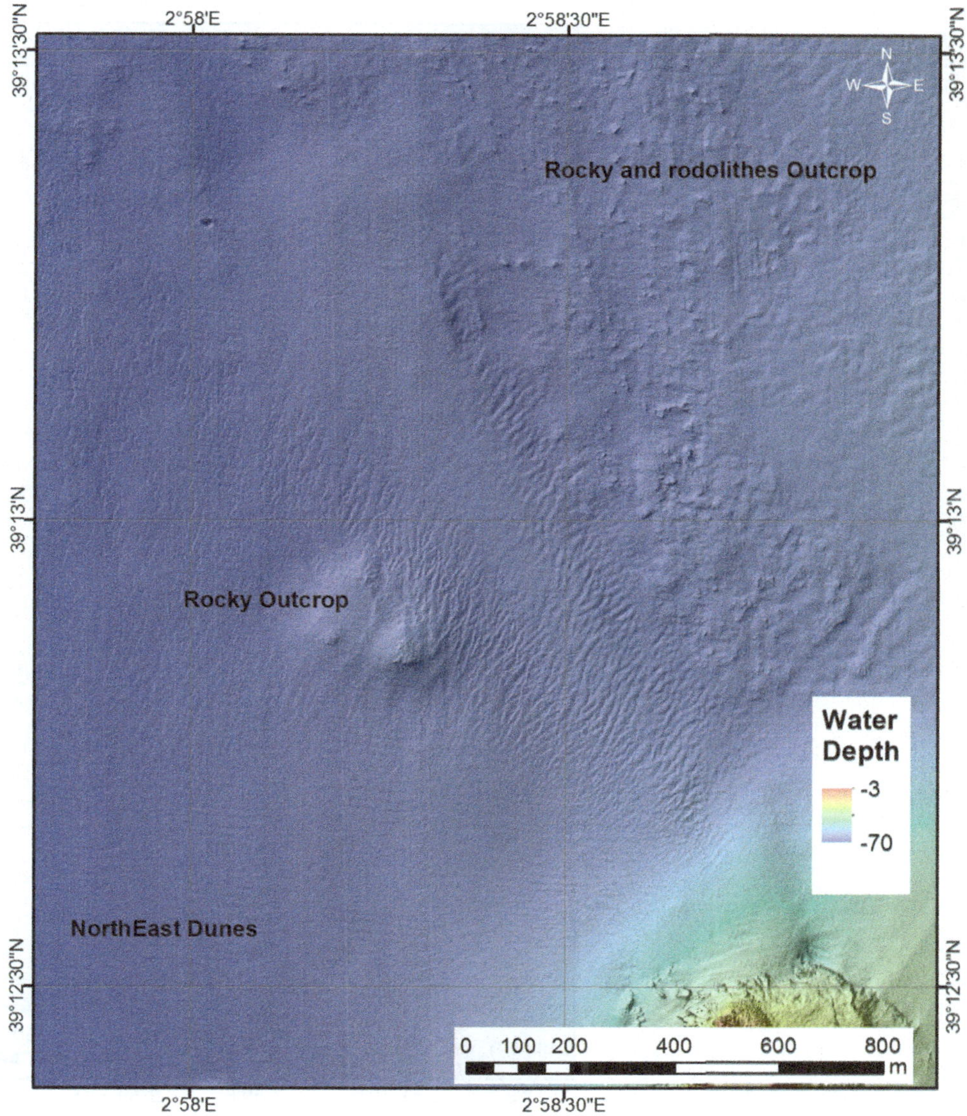

Fig. 28.2 Bathymetric map of the northern dune field, which is extends from Na Foradada island towards the Cabrera Channel (see location in Fig. 28.1)

28.2 Methods

The bathymetric data were acquired during two surveys performed in July 2012 and August 2014 on board R/V *Emma Bardan* with a Kongsberg EM3002D multibeam echosounder. The processing was carried out using the Caris Hips 7.0 and ArcGIS 9.3 software packages. The digital terrain model footprint resolution obtained was 1 m. Visualization of bathymetric data was obtained, using the IVS Fledermaus system and ESRI ArcMap 9.3. Surface sediment samples were collected during the survey using a Shipeck grab and the sediment grain size, organic content and carbonate content of the surficial sediment were analysed.

28.3 Results

28.3.1 The Northern Dune Field (Na Foradada Island)

This field of small dunes occurs on the slope from Na Foradada Island to the Mallorca-Cabrera shelf, over a depth range of 52–66.5 m. The field covers an area of 0.8 km^2, which is limited to the east and west by gravel, rhodoliths and rocky bottoms and to the south by hard bottoms and meadows of *Posidonia oceanica*. The grain size of the bottom sediment in the dune field area ranges from medium to coarse sand with low percentages of gravel (5 %) and mud (<2 %).

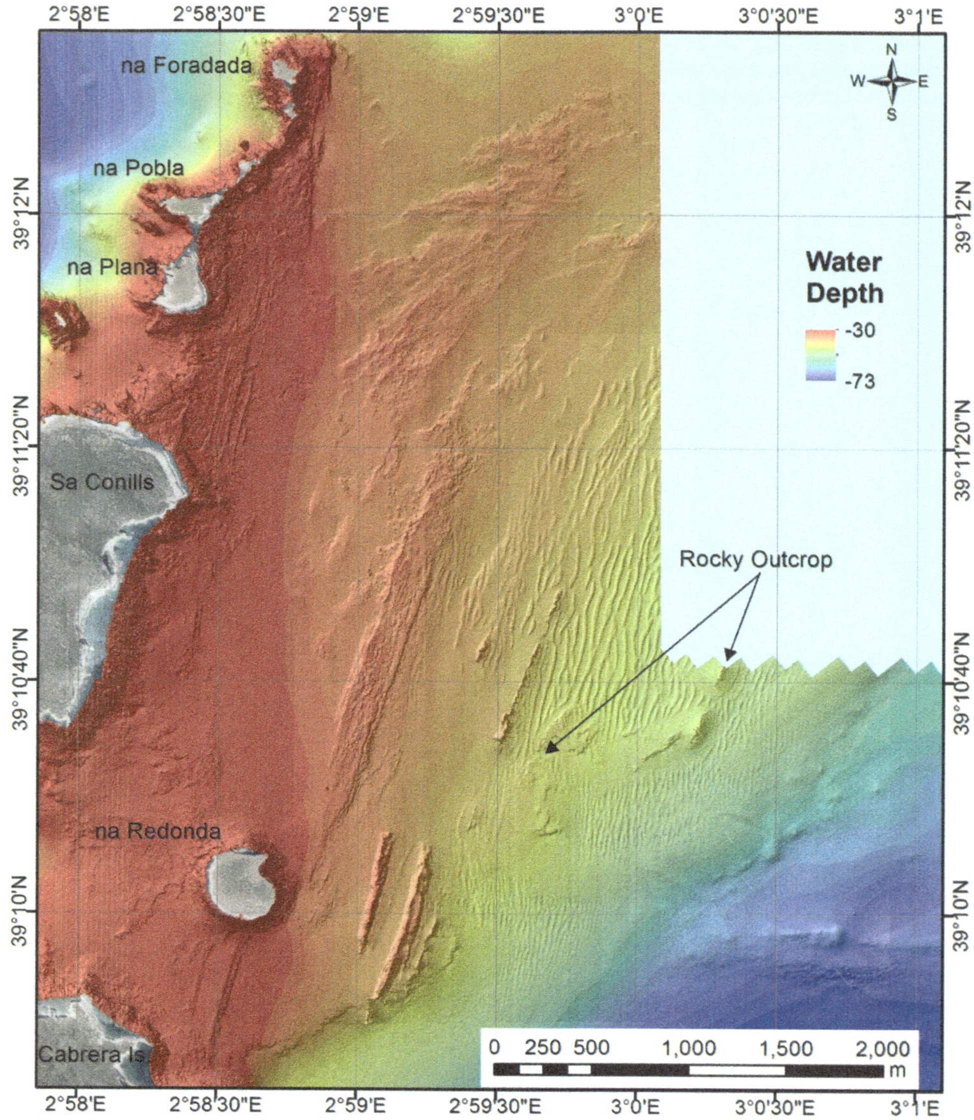

Fig. 28.3 Bathymetric map of the eastern dune field (Sa Conills and Na Redona islands) (see location in Fig. 28.1)

A total of 223 crestlines were identified. They are mainly oriented NE-SW, displaying a spacing ranging from 10 to in excess of 25 m and ranging in height from 0.1 to 0.2 m. These bedforms are characterized by their asymmetrical profile, with the lee side located upslope (SW side of the crestline) and partially rounded crests (Figs. 28.2 and 28.5).

28.3.2 The Eastern Dune Field (Sa Conills and Na Redona Islands)

This dune field is interrupted by the presence of a rocky outcrop in the middle zone (Fig. 28.3). In the northern sector the subaqueous dunes are located between 37 and 47 m depth, occupying an area of 1.83 km^2. The field is composed of 225 dunes with a symmetrical profile, a fairly peaked crest and broad trough, and straight to bifurcating crestlines oriented NE-SW (Figs. 28.3 and 28.5). The spacing ranges from 30 to 60 m and the height ranges from 0.2 m to about 0.6 m (Fig. 28.3). The length of the dunes along the crestline ranges between 130 and 550 m.

In the southern sector, off Cabrera and Na Redona Islands, the field is located between 41 and 55 m water depth and covers an area of 1.60 km^2. A total of 240 crestlines were identified (Fig. 28.3). The dune height ranges between 0.25 and 0.50 m, the wavelength between 20 and 50 m and the profile is symmetrical. The main differences from the northern sector are the N-S crestline orientation and the shorter length (90–240 m) in the southern sector (Fig. 28.3).

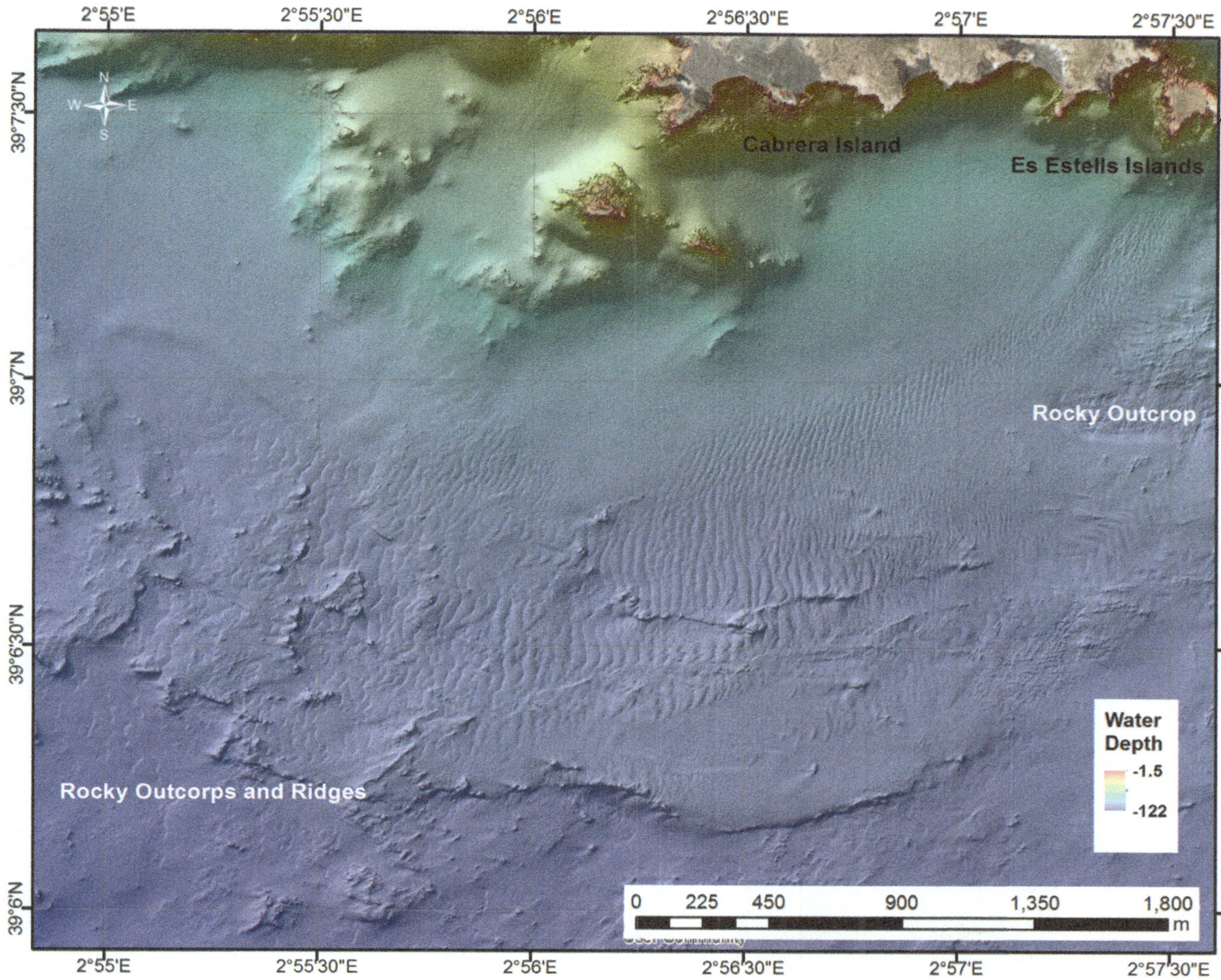

Fig. 28.4 Bathymetric map of the southern dune field (Es Estells islands) (see location in Fig. 28.1)

28.3.3 The Southern Dune Field (Es Estells islands)

The main dune field of this area has developed between 90 and 108 m depth and occupies an area of 4.73 km^2, although smaller dunes are observed at shallower depths (Fig. 28.4). The dune field is composed of 530 symmetrical dunes oriented NNE-SSW (Fig. 28.4). The spacing between crestlines ranges from 20 to 50 m and the height from 0.1 to 0.8 m. The field is limited by the Estels and Cabrera islands in the north, the rocky outcrops in the east and west, and small ridges and escarpments in the south.

The morphological characteristics of the dunes are modified in the proximity of the islands. For instance, near the Estels island the crestline of the dunes has a NE-SW orientation with lower heights and shorter lateral continuity. In general, the height and wavelength increases towards the southern part of the crests (deeper and farther from the islands).

28.4 Discussion and Concluding Remarks

The Cabrera Archipelago is surrounded by meadows of *Posidonia oceanica* that extend to 35 m water depth, and by maërl and coralligenous facies. In the absence of significant continental sedimentary inputs, the bottom sediment on the continental shelf is mainly composed of

Table 28.1 Morphological parameters of the dune fields around the Cabrera National Park

	Area km²	No. dunes	Spacing (m)	Height (m)	Length (m)	Depth (m)
Northern dunefield	0.8	223	10–25	0.1–0.2	20–140	52–66.5
Eastern dunefield N	1.83	225	30–60	0.2–0.6	130–550	37–47
Eastern dunefield S	1.6	240	20–50	0.25–0.50	90–240	41–55
Southern dunefield	4.73	530	20–50	0.1–0.8	20–340	90–108

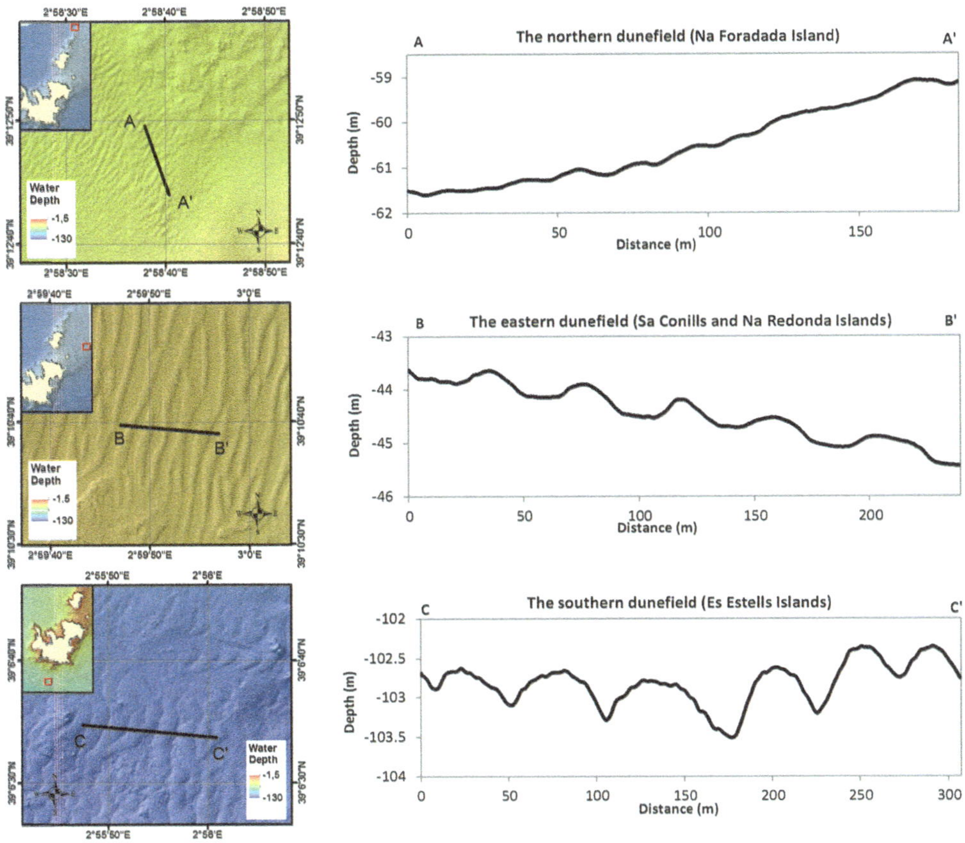

Fig. 28.5 Bathymetric profiles crossing the different dune fields

biogenic fragments derived from coastal meadows and shelf habitats. In these sand-starved environments bedforms have developed from 35 to more than 100 m water depth. The subaqueous dunes of the Cabrera shelf are small to medium in height based on the classification of Ashley (1990), but they have wavelengths in the range of large dunes (Table 28.1).

The northern dune field displays 2-D asymmetric bedforms with the lee side upslope, which are probably related to easterly currents that originated in the channel between the Cabrera Archipelago and the Mallorca Island.

The crestline orientation in the eastern and southern dune fields suggests the dominance of east/west-directed currents altered by the complex bottom topography because of the presence of islands, outcrops and ridges. The morphological appearance and the low height/length ratio in the dunes of the eastern field is probably due to the limited sediment availability and a progressive degradation of dune crests in this area, a finding which is supported by the observed rounding of the crests (Fig. 28.5b). The southern dune field displays a variety of dunes that must be related to the complex circulation pattern in the area, where interaction between oscillatory flows produced by wind-forced, island-trapped waves and the island generates ephemeral recirculation patterns (Basterretxea et al. 2013).

Acknowledgments The data were obtained in the framework of the ESPACE Spanish continental shelf mapping funded by the Secretaría General de Pesca (MAGRAMA). We also wish to thank the SGP-TRAGSA Cartography Group, who participated in the cruises.

References

Acosta, J., Canals, M., López-Martínez, J., Muñoz, A., Herranz, P., Urgeles, R., Palomo, C., Casamor, J.L. (2002). The Balearic Promontory geomorphology (western Mediterranean): morphostructure and active processes. Geomorphology, 1203: 1–28.

Alonso, B., Guillén, J., Canals, M., Serra, J., Acosta, J., Herranz, P., Sanz, J.L., Calafat, A., Catafau, E. (1988). Los sedimentos de la plataforma balear. Acta Geol. Hisp. 23 (3): 185–196.

Ashley, G.M., Chairperson and others. (1990). Classification of large-scale subaqueous bedforms: a new look at an old problem. SEPM bedforms and bedding structures. Journal of Sedimentary Petrology, 60 (1): 160–172.

Basterretxea, G., Catalán, I.A., Jordi, A., Álvarez, I., Palmer, M. and Sabatés, A. (2013). Dynamic regulation of larval fish self-recruitment in a marine protected área. Fisheries Oceanography, 22, 6: 477–495.

Casier, R. (2011). Marine Protected Areas in the Mediterranean Sea. Alfred Toepfer Natural Heritage Scholarship 2011. Awarded on the EUROPARC Conference 2011 in Bad Urach, made possible by the Alfred Toepfer Stiftung F.V.S.

Fornós, J.J., Ahr, W.M. (1997). Temperate carbonates on a modern low-energy, isolated ramp: the Balearic platform, Spain. J. Sediment Res. 67 (2): 364–373.

Jordi, A., Basterretxea, G., Wang, D.P. (2009). Evidence of sediment resuspension by island trapped waves. Geophys. Res. Lett. 36: L18610.

Jordi, A. and Wang, D.P. (2009). Mean dynamic topography and eddy kinetic energy in the Mediterranean Sea: Comparison between altimetry and a 1/16 degree ocean circulation model. Ocean Modell. 29: 137–146.

López-Jurado, J.L. (1990). Masas de agua alrededor de las Islas Baleares. Bol. Inst. Esp. Oceanogr. 6: 3–20.

MAGRAMA. (2013). Memoria de la Red de Parques Nacionales 2013. Ministerio de Agricultura, Alimentación y Medio Ambiente. 166 pp.

Millot, C., Benzohra, M., Taupier-Letage, I. (1997). Circulation off Algeria inferred from the Mediprod-5 current meters. Deep-Sea Res. 44: 1467–1495.

Rio, M.H., Poulain, P.M., Pascual, A., Mauri, E., Larnicol, G., Santoleri, R. (2007). A Mean Dynamic Topography of the Mediterranean Sea computed from altimetric data, in-situ measurements and a general circulation model. J. Mar. Sys., 65: 484–508.

Werner, F., Viúdez, A., Tintoré, J. (1993). An exploratory numerical study of the currents off the southern coast of Mallorca including the Cabrera Island complex. J. Mar. Sys, 4: 45–66.

Subaqueous Dunes Over Sand Ridges on the Murcia Outer Shelf

29

R. Durán, J. Guillén, J. Rivera, A. Muñoz, F.J. Lobo, L.M. Fernández-Salas, and J. Acosta

Abstract

Multibeam swath bathymetry, high-resolution seismic data and sediment samples were used to characterize a field of sand ridges and subaqueous dunes on the outer Murcia continental shelf (western Mediterranean Sea). Sand ridges are 1.5–3 m high and show a predominant E-W orientation oblique to the present-day shoreline. High-resolution seismic data reveal a backstepping stacking pattern of high-angle clinoforms dipping towards the southwest, interpreted as buried sand bodies. Subaqueous dunes have a mean height of 0.3 m and appear superimposed on the sand ridges showing a NW-SE orientation oblique

R. Durán (✉) · J. Guillén
Insituto de Ciencias del Mar (CSIC), Pg. Marítim de la Barceloneta 37-49, 08003 Barcelona, Spain
e-mail: rduran@icm.csic.es

J. Rivera · J. Acosta
Instituto Español de Oceanografia, IEO, Corazón de María 8, 28002 Madrid, Spain

A. Muñoz
Tragsa-SGP, C/Julián Camarillo 6B, 28037 Madrid, Spain

F.J. Lobo
Instituto Andaluz de Ciencias de la Tierra (CSIC-Universidad de Granada), Avenida de las Palmeras no 4, 18100 Armilla Granada, Spain

L.M. Fernández-Salas
Instituto Español de Oceanografía, Centro Oceanográfico de Cádiz, Puerto Pesquero s/n. Muelle de Levante, 11006 Cádiz, Spain

J. Guillén et al. (eds.), *Atlas of Bedforms in the Western Mediterranean*,
DOI 10.1007/978-3-319-33940-5_29

to the ridges. They are composed of sandy sediments and display asymmetric morphology, with the lee side towards the southwest. Ridge and dune asymmetry and internal structure are indicative of long-term sediment transport towards the southwest. At present, dune migration rates deduced from repeated bathymetric surveys indicate that the dunes remain stationary or migrate at very low rates on a decadal scale.

Keywords
Sand ridges • Subaqueous dunes • Tideless continental shelf • Western mediterranean sea

29.1 Introduction

Sand ridges are pervasive bedforms on the middle and outer continental shelf worldwide (Ashley 1990; Goff et al. 2005), as well as in the Mediterranean (Lo Iacono et al. 2010; Simarro et al. 2015). Sand ridges form in the shoreface environment through similar mechanisms to those proposed for shoreface-connected ridges (Goff et al. 2005; Snedden et al. 2011), which grow up during storms and with a strong steady flow component (Calvete et al. 2001). During the sea-level rise, the shoreface-connected ridges became progressively detached from the shoreface, slowing their growth and decreasing their migration rate until they eventually drowned when the near-bed orbital velocity dropped below the critical velocity for erosion of sediment (Nnafie et al. 2014). Subaqueous dunes are ubiquitous bedforms on the continental shelf that can appear superimposed on larger-scale bedforms, such as large and very large dunes or sand ridges (Lobo et al. 2000; Goff et al. 2005; Li and King 2007). Dunes can be stationary bedforms or migrate at different rates, providing valuable information about the local and regional current patterns and sediment transport.

A large field of sand ridges and subaqueous dunes was identified on the outer Murcia continental shelf (western Mediterranean Sea). This chapter provides new evidence about the recent evolution and present-day activity of these bedforms.

The Murcia continental shelf is 8 km wide but it narrows off Cape Cope, where it is less than 3 km wide (Fig. 29.1). It comprises a seaward-dipping platform that is cut by shelf-indenting submarine canyons with their heads located at short distances (5 km) from the coastline (Acosta et al. 2013). Quaternary deposits show a moderate development due to the narrowness of the shelf, and are locally eroded at the shelf break by the submarine canyons (Lobo et al. 2015). Shelf sediments are mainly composed of gravelly and sandy sediments with small areas of fine sands concentrated off small rivers and streams (Fernández-Salas et al. 2015). The Murcia continental shelf is a micro-tidal shelf with a maximum tidal amplitude of 0.2 m. Wave climate follows a marked seasonal pattern, with the most intense events usually occurring from September to May. Statistical analysis of wave records from 2006 to 2013 shows a mean significant wave height (Hs) of 1 m, with Hs maxima of 5.5 m and maximum wave heights of 7.80 m, and predominant NE and SW directions. The circulation on the shelf is dominated by the wind stress and the Northern Current, which carries old Atlantic Waters towards the southwest along the shelf break (Millot et al. 1999).

29.2 Methods

The present study is based on the analysis of high-resolution multibeam data (bathymetry and backscatter), widely-spaced sub-bottom seismic profiles and sediment samples. Data were gathered during the Copesands cruise conducted on board the R/V *Angeles Alvariño* in February 2013. Swath bathymetry data were collected using a Kongsberg EM-710 system (70–100 kHz). Information on the shallow structure of the seabed was obtained using a TOPAS PS18 sub-bottom parametric profiler from Kongsberg. In addition, 20 sediment samples were collected on the dune field using a box-corer grab. A dynamic positioning system combined with the differential global positioning system allowed us to take samples with pinpoint accuracy on the slope, crest and trough of the dunes (position error <0.5 m with a confidence interval of 95 %).

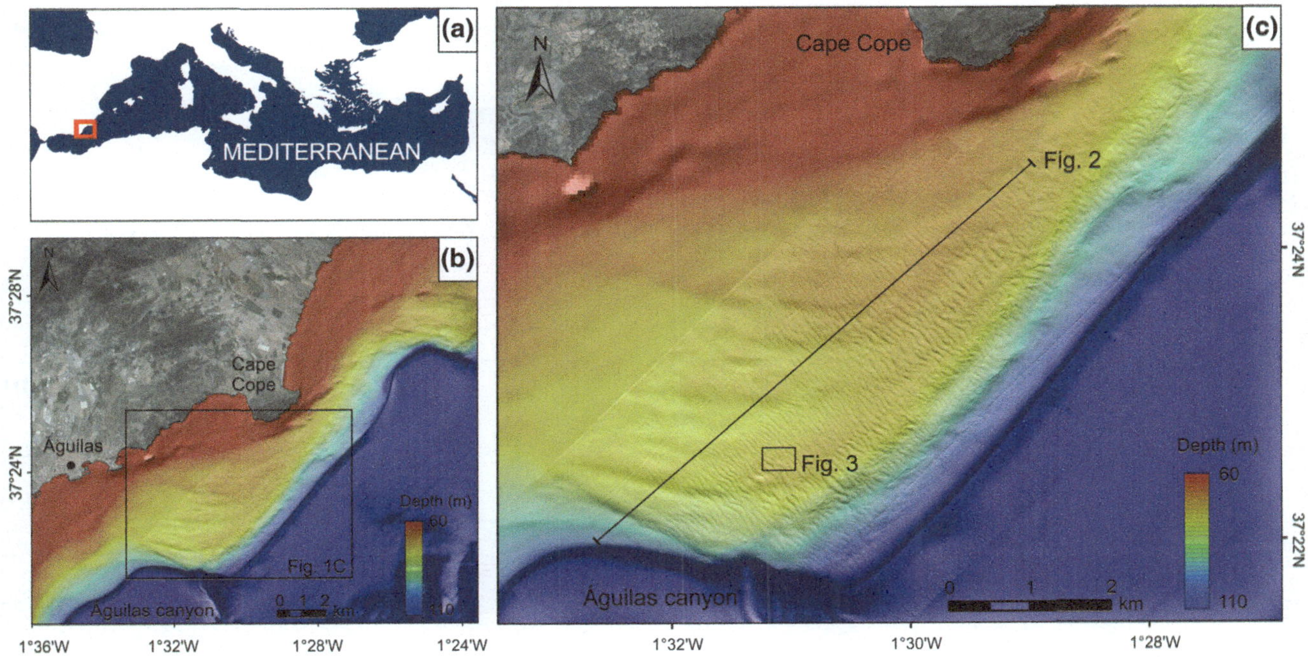

Fig. 29.1 **a** Location map. **b** Shaded relief bathymetric map of the Murcia continental shelf off Cape Cope. **c** Detailed bathymetry of the sand ridges and subaqueous dunes, including the location of Figs. 29.2 and 29.3. Onland background information from the *Instituto Geográfico Nacional de España* (www.ign.es)

Short-term (decadal) morphological changes in the subaqueous dunes were explored by comparing the 2013 bathymetry with a previous survey carried out in 2003 using a Simrad EM-3000D system. Post-processing of the multibeam data from both surveys (including correction for heading, depth, pitch, heave and roll) was conducted using the CARIS HIPS and SIPS Hydrographic Data Processing System. Dune migration rates were determined from the net displacement of the crest positions deduced from a large set of cross-sectional bathymetric profiles, following the methodological approach proposed by Knaapen et al. (2005).

29.3 Results

29.3.1 Sand Ridges

Sand ridges were observed on the outer shelf between 65 and 76 m water depth (Fig. 29.1). They show a predominant E-W orientation that gradually changes to NE-SW with increasing water depth and displays a curved morphology. The height of these bedforms ranges from 1.5 to 3 m, the spacing between 300 and 600 m and the length from 1500 to 3500 m. Very high-resolution seismic profiles across the ridges display a backstepping stacking pattern of prograding seismic units with high-angle reflectors dipping towards the southwest that are interpreted as buried sand bodies (Fig. 29.2a, b). Prograding seismic units have variable thickness from 3 to 5 m in the lower units to 1.5–3 m in the uppermost ones (Fig. 29.2b). They have a limited areal distribution and an asymmetric transverse profile with the steepest slope facing southwards. Prograding units overlie a strong, continuous reflector that can be traced across the study area and merges at the seabed on the landward portions of the shelf, particularly in the vicinity of Cape Cope (Fig. 29.2a). This basal reflector is tilted towards the southwest, and shows evidence of channel incisions. Maximum thickness of the sedimentary infill overlying this reflector is ~12 m (assuming a sound velocity in sediments of 1650 m s^{-1}) in the southern sector of the survey area, but it decreases significantly towards the north.

29.3.2 Subaqueous Dunes

A large field of subaqueous dunes extends from Cape Cope to the Águilas submarine canyon, between 60 and 110 m depth (Fig. 29.1). The dunes show a predominant NW-SE orientation, oblique to the present shoreline. The mean height is 0.3 m but it can be up to 1.3 m. Dunes have asymmetrical profiles (average asymmetry ratio of 0.065) with the lee slope facing towards the southwest, indicating long-term migration in this direction (Fig. 29.2c). The area occupied by the subaqueous dunes is mainly composed of

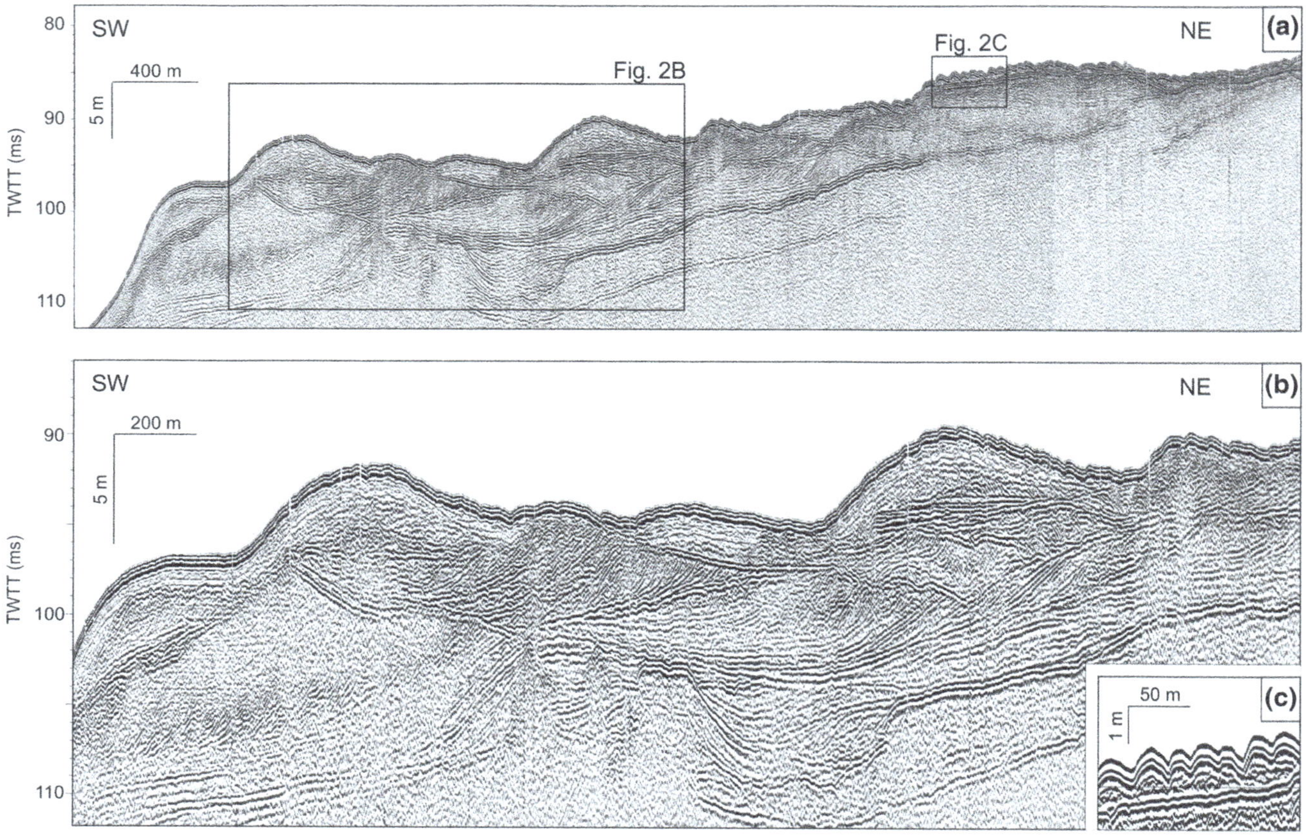

Fig. 29.2 **a** Very high-resolution seismic profile along the Murcia continental shelf. Detailed sections of the seismic profile showing **b** the internal structure of sand ridges, and **c** the subaqueous dunes. Vertical scale in milliseconds two-way travel time (TWTT). Thickness in metres is calculated assuming a sound velocity in sediments of 1650 m s^{-1}. See Fig. 29.1 for location

fine to coarse sand (median grain size of 0.131–0.479 mm), coarser than the surrounding areas (0.025–0.115 mm). Backscatter imagery displays higher intensity values on the crests and lower intensity values on the troughs (Fig. 29.3). High backscatter areas consist of coarse sand (79 % in average) with a low content of fines (16 % on average), whereas the lower backscatter areas correspond to fine to medium sand (58 % on average) with a higher content of fines (34 % in average). Dunes appear superimposed on the sand ridges in the southern sector of the dune field, whereas

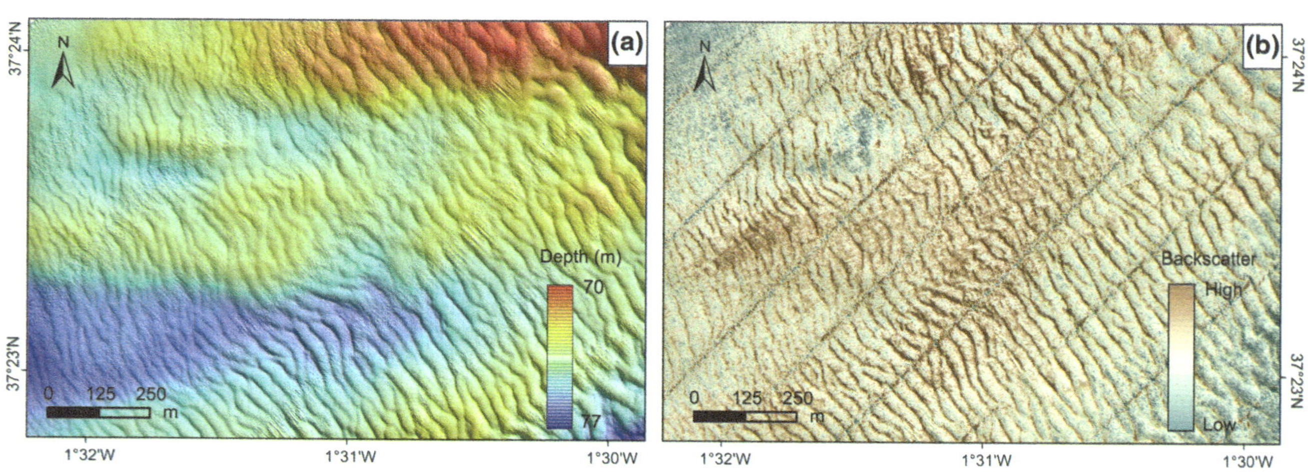

Fig. 29.3 **a** Shaded relief bathymetric map illustrating the subaqueous dunes. **b** Backscatter imagery of the same area. See Fig. 29.1 for location

in the vicinity of Cape Cope, where no sand ridges are present, they overlie a strong basal reflector that truncates the underlying reflectors (Fig. 29.2a, c). Internally, the dunes display southwest-dipping oblique reflectors, consistent with the migration in that direction. To assess the potential mobility of these bedforms, dune migration rates were estimated from the net displacement of the crest and trough positions derived from two bathymetric surveys. Results revealed slow dune migration over the study period with a mean rate of 0.5 m y^{-1}.

29.4 Discussion and Conclusions

The detailed analysis of the sand ridges and subaqueous dunes identified on the Murcia continental shelf provides new insights into the recent evolution and present activity of these bedforms.

The sand ridges of the Murcia continental shelf show a similar morphology and internal structure to that of other storm-dominated sand ridges described on the NW Atlantic and Mediterranean shelves (Goff et al. 2005; Bassetti et al. 2006; Li and King 2007, Lo Iacono et al. 2010; Snedden et al. 2011; Goff and Duncan 2012; Durán et al. 2015). They are asymmetrical and oblique to the shoreline, and migrate in the direction of their steep face (towards the SW). Internally, the sand ridges display SW-dipping oblique reflections indicating episodic ridge migration in that direction. The presence of dipping reflections within the ridges has also been described in other storm-dominated shelves (Goff et al. 2005; Durán et al. 2015). Snedden et al. (2011) associated these internal reflections with mud beds, indicating cessation of migration followed by reactivation of the sand ridge movement during major storms.

The stratigraphic record of the Murcia continental shelf reveals a backstepping stacking pattern of sand bodies. As the sand bodies migrated towards the southwest, new sand bodies were superimposed on the old ones, resulting in a stacking pattern of sand bodies in the study area. A complicated, multilayer internal architecture was also observed in tidal sand ridges of the East China Sea as a result of repeated accumulation–erosion–accumulation cycles during the last transgression (Liu et al. 2003). The pattern of sand ridges on the Murcia shelf is also suggestive of generation during relative sea-level rise conditions, in agreement with previous observations in the western Mediterranean that proposed that sand ridge formation occurred during the last Holocene transgression, favoured by a deceleration of sea-level rise around the Younger Dryas (Bassetti et al. 2006). Though most active ridge formation occurred during the last transgression, sand ridges could still have evolved under episodic high-energy events during the Holocene highstand (Bassetti et al. 2006; Simarro et al. 2015).

The asymmetry and internal structure of the sand ridges and subaqueous dunes indicate long-term migration towards the southwest. At present, the subaqueous dunes over the sand ridges display a very low migration rate that lies at the limit of the measurement method, indicating that the dunes on the Murcia outer shelf remain almost stationary or migrate at very low rates on a decadal scale. However, it cannot be ruled out that longer monitoring of the dune field would have recorded dune migration. Textural changes across the subaqueous dunes are consistent with higher near-bottom shear stresses on the crest than the trough, favouring the winnowing of finer fractions on the crest and their deposition on the trough and thus contributing to the maintenance or migration of these morphologies. This is supported by recent observations on the Valencia outer shelf that evidenced active sediment resuspension and transport episodes during storms (Simarro et al. 2015).

Acknowledgments This work is a contribution to the Spanish projects FORMED (CGL2012-33989). The authors wish to thank the Secretaría General de Pesca and Tragsa for the 2003 Espace Project dataset and the Spanish Oceanographic Institute for the ship time of the Copesands cruise. R. Durán is supported by a CSIC JAE-Doc contract co-funded by the FSE.

References

Acosta, J., Fontán, A., Muñoz, A., Muñoz-Martín, A., Rivera, J., Uchupi, E. (2013). The morpho-tectonic setting of the Southeast margin of Iberia and the adjacent oceanic Algero-Balearic Basin. Mar. Pet. Geol. 45, 17–41.

Ashley, G.M. (1990). Classification of large-scale subaqueous bedforms: a new look at an old problem. SEMP Bedforms and Bedding Structures Research Symposium. J. Sediment. Petr. 60, 160–172.

Bassetti, M.A., Jouet, G., Dufois, F., Berné, S., Rabineau, M., Taviani, M. (2006). Sand bodies at the shelf edge in the Gulf of Lions (western Mediterranean): deglacial history and modern processes. Mar. Geol. 234, 93–109.

Calvete, D., Falqués, A., de Swart, H. E., Walgreen, M. (2001). Modelling the formation of shoreface-connected sand ridges on storm-dominated inner shelves. J. Fluid Mech. 441,169–193.

Durán, R., Guillén, J., Simarro, G., Ribo, M., Puig, P., Muñoz, A., Palanques, A. (2015). Sand ridges in the mid-outer shelf as potential sand borrows areas (NE Mediterranean). Proceedings of the Coast. Sediments 2015, 1–13.

Fernández-Salas, L. M., Durán, R., Mendes, I., Galparsoro, I., Lobo, F. J., Bárcenas, P., Rosa, F., Ribó, M., García-Gil, S., Ferrín, A., Carrara, G., Roque, C., Canals, M. (2015). Shelves of the Iberian Peninsula and the Balearic Islands (I): Morphology and sediment types. Boletín Geológico y Minero 126, 327–376.

Goff, J.A., Austin, Jr., Gulick, S., Nordfjord, S., Christensen, B., Sommerfield, C., Olson, H., Alexander, C. (2005). Recent and modern marine erosion on the New Jersey outer shelf. Mar. Geol. 216, 275–296.

Goff, J.A., Duncan, L.S. (2012). Re-examination of sand ridges on the middle and outer New Jersey shelf based on combined analysis of multibeam bathymetry and backscatter, seafloor grab and chirp seismic data. Int. As. Sed. 44, 121–142.

Knaapen, M. A F. (2005). Sandwave migration predictor based on shape information. J. Geophys. Res. Earth Surf. 110, 1–9.

Li, Z., King, E.L. (2007). Multibeam bathymetric investigations of the morphology of sand ridges and associated bedforms and their relation to storm processes, Sable Island Bank, Scotian Shelf. Mar. Geol. 243, 200–228.

Liu, Z.X., Yin, P., Xiong, Y.Q., Berne, S., Trentesaux, A., Li, C.X. (2003). Quaternary transgressive and regressive depositional sequences in the East China Sea. Chinese Science Bulletin 48 (Suppl.), 81–87.

Lo Iacono, C., Guillén, J., Puig, P., Ribó, M., Ballesteros, M., Palanques, A., Farrán, M., Acosta, J. (2010). Large-scale bedforms along a tideless outer shelf setting in the western Mediterranean. Cont. Shelf Res. 30, 1802–1813.

Lobo, F.J., Hernández-Molina, F.J., Somoza, L., Rodero, J., Maldonado, A., Barnolas, A. (2000). Patterns of bottom current flow deduced from dune asymmetries over the Gulf of Cadiz shelf (southwest Spain). Mar. Geol. 164, 91–117.

Lobo, F.J., Durán, R., Roque, C., Ribó, M., Carrara, G., Mendes, I., Ferrín, A. (2015). Shelves around the Iberian Peninsula (II): Evolutionary sedimentary patterns. Boletín Geológico y Minero 126, 377–408.

Millot, C. (1999). Circulation in the Western Mediterranean sea. J. Marine Syst. 20, 423–442.

Nnafie, A., de Swart, H.E., Calvete, D., Garnier, R. (2014). Effects of sea level rise on the formation and drowning of shoreface-connected sand ridges, a model study. Cont. Shelf. Res. 80, 32–48.

Simarro, G., Guillén, J., Puig, P., Ribó, M., Lo Iacono, C., Palanques, A., Muñoz, A., Durán, R., Acosta, J. (2015). Sediment dynamics over sand ridges on a tideless mid-outer continental shelf. Mar. Geol. 361, 25–40.

Snedden, J.W., Tillman,R. W., Culver, S.J. (2011). Genesis and evolution of a mid-shelf, storm-built sand ridge, New Jersey continental shelf, U.S.A. J. Sediment. Res. 81, 534–552.

Part V

Bedforms and Benthos

Bedforms as Benthic Habitats: Living on the Edge, Chaos, Order and Complexity

30

V. Van Lancker

Abstract

Bedforms as benthic habitats are studied increasingly as acquisition and analysis of acoustic data improve in capturing, visualizing and quantifying terrain variables on various scales. However, feedback mechanisms between geomorphology and benthos are not always clear and complexity increases where humans also affect the benthos-landscape relationship. Based on research-oriented seabed mapping along the Belgian part of the North Sea (BPNS), a synthesis is provided on where increased biodiversity has been observed in relation to active bedforms.

Keywords

Sandbank • Dunes • Substrate • Sediment dynamics • Biogenic reefs • Marine Strategy Framework Directive

30.1 Study Area

The BPNS is a siliciclastic macro-tidal environment (tidal range of 4.5 m) comprising several groups of sandbanks (Fig. 30.1). Depths range from 0 to −50 m Mean Lowest Low Water at Spring (MLLWS). Mean grain sizes range from fine to medium sands varying along a subtle gradient. Sediment transport is mainly driven by tidal currents (max. 1.5 m s^{-1}), though wind-induced currents may have a direct effect on sediment resuspension and bedform morphology. Along the sandbanks, bedform patterns are simple to complex, with sand dunes on average 2–4 m high (100–200 m wavelength) having varying rates and directions of migration (on average 20 m $year^{-1}$ and oscillating). The BPNS is one of world's busiest continental shelf areas, with 30 years of disposal of dredged material and extraction of aggregates, as well as more than 150 years of bottom-disturbing fisheries. Human activities on the BPNS influence natural sediment fluxes, e.g., by dispersion and deposition of plumes originating from the disposal of dredged material, marine aggregate extraction, wind mill farms and fishing activities.

V. Van Lancker (✉)
Operational Directorate Natural Environment, Royal Belgian Institute of Natural Sciences, Gulledelle 100, 1200 Brussels, Belgium
e-mail: vera.vanlancker@naturalsciences.be

30.2 Methodology

Aggregations of ecosystem engineering species (e.g. tubeworms, clams) in bedform areas were derived from side-scan sonar (GeoAcoustics side-scan sonar; 410 kHz, Fig. 30.2) and multibeam imagery (Kongsberg Simrad EM1002/99 kHz (Fig. 30.3) and EM3002/300 kHz (Fig. 30.4). To understand the relationships between dense species aggregations and their physical habitat, current and backscatter profiling (acoustic Doppler current profiling) were also carried out. Sediment and biological samples were obtained for validation (Degraer et al. 2008; Rabaut et al. 2009; Houziaux et al. 2012). Here, only results from acoustic imagery are shown.

30.3 Results

Three cases (A-B-C) are presented in which local higher relief structures ('mounds' sensu Van Lancker et al. 2012), as depicted by detailed terrain analyses of acoustic imagery,

J. Guillén et al. (eds.), *Atlas of Bedforms in the Western Mediterranean*,
DOI 10.1007/978-3-319-33940-5_30

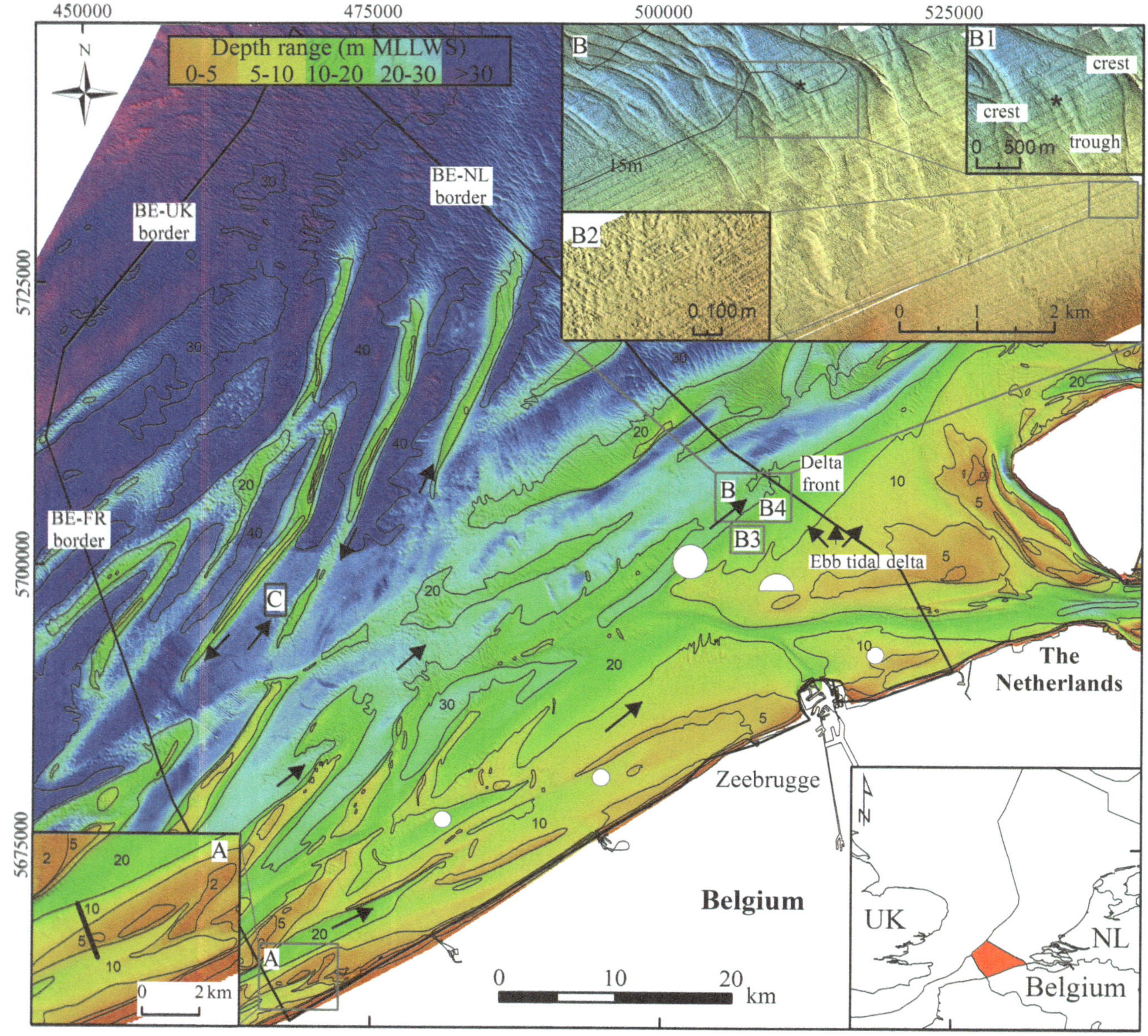

Fig. 30.1 Sandbanks along the Belgian part of the North Sea, with the locations of the three study areas: *A* Shallow sandbank off the French-Belgian border. *B* Front of an ebb-tidal delta where a large dune field occurs at the end of a flood channel. B1 shows a biogenic reef (*) in the trough of large dunes inhabited mainly by the tubeworm *O. fusiformis*; Facies B2 is also indicative of dense aggregations of the latter, but high abundances of the razor clam *E. directus* were also observed; for B3 and B4, see Fig. 30.3. *C* Barchan dune area attached to a linear sandbank, with ecologically valuable gravel beds in their troughs. *(Half) circles* indicate the position and size of disposal grounds of dredged material. *Arrows* represent the main sand transport directions

corresponded with local biodiversity increases. These zones were found (1) on the lee side of topzones of shallow sandbanks (Fig. 30.1, Location A; Fig. 30.2); (2) in dune areas with complex sediment transport pathways (Fig. 30.1, Location B; Fig. 30.3); and (3) near the lee side of complex dunes (Fig. 30.1, Location C; Fig. 30.4). Common explanatory denominators for the benthos occurrences were order, complexity, and also chaos.

Order relates to the fact that in the three cases a steeper morphological gradient associated with enhanced tide-topography effects was the first predictor of increased biodiversity. Acoustically, this translated into circular to elongated mounds 10–40 cm in diameter. These were seen mostly on the lee side of sandbanks (Figs. 30.1A and 30.2) or large dunes (Figs. 30.1B, C and 30.4), where species can shelter from the overall higher sediment dynamics (e.g.,

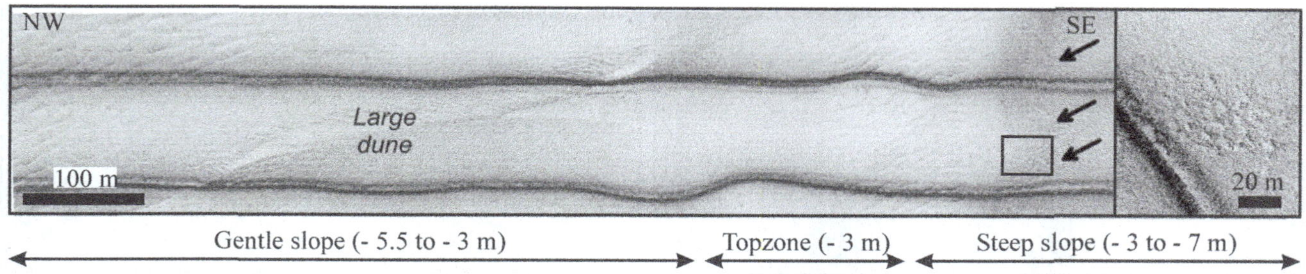

Fig. 30.2 Side-scan sonar imagery showing occurrences of circular to elongated mounds (15–40 cm in height) (*inset*) corresponding to colonies of the tube worm and the ecosystem engineer *L. conchilega* along the lee side of topzones of shallow sandbanks (profile Fig. 30.1 A). From west to east, mean grain sizes ranged from medium sands (280 µm) to fine sands (168 µm), with 5 % mud enrichment. See also Degraer et al. (2008) and Van Lancker et al. (2012)

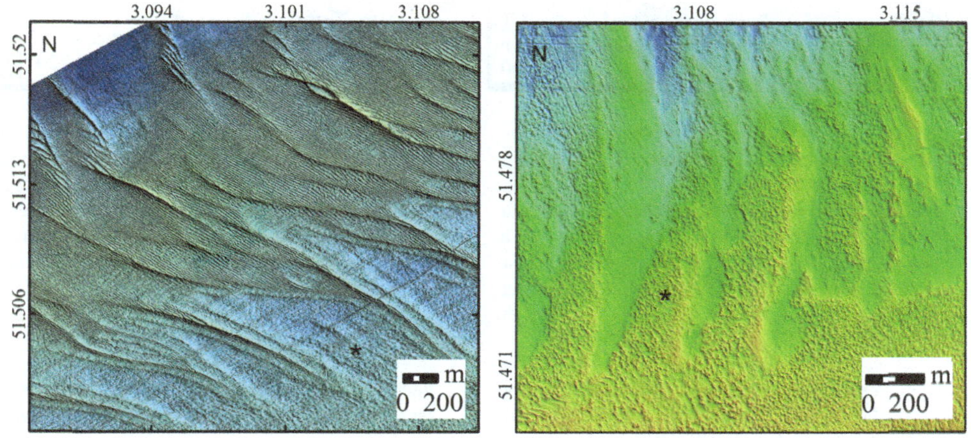

Fig. 30.3 Bedform occurrences along the delta front of the Westerschelde ebb-tidal delta (Fig. 30.1B). *Left* Very large NW–SE oriented dunes (−22 m MLLWS) with superimposed smaller dunes. In the troughs of the very large dunes, circular to elongated mounds (*) (40–50 cm in height) occurred, corresponding to colonies of the tube worm and ecosystem engineer *O. fusiformis*. *Right* Sandy large dunes (−12 m MLLWS) overtopped with muddy sediments (*). The latter comprised dense aggregations of the tube worm *O. fusiformis* (>12000 ind. m^{-2}) and the razor clam *E. directus* (>500 ind. m^{-2}). Here, deposition occurred from along-delta sand fluxes that combined with a cross-delta flux of muddy sediments, both naturally and anthropogenically-induced (e.g. from disposal of dredged material, Fig. 30.1B). In this area, the habitat changed from sand to sandy mud (up to 40 % mud). See also Van Lancker et al. (2013)

Fig. 30.1B1). However, some form of complexity was necessary for species to thrive. The dunes were mostly coarser-grained, and their morphology was more complex than that of their surroundings (e.g. 3D dunes, or very steep dunes), but above all fine-grained sediment flows were measured interacting with the coarser substrate. The transit and/or deposition of this flow, comprising also food and larvae, is attractive to suspension feeders (e.g. ecosystem engineering tube worms and clams). These profit from the robust substrate, whose coarseness is favourable for detritus trapping and for better oxygenation. Species aggregations were most dense where, in the longer term, the interaction of the finer-grained material with the coarser substrate led to chaotic or patchy sediment distributions (Fig. 30.3, right). This is most likely to occur in sand dune areas, where trapping of fine-grained material from sediment plumes, often generated by anthropogenic activities, is most effective (Fig. 30.3). This is not necessarily adverse for benthic communities, but a threshold likely exists up to which species can cope with the excess of fine sediments (i.e. living on the edge).

Figure 30.4 is another example in which steep dunes ('complex') with topography-induced sediment fluxes ('order'), in combination with sedimentation from fine sediments ('chaos') resulted in higher biodiversity in their troughs. Here, the focus is on gravel beds that locally host rich epifauna communities.

Since the geological substrate, as well as sediment dynamics, are important predictors of biodiversity, these are now revisited systematically (Van Lancker et al. In Press) and will assist in assessing 'Good Environmental Status' within Europe's Marine Strategy Framework Directive.

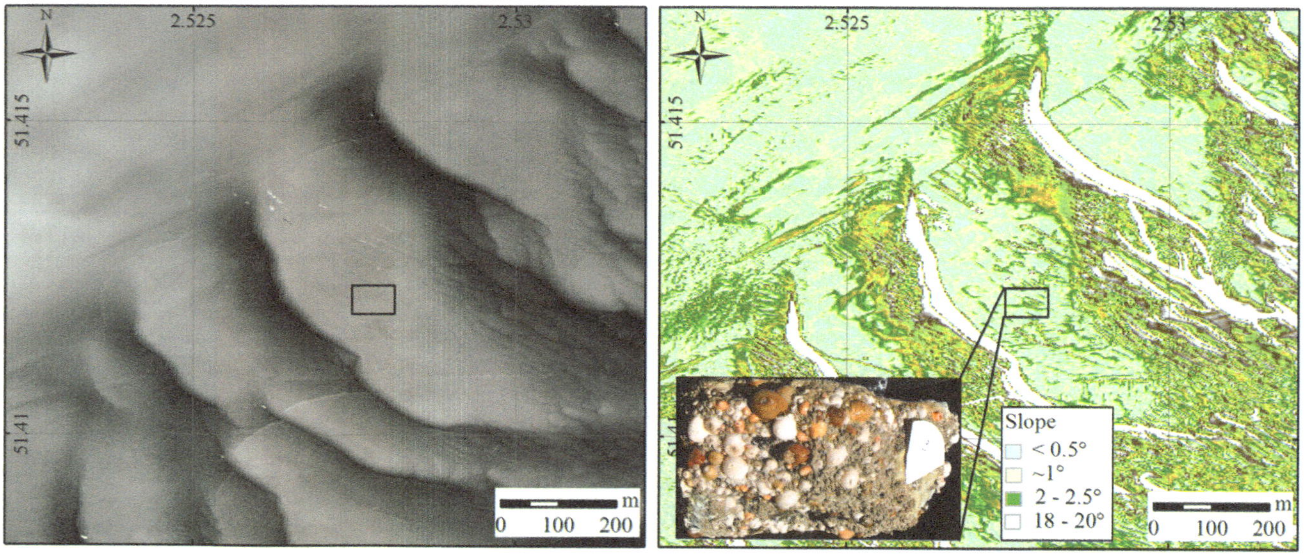

Fig. 30.4 Bathymetry (*left*) and slope (*right*) map of barchan dunes, where the dune troughs host gravel beds overgrown with epifauna (−30 m MLLWS), and richer species compositions. The dunes are 6–7 m in height with wave lengths of around 200 m. The complex dune morphology protects the fauna from abrasion from fisheries, though it is hypothesized that the nature of biodiversity in the dune troughs is also related to acceleration and deceleration of the current over the steep dunes (~20°), resulting in trapping of muddy sediments near the lee side. The source of the fine-grained material may be natural, or anthropogenically-induced (e.g. aggregate extraction). In the longer term, depending on the nature of the flora and fauna, the addition of the fine-grained material may be beneficial (nutrient input) or adverse (smothering) for biodiversity. The slope variations in the troughs (up to 2°–2.5°) are indicative of the occurrence of gravel beds. The inset is a sampled gravel block (20 cm)

Acknowledgments This chapter contributes to the Belspo (www.belspo.be) projects HABITAT (MN/02/89), MAREBASSE (EV/02/18A), QUEST4D (SD/NS/06B), EnSIS (SD/NS/09A) and TILES (BR/121/A2/TILES), as well to contracts from the Agency of Maritime and Coastal Services (HABITAT/99380/200.455; MOZ4/211.177) and from private funds (ZAGRI). Special thanks are due to the officers and crew of the R/V *Belgica*. Shiptime was provided by Belspo and the Royal Belgian Institute of Natural Sciences.

References

Degraer, S., Moerkerke, G., Rabaut, M., Van Hoey, G., Du Four, I., Vincx, M., Henriet, J.P. & Van Lancker, V. (2008). Very high resolution side-scan sonar mapping of biogenic reefs of the tube-worm *Lanice conchilega*. Remote Sensing of Environment 112, 3323–3328.

Houziaux, J.-S., Craeymeersch, J., Merckx, B., Kerckhof, F., Van Lancker, V., Courtens, W., Stienen, E., Perdon, K.J., Goudswaard, P.C., Van Hoey, G., Virgin, L., Hostens, K., Vincx, M. & Degraer, S. (2012). 'EnSIS' - Ecosystem Sensitivity to Invasive Species. Final Report. Belgian Science Policy Office: Brussels. 105 pp.

Rabaut, M., Du Four, I., Nakas, G., Van Lancker, V.R.M., Degraer, S. & Vincx, M. (2009). Ecosystem engineers stabilize sand bank systems: *Owenia fusiformis* aggregations as ecologically important microhabitat, *in*: Rabaut, M. (2009). *Lanice conchilega, fisheries and marine conservation: Towards an ecosystem approach to marine management*. pp. 273–297.

Van Lancker, V., Moerkerke, G., Du Four, I., Verfaillie, E., Rabaut, M. & Degraer, S. (2012). Fine-scale geomorphological mapping for the prediction of macrobenthic occurrences in shallow marine environments, Belgian part of the North Sea, ***in***: Harris, P. and E. Baker (Eds.) 2012. Seafloor Geomorphology as Benthic Habitat: GeoHab Atlas of seafloor geomorphic features and benthic habitats: 251–260. Elsevier Insights.

Van Lancker, V., Houziaux, J.S., Baeye, M., Van den Eynde, D., Rabaut, M., Troost, K., Vermaas, T., van Dijk, T.A.G.P. (2013). Biogeomorphology in the field: bedforms and species, a mystic relationship, ***in:*** Van Lancker, V. et al. (Ed.) (2013). MARID 2013: Fourth International Conference on Marine and River Dune Dynamics. Bruges, Belgium, 15–17 April 2013. VLIZ Special Publication, 65: pp. 277–283.

Van Lancker, V., Francken, F., Kint, L., Terseleer, N., Van den Eynde, D., De Mol, L., De Tré, G., De Mol, R., Missiaen, T., Chademenos, V., Bakker, M., Maljers, D., Stafleu, J. & van Heteren, S. in press. Building a 4D Voxel-Based Decision Support System for a Sustainable Management of Marine Geological Resources. In: Diviacco, P., Leadbetter, A. & Glaves, H. (eds.). Oceanographic and Marine Cross-Domain Data Management for Sustainable Development. IGI Global. DOI: 10.4018/978-1-5225-0700-0

Benthic Communities on Shallow Sedimentary Bottoms in the Western Mediterranean

31

J. Mas, I. Franco, M. Demestre, J. Guillén, F.J. Murcia, and J.M. Ruiz

Abstract

Benthic communities on shallow sedimentary bottoms in the western Mediterranean can play a crucial role in the development, maintenance and evolution of bedforms. This chapter presents an overview of the most characteristic benthic communities in shallow waters classified by the type of substrate: coarse bottoms, sandy bottoms and fine-grained bottoms. Examples from the Murcia continental shelf are provided to illustrate the main communities, giving special emphasis to the seagrass beds (some of them endemic to the Mediterranean) and the scarce maërl community, both with high biodiversity and ecological value.

Keywords

Bedforms • Coastal zone • Benthic communities • Maërl • *Posidonia* meadows • "Soft" and "hard" bottoms

J. Mas (✉) · I. Franco · J.M. Ruiz
Centro Oceanográfico de Murcia (IEO), c/Varadero 1, 30740 San Pedro del Pinatar-Murcia, Spain
e-mail: julio.mas@mu.ieo.es

M. Demestre · J. Guillén
Institut de Ciències del Mar (ICM-CSIC), Passeig Maritim de la Barceloneta 37-49, 08003 Barcelona, Spain

F.J. Murcia
http://www.javiermurcia.com.es

J. Guillén et al. (eds.), *Atlas of Bedforms in the Western Mediterranean*,
DOI 10.1007/978-3-319-33940-5_31

31.1 Introduction

The formation and maintenance of shallow marine bedforms are highly influenced by the benthic community that has developed on the sedimentary seabed. Conversely, the dynamics of bottom sediments influence the development of benthic communities, although the feedback mechanisms between geomorphology and benthos are not always clear (Van Lancker et al. 2013). In general, benthic communities modify the interaction between the seabed and hydrodynamics by reworking the seabed, thus changing the bottom roughness or sediment erodibility (Guillén et al. 2008). Therefore, the characterization of the benthic communities in the western Mediterranean provides the background information required for a better understanding of bedform dynamics.

Infralittoral sedimentary bottoms are characterized by the grain size of the particles of which they are composed, which are in turn affected by local hydrodynamics and the source of the sediment (continental or marine). These bottoms determine the composition, diversity and abundance of the various species that make up their communities. Seabed bottoms can be composed of mud and sand ("soft substrate") or of cobbles and pebbles ("hard substrate"). Different substrate characteristics give rise to a variety of habitats, communities and biocenoses with a prevalence of endobiotic and epibiotic organisms, algae and marine angiosperms, bivalves, gastropods, cephalopods, crustaceans and annelids, with numerous polychaetes, echinoderms and species of fish, among other taxonomic groups.

Benthic communities settled in sedimentary environments in western Mediterranean coastal areas have been the subject of a wealth of scientific literature dealing with species and habitats, especially those that generate bioconstructions (Pérès and Picard 1964; Bellán-Santini 1994, 2002). The environmental importance of these communities has also been clearly stated (Council Directive 1992; Cochrane et al. 2010). The coast of Murcia (SE Spain) is characterized by a narrow continental platform leading to many submarine canyons reaching a depth of 2500 m (Fig. 31.1). In the shallow part between 0 and 30 m, meadows of *Cymodocea nodosa* and *Posidonia oceanica* are generally well conserved. At increasing depth we find infralittoral and circalittoral sedimentary seabeds with biogenic sources of large numbers of free-living coralline algae forming maërl beds. The present paper reviews the characteristics of the benthic communities of the shallow bottoms in the SE of the Iberian Peninsula, paying special attention to meadow and maërl seabeds as an example of the benthic communities living in shallow areas of the western Mediterranean.

31.2 Benthic Communities

31.2.1 Very Coarse Bottoms

Cobble and pebble bottoms are typically associated with strong currents that winnow finer sediment (Fig. 31.2). They usually lack a permanent vegetation cover, with the exception of certain encrusting species, but algae such as *Acetabularia* can take root when currents are of lower intensity. Sessile fauna is similarly sparse, consisting principally of anemones and sea squirts. The mobile species observed include chitons, gastropods such as green ormer (*Haliotis tuberculata*), decapods (*Palaemon serratus, Xanto poresa*), echinoderms such as starfish (*Asterina gibbosa*), and the common brittlestar (*Ophiothrix fragilis*). Fishes are represented by various species of clingfish (genus *Lepadogaster*), as well as blennies and gobies.

Gravel substrates and coarse sands also lack stable vegetation and show a prevalence of endofauna, including bivalves (*Venus verrucosa, Callista chione*), polychaetes and other kinds of sea worm, gastropods, sea squirts and bryozoans (Fig. 31.2).

31.2.2 Sandy Bottoms

The sandy infralittoral seabed usually extends from the coastline to a depth of 10 m, where the silt fraction is incorporated into the sediment and progressively increases its content with water depth (Fig. 31.3). Bivalves are the dominant species in the first few metres, in the form of different kinds of clam, including the striped Venus shell (*Chamelea gallina*), tellins (*Tellina spp.*) and gastropods of the genus Nassarius. As the seabed drops away, the scarcity of vegetation is maintained, as is the predominance of bivalves (*Ruditapes decussatus, Callista chione* and *Spisula subtruncata*), gastropods (*Nassarius reticulatus*), polychaetes, crustaceans (*Diogenes pugilator, Crangon crangon*), isopods, amphipods, and echinoderms such as starfish of the genus *Astropecten*.

Other common species that find both food and shelter on these sandy bottoms are irregular sea-urchins such as the sea potato (*Echinocardium cordatum*), flat fish such as the wide-eyed flounder (*Bothus podas*), pearly razorfish (*Xyrichthys novacula*), sea spiders, the marbled electric ray (*Torpedo marmorata*) (Fig. 31.3) and the common eagle ray (*Myliobatis aquila*).

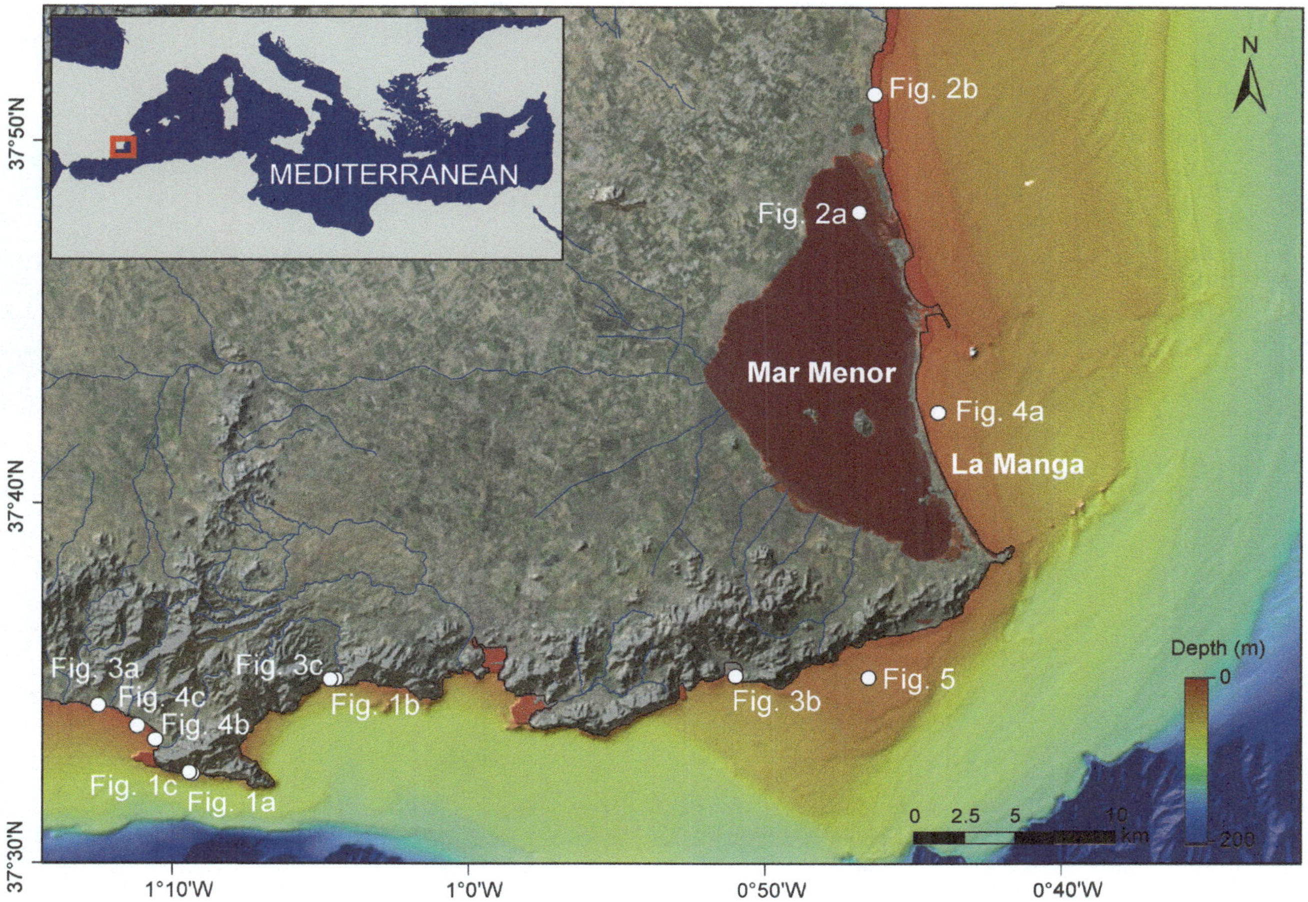

Fig. 31.1 Location of the study area. The *shaded relief* bathymetry of the continental shelf from the Secretaría General del Mar and Tragsa (ESPACE Project). Onland background information from Instituto Geográfico Nacional de España and ESRI World Imagery

31.2.3 Fine-Grained Bottoms

Silty sand and muddy bottoms are mainly found in protected shallow coastal areas such as coves or enclosed bays, coastal lagoons, deltas and ports. Among the coastal lagoons, the Mar Menor (SE Iberian Peninsula) displays a series of characteristics that set it apart from the rest, with a predominance of fine sedimentary bottoms and only the occasional igneous rock outcrop. Here the seabed is covered with thick meadows of blade algae (*Caulerpa prolifera*), sometimes as a single species and sometimes combined with the phanerogams slender seagrass (*Cymodocea nodosa*) and dwarf eelgrass (*Zostera noltii*), populated by gastropods such as the horn nassa (*Nassarius corniculum*), the white furrow shell (*Abra alba*), the green bubble snail (*Oxynoe olivacea*), the needle whelk Bittium reticulatum), the grooved carpet shell (*Tapes decussates*), the banded dye-murex (*Hexaplex trunculus*) and the mud snail (*Cyclope neritea*) (Fig. 31.4). Crustaceans include the common prawn (*Palaemon serratus*) and the caramote prawn (*Melicertus kerathurus*), as well as several species of amphipods and polychaetes. Fish species inhabiting the Mar Menor lagoon include various species of mugilids and others characterized by their euryhaline and eurythermal capabilities, such as the gilt-head bream (*Sparus aurata*) and the European seabass (*Dicentrarchus labrax*), gobies, blennies, and small pelagic species such as smelts (genus *Atherina*). Other commonly found populations are those of the European eel (*Anguilla anguilla*), syngnathids such as seahorses and pipefish, and endemic species such as toothcarp of the genus *Aphanius*, in the Cyprinodontidae family.

A singular example of fine-grained bottoms are the silty sand bottoms that are highly influenced by anthropogenic activities, such as mining in the SE Iberian Peninsula, where the materials dumped into the sea contain a high proportion of heavy metals (Baños-Gonzalez et al. 2013) (Fig. 31.4). This

Fig. 31.2 Coarse bottoms. **a** Characteristic bottom of cobbles (Cala Cerrada, Cartagena, 1.5 m depth). **b** Coarse sands and gravels (El Portús, Cartagena, 1.5 m depth). **c** Rippled bottom in coarse sands, with a cnidarian of the genus *Cerianthus* (about 5 cm in diameter) in the foreground (Cala Cerrada, Cartagena, 4 m depth). Location of photographs is in Fig. 31.1

Fig. 31.3 Sandy bottoms. **a** Mollusc *Bittium sp.* and meadow of *Cymodocea nodosa* in the background (San Javier, Mar Menor, 1.5 m depth). **b** Wave generated ripples developed on a well-sorted fine sand bottom free of macroscopic vegetation (El Mojón, San Pedro del Pinatar, 2 m depth). Location of photographs is in Fig. 31.1

Fig. 31.4 Fine-grained bottoms. **a** Rippled silty sand bottom with significant organic matter content. Abundant infauna in the form of molluscs, annelids and other taxonomic groups, generating a specific biological roughness on the seabed (Isla Plana, Cartagena, 3 m depth). **b** Sedimentary seabed covered by fine material coming from mining waste, with a high heavy metal content. The presence of fauna is quite limited and no vegetation has developed. (Portman, La Unión, 4 m depth). **c** Fine-sediment seabed which is the characteristic habitat of the cartilaginous fish *Torpedo marmorata*. (El Portús, Cartagena, 10 m depth). Location of photographs is in Fig. 31.1

pollution clearly reduces the fauna and flora found in areas close to the dumping point, with a marine biodiversity gradient that increases with distance (Benedicto et al. 2012). In this environment, algal flora is scarce, seagrass meadows have largely disappeared as a result of the combined effects of burial under sediment and water turbidity, and only a few bivalves such as the striped Venus clam (*Chamelea gallina*), gastropods such as the banded dye-murex (*Hexaplex trunculus*), and various species of Mediterranean limpets and mussels (e.g. Mytillus galloprovincialis) are able to maintain a permanent presence (Benedicto et al. 2011). However, several species of pelagic invertebrates and vertebrates make sporadic use of the water column and of these polluted seabeds.

31.2.4 Meadows

Marine angiosperms form what are commonly known as phanerogam meadows or lawns, and although some species are able to take root on a hard substrate, they usually do so in sediment. They constitute an ecosystem in themselves, and are considered to be the 'climax' vegetation in the Mediterranean because of their complexity and ecological maturity (Ros et al. 1985). They have a direct impact on coastal dynamics, playing host to numerous species and thereby significantly increasing biodiversity and enriching food chains, as well as providing room for spawning, shelter and feeding for other members of the associated fauna and flora. By occupying large expanses of the seabed in well-lit coastal areas, they produce a large amount of biomass, absorb significant quantities of carbon dioxide, favour oxygen production and increase the transparency of the water as a result of their filtering effect. Their rhizomes also help to stabilize the sediment, while the barrier effect favours wave dissipation and helps reduce wave-induced bottom sediment transport (Bellán-Santini et al. 1994, 2002).

Posidonia oceanica meadows are endemic in the Mediterranean (Fig. 31.5). They colonize the seabed from a

Fig. 31.5 Meadows. **a** Seabed of *Posidonia oceanica*, an endemic species in the Mediterranean and a fine indicator of good environmental conditions (La Manga, Cartagena, 4 m depth). **b** Silty sand seabed colonized by *Zostera noltii* in the foreground and *Posidonia oceanica* in the background, with two female grey wrasse (*Symphodus cinereus*) (La Azohía, Cartagena, 2 m depth). **c** Low-density meadow of *Cymodocea nodosa* developed on a rippled, well-sorted fine sand bottom. A starfish of the genus *Astropecten* in the foreground (La Azohía, Cartagena, 4 m depth). Location of photographs is in Fig. 31.1

few metres from the coastline to depths of 30 m. In areas of especially clean waters such as the Balearic Islands, they can extend up to 40 m depth (Ballesteros et al. 2007). This species shows little tolerance to changes in salinity and turbidity and to high sedimentation rates, thus being considered a marker of unpolluted waters with a high oxygen content and transparency. In the meadow itself we can differentiate between the foliar stratum, corresponding to the leaves of the plants and subject to an ongoing process of renewal, and the rhizome stratum, corresponding to the network of roots that anchor the plants to the substrate. The foliar stratum is home to a highly complex photophilic community containing a large number of environmental niches. Few species actually feed directly on the leaves, the exceptions being the fish known as the cow bream (*Salpa salpa*) and those that graze on their epiphytes, such as the purple sea urchin (*Paracentrotus lividus*). The fish community is particularly well represented, with a prevalence of sparids (*Diplodus annularis*), labrids (*Labrus viridis, Symphodus mediterraneus, Symphodus tinca*) and syngnathids (*Hippocampus hippocampus, Syngnathus acus*) species, although the species best equipped to survive on the leaves of the seagrass meadow is perhaps the clingfish (*Opeatogenys gracilis*).

The species that live or feed among seagrass leaves include the community of brown algae (*Myrionema orbiculare*), coralline algae (*Hydrolithon farinosum*) and filamentous algae. Other species that colonize the foliar stratum are hydroids (*Plumularia obliqua*), sea squirts (*Botryllus schlosseri*), bryozoans (*Electra posidoniae*), sponges, isopods (*Idotea hectica*), molluscs (*Rissoa variabilis, Tricolia speciosa*) and echinoderms (*Asterina pancerii*).

The fauna that moves by swimming around this photophilic layer includes various species of crustaceans, amphipods, mysids, decapods, various kinds of shrimp (*Hippolyte inermis*) and a few cephalopods such as the common cuttlefish (*Sepia officinalis*) and octopus (*Octopus vulgaris*). The above gives us some idea of the enormous biodiversity of the Mediterranean seagrass meadows, some of which have been recorded as containing upwards of 800 different species (Templado et al. 2009, 2012).

Cymodocea nodosa meadows (Fig. 31.5) or lawns not only look very different to those of *Posidonia oceanica* but also show a very different annual cycle: smaller in size and occupying less of the seabed surface, they can lose all of their leaves during the winter period. Unlike *Posidonia*, they have a wide tolerance to environmental conditions, which makes them a colonizing species at depths of up to 30 m. They are commonly found in coastal lagoons, where they combine with the green alga *Caulerpa prolifera* to form meadows. At locations where the sediment is silty and rich in organic material, they combine with the phanerogam *Zostera noltii*. When they coincide spatially with *Posidonia oceanica* meadows, they are able to colonize its areas of dead undergrowth, or to establish a belt above the upper and lower horizons of these meadows.

Neither the flora nor the fauna associated with *Cymodocea nodosa* meadows is as abundant as those found in *Posidonia* meadows. They include epiphytic coralline algae, some of which they share in common, other filamentous algae (*Ectocarpus siliculosus var. confervoides*), sea anemones (*Paranemonia cinerea*), gastropods (*Gibbula leucophaea, Rissoa monodonta*) and clingfish (*Opeatogenys gracilis*). In the rhizome stratum we can find sea cucumbers (*Holoturia tubulosa*), gastropods (*Bulla striata*), bivalves (*Spisula subtruncata*) and irregular sea-urchins (*Echinocardium mediterraneum*).

The Mediterranean littoral marine ecosystem also contains meadows of *Zostera marina* and *Zostera noltii*, sometimes growing together, which show very similar characteristics to those described for *Cymodocea nodosa*.

31.2.5 Maërl Beds

When continental detrital inputs to the continental shelf are limited, the infralittoral and circalittoral seabeds are mainly composed of particles of biological origin. They are frequently covered with large numbers of free-living coralline alga, forming what are commonly known as maërl beds (Fig. 31.6). Each individual specimen of these algae is known as a rhodolith because of its spherical shape, produced at times by the rubbing effect caused by the hydrodynamics of the waters, and in calmer oceanographic conditions by its densely branched structure. In the Mediterranean maërl beds can be found at depths of between 30 and 150 m, and are highly dependent on transparent waters and medium—to high-intensity currents. Different characteristics of the maërl community and their relationship with bedforms are presented in Demestre et al. (this issue).

Maërl beds are made up of calcareous algae (*Phymatolithon calcareum, Lithothamnion coraloides*), combined with algae of the genus *Peyssonnelia*. The floating substrate thus formed provides an anchor for other types of algae of the red Gigartinales and Ceramiales orders and certain green species, such as *Codium bursa*. Other organisms that adhere to these bottoms are sponges (*Clionia viridis*), sea squirts (*Phallusia mamillata*), bryozoans (*Myriapora truncata*) and various kinds of cnidarians and tubicolous polychaetes (Calvín et al. 1998).

Fig. 31.6 Maërl community characterized by the presence of rhodoliths with a red starfish *Echinaster sepositus* (Calblanque, Cartagena, 37 m depth). Scale 80 × 50 cm. Location of the photography is in Fig. 31.1

The mobile epifauna includes sea-urchins (*Genocidaris maculata*), starfish (*Chaetarser longipus*), various species of crab and the pygmy locust lobster (*Scyllarus pygmaeus*). Fish include scorpaenids such as *Scorpaena notata*, gobies and the serranid *Serranus hepatus*.

31.3 Concluding Remarks

This chapter presents a general overview of the characteristic habitats and benthic communities that have developed in the Murcia coastal area as an example of the shallow benthic communities in the western Mediterranean. The communities of bottoms of very coarse sediment, sandy, fine-grained meadows and maërl are described. Special attention is given to the most representative organisms and their complexity and ecological importance because they provide a significant increase in biodiversity. Furthermore, the habitats of *Posidonia* and maërl give either food or shelter to important target fishing species such as bivalves, crustaceans and fishes. This information reveals the importance of studying interactions between benthos and geomorphology for a better understanding of bedform dynamics because, except in densely vegetated meadows, bedforms develop in all these habitats (see examples in this book).

References

Baños-Gonzalez, I. Baños, P. (2013). Portman: De El Portus Magnus del Mediterraneo Occidental a la bahía aterrada. Ed. Universidad de Murcia: 502 pp.

Ballesteros, E. Torras, X., Pinedo, S. García, M., Mangialajo, L., de Torres, M. (2007). A new methodology based on littoral community cartography dominated by macroalgae for the implementation of the European Water Framework Directive. *Marine Pollution Bulletin* 55: 172–180.

Bellan-Santini D., Bellan G., Bitar G., Harmelin J.G., Pergent G. (2002). Manuel d´interprétation des types d´habitats marins pour la sélection des sites à inclure dans les Inventaires Nationaux de Sites Naturels d´Intérêt pour la Conservation. *Programme des Nations Unies pour l´Environnement, Plan d´Action pour la Méditerranée.* 197 pp.

Bellan-Santini D., Lacaze J.C., Poizat C. (1994). Les biocénoses marines et littorales de Méditerranée, synthèse, menaces et perspectives. *Collection Patrimoines Naturels, Secrétariat Faune et Flore, M.N.H.N.*, Paris. 246 pp.

Benedicto, J.; Andral, B.; Martínez-Góomez, C.; Guitart, C.; Deudero, S.; Cento, A.; Scatrppato, A.; Caixach, J.; Benbrahim, S.; Chouba, L.; Boulahdidi, M., Galgani, F. (2011). A large scale survey of trace metals levels in coastal waters of the Western Mediterranean basin using caged mussels. *Journal of Environmental Monitoring*, 13: 1495–1505.

Benedicto, J.; Fernández, J.M.; Martínez-Gómez, C.; Fernández, B.; Mas, J.; Campillo, J.A., León, V. (2012). Informe del estado preoperacional del medio marino de la Bahía de Portman (La Unión-Murcia) y su entorno para la obra de regeneración y adecuación ambiental. Informe Interno del Instituto Español de Oceanografía.

Calvín, J.C., Franco, I. Marín, A., Martínez-Ingles, A. Belmonte, J.M. Ruiz, J.M. (1998). El litoral sumergido de la Región de Murcia: Cartografía bionómica y valores ambientales. *Dirección General de Medio Ambiente, Agricultura y Agua. Región de Murcia.* 128 pp.

Cochrane, S.K.J., Connor, D.W, Nilsson, P., Mitchell, I., Reker, J., Franco, J., Valavanis, V, Moncheva, S., Ekebom, J., Nygaard, K., Serrão Santos, R., Narberhaus, I., Packeiser, T., van de Bund W., Cardoso, A.C. (2010). Marine Strategy Framework Directive Task Group 1, *Report Biological diversity*. Ed: N.Zampoukas. EUR Scientific and Technical Research series, ISSN 1018-5593 ISBN 978-92-79-15650-2, doi: 10.2788/86653: 120 pp.

Demestre, M., Muntadas, A., Sánchez, P., García de Vinuesa, A., Franco, I., Más, J., Durán, R. and Guillén, J. (2016). Bio and anthropogenic disturbance on maërl communities in subaqueous dunes of the Mar Menor shelf (SW Mediterranean). In Guillén et al (eds): Atlas of bedforms in the western Mediterranean. Springer. (this issue).

Council Directive 92/43/EEC of 21 May 1992 on the conservation of natural habitats and of wild fauna and flora. (DO L 206 de 22.7 1992). 7 pp.

Guillén, J., Soriano, S., Demestre, M., Falqués, A., Palanques, A., Puig, P. (2008). Alteration of bottom roughness by benthic organisms in a sandy coastal environment. *Continental Shelf Research*, 28: 2382–2392.

Pérès, J.M., Picard J. (1964). Nouveau manuel de bionomie benthique de la Mer Méditerranée. *Recueil des Travaux de la Station Marine d´Endoume*. Bulletin nº34, fasc. Nº7. 137 pp.

Ros, J., Romero, J., Ballesteros, E., Gili, J.M. (1985). Diving in blue water: the benthos. In Western Mediterranean, R. Margalef (ed.), *Oxford: Pergamon*, 233–295.

Templado, J. Guallart, J., Capa, M., Luque, A. (2009). 1170 Arrecifes. In: W.AA., Bases ecológicas preliminares para la conservación de los tipos de hábitats de interés comunitario en España. Madrid: *Ministerio de Medio Ambiente, y Medio Rural y Marino*. 142 pp.

Templado, J. Ballesteros, E., Galparsoro, I., Borja, A., Serrano, A., Marín, L., Brito, A. (2012). Guía interpretativa. Inventario español de hábitats marinos. *Ministerio de Agricultura, Alimentación y Medio Ambiente*. 229 pp. http://www.magrama.gob.es/es/costas/publicaciones/GUIA_INTERP_HABITATS_WEB_tcm7-270736.pdf.

Van Lancker, V., Houziaux, J.S, Baeye, M., Van den Eynde, D., Rabaut, M., Troost, K., Vermaas, T., van Dijk, T.A.G.P. (2013). Biogeomorphology in the field: bedforms and species, a mystic relationship. Marine and River Dune Dynamics, MARID IV, April 2013, Bruges, Belgium: 277–283.

32 Characterization of Benthic Communities in a Subaqueous Dune Field on the Continental Shelf (Mar Menor, Western Mediterranean)

M. Demestre, A. Muntadas, R. Duran, A. García-de-Vinuesa, P. Sánchez, J. Mas, I. Franco, A. Muñoz, and J. Guillén

Abstract

Multibeam swath bathymetry, sediment and benthos samples (including both epifauna and infauna) and submarine images were explored to characterize benthic communities in a subaqueous dune field on the Mar Menor middle shelf (western Mediterranean). The dunes are 2–5 m high and spaced 400–1100 m apart, with a predominant NE-SW orientation oblique to the shoreline. The benthic communities show high spatial variability in diversity. Three types of habitat showing a different species composition are distinguished in the subaqueous dune field.

Keywords

Benthic communities • Bedforms • Epifauna • Infauna • Middle shelf • Mediterranean

32.1 Introduction

The Mar Menor is a coastal lagoon located in the western Mediterranean (Fig. 32.1a). It is separated from the open sea by a 20 km-long and 100–900 m-wide barrier island (Conesa and Jiménez-Cárceles 2007). The continental shelf off the Mar Menor lagoon has a mean width of 38 km, with the

M. Demestre (✉) · A. Muntadas · R. Duran · A. García-de-Vinuesa · P. Sánchez · J. Guillén
Institut de Ciències del Mar (ICM-CSIC), Passeig Marítim de ;la Barceloneta, 37-49, 08003 Barcelona, Spain
e-mail: montse@icm.csic.es

J. Mas · I. Franco
Centro Oceanográfico de Murcia (IEO), C/Varadero 1, Lo Pagán, San Pedro del Pinatar, Murcia, Spain

A. Muñoz
Tragsa-SGP, C/Julián Camarillo 6B, 28037 Madrid, Spain

J. Guillén et al. (eds.), *Atlas of Bedforms in the Western Mediterranean*,
DOI 10.1007/978-3-319-33940-5_32

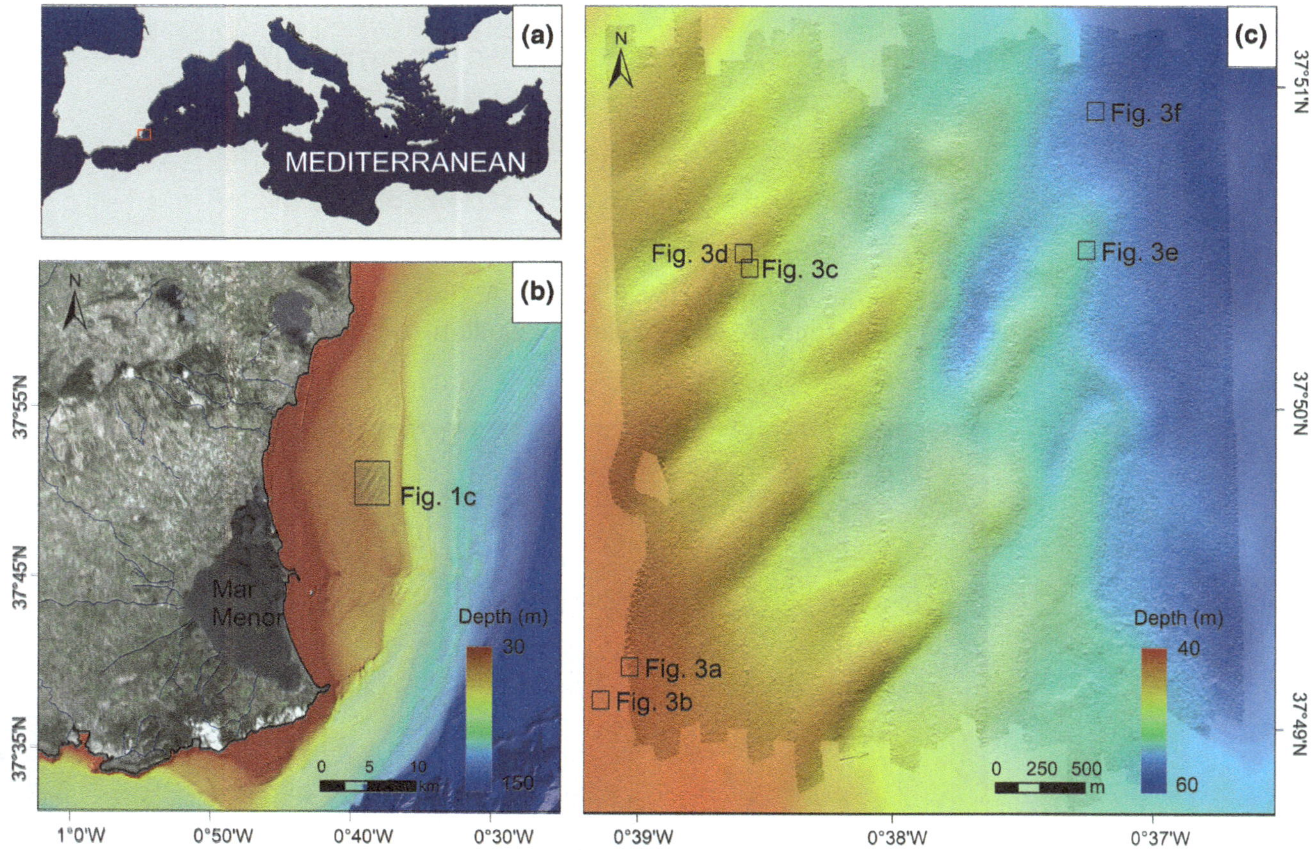

Fig. 32.1 **a** Location map. **b** Shaded relief colour bathymetric map of the continental shelf off the Mar Menor lagoon. **c** Detailed bathymetry of the subaqueous dunes, including the location of the images shown in Fig. 32.3. Onland background information from ©Instituto Geográfico Nacional de España (www.ign.es)

shelf edge located at a water depth of 150–200 m (Acosta et al. 2013), and most of the middle shelf is occupied by a subaqueous dune field (about 200 km^2) (Fig. 32.1b). The seabed is characterized by biogenic coarse-grained sediments and rocky outcrops covering the middle and outer shelf and a well-developed coastal belt of seagrass meadows covering the inner shelf (Fernández-Salas et al. 2015). Muddy sediments dominate the middle and outer shelf only in the northernmost sector (Catafau et al. 1994). The continental shelf is characterized by heterogeneous habitats, with maërl dominating the biocenosis (de Juan et al. 2013). This highly structured habitat characterized by high species richness is threatened by trawling activities performed in the area (Demestre et al. 2015; Muntadas et al. 2015).

Though trawl fishing may alter benthic habitats, the distribution and abundance of benthic communities vary spatially in relation to seabed morphology, water depth and sediment composition (Bouret et al. 1994). The seabed heterogeneity introduced by large-scale bedforms also influences the distribution of benthic habitats, and significant differences have been described between the crest and trough of sand ridges (Baptist et al. 2006; van Dijk et al. 2012). Furthermore, the presence of bedforms on the continental shelf indicates specific environmental conditions (grain size, current intensity, sediment availability, etc.) that determine the development of benthic assemblages. The aim of this chapter is to characterize the benthic community structure in and around a large subaqueous dune field (about 200 km^2) located on the Mar Menor middle continental shelf in order to evaluate the influence of these bedforms on the benthic community structure.

32.2 Methods

The detailed morphology, sediment characteristics and benthic community structures of these subaqueous dunes were based on multibeam bathymetry and sediment and benthos samples gathered during the FORMED 5 cruise onboard the R/V *García del Cid* in June 2014. Swath bathymetry was acquired using the Elac Seabeam 1050D 180 kHz Multibeam System. The multibeam data were post-processed using the CARIS HIPS and SIPS Hydrographic Data Processing System, including correction for heading, depth, pitch, heave and roll. A grid of the study area dataset was generated at 5 m resolution. Additionally, 20 sediment samples were collected from the dune field (on the slope, crest and trough of dunes) and the surrounding area using a HAPS box corer.

Regarding the benthic component, the epifauna (macrofauna >1 cm that lives on the sediment and occasionally within the first centimetres, e.g. sea stars and bivalves) and the infauna (organisms that live within the sediment, e.g. polychaetes, small crustaceans and bivalves) were sampled across the subaqueous dune field. Six epifaunal samples were collected using an experimental benthic dredge with a 2 m iron-frame aperture and a 1-cm cod-end. Three van Veen grabs were randomly collected to survey the infauna. Five grabs were collected at each sampling point to obtain the minimum sample size (de Juan et al. 2007). Moreover, three video transects with a GoPro camera and zenithal images with a NIKON D3200 camera were obtained using an experimental benthic sledge.

32.3 Results

32.3.1 Subaqueous Dunes

Dunes were identified between 40 and 70 m water depth, showing a predominant NE-SW orientation oblique to the shoreline (Fig. 32.1c). The dune height (H) varies from 2 to 5 m, spaced by distances (L) of between 400 and 1100 m, and the length varies between 800 and 3000 m. The dunes are slightly asymmetric, with the lee face dipping to the south. Crestline orientation is almost perpendicular to the general current towards the SE, which is probably accelerated by the influence of Cape Palos. However, the highly rounded dune crests suggest some degradation of the morphology of these bedforms caused by hydrodynamic conditions or by intense trawling activities over the dune field (see Demestre et al. this issue). Surficial sediment is mostly composed of sand (30–76 %) and mud (23–70 %), with a low content of biogenic gravel (0–4 %). Seabed sediment displays high spatial variability, as revealed by video sequences, with small areas of mud (up to 70 %) within larger areas dominated by coarse sediment. In spite of this variability, there is a trend towards a decreasing sand fraction and an increasing mud fraction of the sediment with increasing depth.

32.3.2 Habitat Types and Benthic Communities

Three habitat types distributed in different areas of the field of subaqueous dunes were identified: (a) a habitat of maërl composed of coarse sediment with biogenic components in the southwestern area; (b) a transition habitat characterized by a transition from maërl to mud habitats in the northwestern area; and (c) a muddy sand habitat located in the northeastern area, adjacent to the dune field.

Benthic community structure depends, among other variables, on sediment characteristics (de Juan et al. 2013). Therefore, as the three habitat types have different sediment compositions, each habitat exhibits a characteristic set of species abundance and composition that clearly represents three distinct benthic communities (Fig. 32.2, MDS). According to a SIMPER analysis, the echinodermata and polychaete group abundance clearly distinguishes the three habitat types. Polychaetes are the species group that contribute most to the dissimilarity between mud and the other two habitat types: 24 % for mud versus transition and 20.9 % for mud versus maërl. Echinodermata are the second species group contributing most to the dissimilarity between maërl and the other two habitat types: 19.5 % for maërl versus mud and 18.48 % for maërl versus transition.

In general, the epifaunal community is dominated by invertebrates in the three habitat types. The presence of fish is slightly higher in the fine-sediment community habitat adjacent to the dune field (3 % vs. less than 1 %), due to the occurrence of flat fish such as *Arnoglosus laterna* that typically inhabit muddy areas.

The infauna community is mainly composed of polychaetes, which constitute an average of 50 % of the abundance in all three habitats, followed by crustaceans and bivalves. It is worth noting the decrease in abundance of the family Syllidae from the maërl community (479 ind/m^2), in which this family dominates the polychaete community, to the finer sediment community (133 ind/m^2), in which no polychaete family stands out from the rest. This is consistent with the fact that the family Syllidae is usually associated with hard substrata (Rose and Pleijel 2001). On the other hand, infaunal crustaceans (mainly amphipods) are more abundant than bivalves in the habitat composed of coarse sediment within the dune field, while the opposite occurs in the fine-sediment habitat. The higher presence of crustaceans within the coarse habitat is due to the presence of maërl in this habitat, as rhodoliths' interstices provide shelter for these organisms (Sciberras et al. 2009).

32.3.2.1 The Maërl Community

The maërl community is characterized by the presence of the coralline algae *Lithothamnion corallioides* and *Phymatolithon calcareum*, along with the non-coralline algae of the genus *Peysonnelia* (Fig. 32.3a, b), and it is distributed between 40–50 m water depth along the southwestern area of the dune field.

The maërl habitat is the one showing the highest organism abundance overall for both infauna and epifauna in the study area. The epifauna community is mainly composed of

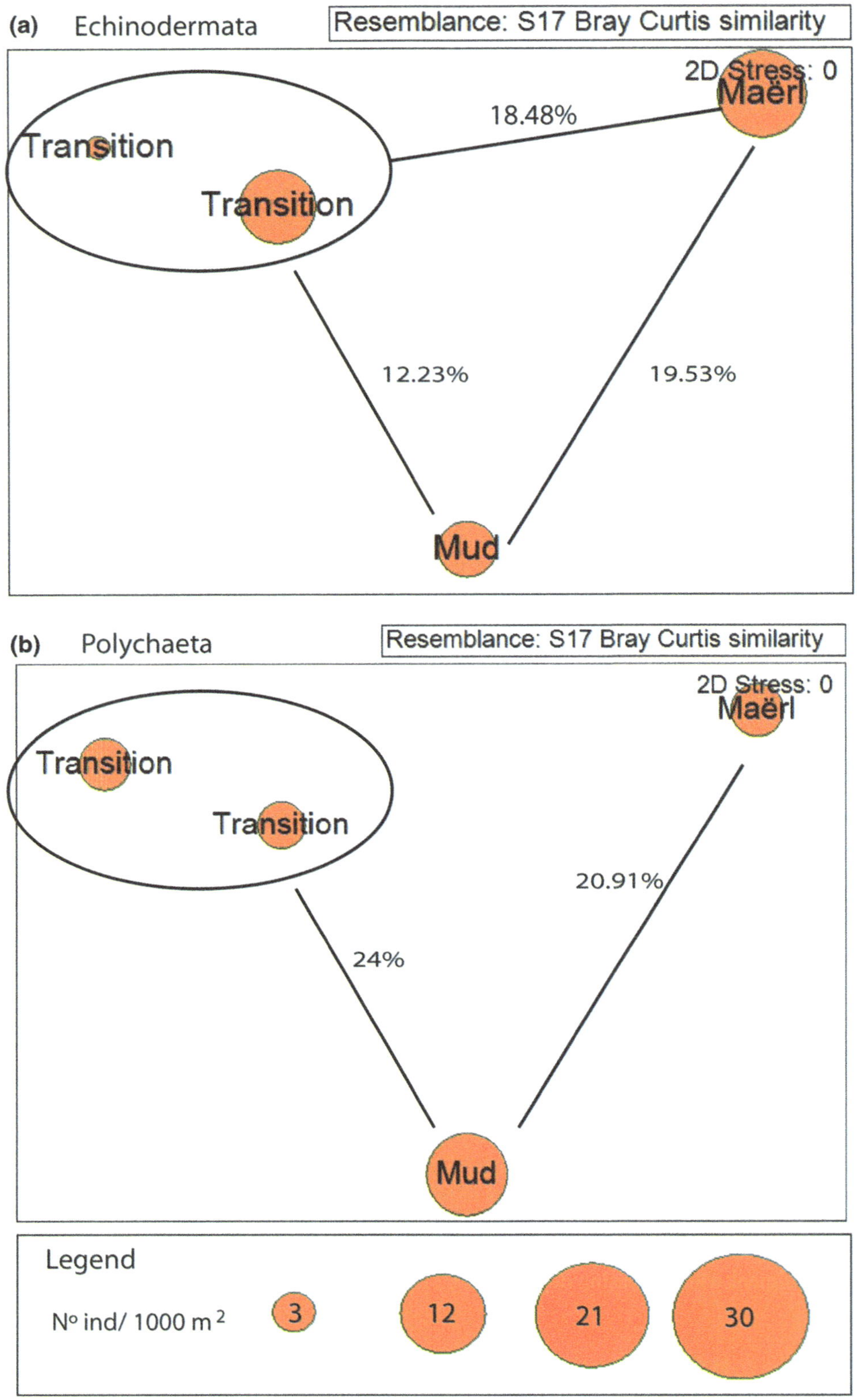

Fig. 32.2 MDS showing the different community composition of the three habitat types. The Echinodermata and polychaete groups are chosen as an example because their abundance clearly distinguishes the maërl and mud areas, whereas they show an intermediate abundance in the transition area (see text for further details)

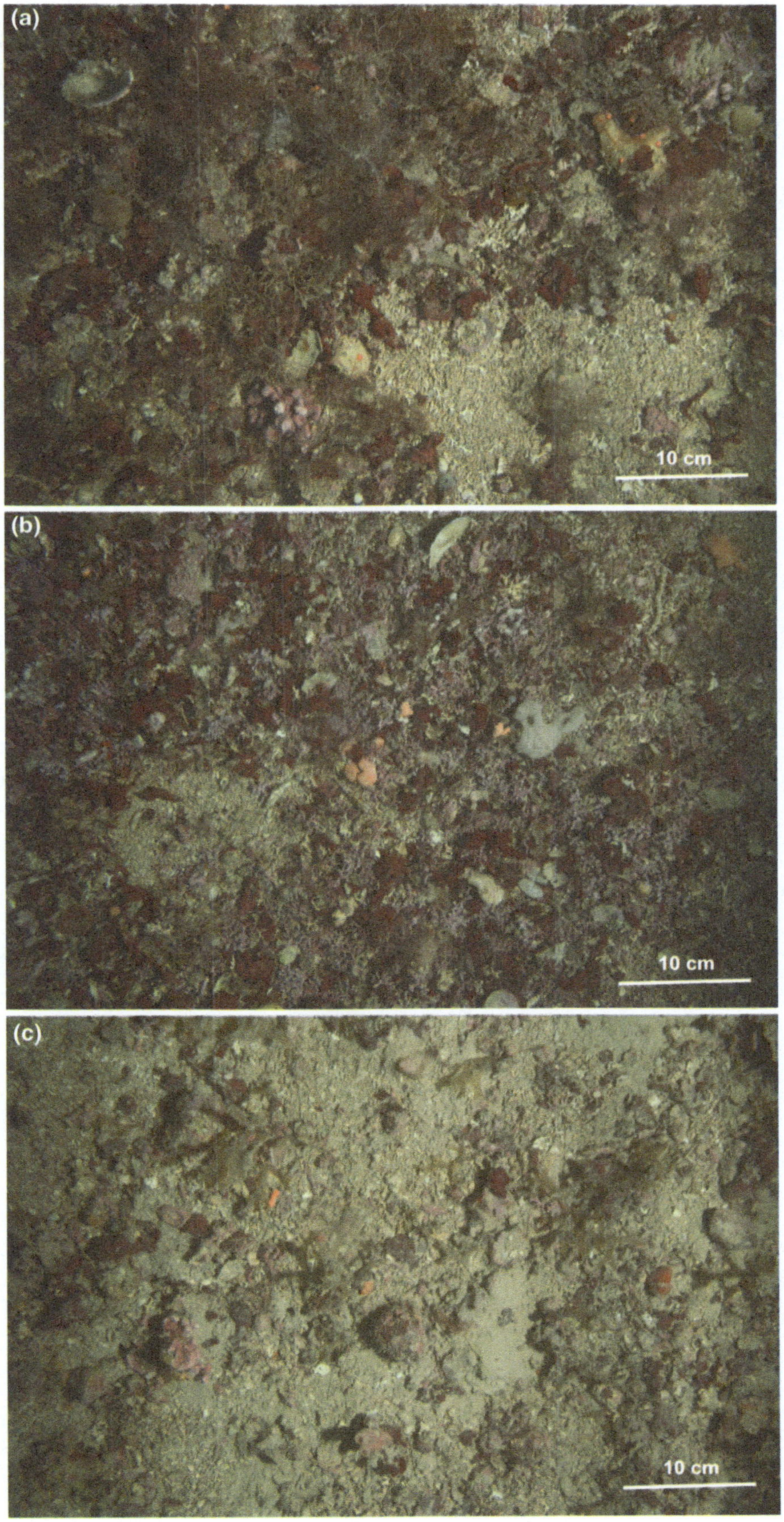

Fig. 32.3 Images showing the characteristics of the three habitats types: a maërl habitat in the southwestern dune sector (**a**, **b**), a transition habitat between maërl and mud in the northeastern dune sector (**c**, **d**), and a muddy habitat in the northwestern sector adjacent to the dune field (**e**, **f**)

Fig. 32.3 (continued)

ascidians (28 %) and Echinodermata (26 %) (Fig. 32.2a, MDS), being sea urchins the most abundant Echinodermata group (72 %). The sea urchins, although dominant in abundance, tend to be of small size and the Echinodermata biomass is dominated by sea stars such as *Echinaster sepositus,* a species showing a preference for maërl beds (Barberá et al. 2012). However, large sea urchins such as *Spatangus purpureus* are also found in this habitat (see Demestre et al. this issue).

32.3.2.2 The Transition Community

The northwestern dune field area from 43 to 47 m water depth has a mixed sediment composition characterized by two grain-size populations (medium-coarse sand and mud), and the median grain size of the sediment is around 300 microns. This transition community shows a higher proportion of non-coralline algae, such as the genus *Peysonnelia,* than the maërl communities (Fig. 32.2c, d). It is suggested that trawling perturbation in this zone degrades the maërl community and might have caused the emergence of this transition habitat between maërl and mud (see Demestre et al. this issue). In general, maërl habitats altered by trawling activity show a higher sand-mud content than undisturbed maërl bottoms, where the sediment is coarser (Bordehore et al. 2003).

The benthic community in this sector is dominated by crustaceans (30 %), most of them hermit crabs (*Dardanus arrosor, Pagurus spp.*). Hence, this epifaunal community is dominated by opportunistic/scavenger organisms, a typical characteristic of fished sites (de Juan et al. 2007). Gastropoda, mainly represented by the deposit feeder *Turritella monterosatoi,* are also abundant (17 %). In this case Echinodermata represent only 10 % of the epifaunal community. As in the maërl area, sea urchins are the most abundant Echinodermata group, but in the transition area larger sea urchins such as *Spatangus purpureus* dominate in abundance and biomass. The finer bottom sediment favours the presence of this burrowing sea urchin species.

32.3.2.3 The Fine-Sediment Community

Finally, in the northeastern sector of the study area, and properly outside the dune field, there is a muddy habitat at around 53–55 m depth. The main epifaunal components from this muddy habitat are crustaceans (27 %) and polychaetes (27 %), the later represented mainly by the sedentary family Onuphidae (Fig. 32.2b, MDS). Onuphidae organisms are considered to be primarily omnivores and scavengers (Rose and Pleijel 2001), and most crustaceans are also scavengers. Similar to the transition habitat in the northeastern dune section, this is a scavenger-dominated community. As in the transition habitat, crustaceans are mainly represented by hermit crab species, but a more diverse crustacean community is found in this muddy area. Actually, this is the area holding the highest epifaunal species number. Although not as evident as in the habitats within the dune field, muddy habitats also hold structured communities with many organisms living in them, mainly buried within the sediment. The presence of these organisms is indicated by the many holes and tracks observed in the sediment surface (Fig. 32.3e, f). In general, this area exhibits the characteristics of the muddy soft-bottom communities of trawling grounds on the continental shelf (Demestre et al. 2008) (Fig. 32.3 e, f).

32.4 Concluding Remarks

The presence of diverse bottom sediments on the middle shelf off the Mar Menor lagoon favours the development of different benthic communities. In the area where the dune field is better developed, the bottom sediment is typically heterogeneous, with gravel (mainly associated with rhodoliths), sand and mud grain-size fractions varying in different proportions, and a typical maërl community dominates. The presence of a coarse bottom probably favours the preservation of the bedform morphology due to hydrodynamic reworking. However, bottom reworking by trawling activities produces a fining of the bottom sediment and develops a "transition" habitat in the dune field with a significantly different benthic community. Finally, a characteristic fine-sediment benthic community develops in the deeper areas adjacent to the dune field.

Acknowledgments This work was funded by the FORMED project (CGL2012-33989). The crew of the R/V *García del Cid* are thanked for their help and enthusiasm during the sample cruise. The authors thank S. de Juan for useful comments on an early draft. R. Durán is supported by a CSIC JAE-Doc contract co-funded by the FSE.

References

Acosta, J., Fontán, A., Muñoz, A., Muñoz-Martín, A., Rivera, J., Uchupi, E. (2013). The morpho-tectonic setting of the Southeast margin of Iberia and the adjacent oceanic Algero-Balearic Basin. Mar. Pet. Geol. 45, 17–41.

Baptist, M.J., van Dalfsen, J., Weber, A., Passchier, S., van Heteren, S. (2006). The distribution of macrozoobenthos in the southern North Sea in relation to meso-scale bedforms. Estuar. Coast. Shelf Sci. 68, 538–546.

Barberá C, Moranta J, Ordines F, Ramón M, de Mesa A, Díaz-Valdés M, Grau AM, Massutí E. (2012). Biodiversity and habitat mapping of Menorca Channel (western Mediterranean): implications for conservation. Biodivers Conserv. 21:701–728 DOI: 10.1007/s10531-011-0210-1

Bordehore, C., Ramos-Esplá, A.A, Riosmena-Rodrigues, R. (2003). Comparative study of two mäerl beds with different otter trawling history, southeast Iberian Peninsula, Aquat. Conserv., 13. S43-S54.

Bourget, E., De Guise,J., Daigle, G. (1994). Scales of substratum heterogeneity, structural complexity and the early establishment of a marine epibenthic community. J.Exp. Mar. Biol.Ecol.181, 31–51.

Conesa, H.M., Jiménez-Cárceles, F.J. (2007). The Mar Menor lagoon (SE Spain): a singular natural ecosystem threatened by human activities. Mar. Pollut. Bull. 54, 839–849.

Catafau, E., Gaytán, M., Pereda, I., Vázquez, J.T. and Wandossell, J. (1994). Mapa geológico de la plataforma continental española y zonas adyacentes. Escala 1:200.000. Hojas 72–73 (Elche-Alicante). Madrid, Instituto Geológico y Minero de España.

de Juan, S., Thrush, S., Demestre, M. (2007). Functional changes as indicators of trawling disturbance on a benthic community located in a fishing ground (NW Mediterranean Sea). Mar. Ecol. Prog. Ser. 334, 117–129.

De Juan, S., Lo Iacono, C., Demestre, M. (2013). Benthic habitat characterisation of soft-bottom continental shelves: Integration of acoustic surveys, benthic samples and trawling disturbance intensity. Estuar. Coast. Shelf Sci. 117, 199–209.

Demestre, M., de Juan, S., Sartor, P., Ligas, A. (2008). Seasonal closures as a measure of trawling effort control in two Mediterranean trawling grounds: effects on epibenthic communities. Mar. Pollut. Bull. 56 (10), 1765–1773.

Demestre, M., Muntadas, A., de Juan, S., Mitilineou, C., Sartor, P., Mas, J., Kavadas, S., Martín, J. (2015). The need for fine-scale distribution of fishing effort to inform on an ecosystem-based management approaches: exploring three data sources in Mediterranean trawling grounds. Mar. Pol. 62, 134–143.

Demestre M., Muntadas A., Sanchez P., Garcia-de-Vinuesa A., Mas J., Franco I. Duran R., Guillén J. (2016). Bio and anthropogenic disturbance on maërl communities settled on subaqueous dunes in the Mar Menor continental shelf (Western Mediterranean), this issue.

Fernández-Salas, L.M., Durán, R., Mendes, I., Galparsoro, I., Lobo, F. J., Bárcenas, P., Rosa, F., Ribó, M., García-Gil, S., Ferrín, A., Carrara, G., Roque, C., Canals, M. (2015). Plataformas de la Península Ibérica y las Islas Baleares (I): Morfología y tipos de sedimentos. Bol. Geol. y Min. 126, 327–376.

Muntadas, A., de Juan, S., Demestre, M. (2015). The delivery of ecosystem services in trawling grounds: an approach based on Ecosystem Service Providers. Science of the total Environment 506-507:594–603. DOI:10.1016/j.scitotenv.2014.11.042

Rose, G. and Pleijel, F. (2001). Polychaetes. Oxford University Press. 354 pp.

Sciberras, M., Rizzo, M., Mifsud, J.R., Camilleri, K., Borg, J. A., Lanfranco, E., Schembri, P. J. (2009). Habitat structure and biological characteristics of a maerl bed off the northeastern coast of the Maltese Islands (central Mediterranean).Mar. Biodiv. 39, 251–264.

Van Dijk, T.A.G.P., Van Dalfsen, J.A., Van Lancker, V., Van Overmeeren, R.A., Van Heteren, S., Doornenbal, P.J. (2012). Seafloor Geomorphology as Benthic Habitat, Seafloor Geomorphology as Benthic Habitat. Elsevier.

Bio and Anthropogenic Disturbance of Maërl Communities Settled on Subaqueous Dunes on the Mar Menor Continental Shelf (Western Mediterranean)

33

M. Demestre, A. Muntadas, P. Sanchez, A. Garcia-de-Vinuesa, J. Mas, I. Franco, R. Duran, and J. Guillén

Abstract

Maërl habitats are of great biological value due to their high biodiversity and productivity, but they are also very fragile. The subaqueous dune field on the continental shelf off the Mar Menor exhibits sediment characteristics that favour the presence of maërl habitats. However, this seabed is altered by two kinds of disturbance: biogenic disturbance caused mainly by the feeding and burrowing activities of the sea urchin *Spatangus purpureus*, and anthropogenic disturbance caused by trawling activities.

Keywords

Bedforms • Habitat stability • Biodiversity conservation • Bioturbation • Trawling • Mediterranean

M. Demestre (✉) · A. Muntadas · P. Sanchez · A. Garcia-de-Vinuesa · R. Duran · J. Guillén
Institut de Ciències del Mar ICM-CSIC, P. Marítim de la Barceloneta, 37-49, 08003 Barcelona, Spain
e-mail: montse@icm.csic.es

J. Mas · I. Franco
Instituto español de Oceanografía, Centro Oceanográfico de Murcia, C/Varadero, 1, 30740 San Pedro del Pinatar, Spain

J. Guillén et al. (eds.), *Atlas of Bedforms in the Western Mediterranean*,
DOI 10.1007/978-3-319-33940-5_33

33.1 Introduction

Maërl habitats distributed across the Mediterranean harbour a high diversity of benthic species, mostly epifaunal and infaunal invertebrates. Organisms live among or are attached to the surface of maërl, or burrow in the coarse gravel of dead maërl beneath the top living layer. The main characteristic of the maërl habitat is the presence of species of calcified red seaweed, a coralline algae, which grow slowly as unattached nodules (rhodoliths) on the seabed (Pardo et al. 2014). Furthermore, maërl habitats are important carbon producers and stores (Barberá et al. 2003).

These habitats are fragile and scarce because maërl algal components are very slow-growing organisms and renewal is highly limited. Because of these attributes, maërl seabeds are considered habitats with an important ecological value. Due to their vulnerability and high ecological value, maërl habitats were designated a Special Area of Conservation (SAC) under the Habitats Directive (92/43/EEC). Consequently, the two most characteristic coralline algae in the Mediterranean, *L. corallioides* and *P. calcareum*, have legal protection at European level. However, there are other species of red coralline algae in Mediterranean maërl habitats that are not protected (e.g. *Lithothamnion corallioides, L. glaciale, Phymatolithon calcareum, Lithophyllum, Neogoniolithon brassica-florida* and *Mesophyllum lichenoides*) (Barberá et al. 2003; Bordehore et al. 2003).

In this work we study a maërl community settled on a subaqueous dune field. The bedforms are located between 40 and 70 m depth on the continental shelf off the Mar Menor (Murcia) (Fig. 33.1). It is assumed that they developed in shallower waters during transgressive conditions but the present-day dynamics are under discussion (Simarro et al. 2015).

The southeastern sector of the dune field is colonized by a maërl benthic community (Demestre et al. this issue). The bottom sediment of this area is composed of sand (90 %) with a very low content of biogenic gravel (2 %) and mud (8 %) (Fig. 33.2). The sediment grain size and composition facilitate a high level of oxygenation throughout the sediment (Martin et al. 2014) and allow the settlement of a maërl community.

We examine in this work the natural and anthropogenic disturbances on the maërl bottoms on the Mar Menor continental shelf as a proxy to infer the processes of alteration and degradation of the dune field after it was occupied by the maërl habitats.

The study is based on epifaunal and infaunal samples collected during the FORMED 5 oceanographic cruise with the R/V *García del Cid* in June 2014. Six epifaunal samples were collected using an experimental benthic dredge with a 2 m iron-frame aperture and a 1 cm cod-end. Three random samples of infauna were collected with a 0.1 m^2 Van Veen grab. Each sample was composed of five grabs to obtain the minimum sample size (de Juan et al. 2007). Moreover video

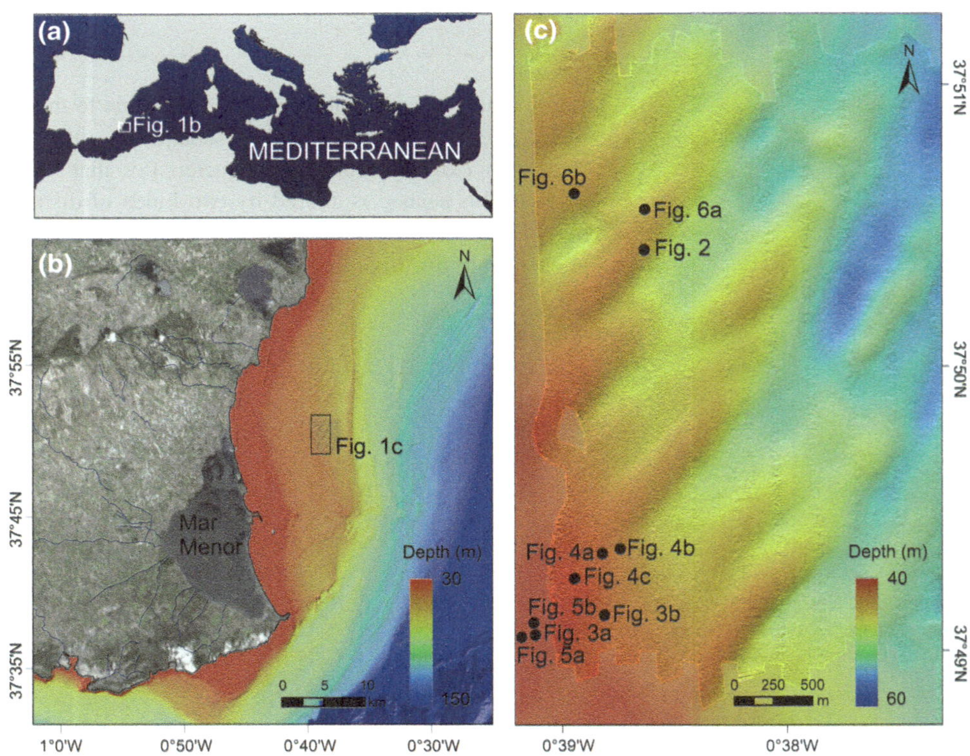

Fig. 33.1 **a** Location map. **b** Shaded relief colour bathymetry of the continental shelf off the Mar Menor. **c** Detailed bathymetry of the subaqueous dunes, including the location of Figs. 33.4–33.6. Onland background information from ©Instituto Geográfico Nacional de España (www.ign.es)

Fig. 33.2 Bottom sediment composed by biogenic components from the southeastern area of the dune field

transects with a GoPro camera and zenithal images with a NIKON D3200 camera were carried out using an experimental benthic sledge.

33.2 Maërl Beds

Maërl beds are composed of the coralline algae *Lithothamnion corallioides* and *Phymatolithon calcareum* together with non-coralline algae of the genus Peysonnelia (Fig. 33.3a, b). The epibenthic community of epifauna inhabiting maërl habitat types on Mar Menor dunes is mainly composed of ascidians (28 %) and Echinodermata (26 %) but Crustaceans (16 %) and Gastropoda (12 %) are also important epifaunal components. On the other hand, polychaetes dominate the maërl infaunal community (51 %) and members of the family Syllidae are the most abundant organisms (32 %). Infaunal crustaceans (mainly amphipods) are also abundant (13 %).

33.3 Biogenic Disturbance

In the maërl community of the Mar Menor dunes, sea urchins are an important component of the epifauna, the most abundant being the Echinodermata group (72 %). Seabed photographs show the irregular sea urchin *Spatangus purpureus*, which bioturbates maërl habitats by mixing and displacing the surface soft sediment and creating furrows across algae communities on the seabed (Fig. 33.4). This omnivorous species feeds on phytodetritus found within maërl algal beds, which also contribute to the creation of seabed furrows (Barberá et al. 2011). From previous studies, it is estimated that these sea urchins are capable of moving along about 30 cm of sediment in a period of 5–6 h (Lohrer et al. 2004). Depending on the abundance and densities of sea urchins, the upper 5 cm of sediment could be reworked every 3–20 days, and large volumes of sediment per day can be displaced across the sea urchin tracks (Lohrer et al. 2004).

The high productivity of these habitats generates aggregation of other sea urchin species such as *Spaerechinus granularis*, which grazes on algae, especially encrusting coralline algae, to feed on their epiphytic organisms and detritus (Guillou et al. 2002). The individuals aggregate to facilitate the fertilization during the spawning period of *S. granularis* (Fig. 33.5). The peak spawning period takes place in spring and early summer (Guillou and Lumingas 1998), when individuals increase their ingestion rate,

Fig. 33.3 Seabed photographs illustrating the composition of maërl habitats; **a** *pink-coloured* rhodolith (*full white arrow*) and genus *Peysonnelia*, a non-coralline algae (*dashed white arrow*); **b** *Peysonnelia*-dominated maërl area (*reddish colour*)

Fig. 33.4 Seabed disturbance caused by sea urchins: **a** panoramic image of bioturbation on maërl habitat produced by *Spatangus purpureus* (*arrow* indicates presence of sea urchin) (approximate scale); **b** and **c** detail of furrows created by the bioturbation action of S. purpureus

Fig. 33.5 Aggregations of the sea urchin *Spaerechinus granularis* during its spawning period. Panoramic vision of sea urchin aggregation (**a**) and detail of the aggregation (**b**)

devouring everything in their path (Guillou et al. 2002) and also creating bioturbation.

33.4 Anthropogenic Disturbance

Maërl habitats are subjected to demersal fisheries because they are very productive and represent an Essential Fish Habitat (EFH) for nursery, spawning and feeding areas for a great number of commercial and non-commercial species (Kamenos et al. 2004a, b). Towed gears such as trawls and dredges have one of the biggest impacts on maërl beds (Hall-Spencer and Moore 2000; Bordehore et al. 2003), disturbing and breaking the layer of coralline algae and causing an increase in sedimentation and turbidity. Consequently, the maërl community and the associated species are damaged and the diversity of flora and fauna is dramatically reduced, leading to fragmentation of the maërl community. The maërl study area exhibits chronic anthropogenic disturbance caused by trawling activity. Perturbation of the maërl community is caused by the groundrope and doors during the passage of the trawl gear over the seabed (Fig. 33.6). Chronic trawling disturbance causes severe changes in the benthic community, transforming the coarse heterogeneous maërl habitats into homogeneous habitats of muddy-sand sediment.

Thus, the cover of coralline algae is less dense in maërl habitats subjected to fishing disturbance than in protected or unexploited habitats, leading to an increase in non-coralline algae such as the genus *Peysonnelia* (Fig. 33.3b) (Bordehore et al. 2003). Furthermore, collateral perturbation that damages slow-moving and sessile organisms of the benthic fauna attracts both scavengers and opportunistic species, altering the benthic community structure and modifying the

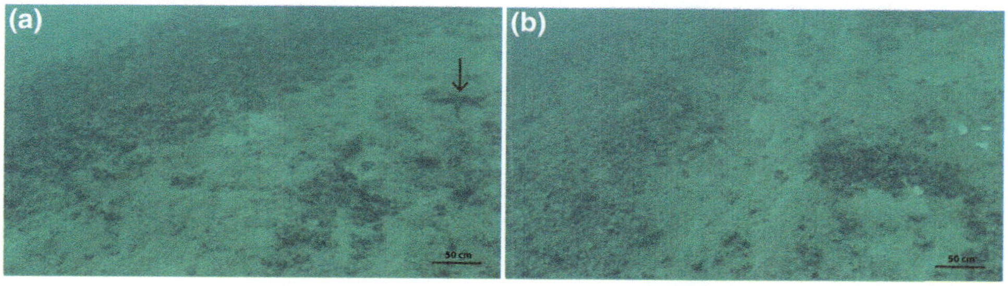

Fig. 33.6 Images of trawl disturbance on the maërl seabed: **a** presence of an opportunistic sea star; **b** tracks of otter boards ("doors")

predator/prey relationship (Demestre et al. 2000; de Juan et al. 2007).

33.5 Concluding Remarks

The development of a maërl habitat on a subaqueous dune field modifies the environmental conditions of bedforms in different ways. Coarser bioclastic sediment covers the seabed and new and very active benthic communities rework the surficial sediment layer. Moreover, trawl fishing activities favoured by the high productivity of this habitat cause strong disturbance on maërl communities and on surficial bottom sediment. Since it is assumed that the initial development of the dune field occurred in coastal waters during a period of lower sea-level, the settlement of the coarse maërl community on the dune field could favour the preservation of bedform morphology. Conversely, the seabed alteration due to biological or anthropogenic disturbance may lead to the degradation of the dune, but this alteration has not been sufficient to destroy the dune morphology.

Acknowledgments This work was funded by the FORMED project (CGL2012-33989). The crew of the RV *García del Cid* are thanked for their help and enthusiasm during the sample cruise. The authors thank S. de Juan for useful comments on an early draft. R. Durán is supported by a CSIC JAE-Doc contract co-funded by the FSE.

References

Barberá, C., Bordehore, C., Borg, J.A., Glémarec, M., Grall, J., Hall-Spencer, J.M., de la Huz, C., Lanfranco, E., Lastra, M., Moore, P.G., Mora, J., Pita, M.E., Ramos-Esplá, A.A., Rizzo, M., Sánchez-Mata, A., Seva, A., Schembri, P.J., Valle, C. (2003). Conservation and management of northeast Atlantic and Mediterranean maerl beds. Aquat. Conserv. Mar. Freshw. Ecosyst. 13, S65–S76.

Barberá C, Fernández-Jover D, López Jiménez JA, González Silvera D, Hinz H, Moranta J. (2011). Trophic ecology of the sea urchin Spatangus purpureus elucidated from gonad fatty acids composition analysis. Mar. Env. Res. 71:235–246

Bordehore C, Ramos-Espla AA, Riosmena-Rodriduez R. (2003). Comparative study of two maerl beds with different otter trawling history, southeast Iberian Peninsula. Aquat Conserv Mar Freshw Ecosyst. 13:S43–S54

Demestre, M., Sanchez, P., Kaiser, M.J. (2000). The behavioral response of benthic scavengers to otter-trawling disturbance in Mediterranean. In *Effects of Fishing on Non-target Species and Habitats*. Eds Kaiser and de Groot. Blackwell Science Ltd.

de Juan S., S. Thrush, M. Demestre. (2007). Functional changes as indicators of trawling disturbance on benthic community from a fishing ground (NW Mediterranean). Mar. Ecol. Progr. Ser. 334:117-129.

Guillou, M., Grall, J., Connan, S. (2002). Can low sea urchin densities control macro-epiphytic biomass in a north-east Atlantic maërl bed ecosystem (Bay of Brest, Britanny, France)? J. Mar. Biol. Assoc. UK. *82*(05), 867–876.

Guillou, M., Lumigas, L.J.L. (1998). The reproductive cycle of the 'blunt' sea urchin. Aquacult. Int. 6, 147–160.

Hall-Spencer, J.M., Moore, P.G. (2000). Impact of scallop dredging on maerl grounds. In Effects of Fishing on Non-target Species and Habitats. Eds Kaiser and de Groot. Blackwell Science Ltd.

Kamenos, N.A., Moore, P.G., Hall-spencer, J.M. (2004a). Nursery-area function of maerl grounds for juvenile queen scallops Aequipecten opercularis and other invertebrates. Mar. Ecol. Prog. Ser. 274, 183–189.

Kamenos, N.A., Moore, P.G., Hall-spencer, J.M. (2004b). Small-scale distribution of juvenile gadoids in shallow inshore waters; what role does maerl play? ICES J. Mar. Sci. 61, 422–429.

Lohrer AM, Thrush SF, Gibbs MM. Bioturbators enhance ecosystem function through complex biogeochemical interactions. (2004). NATURE, VOL 431, 28 OCT.

Martin, C.S., Giannoulaki, M., De Leo, F., Scardi, M., Salomidi, M., Knittweis, L., Pace, M.L., Garofalo, G., Gristina, M., Ballesteros, E., Bavestrello, G., Belluscio, A., Cebrian, E., Gerakaris, V., Pergent, G., Pergent-Martini, C., Schembri, P.J., Terribile, K., Rizzo, L., Ben Souissi, J., Bonacorsi, M., Guarnieri, G., Krzelj, M., Macic, V., Punzo, E., Valavanis, V., Fraschetti, S. (2014). Coralligenous and maërl habitats: predictive modelling to identify their spatial distributions across the Mediterranean Sea. Sci. Rep. 4, 5073.

Muntadas, A., de Juan, S., Demestre, M. (2015). The delivery of ecosystem services in trawling grounds: an approach based on Ecosystem Service Providers. Science of the total Environment 506-507:594–603. DOI: 10.1016/j.scitotenv.2014.11.042

Pardo C, Lopez L, Peña V, Hernández-Kantún J, Le Gall L, Bárbara I, et al. (2014). A Multilocus Species Delimitation Reveals a Striking Number of Species of Coralline Algae Forming Maerl in the OSPAR Maritime Area. PLoS ONE 9(8): e104073. doi:10.1371/journal.pone.0104073

Simarro, G., Guillén, J., Puig, P., Ribó, M., Lo Iacono, C., Palanques, A., Muñoz, A., Durán, R. Acosta, J. (2015). Sediment dynamics over sand ridges on a tideless mid-outer continental shelf. Mar. Geol. 361: 25–40.

Part VI

Beforms in Submarine Canyons

34 Bedforms in the Southern Submarine Canyons of the Balearic Islands (Western Mediterranean) Interpreted as Cyclic Steps

Matthieu Cartigny, Claudio Lo Iacono, Roger Urgeles, Maria Druet, and Juan Acosta

Abstract

Multibeam bathymetric data collected along the Menorca Channel (Balearic Islands) for a depth range of 70–850 m revealed the occurrence of bedforms along some of the submarine canyons of the Balearic southern slope. The four main canyons presented in this study display a linear to sinuous geometry, incising the slope for up to 170 m and carving the shelf-edge at a depth of 110 m. Bedforms along these canyons are more developed along the two northernmost canyons, particularly within the 300–700 m depth range. The crescentic shape of the bedforms and their wavelengths of between 120 and 210 m resemble the morphology of bedforms found in other submarine channels and canyons. In line with these previously observed bedforms, we here explore the possibility that the bedforms in the canyons could be interpreted as cyclic steps formed by supercritical turbidity currents. Assuming this interpretation holds, then some simple modelling would

M. Cartigny (✉) · C. Lo Iacono
Marine Geosciences, National Oceanography Centre, Southampton, UK
e-mail: m.cartigny@noc.ac.uk

C. Lo Iacono
e-mail: cllo@noc.ac.uk

R. Urgeles
Departament de Geociencies Marines, Institut de Ciencies del Mar (CSIC), Barcelona, Spain

M. Druet · J. Acosta
Instituto Español de Oceanografía, Madrid, Spain

J. Guillén et al. (eds.), *Atlas of Bedforms in the Western Mediterranean*,
DOI 10.1007/978-3-319-33940-5_34

allow us to give rough estimates of velocity and thickness of the flows that formed the cyclic steps.

Keywords
Cyclic steps • Supercritical-flow bedforms • Menorca channel • Western mediterranean

34.1 Introduction

Trains of bedforms have been mapped along the southern canyons of the Menorca Channel (Balearic Islands, western Mediterranean; Fig. 34.1). The bedforms are several hundred meters long and have a distinct crescentic shape (Fig. 34.2). In geometry and size these bedforms resemble trains of sediment waves as observed on the floor of many other submarine canyons and channels (Smith et al. 2005; Paull et al. 2010; Hughes Clarke et al. 2012; Conway et al. 2012; Casalbore et al. 2014; Normandeau et al. 2014). Monitoring studies have shown that these sediment waves are migrating upslope in response to downslope-moving turbidity currents (Hughes Clarke et al. 2012, 2014). Recently a review of ocean floor sediment waves has shown that these upstream migrating bedforms are very abundant and occur over a large range of sizes (Symons et al. 2016). In general upslope migrating bedforms such as antidunes (Kennedy 1963) and cyclic steps (Taki and Parker 2005) are characteristic of Froude supercritical flows (Simons 1960; Cartigny et al. 2014). Supercritical-flow conditions are likely to occur in turbidity currents as a result of the low density differences between these flows and the ambient water (Cartigny and Postma 2016). Numerical and experimental studies have shown that cyclic steps are controlled by alternating subcritical and supercritical conditions within turbidity currents (Kostic and Parke 2006; Fildani et al. 2006; Spinewine et al. 2009; Postma and Cartigny 2014).

If such canyon floor bedforms can be confidently interpreted as supercritical-flow features, then they can be used to estimate the size of the flows that controlled their evolution. This link would then in turn allows to roughly estimate the

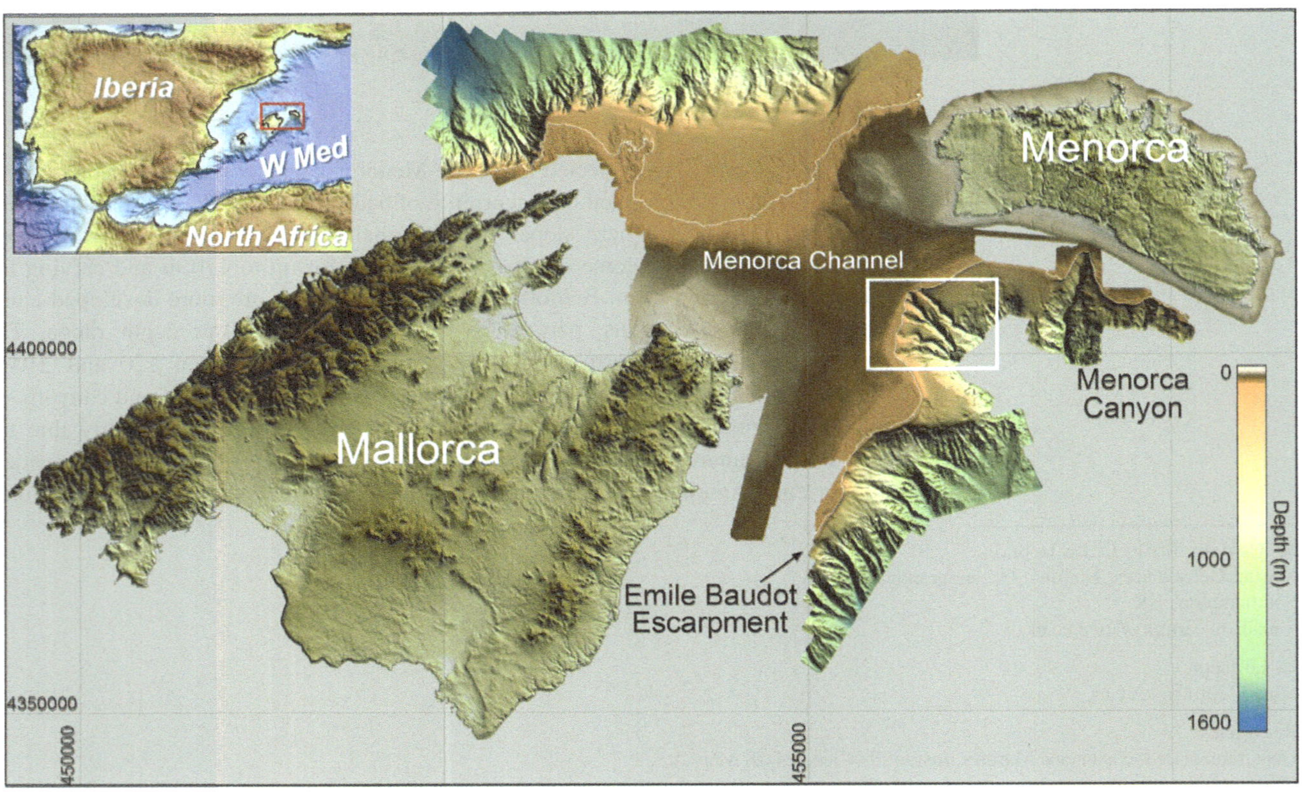

Fig. 34.1 Bathymetric model of the Menorca Channel. The *white rectangle* corresponds to the area under study

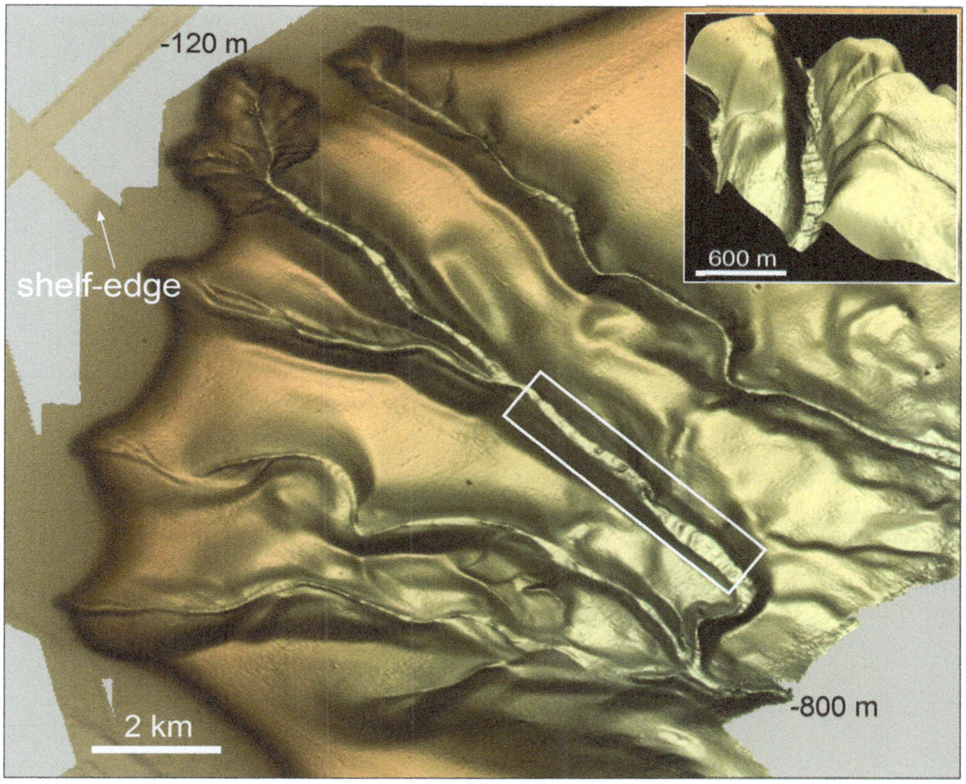

Fig. 34.2 The four submarine canyons where the studied bedforms were mapped. The *white rectangle* corresponds to the area illustrated in the 3D bathymetric model in the *top-left inset* and to the cyclic steps where numerical models were applied to reconstruct their genetic turbidity currents (Fig. 34.3)

dynamics of turbidity currents down submarine canyon systems just on the basis of the bedform morphologies, representing then a big step forward in our understanding of turbidity currents, which are otherwise so difficult to measure. However, to confidently interpret these bedforms as cyclic steps it would be desirable to confirm the upslope migration of the bedform over time, either by repeat mapping (Smith et al. 2005; Conway et al. 2012; Hughes Clarke et al. 2014) or by recognizing upslope migrating backsets patterns (Ventra et al. 2015) in the subsurface expression (Covault et al. 2014; Zhong et al. 2015). As such data is lacking for the canyons of the Balearic Islands, we have to base our cyclic step interpretation solely on the typical crescentic shape and common wavelength of the bedforms.

Here we aim to describe the main morphological characteristics of the mapped bedforms in the submarine canyons of the Balearic Islands. Based on this morphological character we interpret these bedforms as cyclic steps, and we use some basic numerical modelling to estimate the flow characteristics that could have formed cyclic steps of this geometry and scale.

34.2 Methods

Multibeam bathymetry data were acquired in 2012 onboard the RV *Miguel Oliver* using a Simrad EM302, in the framework of the INDEMARES-LIFE Project (www.indemares.es) of the Instituto Español de Oceanografía (IEO). Digital Terrain Models were produced with a footprint resolution of 20 m. Global Mapper and Golden Software Surfer9 were used to map the train of bedforms and calculate their main morphometric characteristics. For further details, see Lo Iacono et al. (2013). Assuming the bedforms were formed by supercritical turbidity currents, then rough estimations of the flow characteristics of the flows in the canyons can be made by comparing the bedform geometries on the canyon floor with bedform geometries predicted by a numerical model for a given range of flow characteristics (Cartigny et al. 2011). The numerical model strongly simplifies the flows by averaging all flow parameters over the depth, by excluding any exchange of sediment in between the flow and the bed, and by limiting the downstream evolution of the flow to only include small variations as described by gradual varying flow theory. More details on the model and its assumptions can be found in Cartigny et al. (2011). The model input consist of an average stoss and lee side slope of the bedforms (Table 34.1) and an average grain size, which is here set at 1 mm (Lo Iacono et al. 2013). Based on these inputs the model then runs several thousands of simulations for flows with different combinations of discharges, Froude numbers and sediment concentration, and calculates the bedform wavelength and amplitude that would result from the combination of flow parameters and the observed canyon floor bedform slopes.

Table 34.1 Morphometric characteristics of the cyclic steps of Fig. 34.3

	Slope stoss side (–)	Slope lee side (–)	Length (m)	Amplitude (m)
Step 1	0.049	0.043	215	4.5
Step 2	0.064	0.066	230	7.2
Step 3	0.013	0.046	269	4.0
Step 4	0.031	0.052	189	3.0
Step 5	0.027	0.061	199	3.4
Step 6	0.044	0.051	235	4.0
Average	0.038	0.053	223	4.3

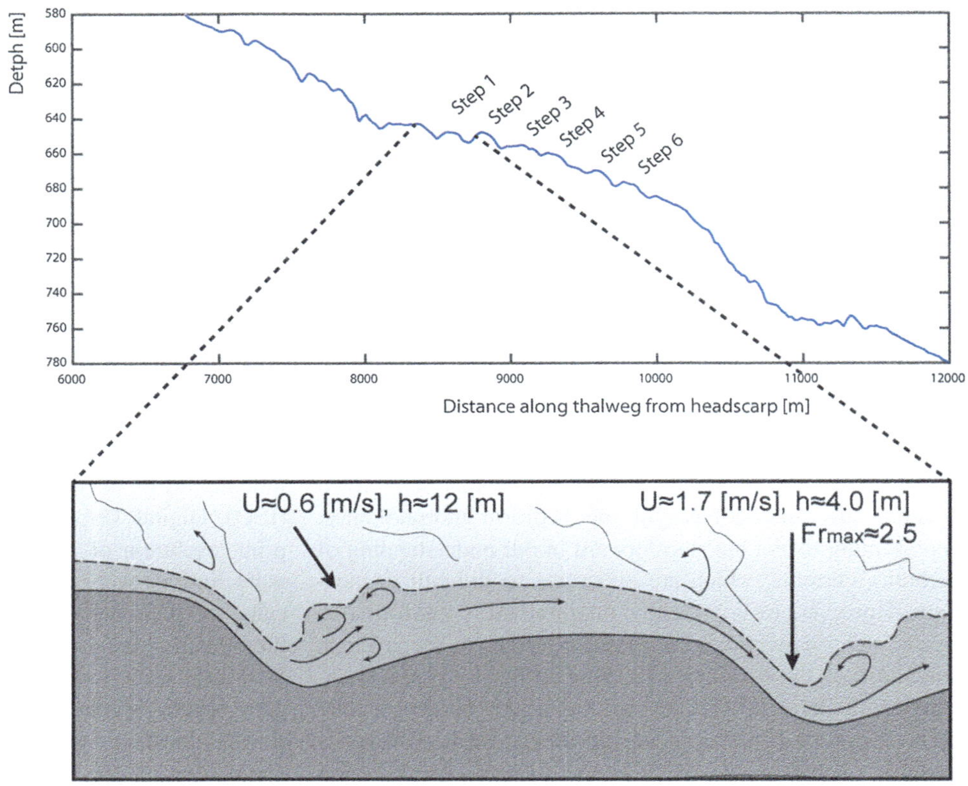

Fig. 34.3 Longitudinal bathymetric transect of the most pronounced cyclic steps in the study area (see Fig. 34.2 for location) and estimation of intensity (*U*) and thickness (*h*) of the corresponding genetic currents based on numerical modelling on an idealised cyclic step drawing

By comparing the predicted bedform wavelength and amplitude with the observed wavelength and amplitude, the most appropriate flow characteristics can be selected.

34.3 Results-Discussion

The studied bedforms were observed along four linear canyons of the southern Balearic slope. These canyons develop east of the Emile Baudot Escarpment (EBE), a passive tectonic feature which bounds the slope region offshore of Mallorca. The canyons breach the paleo/coastal sandy deposits of the southern edge of the Menorca Channel at an average depth of 130 m, and do not presently have a direct connection with the coastal environment (Lo Iacono et al. 2013) (Fig. 34.1). The two northernmost canyons breach the shelf margin the furthest, corresponding to an increase in their width and incision and suggesting a more mature evolutionary stage compared with the two southernmost canyons, which barely incise the shelf (Lo Iacono et al. 2013). The canyons have been mapped to a maximum depth of ~800 m. The bedforms within the canyons display wavelengths ranging from 120 to 300 m and amplitudes ranging from 4 to 7 m. The largest and most pronounced bedforms mapped along the downstream part of the northernmost canyons occur along a thalweg stretch with an average gradient of 2° (Fig. 34.2).

Given the crescentic shape and the scale bedforms on the canyon floor we here interpret these bedforms as cyclic steps. Assuming this interpretation holds then we can estimate the velocity and depth of the flows that have formed the cyclic steps. As input for the model, we used the most prominent bedforms, which are observed at depths of 600–780 m (Figs. 34.2 and 34.3). The average characteristics for the bedforms are summarized in Table 34.1. The model indicates that the flows in the canyon might have had average velocities of 1.3 m/s and a flow depth of around 7 m. The maximum velocity at the toe of the steep lee sides of the steps attains higher values of ~ 1.7 m/s, while on the flatter stoss sides the flow thickness reaches a maximum value of about 12 m combined with a minimum velocity of ~0.6 m/s. These flows are likely related to periods of strong hydrodynamics during the winter storms occurring in the area. These storms enhance the sediment transport processes along the outer shelf and trigger sandy turbidity currents flowing from the shelf-edge down the canyon (Lo Iacono et al. 2013).

As stated before these velocity and flow thickness estimates are only valid if our cyclic steps interpretation holds. Additionally, it should be kept in mind that the modelling results are only rough estimations due to the assumption that are made in the model.

34.4 Conclusions

Recent multibeam mapping along the Menorca Channel has revealed the occurrence of trains of bedforms along four submarine canyons incising the southern slope of the Balearic Margin. Observed bedforms display maximum lengths of 300 m. If we interpret these bedforms as cyclic steps, then some basic numerical modelling shows that these bedforms might have formed by roughly 10 m thick turbidity currents whose velocities can be estimated in the range of ~0.5–2 m/s.

Acknowledgments Data acquisition was made possible thanks to the European project INDEMARES (Life-NAT/E/000732). Constructive anonymous reviewers greatly improved the submitted version of the manuscript.

References

Cartigny, M. J., Postma, G., van den Berg, J. H., & Mastbergen, D. R. (2011). A comparative study of sediment waves and cyclic steps based on geometries, internal structures and numerical modeling. Marine Geology, 280(1), 40–56.

Cartigny, M. J., Ventra, D., Postma, G., & Den Berg, J. H. (2014). Morphodynamics and sedimentary structures of bedforms under supercritical-flow conditions: New insights from flume experiments. Sedimentology, 61(3), 712–748.

Cartigny, M.J.B., Postma, G. (2016). Turbidity current bedforms. In J. Guillén et al. (eds), Atlas of bedfoms in the western Mediterranean, Springer.

Casalbore, D., Romagnoli, C., Bosman, A., & Chiocci, F. L. (2014). Large-scale seafloor waveforms on the flanks of insular volcanoes (Aeolian Archipelago, Italy), with inferences about their origin. Marine Geology, 355, 318–329.

Conway, K. W., Barrie, J. V., Picard, K., & Bornhold, B. D. (2012). Submarine channel evolution: active channels in fjords, British Columbia, Canada. Geo-Marine Letters, 32(4), 301–312.

Covault, J. A., Kostic, S., Paull, C. K., Ryan, H. F., & Fildani, A. (2014). Submarine channel initiation, filling and maintenance from sea-floor geomorphology and morphodynamic modelling of cyclic steps. Sedimentology, 61(4), 1031–1054.

Fildani, A., Normark, W. R., Kostic, S., & Parker, G. (2006). Channel formation by flow stripping: Large-scale scour features along the Monterey East Channel and their relation to sediment waves. Sedimentology, 53(6), 1265–1287.

Hughes Clarke, J. E., Brucker, S., Muggah, J., Church, I., Cartwrigght, D. & Kuus, P., Eisan, B. (2012). The Squamish ProDelta: monitoring active landslides and turbidity currents. In Canadian Hydrographic Conference 2012, Proceedings.

Hughes Clarke, J. E., Vidiera Marques, C. & Pratomo, D. (2014). Imaging active mass-wasting and sediment flows on a fjord delta, Squamish, British Columbia. In S. Krastel et al. (eds.), Submarine Mass Movements and Their Consequences, Advances in Natural and Technological Hazards Research 37, Springer.

Kennedy, J. F. (1963). The mechanics of dunes and antidunes in erodible-bed channels. Journal of Fluid Mechanics, 16(04), 521–544.

Kostic, S., & Parker, G. (2006). The response of turbidity currents to a canyon–fan transition: internal hydraulic jumps and depositional signatures. Journal of Hydraulic Research, 44(5), 631–653.

Lo Iacono C., Urgeles R., Polizzi S., Grinyo, J., Druet M., Agate M., Gili J.M. & Acosta J. (2013). Submarine Mass Movements Along a Sediment Starved Margin: The Menorca Channel (Balearic Islands –Western Mediterranean). In S. Krastel et al. (eds.), Submarine Mass Movements and Their Consequences, Advances in Natural and Technological Hazards Research 37, Springer.

Normandeau, A., Lajeunesse, P., St-Onge, G., Bourgault, D., Drouin, S. S. O., Senneville, S., & Bélanger, S. (2014). Morphodynamics in sediment-starved inner-shelf submarine canyons (Lower St. Lawrence Estuary, Eastern Canada). Marine Geology, 357, 243–255.

Paull, C. K., Ussler III, W., Caress, D. W., Lundsten, E., Covault, J. A., Maier, K. L., & Augenstein, S. (2010). Origins of large crescent-shaped bedforms within the axial channel of Monterey Canyon, offshore California. Geosphere, 6(6), 755–774.

Postma, G. & Cartigny, M. J. (2014). Supercritical and subcritical turbidity currents and their deposits—A synthesis. Geology, 42(11), 987–990.

Simons, D. B. (1960). Sedimentary structures generated by flow in alluvial channels.

Symons, W. O., Sumner, E. J., Talling, P. J., Cartigny, M. J., & Clare, M. A. (2016). Large-scale sediment waves and scours on the modern seafloor and their implications for the prevalence of supercritical flows. *Marine Geology, 371*, 130–148.

Smith, D. P., Ruiz, G., Kvitek, R., & Iampietro, P. J. (2005). Semiannual patterns of erosion and deposition in upper Monterey Canyon from serial multibeam bathymetry. Geological Society of America Bulletin, 117(9–10), 1123–1133.

Spinewine, B., Sequeiros, O. E., Garcia, M. H., Beaubouef, R. T., Sun, T., Savoye, B., & Parker, G. (2009). Experiments on wedge-shaped deep sea sedimentary deposits in minibasins and/or on channel levees emplaced by turbidity currents. Part II. Morphodynamic evolution of the wedge and of the associated bedforms. *Journal of Sedimentary Research*, *79*(8), 608–628.

Taki, K., & Parker, G. (2005). Transportational cyclic steps created by flow over an erodible bed. Part 1. Experiments. Journal of Hydraulic Research, 43(5), 488–501.

Ventra, D., Cartigny, M. J., Bijkerk, J. F., & Açikalin, S. (2015). Supercritical-flow structures on a Late Carboniferous delta front: Sedimentologic and paleoclimatic significance. *Geology*, *43*(8), 731–734.

Zhong, G., Cartigny, M.J.B., Kuang, Z., & Wang, L. (2015). Cyclic steps along the South Taiwan Shoal and West Penghu submarine canyons on the northeastern continental slope of the South China Sea. Geological Society of America Bulletin, B31003-1.

Cyclic Steps at the Head of Channelized Features Along the Calabrian Margin (Southern Tyrrhenian Sea, Italy)

35

Alessandro Bosman, Daniele Casalbore, and Rocco Dominici

Abstract

High-resolution multibeam bathymetry have revealed coaxial trains of crescent-shaped bedforms within the heads of channelized features lying in shallow-water sectors along the tectonically-controlled Calabrian Margin. These bedforms have wavelengths of tens or a few hundreds of metres and wave heights of a few metres, and their crest-lines trend perpendicular to the maximum slope gradients. Repeated multibeam surveys carried out in 2007, 2008, 2012 and 2013 showed a rapid and significant morphological evolution of the channel's floors, with the generation or upslope migration of the bedforms. Based on their size, upslope migration and similarities to crescent-shaped bedforms recognized in other active canyon heads, these features can be interpreted as cyclic steps. The bedforms are, in fact, formed or modified by frequent slope failures and related sedimentary flows, whose occurrence is favoured by the concurrent presence of several predisposing and triggering mechanisms, such as a high sedimentation rate due to a steep coastal creek, severe storms and seismic events.

A. Bosman (✉) · D. Casalbore
Istituto di Geologia Ambientale e Geoingegneria, Consiglio Nazionale delle Ricerche, Rome, Italy
e-mail: alessandro.bosman@uniroma1.it

R. Dominici
Dipartimento di Biologia, Ecologia e Scienze della Terra, Università della Calabria, Arcavacata di Rende, Cosenza, Italy

J. Guillén et al. (eds.), *Atlas of Bedforms in the Western Mediterranean*,
DOI 10.1007/978-3-319-33940-5_35

Keywords
Crescent-shaped bedforms • Seafloor mapping • Repeated multibeam bathymetry • Geohazard

35.1 Introduction and Study Area

In the last decade, the development of high-resolution multibeam and high-precision positioning systems have allowed the morphology of coastal and internal waters to be mapped with unprecedented detail, encompassing canyon heads, insular volcanoes, rivers, lakes and lagoons (e.g. Bosman et al. 2014). These studies are rapidly expanding our knowledge on geological and oceanographic processes acting in these areas. When repeated in time, high-resolution multibeam surveys may also depict small variations in seafloor morphology, thus giving hints on the reconstruction of ongoing processes. Specifically, significant short-term evolution of the seafloor has been observed within canyon heads lying in very shallow-water areas (e.g. Casalbore et al. 2011; Paull et al. 2010), where crescent-shaped bedforms with wavelengths of tens or hundreds of metres and heights of a few metres are ubiquitous. Despite their small size, the study of these bedforms is rapidly expanding, because they may be considered a proxy of recent sedimentary activity. In this paper, we show the presence and migration of similar crescent-shaped bedforms within shallow-water canyon heads that have developed along the Calabrian Margin. Since the middle Pleistocene, this area has been affected by rapid uplift rates (about 1 mm/a), witnessed by a flight of raised marine terraces (e.g. Antonioli et al. 2006). This uplift produced mountainous relief very close to the coast, with short, steep subaerial valleys periodically affected by flash-flooding. Moreover, this area is largely affected by historical and pre-instrumental strong earthquakes as well as by severe storms during the winter months. Largely due to its tectonic evolution as a rifted margin during the spreading of the southern Tyrrhenian Sea, the facing continental margin is characterized by a very narrow to nearly absent continental shelf changing abruptly to steep slopes (Fig. 35.1). All these conditions led to a general instability of the submarine sectors, as witnessed by the recognition of a large number of erosive features, such as canyons and landslide scars (e.g. Casalbore et al. 2014).

35.2 Methods

This work relies on the interpretation of high-resolution multibeam data collected during several cruises carried out over the last 10 years aboard R/Vs *Urania* and *Universitatis* and small boats for shallow-water surveys. Data were acquired using multibeam echo-sounder systems (Teledyne Reson and Kongsberg) operating at different frequencies (from 50 to 455 kHz) and they are DGPS- or RTK-positioned to ensure maximum accuracy at all the water depths. Repeated multibeam surveys were used to generate residual maps for monitoring rapid morphological changes in the highly dynamic canyon heads.

35.3 Results and Discussion

The Gioia Canyon is a 70-km-long erosive feature draining the entire Calabrian Continental Margin (Fig. 35.1). At depths of less than 250 m, the canyon divides into two main branches that deeply indent the shelf up to −10 m, isolating the main entrance of the Gioia Tauro harbour. Crescent-shaped bedforms are recognized in both canyon heads, although they are mainly developed in the southern one (Fig. 35.2).

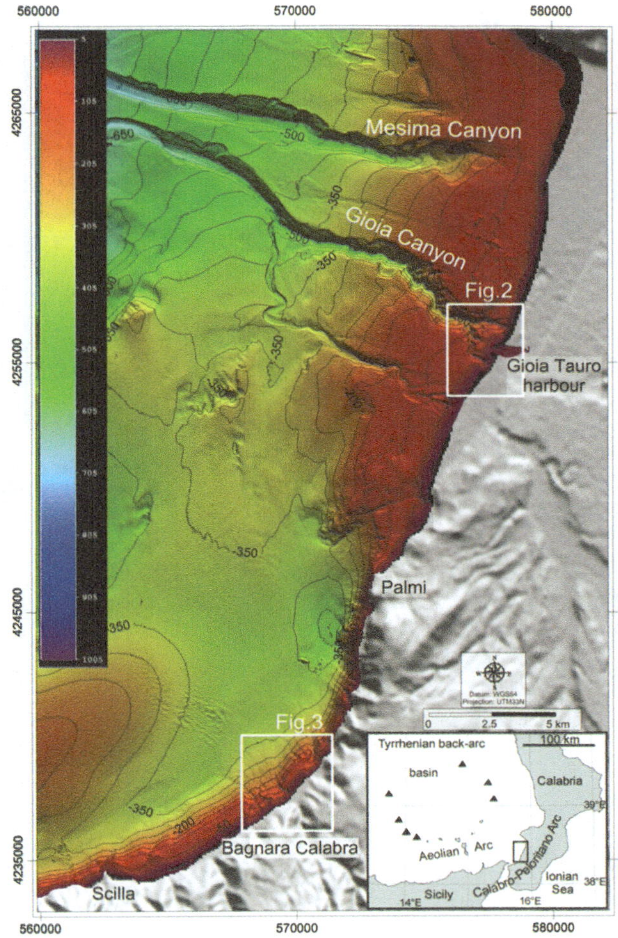

Fig. 35.1 Shaded relief map of the study area, with the indication of the main submarine canyons and the following figures (*white boxes*); the *inset* shows the regional setting

Here, bedforms have wavelengths of 10–30 m and wave heights of 1–3 m; bedform size generally increases downslope. Multibeam surveys collected in August 2012 (Fig. 35.2a) and February 2013 (Fig. 35.2b) show significant morphological variations in the canyon floor from the coast down to −130 m, with the formation of a coaxial train of new crescent-shaped bedforms on the upper part of the slope and the upslope migration of larger and previously-formed bedforms on the lower part of the slope (Fig. 35.2c, d).

The Bagnara Calabra channels are 1-km-wide erosive features located between −10 and −130 m that feed underlying depositional fan-shaped features that extend down to −400 m (Fig. 35.1). The heads of the channels have an overall amphitheatre-like shape, formed by coalescing landslide scars. The scars are 50–200 m wide and deeply indent the littoral wedge up to −10 m and a few hundred metres from the coastline (Fig. 35.3). Coaxial trains of crescent-shaped bedforms are recognized within the channel head on both the thalweg and the underlying fan (Fig. 35.3a). The bedforms have wavelengths ranging from tens to hundreds of metres and wave heights of a few metres; their wave dimension tends to increase downslope. Repeated bathymetric surveys at the head of the Bagnara Calabra channels show significant morphological variations (Fig. 35.3b, c). Specifically, the residual map obtained as a difference between the 2007 and 2008 bathymetries shows in the northern channel the formation of a semi-circular scar at −15 m, whereas downslope an alternation of seafloor erosion and accretion is recognizable (Fig. 35.3b), giving rise to the formation of crescent-shaped bedforms. The residual map obtained as a difference between the 2012 and 2013 bathymetries shows, instead, seafloor erosion and accretion mainly along the southernmost channel, where an upslope migration of the bedforms is observed (Fig. 35.3c).

In both study areas, we identified coaxial trains of small-scale crescent-shape bedforms within the heads of channelized features lying in shallow-water areas. Based on their size, upslope migration and similarities to crescent-shaped bedforms observed in other active channel headwalls (e.g. Paull et al. 2010; Hughes-Clarke et al. 2014), these features can be interpreted as cyclic steps, i.e. a class of upslope-migrating turbidite sediment waves (Kostic 2011). The formation and migration of these bedforms can be related to the occurrence of minor slope failures and/or related gravity flows from the upper part of the channel head, as witnessed by repeated multibeam surveys. In the study areas, multiple predisposing factors for the occurrence of slope failures are present, including high-rate and rapid sedimentation supplied both by coastal creeks directly discharging throughout the narrow shelf and onto steep slope settings and by sediment transported by long-shore currents that can be trapped within canyon heads. The triggering mechanisms may be load exerted by hyperpycnal flows generated during flash-flood events (e.g. Casalbore et al. 2011) or cyclic load due to earthquakes and storm-waves.

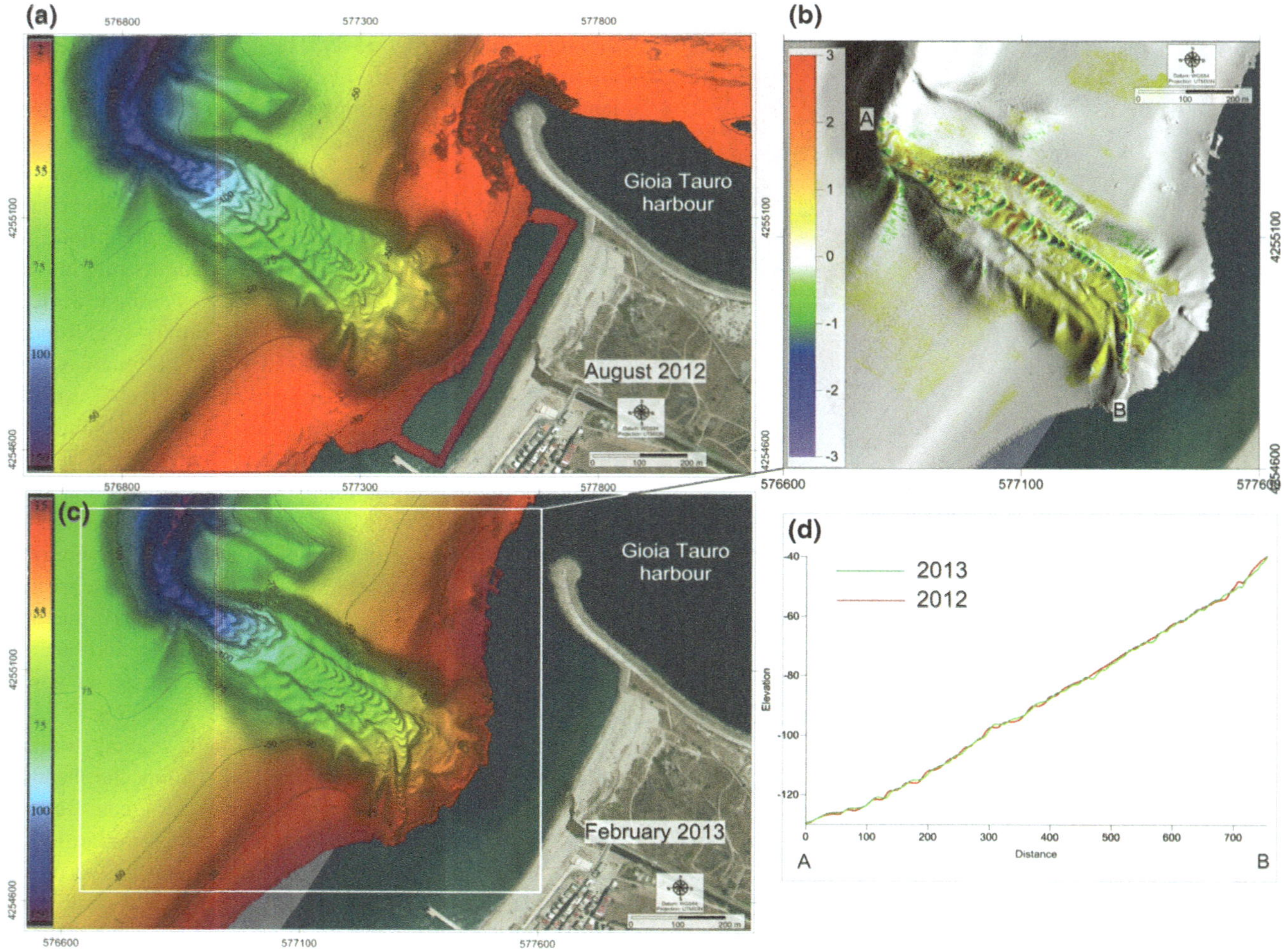

Fig. 35.2 Shaded relief map of the southern head of Gioia Tauro Canyon from the 2012 (**a**) and 2013 (**b**) multibeam bathymetries, on which morphological changes are evident. **c** Residual map shows the morphological changes in terms of seafloor erosion and accretion; scale bar is in metres. **d** 2012 and 2013 bathymetric cross-sections along the bedforms on the thalweg (for location see *dashed line* in Fig. 35.2c)

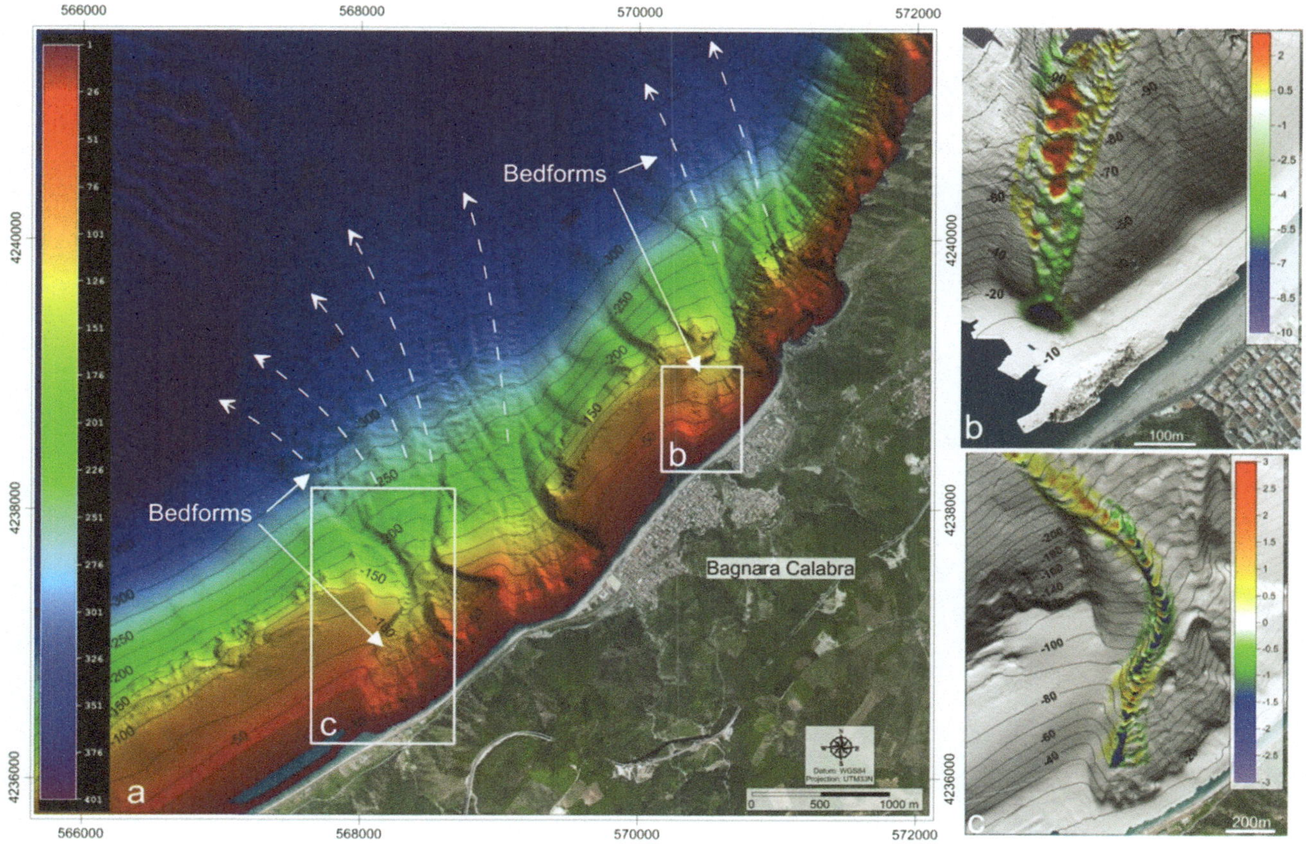

Fig. 35.3 a Shaded relief map of submarine channels and downslope fan-shaped features offshore of Bagnara Calabra village, where crescent-shaped bedforms are present. Residual maps obtained as a difference between the 2007 and 2008 bathymetries in the northern channel (**b**) and between the 2012 and 2013 bathymetries in the southern channel (**c**)

Acknowledgments This research was funded by the Italian MaGIC (Marine Geohazards along the Italian Coasts) Project and carried out in the framework of the RITMARE (Ricerca ITaliana per il MARE) Project. Crews of R/Vs *Urania*, *Universitatis* and *Thetis* are gratefully acknowledged along with the fellow researchers and students who took part in the sea surveys. We also thank Alan Lounds for the English revision.

References

Antonioli F, Ferranti L, Lambeck K et al (2006). Late Pleistocene to Holocene record of changing uplift rates in southern Calabria and northeastern Sicily (southern Italy, Central Mediterranean Sea). Tectonophysics 422: 23–40.

Bosman A, Casalbore D, Romagnoli C, Chiocci FL (2014). Formation of an ‘a’ā lava delta: insights from time-lapse multibeam bathymetry and direct observations during the Stromboli 2007 eruption. Bulletin of Volcanology, 76(7): 1–12.

Casalbore D, Chiocci FL, Scarascia Mugnozza G et al (2011). Flash-flood hyperpycnal flows generating shallow-water landslides at Fiumara mouths in Western Messina Straits (Italy). Mar. Geophys. Res. 32: 257–271.

Casalbore D, Bosman A, Ridente D, Chiocci FL (2014). Coastal and submarine landslides in the tectonically-active Tyrrhenian Calabrian Margin (Southern Italy): examples and geohazard implications. In: Krastel, S., et al. (Eds.), Submarine Mass Movements and Their Consequences. Advances in Natural and Technological Hazards Research, 37: 261–269.

Hughes Clarke JE, Marques CR, Pratomo D (2014). Imaging active mass-wasting and sediment flows on a Fjord Delta, Squamish, British Columbia. In: Krastel, S., et al. (Eds.), Submarine Mass Movements and Their Consequences, Advances in Natural and Technological Hazards Research, 37: 249–260. doi: 10.1007/978-3-319-00972-8.

Kostic S (2011). Modeling of submarine cyclic steps: controls on their formation, migration and architecture. Geosphere 7: 294–304.

Paull CK, Ussler W, Caress DW, Lundsten E, Covault JA, Maier KL, Xu J, Augenstein S (2010). Origins of large crescent-shaped bedforms within the axial channel of Monterey canyon, offshore California. Geosphere 6: 1–20.

Potential Cyclic Steps in a Gully System of the Gulf of Palermo (Southern Tyrrhenian Sea)

36

Claudio Lo Iacono, Matthieu Cartigny, Elisabetta Zizzo, Mauro Agate, and Attilio Sulli

Abstract

Multibeam bathymetric data revealed the occurrence of a train of bedforms along a gully system in the Gulf of Palermo, southern Tyrrhenian Sea. The observed gullies, located in the westernmost sector of the Gulf of Palermo, incise the outer shelf at a depth of 120 m and converge at the Zafferano Canyon, connecting to the Palermo Basin at a depth of 1300 m. Bedforms develop along these gullies and along the thalweg of the canyon, displaying an average wavelength of 200 m, with maximum values of 340 m. Their gully floor location combined with their wave length, upslope asymmetry and crescent shape point to a possible cyclic step origin of these bedforms. Preliminary numerical modelling suggests that, assuming that these bedforms were formed by cyclic steps in turbidity currents, these flows might have been few meters thick and have had velocities in the range of 0.2–1.5 m/s.

Keywords

Cyclic steps • Gullies • Submarine canyons • Turbidity currents • Gulf of palermo • Tyrrhenian sea

C. Lo Iacono (✉) · M. Cartigny
Marine Geoscience, National Oceanography Centre, Southampton, European Way, Southampton, SO14 3ZH, UK
e-mail: cllo@noc.ac.uk

E. Zizzo · M. Agate · A. Sulli
Dipartimento delle Scienze della Terra d del Mare, Universita' di Palermo, Via Archirafi, 90100 Palermo, Italy

J. Guillén et al. (eds.), *Atlas of Bedforms in the Western Mediterranean*,
DOI 10.1007/978-3-319-33940-5_36

36.1 Introduction-Study Area

The submarine canyons of the Gulf of Palermo (northwestern Sicily, southern Tyrrhenian Sea) deeply carve the sedimentary sequence of the Palermo slope and basin system, most of them breaching the shelf margin, and extend to a depth of 1500 m, coinciding with the deepest area of the Palermo intraslope Basin (Lo Iacono et al. 2011, 2014). A train of bedforms occurs along a gully network in front of Cape Zafferano, in the easternmost sector of the gulf, where the continental shelf is no wider than 3 km (Fig. 36.1). The channel floor location of the bedforms, together with their wave length, crescent shape and upslope asymmetry, suggests that they might have developed as a result of a cyclic step process. In such a process, the downstream side of the bedform (the lee side of the bedform) is continuously eroded by a supercritical flow (Froude Number [Fr] > 1), while the flatter upstream side of the bedform is shaped by depositional subcritical flows (Fr < 1) (Parker 1996). This process leads to the formation of a series of asymmetrical upslope-migrating (often crescent-shaped) bedforms (Clarke et al. 2014). Similar bedforms have been observed in open slope environments (Migeon et al. 2000; Pratson et al. 2000; Hill et al. 2008; Urgeles et al. 2011) and more frequently along the thalweg of several submarine canyons and gullies (Fildani et al. 2006; Clark et al. 2014; Covault et al. 2014; Zhong et al. 2015) and could be interpreted as cyclic steps (Cartigny et al. 2011; Kostic 2011). Unfortunately, there are no available data to definitely confirm a cyclic step origin of the bedforms presented here, or even to proof their upslope migration over time. These bedforms are interpreted here as cyclic steps purely on the basis of their geometry and location. We aim to present their morphologic interpretation and in second instance to roughly estimate the main characteristics of the flow potentially responsible for their formation, applying an hydraulic flow model.

36.2 Methods

Swath-bathymetry multibeam (MB) data available for this study were acquired during two different oceanographic cruises in 2001 (CARG Project) and 2009 (MaGIC Project) (Chiocci and Ridente 2011). The MB system of the 2001 cruise was a Reson SeaBat 8111 generating 105 beams at a frequency of 100 kHz. The MB system of the 2009 cruise was a Reson SeaBat8160 generating 126 beams at a frequency of 50 kHz. The MB data were post-processed with the PDS-2000 system. Digital terrain models were produced with a footprint resolution of 20 m. For further details, see Lo Iacono et al. (2011, 2014). Global Mapper and Golden Software Surfer 9 were used to map the trains of bedforms and to calculate their main morphometric characteristics (Table 36.1). The bedform steepness has been defined as the step heights divided by the step lengths (h/L). The asymmetry index (AI) (Knaapen 2005) is defined as L2-L1/L, where L is the distance between two troughs, L1 is the distance between upslope trough and crest and L2 is the distance between the crest and the downslope trough. An AI > 0.02 indicates the presence of asymmetric bedforms. Negative AI values indicate a downslope asymmetry and positive AI values indicate an upslope asymmetry. The applied numerical model strongly simplifies the flows by averaging all flow parameters over the depth, by excluding any exchange of sediment in between the flow and the bed, and by limiting the downstream evolution of the flow to only include small variations as described in gradual varying flow theory. The numerical model uses an average grain size (medium sands) and the stoss and lee side slopes of observed bedforms as input data. The model runs several thousands of simulations for flows combining different discharges, Froude numbers and sediment concentrations. The synthetic bedform wavelengths and amplitudes predicted by these simulations are finally compared with the dimensions of the observed cyclic steps, and the most appropriate characteristics of their genetic flow are then fitted. More details on the model and its assumptions can be found in Cartigny et al. (2011).

36.3 Results-Discussion

The cyclic steps of the Gulf of Palermo were mapped in a depth range of 125–1050 m along a network of 9 gullies breaching the shelf-edge in front of Cape Zafferano, the eastern cape of the gulf (Figs. 36.1, 36.2). The gullies have an average width of 180 m and are up to 20 m deep. Some of the gullies display a smoothed and planed morphology of unclear origin within a depth range of 500–700 m (Fig. 36.2). The mapped bedforms have a wavelength ranging from 110 to 340 m and an amplitude ranging from 0.8 to 5 m. Their steepness ranges from 0.007 to 0.01, this last value corresponding to the most morphologically pronounced bedforms, where steeper lee (downslope) and stoss (upslope) sides occur (Fig. 36.3). These bedforms are also the most (upslope) asymmetric, with an AI of 0.27. The

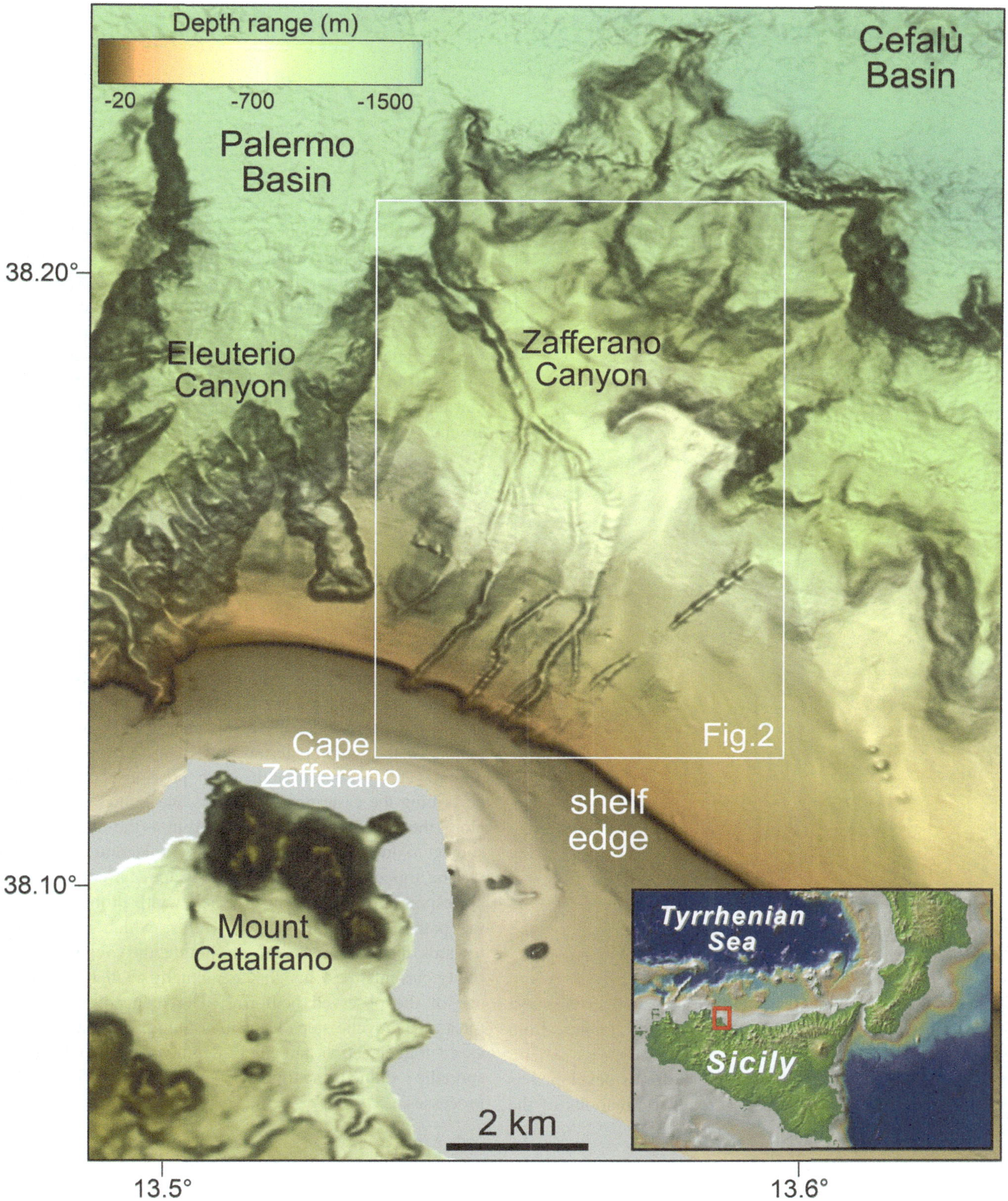

Fig. 36.1 Bathymetric model of the study area. Cape Zafferano is the eastern cape of the Gulf of Palermo

Table 36.1 Bedform characteristics of profile A–A′ in Fig. 36.3

Step #	Slope stoss side (m)	Slope lee side (–)	Length (m)	Amplitude (m)
Step 1	−0.008	0.164	224	5.8
Step 2	0.026	0.217	198	13.2
Step 3	0.003	0.110	249	3.1
Step 4	−0.038	0.063	286	5.0
Average	−0.004	0.138	239	6.8

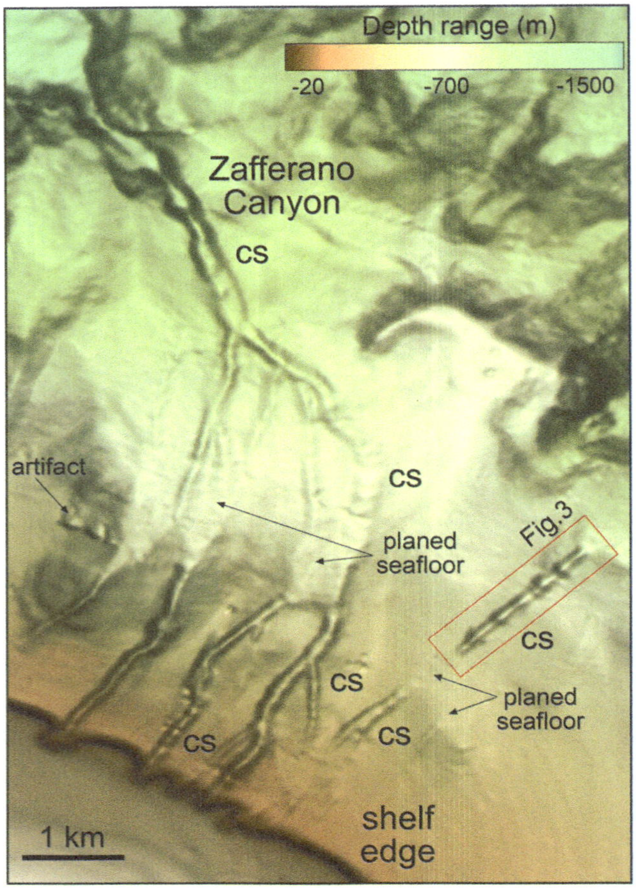

Fig. 36.2 Bathymetric model of the submarine gullies and of the Zafferano Canyon, where interpreted cyclic steps (*CS*) were mapped

general slope gradient of the gullies along which the most asymmetric and steepest bedforms occur ranges from 7° to 10°.

Based on the assumption that the bedforms developed as a result of a cyclic step process, rough estimations of the turbidity currents which generated the cyclic steps can be made using a simple numerical model for a given range of flow characteristics. The bedforms of Cape Zafferano displaying a more pronounced morphology are here used as input for the model (Fig. 36.3). The average characteristics for the selected bedforms (700–800 m water depth) are summarized in Table 36.1. The model calculations indicate that the observed cyclic steps are likely generated by flows around 1 m thick, with average velocities of exceeding 1.0 m/s. The maximum velocities at the toe of the steep lee sides could reach values of ~1.5 m/s, whereas on the flatter stoss sides the flow reaches a maximum thickness exceeding 2 m combined with a minimum velocity of ~0.2 m/s (Fig. 36.3). As the model makes several assumptions to simplify the flow dynamics, it is necessary to point out that these values are only very rough estimates, and are fully dependent on our cyclic step interpretation.

The occurrence of the studied bedforms along the Capo Zafferano Canyon and associated gullies likely suggests intense turbidity current or a variation in morphology and grain size along these incisions compared to the other canyons mapped in the Gulf of Palermo, where bedforms are apparently absent. This observation fits with previous considerations describing the easternmost canyons and gullies of each gulf along the northwestern Sicilian margin as the most intensely subject to downslope turbidity flows (Lo Iacono et al. 2014).

The topography of the eastern Cape Zafferano and the corresponding decrease in the shelf width probably control the path of the along-shelf currents, which are diverted towards the canyon heads, promoting the creation of turbidity currents. Few insights are actually available about the age of the observed bedforms. The reduced steepness of most of the cyclic steps leads us to interpret these bedforms as no longer active and likely degraded in their height by sporadic diluted sedimentary flows and bioturbation processes.

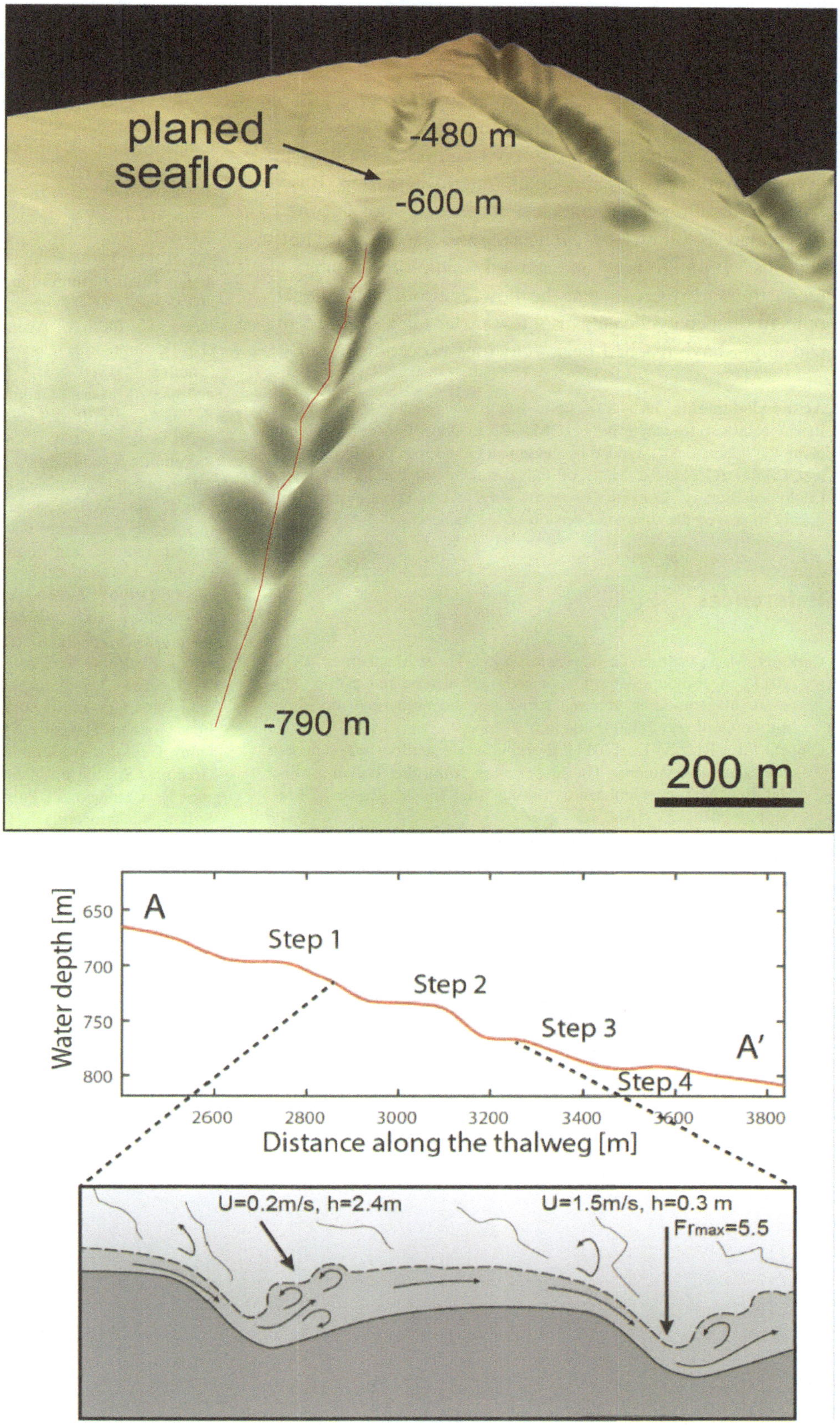

Fig. 36.3 3D bathymetric model of the most pronounced cyclic steps in the area, where numerical models were applied to reconstruct the intensity (U) and thickness (h) of the corresponding turbidity currents (section A–A′)

36.4 Conclusions

Swath-bathymetry mapping along the Gulf of Palermo (southern Tyrrhenian) revealed the occurrence of trains of bedforms along a set of shelf incising gullies connected to the Zafferano Canyon. The observed bedforms are upslope asymmetric and display maximum lengths of 340 m. Based on their location and geometry, these bedforms are here interpreted as formed by cyclic step processes in turbidity currents. A preliminary and strongly simplified numerical reconstruction suggests that the flow controlling the development of such cyclic steps is a few meter thick (0.3–2.5 m) and might have reached peak velocities exceeding 1 m/s.

Acknowledgments Data acquisition was made possible thanks to the Italian National Research Projects MaGIC (Marine Geological Hazard along the Italian Coast), funded by the Italian Civil Protection Department, and CARG (Geological Maps of Italy), funded by the ISPRA-Italian Geological Survey. Constructive reviews by two anonymous reviewers greatly improved the submitted version of the manuscript

References

Cartigny, M. J., Postma, G., van den Berg, J. H., & Mastbergen, D. R. (2011). A comparative study of sediment waves and cyclic steps based on geometries, internal structures and numerical modeling. Marine Geology, 280(1), 40–56.

Chiocci F.L., Ridente D., (2011). Regional-scale seafloor mapping and geohazard assessment. The experience from the Italian project MaGIC (Marine Geohazards along the Italian Coasts). Mar. Geophys. Res. 32, 13–23.

Clarke, J. E. H., Vidiera Marques, C., Pratomo, D. (2014). Imaging active mass-wasting and sediment flows on a fjord delta, Squamish, British Columbia. In S. Krastel et al. (eds.), Submarine Mass Movements and Their Consequences, Advances in Natural and Technological Hazards Research 37, Springer.

Covault, J. A., Kostic, S., Paull, C. K., Ryan, H. F., & Fildani, A. (2014). Submarine channel initiation, filling and maintenance from sea-floor geomorphology and morphodynamic modelling of cyclic steps. *Sedimentology*, *61*(4), 1031–1054.

Fildani, A., Normark, W. R., Kostic, S., & Parker, G. (2006). Channel formation by flow stripping: Large-scale scour features along the Monterey East Channel and their relation to sediment waves. *Sedimentology*, *53*(6), 1265–1287.

Hill, P. R., Conway, K., Lintern, D. G., Meulé, S., Picard, K., & Barrie, J. V. (2008). Sedimentary processes and sediment dispersal in the southern Strait of Georgia, BC, Canada. *Marine environmental research*, *66*, S39–S48.

Knaapen, M.A.F., (2005). Sandwave migration predictor based on shape information. J. Geophys. Res. 110, F04S11. doi:10.1029/2004JF000195

Kostic, S. (2011). Modeling of submarine cyclic steps: Controls on their formation, migration, and architecture. *Geosphere*, *7*(2), 294–304.

Lo Iacono C., Sulli A., Agate M., LoPresti V., Pepe F., Catalano R., (2011). Submarine canyon morphologies in the Gulf of Palermo (Southern Tyrrhenian Sea) and possible implications for geo-hazard. Marine Geophysical Researches 32, 127–138.

Lo Iacono C., Sulli A., Agate M., (2014). Submarine canyons of north-western Sicily (Southern Tyrrhenian Sea): Variability in morphology, sedimentary processes and evolution on a tectonically active margin. Deep-Sea Research II 104, 93–105.

Migeon S., Savoye B., Faugeres JC., (2000). Quaternary development of migrating sediment waves in the Var deep-sea fan: distribution, growth pattern, and implication for levee evolution. Sedimentary Geology 133, 265–293.

Parker, G. (1996). Some speculations on the relation between channel morphology and channel-scale flow structures. *Coherent flow structures in open channels*, *423*.

Pratson L.F., Imran J., Parker G., Syvitski JPM., Hutton E., (2000). Debris flows versus turbidity currents: A modeling comparison of their dynamics and deposits. In: Bouma A.H. and Stone C.G. eds: Fine-Grained Turbidite Systems: American Association of Petroleum Geologists Memoir 72, SEPM (Society for Sedimentary Geology) Special Publication 68, 57–72.

Urgeles R., Cattaneo A., Puig P., Liquete C., De Mol B., Amblas D., Sultan N., Trincardi F., (2011). A review of undulated sediment features on Mediterranean prodeltas: distinguishing sediment transport structures from sediment deformation. Marine Geophysical Research 32, 1–1, 49–69.

Zhong, G., Cartigny, M.J.B., Kuang, Z., & Wang, L. (2015). Cyclic steps along the South Taiwan Shoal and West Penghu submarine canyons on the northeastern continental slope of the South China Sea. Geological Society of America Bulletin, B31003-1.

Bedforms Feeding and Bedforms Fed by Canyon Activity Around Punta Alice Promontory (Calabria Ionian Margin, Italy)

37

D. Ridente, A. Bosman, D. Casalbore, and F.L. Chiocci

Abstract

Submarine canyons along the Ionian Calabria margin indent large shelf sectors as the coalescing of smaller failure scarps progresses under the control of tectonic deformation and high sedimentation rates, promoted by regional uplift and rejuvenation of the drainage basin. The Punta Alice Canyon and the Cirò Canyon show such complex canyon heads, almost completely eroding the shelf and thus affecting sediment pathways along and across strike. Wavy bedforms of variable shape and size are observed both outside and inside the two canyon systems, providing an indication of how sediment is sourced and transported. We analyse the distribution of bedforms in order to define how both open shelf currents and confined sediment fluxes contribute to their formation outside and inside canyons. We also remark on the importance of the orientation of canyon heads and channels relative to flow direction of currents, and of slope failure as a sediment source.

Keywords

Morpho-bathymetry • Sediment waves • Shelf currents • Sediment failure

37.1 Introduction

The Calabria region is the southernmost part of peninsular Italy and, together with the north-eastern tip of Sicily (Peloritani Mountains), it forms an allochthonous terrain overriding the Apennine domain along the narrow (only

D. Ridente (✉) · A. Bosman · D. Casalbore
Istituto di Geologia Ambientale e Geoingegneria, Consiglio Nazionale delle Ricerche, Rome, Italy
e-mail: domenico.ridente@cnr.it

F.L. Chiocci
Dipartimento Scienze della Terra, Sapienza Università di Roma, Rome, Italy

J. Guillén et al. (eds.), *Atlas of Bedforms in the Western Mediterranean*,
DOI 10.1007/978-3-319-33940-5_37

200 km wide) Ionian subduction zone (e.g. Royden et al. 1987). Deformation of the Ionian Sea accretionary prism is accompanied by regional uplift along the continental margins at rates of ca. 1 mm/a, at least since the Middle Pleistocene (Ferranti et al. 2006). Among other effects, this tectonic evolution is responsible for shaping narrow continental shelves and steep slopes (Ridente et al. 2013), backed by high-relief coastal areas with a tectonically rejuvenated drainage area. This network of short-steep torrents (*fiumare*) typically yields coarse-grained sediment during frequent high-energy floods. In spite of this supply regime, the sedimentary equilibrium of shores and beaches is unbalanced, since most of the clastic input directly bypasses the narrow shelf during major floods, and is delivered downslope by gravitational processes that, locally, may be conveyed by canyons.

The Punta Alice Promontory, in the northern part of the Ionian Calabria margin, is bordered by a narrow continental shelf (maximum width >1.5 km) where two canyons (among others) develop canyon heads that approach the coastline (ca. 200 m away): these are the Punta Alice Canyon and the Cirò Canyon, north and south of Punta Alice Promontory, respectively (Fig. 37.1). Evidence of active sediment transport along these canyons is provided by the presence of wavy-stepped features blanketing the thalweg of channels. In addition, fields of wavy bedforms of different scale and shape are present on the shelf outside Punta Alice Canyon, immediately north of the canyon head region (Fig. 37.1).

The aim of this study was to describe the distribution of bedforms outside and inside the canyons, in order to define how both open shelf currents and confined sediment fluxes contribute to their formation by interacting with (a) the reciprocal location of canyons; (b) the width/depth of shelf sectors in between canyons; (c) the orientation of canyon heads and channels relative to flow direction of currents; (d) the canyon head morphology; and (e) the length and steepness of channels. We also consider the importance of slope failure as a sediment source.

37.2 Data and Methods

Morpho-bathymetric data were collected during two cruises (in December 2005 and April 2006) using a small vessel equipped with Reson multibeam systems operating at different frequencies (240 and 455 kHz). The survey area extends from near-coast to −300 m depth, offshore of Punta Alice Promontory and the town of Cirò Marina (Fig. 37.1). Data were real-time kinematic-positioned and processed with Caris Hips and Sips 8.1 (including correction for tide effects, vessel stability and water sound velocity); the digital elevation model was generated with a cell size of 0.5–1 m.

37.3 Results and Discussions

Submarine canyons along the Ionian Calabria margin are generally broad and dissect large shelf sectors (e.g. Rebesco et al. 2009), via the coalescing of headscarps that originate through instability processes and later evolve into a more or less canalized and composite canyon head region. The Punta Alice Canyon and the Cirò Canyon underwent a similar geomorphic evolution, and now indent the shelf at a shallower water depth (ca. −5 m) and closer to the coast than the other canyons that develop in between (Fig. 37.1).

The Cirò Canyon extends in front of, and very close to, the harbour of Cirò Marina, where the head region is formed by several tributaries trending approximately W-E and merging downslope into a main channel oriented ca. SSW-NNE (Figs. 37.1 and 37.2). The main channel and its tributaries are characterized by the presence of downslope-concave bedforms, elongated crosswise to the thalweg (Bf1 in Fig. 37.2). They are ubiquitous in the depth interval from −10 to −180 m, and appear as "wavy" features displaying wavelengths of between ca. 10 and 50 m and crest heights of between 1 and 3 m. Their shape is typically asymmetric, with a sub-horizontal stoss side and a steep lee side, thus resulting in a "staircase" canyon floor morphology (Fig. 37.2).

The Punta Alice Canyon shows a general W-E, coast-parallel course and a composite head region with coalescing headscarps more closely spaced on the northern side of the canyon (Fig. 37.3). Wavy bedforms similar in size and shape to those observed in the Cirò Canyon are present, though limited to the northwesternmost canyon head channel, between ca. −20 and −200 m (Bf1 in Fig. 37.3). The area around the northern flank of the canyon head region is covered by seafloor undulations of different scale (Bf2 and Bf3 in Fig. 37.3), the larger ones forming a train of waves trending NW-SE for more than 3 km (Bf3 in Fig. 37.3).

The field of smaller bedforms (bf 2) extends at depths between −10 and −20 m, over an area of about 0.2 km^2. Individual waves have crest heights of less than 1 m (ca 0.4–0.7 m) and wavelengths of up to 15 m. They are symmetric in cross-sections (A–B in Fig. 37.3) and their crest-lines are slightly sinuous and oblique to the coastline. The larger bedforms (bf 3), located between −25 and −35 m, show a wave height of 2–3 m and a wavelength of 60–90 m. They are slightly asymmetric in cross-section, with a longer stoss side and a shorter and steeper lee side (C-D in Fig. 37.3). Crest-lines are rectilinear or slightly sinuous and tend to rotate as they approach the canyon head (likely reflecting a similar rotation of current flow as it is intercepted by the canyon head).

The shelf bedforms north of Punta Alice Canyon are intercepted by the canyon head as they migrate down-current, providing a significant sediment source to

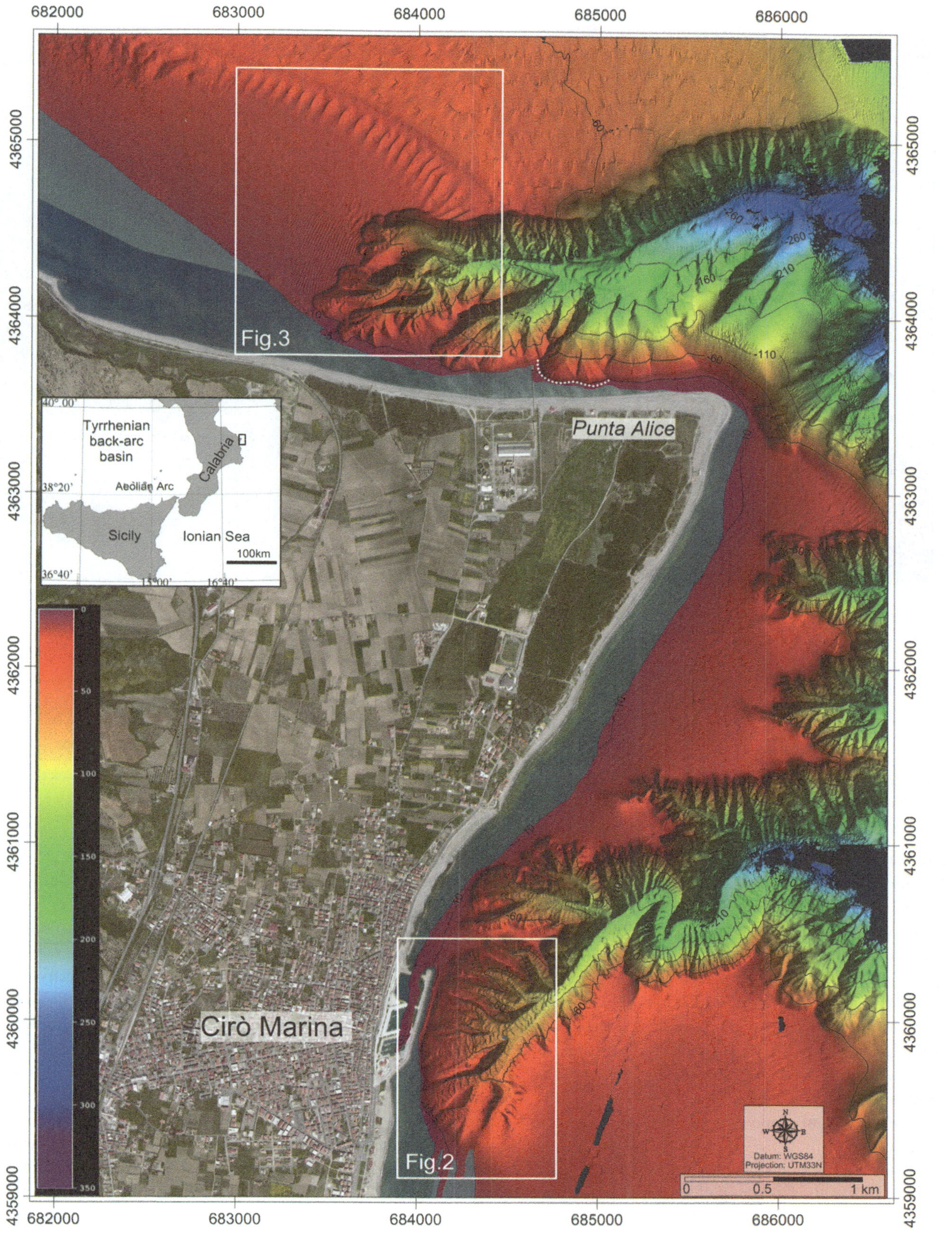

Fig. 37.1 Shaded relief of the shelf and upper slope around Punta Alice Promontory, in the Ionian Sea (see *inset*), where the Punta Alice Canyon (to the north) and the Cirò Canyon (to the south) deeply indent the shelf. The *dashed white line* indicates the scar of the Punta Alice slide, which occurred in December 2005 (Casalbore et al. 2012)

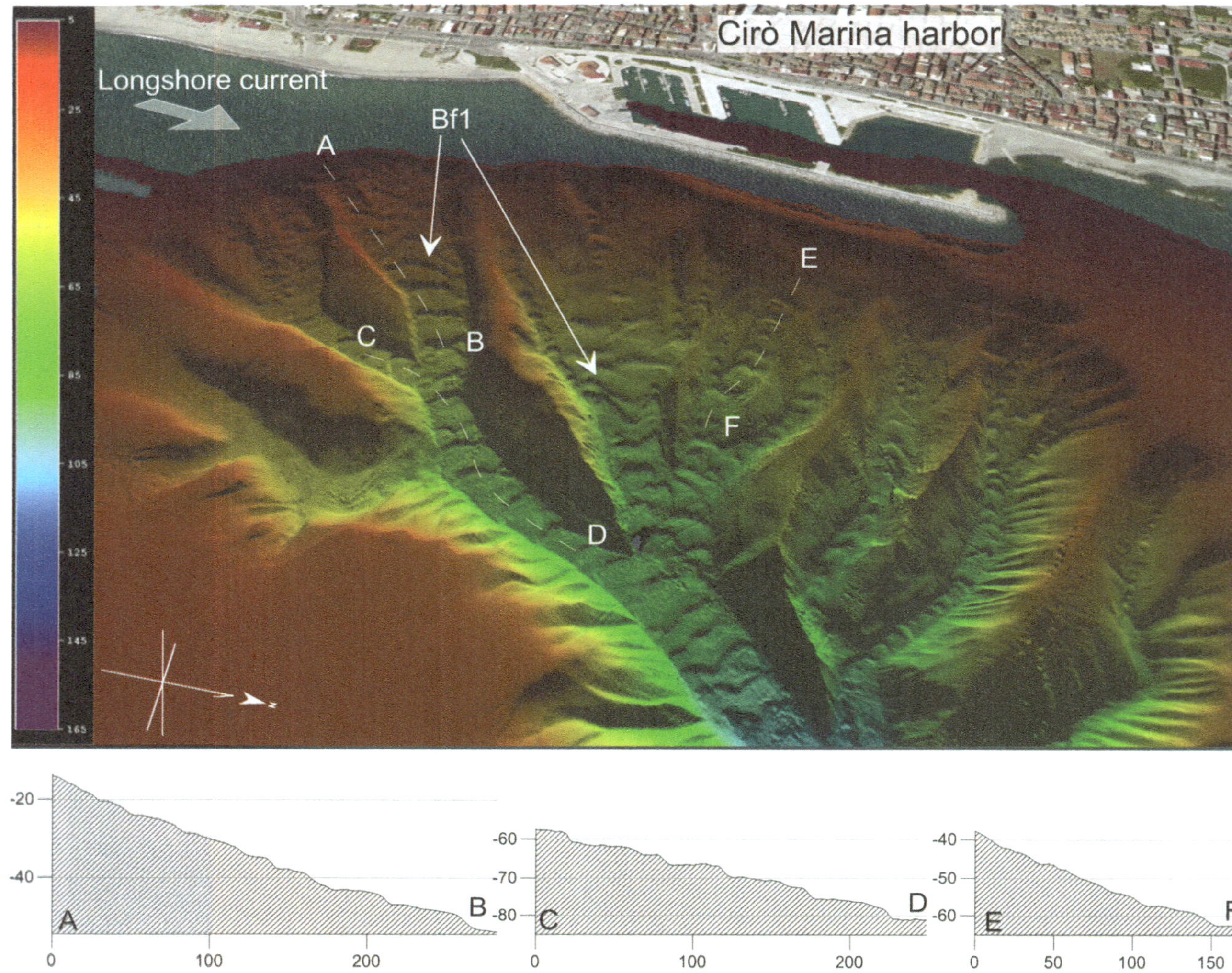

Fig. 37.2 3-D perspective image of the Cirò Canyon head region very close to the Cirò Marina harbour, with channels showing coaxial trains of crescent-shaped bedforms. Below, profiles A–B, C–D, E–F are a bathymetric cross-section along the *white dashed lines*

canyon activity and virtually feeding bedform development within the channels. South of Punta Alice Canyon, a narrowed and shallower shelf may hamper the efficiency of currents, and bedforms attain a size below the resolution of our data; consequently, the feeding of canyons by sediment wave migration is also expected to be less effective. Nevertheless, wavy to stepped bedforms are widespread within the Cirò Canyon. This evidence indicates that (1) sediment is supplied by other means than bedform migration, for instance by mass-wasting processes; and (2) other factors control the development of intra-canyon bedforms, such as the steepness and shape (length, width and sinuosity) of the channels.

Thus, the northern flank of the Punta Alice Canyon, although directly fed by shelf bedform migration, is steep and deprived of well-developed tributary channels, except for the most proximal segment where, indeed, bedforms are observed (Bf1 in Fig. 37.3) and show evidence of upslope migration based on available residual bathymetric maps (unpublished data). The southern canyon flank is less steep but it is sheltered from direct feeding by bedform migration. Although residual bathymetric maps (Casalbore et al. 2012) evidence sediment mass failure along the canyon head (dashed line in Fig. 37.1), there are no well-developed channels where this failed sediment can be conveyed to produce confined bedforms. Instead, well-developed tributary channels dissect the Cirò Canyon head, each showing a thalweg covered by bedforms.

Based on their size, shape and upslope migration, confined bedforms in the study area are comparable with the "cyclic steps" described in active canyon heads (e.g. Paull et al. 2010; Hughes-Clarke et al. 2013), and interpreted as upslope-migrating turbidite sediment waves (Cartigny et al. 2011).

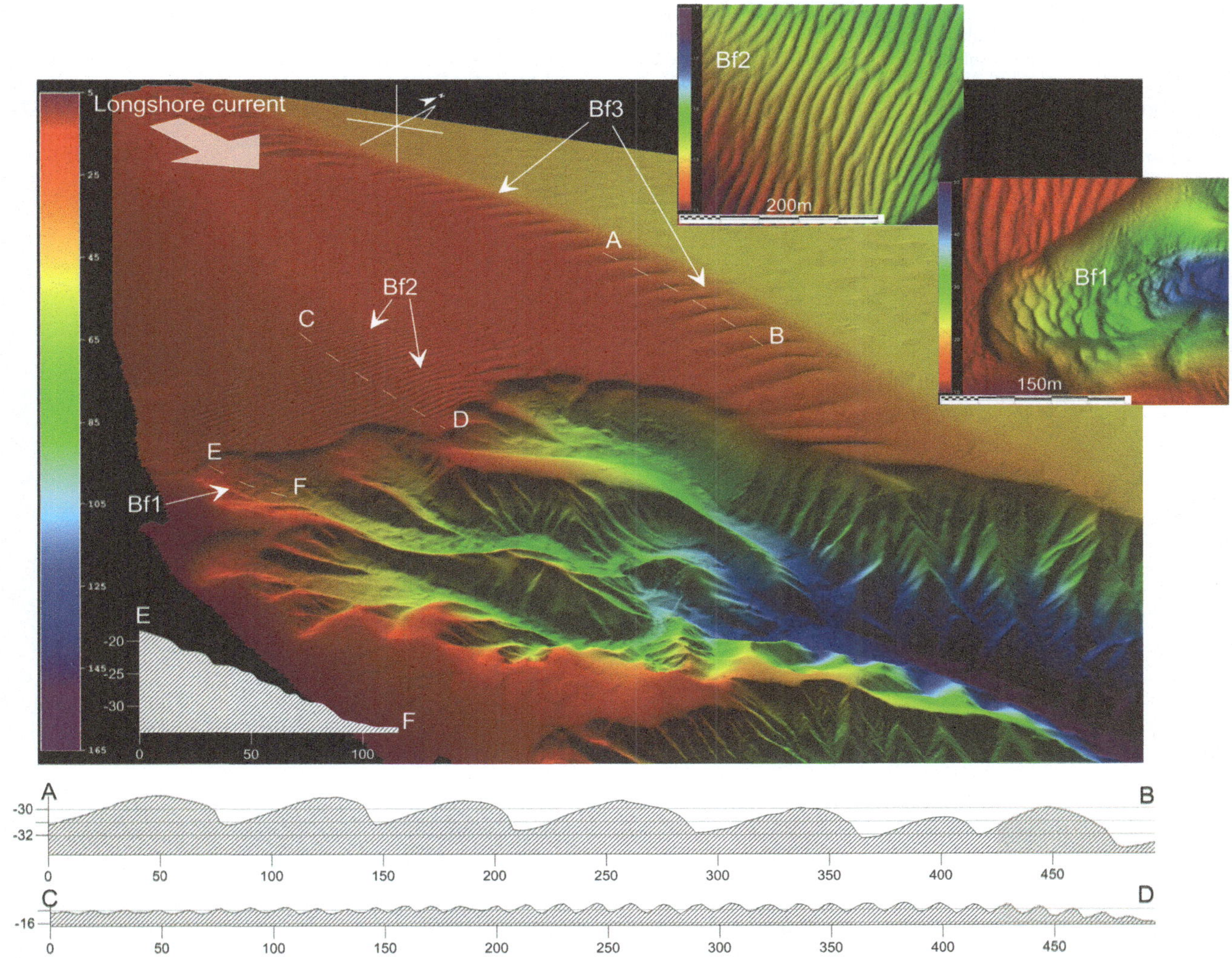

Fig. 37.3 3-D perspective image of Punta Alice Canyon head, where three types of bedform are present (Bf1, Bf2 and Bf3; see text for details); upper right, detail of Bf1 and Bf2. A–B, C–D and E–F are bathymetric cross-sections along the *dashed white line*

Acknowledgments We gratefully thank the Coastal Consulting & Exploration for the acquisition of multibeam data, with special reference to Stefano Mosticchio and Francesco De Giosa. This research was funded by the Italian Civil Protection Department and further developed in the MAGIC Project (MArine Geohazards along the Italian Coasts) and in the framework of RITMARE (Ricerca ITaliana per il MARE).

References

Cartigny MJB, Postma G, Van den Berg JH, Mastbergen DR (2011). A comparative study of sediment waves and cyclic steps based on geometries, internal structures and numerical modeling. Marine Geology 280, 40–56.

Casalbore D, Bosman A, Chiocci FL (2012). Study of recent small-scale landslides in geologically active marine areas through repeated multibeam surveys: examples from the Southern Italy. In: Submarine mass movement and their consequences, “Advances in natural and technological hazards research”, vol 31 Springer International Publishing, pp. 573–582.

Ferranti L, Antonioli F, Mauz B, Amorosi A, Dai Pra G, Mastronuzzi G, Monaco C, Orrù P, Pappalardo M, Radtke U, Renda P, Romano P, Sansò P, Verrubbi V (2006). Markers of the last interglacial sea-level highstand along the coast of Italy: Tectonic implications. Quaternary International 145–146, 30–54.

Hughes Clarke JE, Vidiera Marques CR, Pratomo D (2013). Imaging active mass-wasting and sediment flows on a Fjord Delta, Squamish, British Columbia. In: Krastel, S., et al. (Eds.), Submarine Mass Movements and Their Consequences. Advances in Natural and Technological Hazards Research, 37, pp. 249–260.

Paull CK, Ussler W, Caress DW, Lundsten E, Covault JA, Maier KL, Xu J, Augenstein S (2010). Origins of large crescent-shaped bedforms within the axial channel of Monterey canyon, offshore California. Geosphere 6:1–20.

Rebesco M, Neagu RC, Cuppari A, Muto F, Accettella D, Dominici R, Cova A, Romano C, Caburlotto A (2009). Morphobathymetric analysis and evidence of submarine mass movements in the western Gulf of Taranto (Calabria margin, Ionian Sea). International Journal of Earth Sciences 98(4):791–805.

Ridente D, Martorelli E, Bosman A, Chiocci FL (2013). High-resolution morpho-bathymetric imaging of the Messina Strait (Southern Italy). New insights on the 1908 earthquake and tsunami. Geomorphology 208, 149–159.

Royden LE, Patacca E, Scandone P (1987). Segmentation and configuration of subducted lithosphere in Italy: an important control on thrust-belt and foredeep-basin evolution. Geology 15, 714–717.

Examples of Sediment Waves in and Around Submarine Canyons of the North Alboran Sea

38

Araceli Muñoz, Elena Elvira, César León, Juan Acosta, and Patricia Jiménez

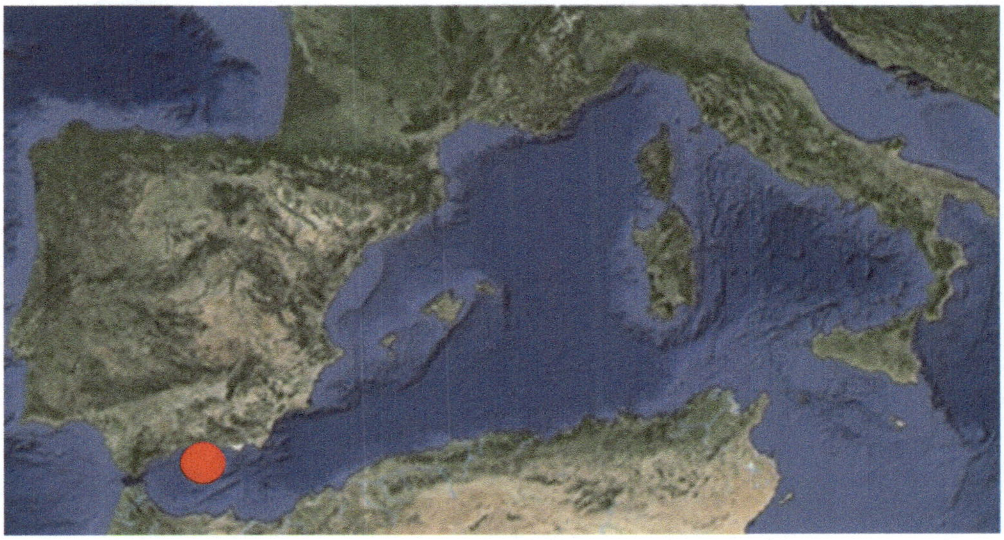

Abstract

Swath bathymetry in the north Alboran Sea shows different types of sediment waves associated with submarine canyons. Bedforms are located in the channel axis, on the levee backslopes of a meander and on the fan lobe developed at the mouth of the canyons. They probably represent different responses to turbidity current flows and suggest active sediment dynamics (at present and/or in the past) in this area.

Keywords

Continental slope • Bedforms • Sediment waves • Cyclic steps • Fan lobe

38.1 Introduction and Methods

The northern Alboran Sea is characterized by an arid climate with small rivers able to supply large amounts of sediment during flood events and a continental margin incised by several submarine canyons (Muñoz et al. 2008) (Fig. 38.1). This chapter provides examples of sediment waves developed along the axis, at the mouth and on bends of the Almuñecar and Motril submarine canyon systems (Fig. 38.1).

The Almuñecar Canyon System is located in front of the Guadalfeo and Rios Verdes rivers. It is composed of five short valleys with their heads below the shelf-break and lengths ranging from 4 to 13 km (Muñoz et al. 2008). These valleys are joined together at depth and finally only two persist (Fig. 38.1). A field of sediment waves was identified along the distal part of the axis of the two main channels and on the fan lobe developed at their mouth.

A. Muñoz (✉) · E. Elvira · C. León · P. Jiménez
Tragsa, C/Valentín Beato, no. 6, 28037 Madrid, Spain
e-mail: amur@tragsa.es

J. Acosta
IEO, C/Corazón de María, no. 8, 28006 Madrid, Spain

J. Guillén et al. (eds.), *Atlas of Bedforms in the Western Mediterranean*,
DOI 10.1007/978-3-319-33940-5_38

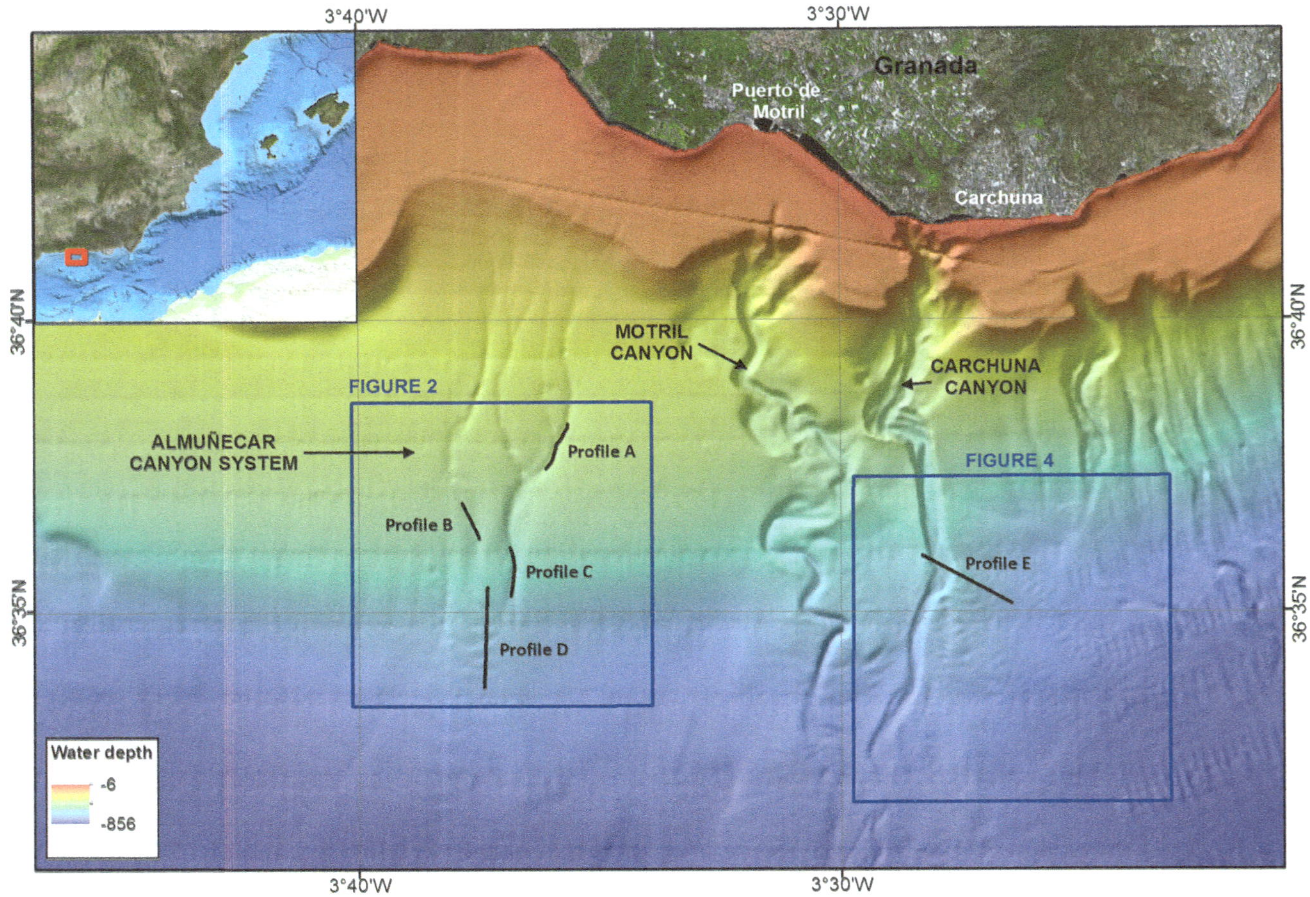

Fig. 38.1 The Almuñecar and Motril-Carchuna Canyon Systems in the northern Alboran Sea (SE Spain). Location of maps and profiles of Figs. 38.2, 38.3 and 38.4

The Motril Canyon System is formed by the meandering Carchuna and Motril canyons, which flow almost in parallel across the continental margin along 23 km and have widths of 1–1.5 km (Alonso and Ercilla 2003). The Carchuna Canyon head is deeply indented in the continental shelf, being located just 200 m from the coastline and in front of an ephemeral stream. This chapter describes a sediment wave field developed on a levee-overbank system located on the east bank of a meander (Fig. 38.1).

The bathymetric data were acquired during two surveys in January 2002 and July 2003 on board the R/V *Vizconde de Eza* with a Kongsberg EM300 multibeam echosounder. The processing was carried out using the Neptune, Caris and ArcGIS 9.3 software packages to obtain a final resolution grid of 20 × 20 m.

38.2 Results and Discussion

Seabed undulations were observed along the axis of submarine canyons in the Almuñecar Canyon System (Figs. 38.2 and 38.3). In the westerly channel axis, undulations are developed at 475 to 522 m depth (Figs. 38.2 and 38.3B). Their height ranges between 0.20 and 1.20 m and the wavelength varies between 190 and 285 m. In the easterly channel axis, two fields of undulations are shown (Figs. 38.2 and 38.3A, C): On the shallow section (390–435 m depth), the height of the undulations is only 0.65–0.80 m and the wavelength ranges from 215 to 330 m. On the deeper section (550–610 m depth), the undulations reach a height of 10 m and the distance between crests ranges from 400 to 550 m. There are no data dealing with sedimentary

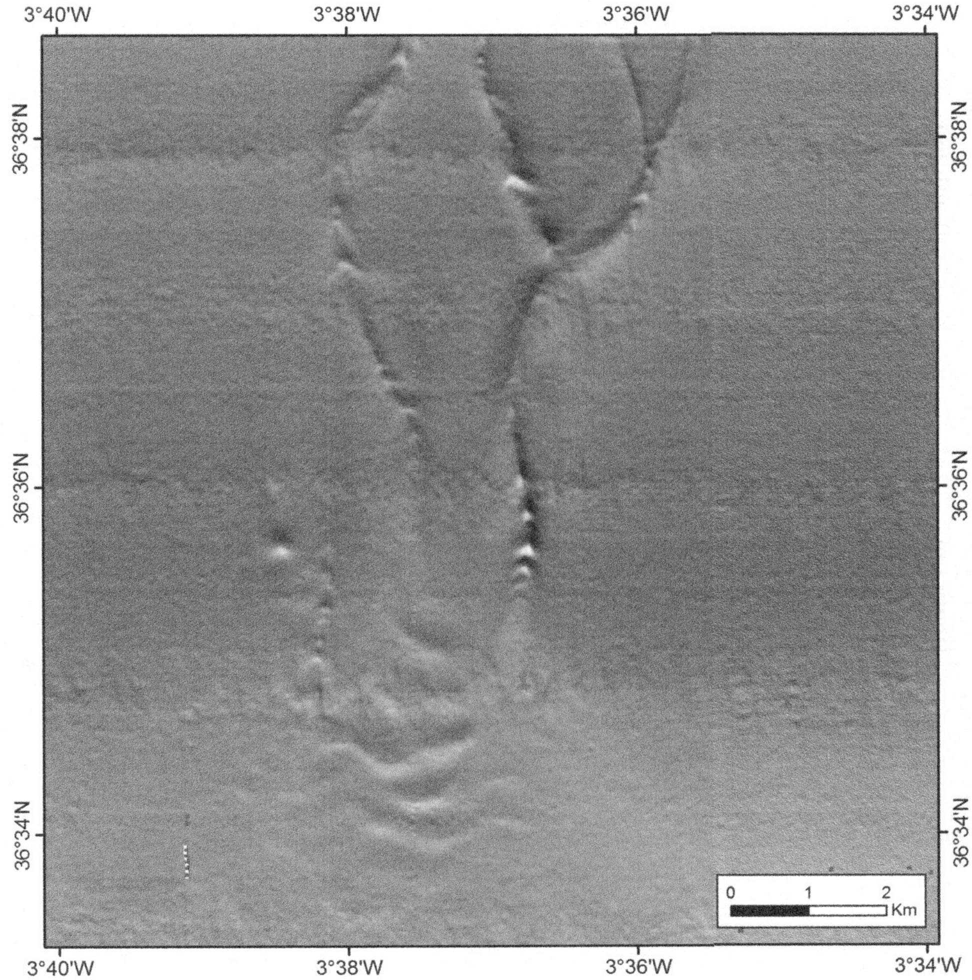

Fig. 38.2 Detail of the Almuñecar Canyon System, showing bedforms along the canyon axis (probably cyclic steps) and the sediment wave field developed on the Almuñecar fan lobe (see Fig. 38.1 for location of the area)

processes in this area that allow the origin of these undulations to be interpreted. However, similar undulations have been described on other canyon axes and were interpreted as cyclic steps, which are up-sloping migrating bedforms that can be generated by hydraulic jumps and supercritical turbidity currents (Smith et al. 2005; Cartigny et al. 2011; Cartigny and Postma this issue).

At the distal part of these canyons a sediment wave field has developed where the relief of the channel vanishes downslope and a lobate deposit has been generated at the mouth of the canyon. This field is composed of 13 sediment waves covering an area of 21.63 km^2 between 605 and 808 m depth (Figs. 38.1 and 38.2). The bedform morphology is characterized by its asymmetrical profile, with a relatively smooth crest, a broad trough and sinuous crestlines (Figs. 38.2 and 38.3D). The spacing of these bedforms ranges from 530 to 750 m and they range in height from 2 to 5 m (Fig. 38.3). Similar bedforms located in the proximal part of submarine fan systems and channel-lobe transitions have been described and interpreted as sediment wave fields formed by unconfined turbidity currents (Wynn et al. 2002).

Finally, a sediment wave field on the slope of the Motril Canyon System is described (Fig. 38.4). It has developed between 650 and 750 m depth and covers an area of 49 km^2. The crestlines display sinuous or crescentic shapes and their orientation varies from 135 to 190°. The spacing between crests ranges from 130 to 250 m and they range in height from 1 to about 5 m. The mean length of the forms is about 500 m. The sediment wave fields are located at the eastern side of the canyon, most of them associated with the meandering of the channel, suggesting that their formation may be related to the overflow of the water and sediment fluxes conducted along the canyon in the meander and the spillover of density currents on the continental slope. Since the heads of some coastal tributaries of the Carchuna Canyon are located at 10 m depth (Ortega-Sánchez et al. 2014), it is plausible to consider this canyon as a direct conduct of active sediment transport towards deeper areas during storms or river floods. Similar bedforms are commonly described on the back slopes of channel levees, where turbidity currents flow unconfined (Fildani et al. 2006).

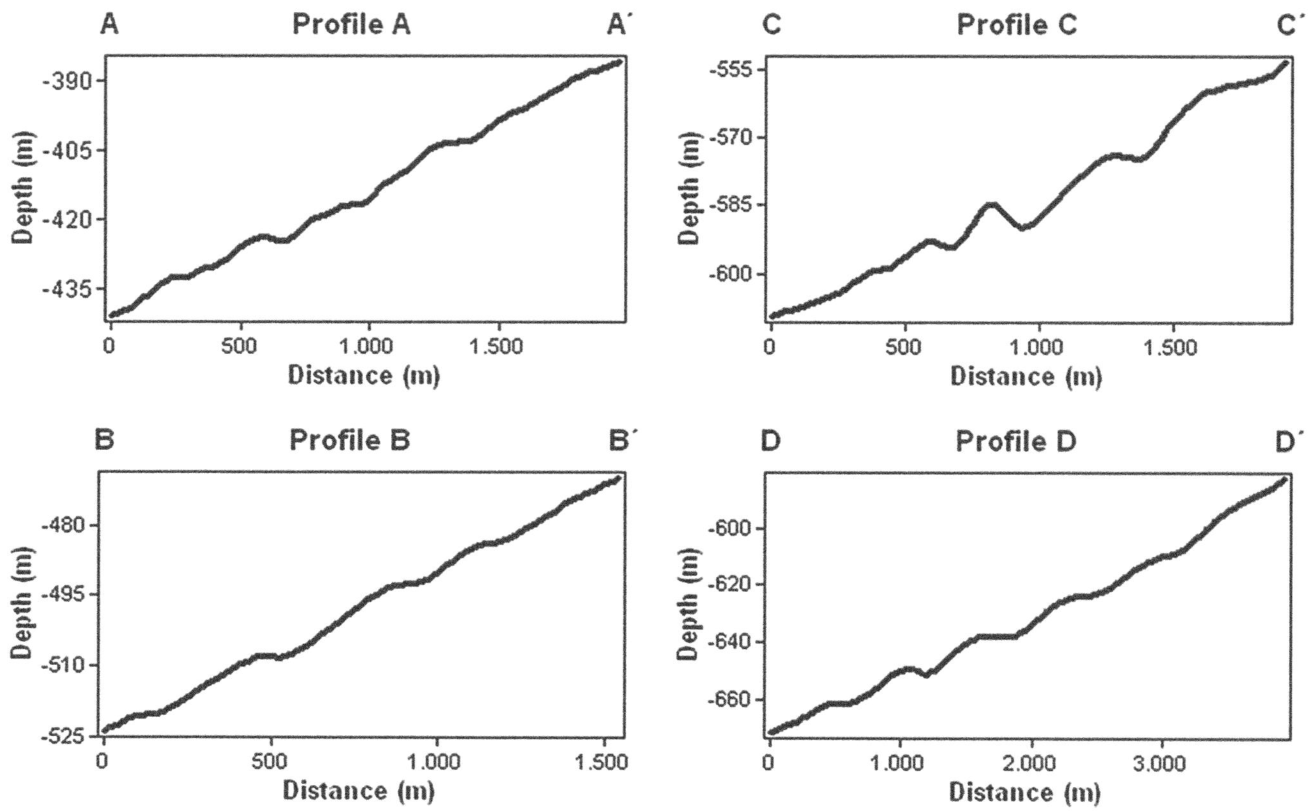

Fig. 38.3 Bathymetric profile across bedforms in the Almuñecar Canyon System: interpreted cyclic steps along the channel axis (profiles *A*, *B* and *C*) and sediment waves at the distal lobe (*D*) (see Fig. 38.1 for location of the profiles)

Examples of different types of bedforms in the channel (interpreted as cyclic steps), on the overbank deposit of a meander and on the fan lobe suggest active transport of sediment at present and/or in the past along the submarine canyon systems on the north Alboran margin. These are examples of different seabed responses to confined and unconfined turbidity currents with varying hydraulic conditions in a submarine canyon environment.

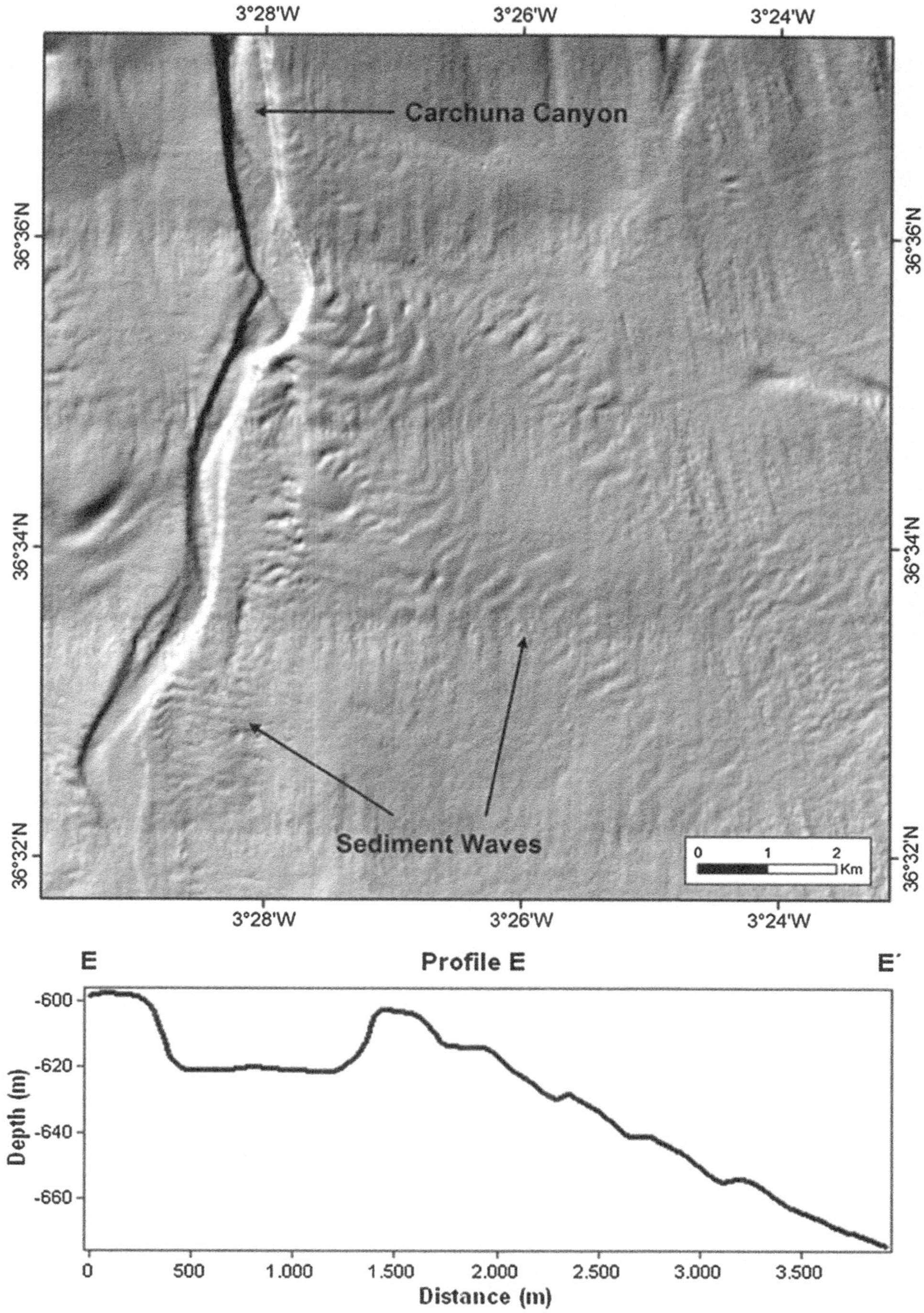

Fig. 38.4 **a** Sediment wave fields associated with meanders in the Carchuna Canyon.
b Bathymetric profile across the channel and sediment waves (see Fig. 38.1 for location of the area and the profile)

Acknowledgments The data were obtained in the framework of the Fisheries Mapping Programme of the Mediterranean, funded by the Secretaría General de Pesca (MAGRAMA). We also wish to thank the SGP-TRAGSA Cartography Group, who participated in the cruises.

References

Alonso, B. and Ercilla, G. (2003). Small turbidite systems in a complex tectonic setting (SW Mediterranean Sea): morphology and growth patterns. Marine and Petroleum Geology, 19, 1225–1240.

Cartigny, M.J.B., Postma, G., van den Berg, J. H., Mastbergen, D.R. (2011). A comparative study of sediment waves and cyclic steps based on geometries, internal structures and numerical modeling. Marine Geology, 280: 40–56.

Cartigny, M.J.B., Postma, G. (2016). Turbidity current bedforms (this issue).

Fildani, A., Normark, W.R., Kostic, S., Parker, G. (2006). Channel formation by flow stripping: large scale scour features along the Monterey East Channel and the relation to sediment waves. Sedimentology, 53: 1–23.

Muñoz, A., Ballesteros, M., Montoya, I., Rivera, J., Acosta, J., Uchupi, E. (2008). Alborán Basin, southern Spain—Part I: Geomorphology. Marine and Petroleum Geology, 25 (1): 59–73.

Ortega-Sánchez, M., Lobo, F.J., López-Ruiz, A., Losada, M.A., Fernández-Salas, L.M. (2014). The influence of shelf-indenting canyons and infralittoral prograding wedges on coastal morphology: The Carchuna system in Southern Spain. Marine Geology, 347: 107–122.

Smith, D.P., Ruiz, G., Kvitek, R., Iampietro, P.J. (2005). Semiannual patterns of erosion and deposition in upper Monterey Canyon from serial multibeam bathymetry. Geological Society of America Bulletin 117 (9–10), 1123–1133.

Wynn, R.B., Kenyon, N.H., Masson, D.G., Stow, D.A.V., Weaver, P.P. E. (2002). Characterization and recognition of deep-water-channel-lobe transition zones. American Association of Petroleum Geologists Bulletin, 86:1441–1462.

39 Sediment Waves at the Mouth of the Almanzora-Alías-Garrucha Canyon

Patricia Jiménez, Elena Elvira, Araceli Muñoz, and Juan Acosta

Abstract

Several tributaries highly incised in the continental margin merge in the Almanzora-Alías-Garrucha Canyon, on the lower slope of the Vera Gulf. This canyon generated a turbidite system and a submarine fan at the mouth, where a field of large sediment waves developed at about 2500 m water depth. These sediment waves reach maximum heights and wavelengths of 80 m and 4.5 km, respectively, and are preferentially orientated N-S. The sediment waves display an upslope migration and are interpreted as being associated with supercritical turbidity currents.

Keywords

Almanzora-Alías-Garrucha canyon • Vera gulf • Large sediment waves • Supercritical turbidity currents

39.1 Introduction

The Vera Gulf is located in the transition zone between the East Alborán Basin and the Algero-Balearic Basin (Acosta et al. 2013). Since the Messinian, the physiography of the Vera Gulf margin has been subjected to strong tectonic control (Perez-Hernández et al. 2009). The continental margin is incised by several valleys (the shallowest head at 64 m water depth) that give rise to four main tributary canyons. These canyons merge at 1000–1200 m water depth into two tributaries, which in turn merge into a single canyon at about 1900 m depth (Pérez-Hernández et al. 2014). This complex physiography leads to some confusion about the name of the canyon because the canyon heads are unrelated to the present location of the river mouths. It has been called the Palomares Canyon (Medialdea et al. 1982), the

P. Jiménez (✉) · E. Elvira · A. Muñoz
Tragsa – SEPI. Calle Valentín Beato 6, 28037 Madrid, Spain
e-mail: pjg@tragsa.es

J. Acosta
Instituto Español de Oceanografía, Calle Corazón de María 8, 28002 Madrid, Spain

J. Guillén et al. (eds.), *Atlas of Bedforms in the Western Mediterranean*,
DOI 10.1007/978-3-319-33940-5_39

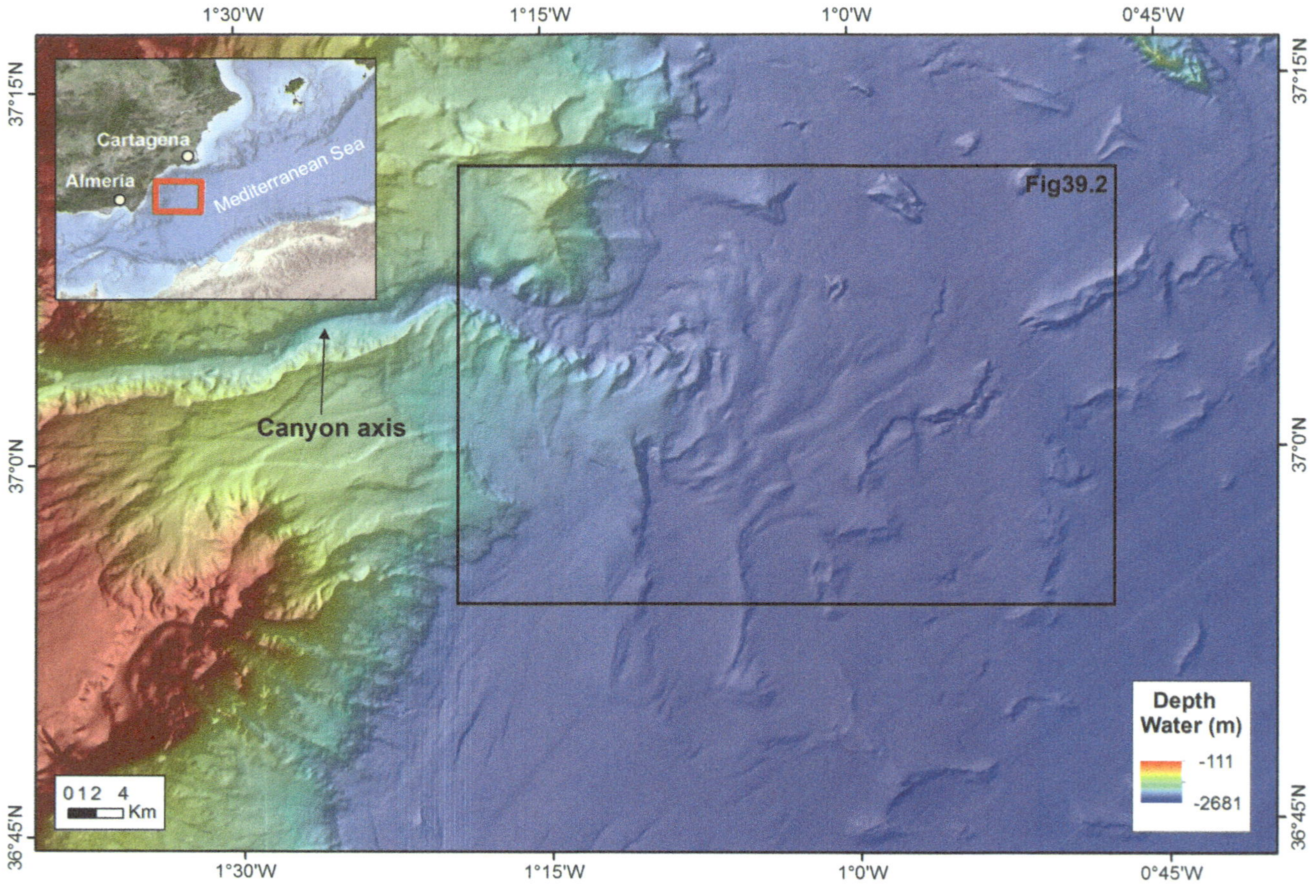

Fig. 39.1 Location of the study area in the western Mediterranean Sea

Aguas-Almanzora Canyon (Comas et al. 2006), the Alías-Almanzora Canyon (Pérez-Hernández et al. 2009, 2014) and the Almanzora-Alías-Garrucha Canyon (Acosta et al. 2012). In this chapter we use the last name because it includes the main rivers discharging in the area and the location of the most incised canyon.

The Almanzora-Alías-Garrucha Canyon is 67 km long and its width increases with depth, reaching about 5 km in the deepest sector, where it becomes meandering and progressively diminishes its relief towards the mouth, developing a submarine fan and several large sediment waves in this zone (Fig. 39.1). This depositional system appears to be dominated by fine-grained sediments (Pérez-Hernández et al. 2014). The objective of this chapter is to describe the morphology and internal structure of these sediment waves as an example of large-scale bedforms associated with the dynamics of submarine canyons in deep water environments.

39.2 Methods

Data were obtained during the CAPESME-06 cruise on board the R/V *Vizconde de Eza* in October 2006 (Fig. 39.1). The bathymetry was acquired with a Kongsberg-Simrad EM300 Multibeam Echo-Sounder and the seismic data using a TOPAS (parametric sub-bottom profiler). The raw bathymetric data were processed with the Neptune software and the shade-bathymetry or Digital Terrain Models (DTM) was generated with CFloor software with a cell size of 50 m and ArcGis.

39.3 Results and Discussion

The sediment wave field occupies an area of 176.55 km^2, with depths ranging from 2437 to 2605 m and a slope ranging from almost flat areas among the sediment waves to

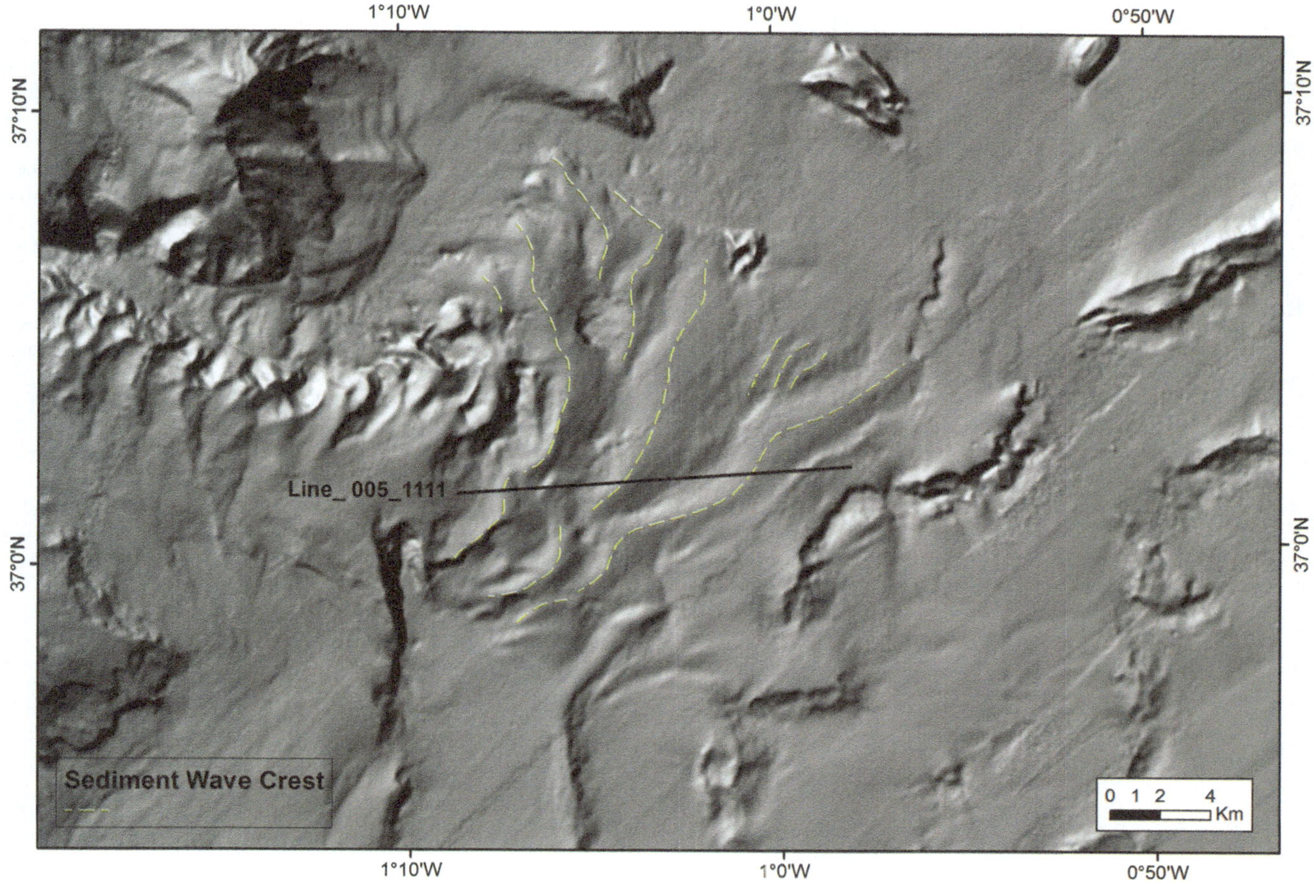

Fig. 39.2 Digital terrain model of the sediment wave field. The *dashed yellow line* represents the crest of the sediment wave and the *black line* shows the location of the seismic profile

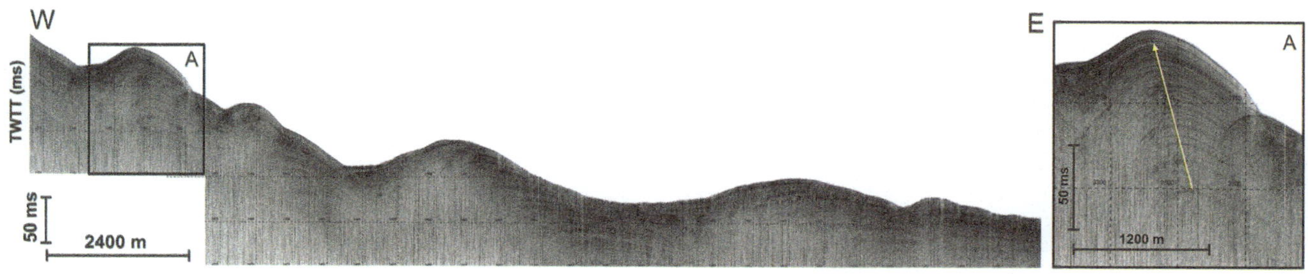

Fig. 39.3 Topas seismic profile across the dune-field (see location in Fig. 39.2). The internal structure displays an aggradation of parallel stratified reflectors with a migration towards *W* (upsloping), as indicated by the *yellow arrow* in the detailed figure *A* (ms is milliseconds and m is metres)

60° on some slipfaces located at the east of the bedform field. The field is constituted by a total of 12 sediment waves with crestlines between 0.7 and 4.8 km apart, heights of between 8.3 and 83.9 m and lengths of between 1.7 and 16.6 km. They appear to be limited by scoured zones, suggesting the reworking of the bottom induced by currents (Fig. 39.2).

The seismic profile across the sediment wave field shows an internal structure mainly characterized by stratified reflectors with convex geometries (Fig. 39.3). The vertical aggradation of sediment waves is accompanied by an upslope migration towards the west, opposite to the direction of the expected flow (Fig. 39.3a).

The zone located at the mouth of the Almanzora-Alías-Garrucha Canyon is interpreted as a turbidite lobe. By comparison with other sediment wave fields developed in turbide lobes, it has been hypothesized that the origin of the sediment waves is related to hydraulic jumps generated by strong sediment flows (Pérez-Hernández et al. 2014). In a regional context of high tectonic activity, instability processes on the lower canyon walls can trigger sediment flows with supercritical turbidity currents, generating hydraulic jumps that are favoured by the changes in the slope, as observed in other areas.

References

Acosta, J., Rivera, J., Muñoz, A., León, C. (2012). Mapa topobatimétrico en relieve del margen continental del sureste Español. Cómo es el fondo marino y su geología en el margen SE Español. Instituto Español de Oceanografía, Madrid, 26 pp plus a 3D map of the area at scale of 1;400, 000.

Acosta, A., Fontan, A., Muñoz, A., Muñoz-Martín, A., Rivera, J., Uchupi, E. (2013). The morpho-tectonic setting of the South East margin of Iberia and the Adjacent Oceanic Algero- Balearic Basin. Marine and Petroleum Geology, 45:17–41.

Comas, M.C., Ivanov, M., Scientific party TTR14 Leg2. (2006). Chapter 2: Southeastern Iberian Margins: The Alboran Basin and the Palomares and Cartagena Margins: Preliminary results of geological and geophysical investigations during the TTR-14 leg 2 cruise of R/V Professor Logachev. IOC Technical Series, 70. UNESCO, pp. 44–61.

Medialdea Vega, J., Baena Pérez, J., García Rodríguez, J., Maldonado, A.E., Uchupi, E., Udías Vallina, A., Wandossell, J. (1982). Mapa geológico de la plataforma Española y zonas adyacentes. E: 200,000. Almería-Garrucha. Chella-Los Genoveses. Instituto Geológico y Minero de España, 105 pp.

Pérez-Hernández, S., Comas. M.C., Escutia, C., Martínez-García, P. (2009). Morphologic characterization of submarine canyons in the Palomares margin (western Mediterranean). Geo-Marine Research on the Mediterranean and European-Atlantic margins. International Conference and TTR-17 Post-Cruise Meeting of the Training-through-Research Programme. Granada, Spain, 2–5 February 2009. IOC Workshop Report No. 222 (English), UNESCO, 2009, 23.

Pérez-Hernández, S., Comas, M.C., Escutia C. (2014). Morphology of turbidite systems within an active continental margin (the Palomares Margin, western Mediterranean). Geomorphology, 219: 10–26.

Part VII

Slope and Deep Bedforms

Large Sediment Waves Over the Gulf of Roses Continental Slope (NW Mediterranean)

40

Marta Ribó, Ruth Durán, Pere Puig, David Van Rooij, and Jorge Guillén

Abstract

Recently acquired swath bathymetry and seismic datasets on the Gulf of Roses continental slope revealed the presence of large sediment waves that have developed between ~200 and ~400 m water depth. Geometric parameters were computed from the multibeam dataset and the obtained results showed mean sediment wave lengths of ~2000 m and maximum heights of ~60 m. The analysis of seismic profiles showed the general architecture of the Late Quaternary deposits over the Gulf of Roses outer shelf and slope. The outer continental shelf is characterized by morphological erosional features affecting the upper part of the most recent seismic unit. Over the continental slope five main unconformity-bounded seismic units are differentiated, with sediment waves developed in all of them. The sediment waves observed over the continental slope are presumably generated by bottom currents intensified during major dense shelf water cascading events flowing downslope at an oblique angle with respect to the main bathymetry.

M. Ribó (✉) · R. Durán · P. Puig · J. Guillén
Institut de Ciències del Mar (ICM-CSIC), Barcelona, Spain
e-mail: mribo@icm.csic.es

D. Van Rooij
Department Geology and Soil Science, Ghent University, Ghent, Belgium

J. Guillén et al. (eds.), *Atlas of Bedforms in the Western Mediterranean*,
DOI 10.1007/978-3-319-33940-5_40

Keywords
Sediment waves • Continental slope • NW Mediterranean sea • Dense shelf water cascading

40.1 Introduction

Previous studies in the northwestern Mediterranean Sea have shown that the Gulf of Lions (GoL) is one of the regions of the world where dense shelf waters (DSW) are generated (Puig, this issue). During wintertime, the northerly (Mistral) and northwesterly (Tramuntana) dry and cold winds favour the cooling and homogenization of the shelf waters in the GoL, triggering the formation of DSW (Béthoux et al. 2002; Durrieu de Madron et al. 2005, 2008; Canals et al. 2006, 2009). When the storm-induced downwelling interacts with dense shelf water cascading (DSWC), this enhances the near-bottom transport of sediment, advecting resuspended sediments towards deeper reaches of the westernmost GoL submarine canyons (Palanques et al. 2006, 2008; Ulses et al. 2008a, b). The Cap de Creus submarine canyon (Fig. 40.1) has been identified as a major sediment transport conduit in the northwestern Mediterranean during high-energy storms and DSWC events. However, it is known that part of the bottom currents and the associated sediment transport are advected southwards towards the Gulf of Roses (GoR), channelized between the Cap de Creus promontory and the submarine canyon flank (Lastras et al. 2011).

The GoR is located on the northern Catalan margin (northwestern Mediterranean Sea), south of the Cap de Creus promontory (Fig. 40.1). Its structure is controlled by NW–SE to NNW–SSE normal faults, which created a system of horsts and grabens reflected in the present configuration of the coastal plain (Muñoz et al. 1986; Maillard et al. 1992; Tassone et al. 1994; Ercilla et al. 1994; Roca et al. 1999). The construction of the modern continental shelf and slope was modulated by the Plio-Quaternary glacio-eustatic sea level changes leading to alternating sediment deposition and erosion stages over the continental shelf (Liquete et al. 2008). The GoR continental shelf width varies from ~2.6 km in the vicinity of the Cap de Creus canyon to ~30 km off the Roses Bay. The shelf gradient ranges from 0.4° to 0.6° and the shelf edge is located approximately at 150 m water depth (Durán et al. 2014). The limit between the GoR inner and outer shelf is evidenced by the presence of a 1.5-km-wide, 15-m-deep channel-like depression at ~125 m water depth (Lastras et al. 2011). This structure, known as the Roses shelf channel, is oriented NNE–SSW and extends for 40 km along the GoR shelf, from the southern rim of the Cap de Creus submarine canyon upper course. The inner and outer shoulders of this shelf channel consist of rocky alignments parallel to the shelf edge (Fig. 40.1), covered by coarse sediments (Lo Iacono et al. 2010). The circulation in the GoR exhibits seasonal variability with significant spatial mesoscale variability, which plays a decisive role in exchange process between shelf and oceanic waters (Font et al. 1995). Over the GoR shelf, waves mainly come from the north (Mendoza et al. 2011) and the northern severe storms are triggered by the strong, cold and persistent Tramuntana wind blowing mainly in winter months.

Most observations of sediment dynamics in the northwestern Mediterranean have focussed on the transfer of sediment from the continental shelf into the deeper areas through the submarine canyons, principally the Cap de Creus canyon. However, as stated above, previous modelling studies have shown that the sediment transfer also occurs across the slope, associated with the overflow of DSWC south of the Cap de Creus coastal promontory (Ulses et al. 2008b). Sediment waves over the GoR continental slope, previously related to slump and creep-like deformation (Ercilla et al. 1994), are characterized in this chapter. Evidence of the sediment transfer across the continental slope is also provided, in agreement with the expected hydrodynamic downslope DSWC conditions and the morphological configuration of the continental margin.

40.2 Methods

The results presented are based on the swath bathymetry acquired using the Elac SeaBeam 1050D 50 kHz multibeam echosounder on board the R/V *García del Cid* in July 2013 during the FORMED project. The multibeam data (including correction for heading, depth, pitch, heave and roll) was post-processed using the CARIS HIPS and SIPS Hydrographic Data Processing System.

In addition, eight high-resolution single-channel seismic profiles were acquired during the same oceanographic cruise, using a SIG Sparker source (600 Joules Sparker source triggered every 3 s, with the vessel velocity maintained at ~3.5 knots.). The seismic lines were recorded directly in SegY-Motorola format with associated navigation files. The seismic data (including bandpass filtering, water noise and burst noise removal, amplitude correction, and trace editing) were processed using the seismic Rad Ex Pro

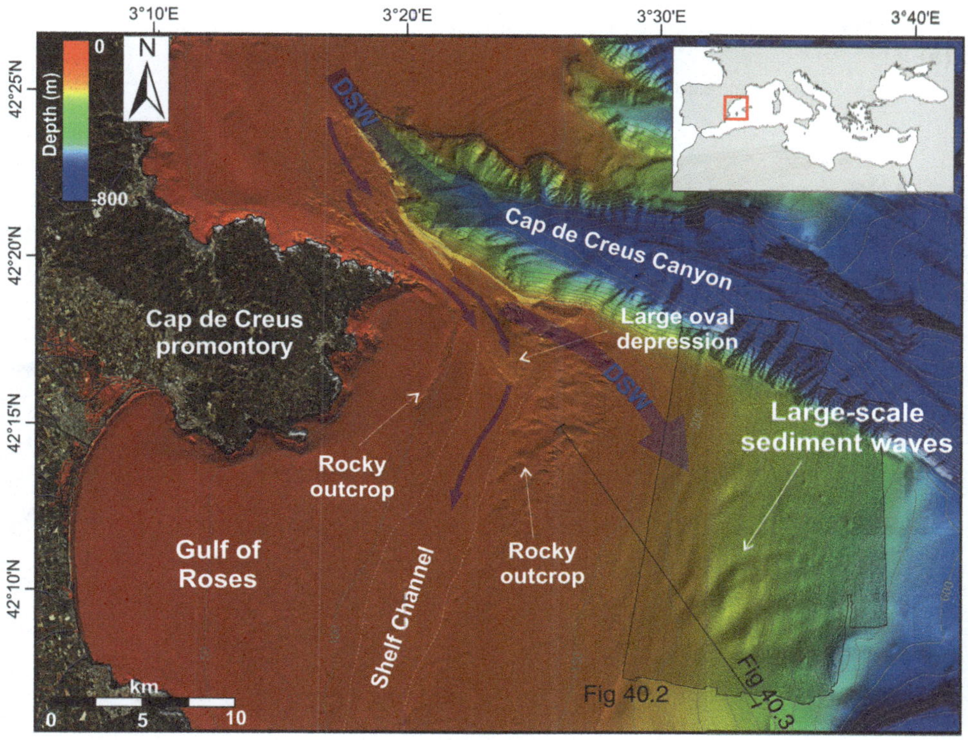

Fig. 40.1 Bathymetric map of the north Catalan margin, including the recently acquired swath bathymetry over the GoR slope. *Arrows* indicate the dense shelf water circulation through the Cap de Creus Canyon and towards the Gulf of Roses. The major morphological features observed are indicated, as well as the location of the single-channel seismic profile

software (Deco Geophysical) and the data were interpreted using the Kingdom Suite software.

40.3 Results and Discussion

Several large sediment waves were observed over the GoR continental slope, extending from ~200 to ~400 m depth, with a NNE–SSW to NE–SE orientation. The continental slope gradient is relatively gentle, ranging from 1° to 4°, and the sediment waves, with a maximum lateral continuity extending to up to 3000 m, tend to disappear towards the south-southeast and seawards, near an abrupt scarp of ~250 m height (Fig. 40.1). Morphological parameters were computed from the multibeam dataset, showing wave lengths of ~2000 m and maximum wave heights of ~60 m (Fig. 40.2).

The single-channel seismic profiles show the general stratigraphic structure over the GoR outer shelf and slope (Fig. 40.3). Five main unconformity-bounded seismic units are differentiated, with sediment waves developed in all of them (Fig. 40.3b). The development of these erosional unconformities has been previously described by Ercilla et al. (1994) and defined as being controlled by eustatic oscillations. These erosional surfaces can be followed downslope into conformable strata of the sediment waves on the upper continental slope. The seismic profiles show continuous internal reflectors downslope and the sediment waves show a clear up-slope migrating pattern (Fig. 40.3). It

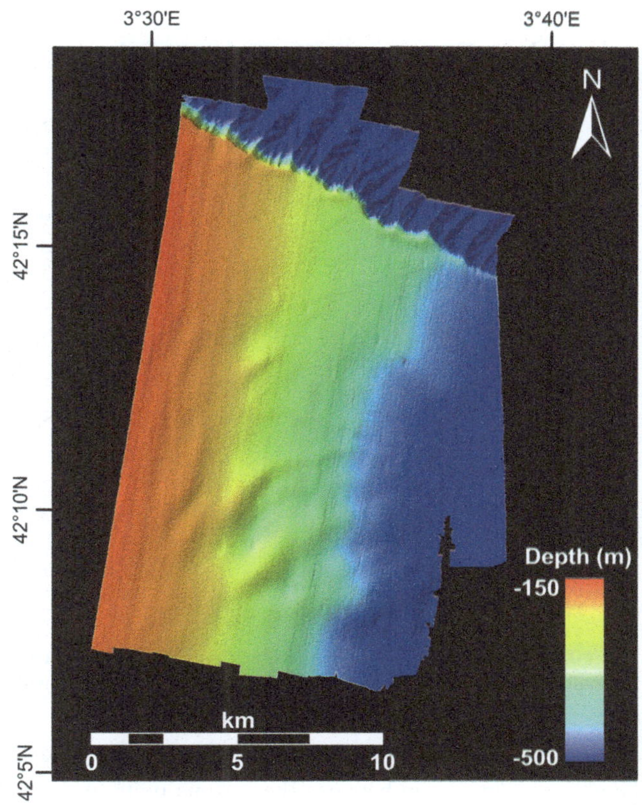

Fig. 40.2 Shaded relief image of the Gulf of Roses (20 × 20 m grid) illustrating the sediment waves observed over the continental slope

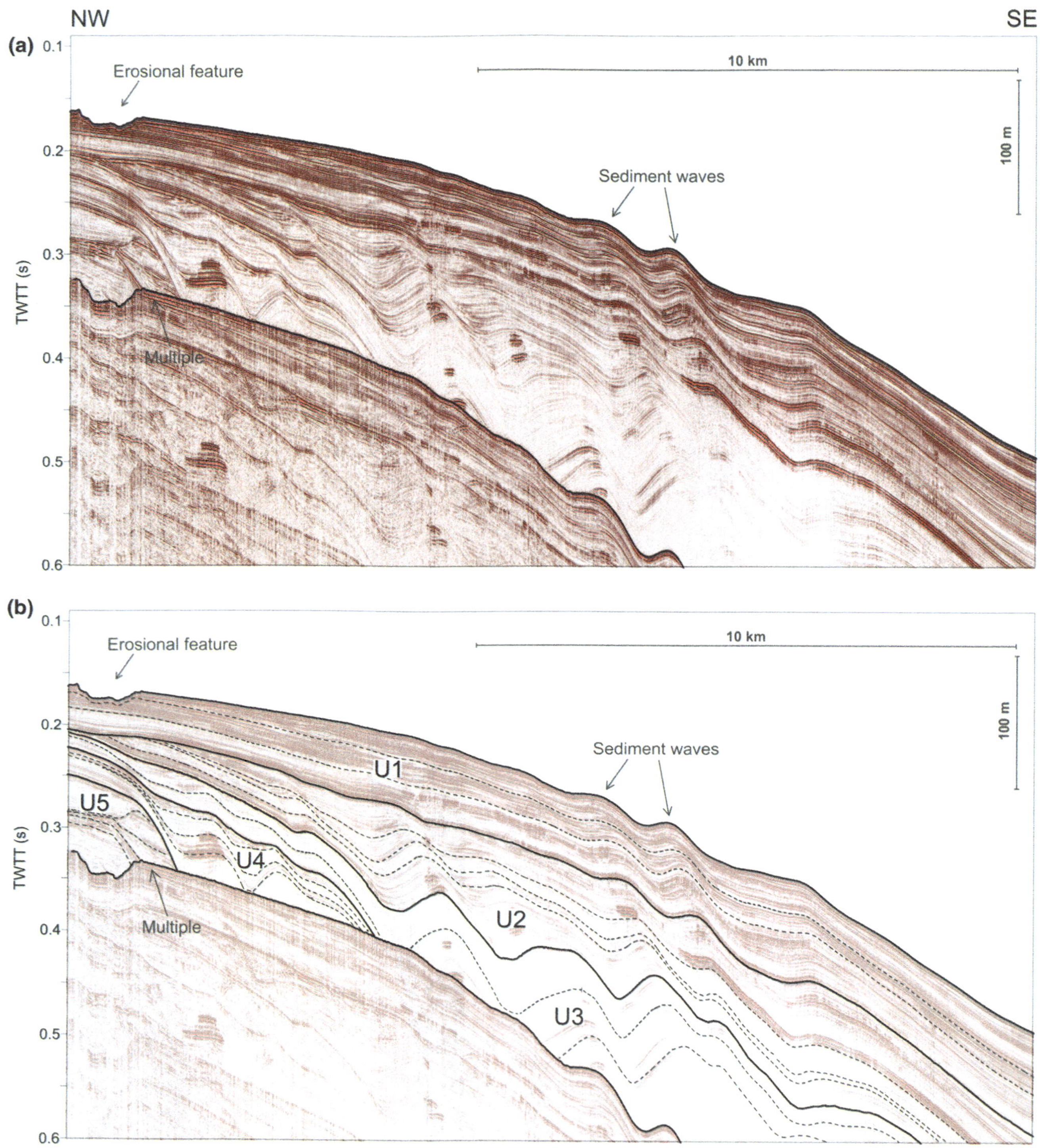

Fig. 40.3 Single-channel (Sparker) seismic profile crossing the sediment waves over the northern field perpendicularly (see location in Fig. 40.1). Unconformity surfaces defining seismic units U1–U5 are marked with *solid lines*, and *dashed lines* follow some internal reflectors within each unit

is worth noting that the sediment waves show different wave heights and lengths in between the seismic units (described here from younger to older units). The youngest unit, U1, is topped by the present seafloor and shows a relatively constant thickness of ~50 m. Over the outer continental shelf, seismic data show local erosional truncation of the subsurface reflectors of U1 at the seafloor (Fig. 40.1). Unit U1 includes the seismic units 6 and 7 defined by Ercilla et al.

(1994). The underlying units U2 and U3 pinch out landwards, increasing their thickness seawards from ~5 to ~70 m and from ~10 to ~100 m, respectively. Sediment waves in unit U2 tend to disappear from base to top and from the outer shelf to the upper slope (Fig. 40.3b). Over the outer shelf, the base of units U1, U2 and U3 truncates the sediment waves of the underlying units U2, U3 and U4, respectively, while sediment waves over the continental slope are well preserved. Sediment waves in unit U3 correspond to the slump features in seismic unit 5 described by Ercilla et al. (1994), and unit U4 correspond to their seismic unit 4. A sharp unconformity defines the boundary between U4 and U5. In unit U5 several sediment deposition events can be differentiated, with the reflectors onlapping the unconformity surface (Fig. 40.3b). These reflectors within unit U5 delimit the seismic units 2 and 3 defined by Ercilla et al. (1994), and the uppermost reflector presumably corresponds to their reflector J, defining the limit between the lower and upper Quaternary (Got 1973).

It has been described that the Pleistocene deposit development on the northern Catalan shelf has no direct relation with the rivers draining this area, as is suggested by the absence of fluvial paleovalleys eroding across the continental shelf (Ercilla 1992). Several studies have demonstrated that the combination of storm-induced downwelling and DSWC events in the GoL increases the bottom currents and induces active sediment resuspension and transport from the shelf to the slope through the submarine canyons, but sediment is also advected southwards along the Cap de Creus promontory shelf and towards the GoR (Ogston et al. 2008; Bourrin et al. 2008; Ribó et al. 2011). The sediment waves observed over the GoR continental slope provide evidence that a large part of the sediment transported within the DSW is also transferred across the slope. The bottom currents enhanced during the DSWC events are probably the main formation process developing the sediment waves, which are oriented oblique to the regional slope and apparently perpendicular to the bottom current flow (Fig. 40.1). The presence of several sediment wave units over the GoR continental slope through the sedimentary record confirms that successive DSWC events transferring sediment across the slope have taken place in the course of time, but direct dating and stratigraphic correlations would be needed to find a relation between these sediment waves and the storm downwelling and DSWC events.

40.4 Conclusions

The recently acquired swath bathymetry and seismic profiles show large sediment waves over the GoR continental slope, which were initially interpreted as slump or creep-like deformation. These sediment waves extend from ~200 to ~400 m water depth and have wave lengths of ~2000 m and maximum wave heights of ~60 m, with a maximum lateral continuity of up to 3000 m. They are oriented NNE–SSW to NE–SE and tend to disappear towards the south-southeast and seawards.

It is suggested that the bottom currents and sediment transport from the shelf to the slope, enhanced during storm-induced downwelling and DSWC events, are the main formation mechanisms for the sediment waves observed over the continental slope.

Acknowledgments This work was funded by the project COSTEM (CMT2009-07806) and FORMED (CGL2012-33989). The authors wish to thank the captain and the crew of the R/V *García del Cid*, and also the UTM technicians and J. Pozo and M. Lloret from the Instrumental Service, for their assistance. The author M. Ribó was supported by a FPI grant (Ref. BES-2010-029949) from the Spanish Ministry of Economy and Competitiveness. R. Durán is supported by a CSIC JAE-Doc contract co-funded by the FSE.

References

Béthoux, J., Durrieu de Madron, X., Nyffeler, F., Tailliez, D. (2002). Deep water in the western Mediterranean: peculiar 1999 and 2000 characteristics, shelf formation hypothesis, variability since 1970 and geochemical inferences. Journal of Marine Systems, 33–34: 117–131.

Bourrin, F., Friend, P. L., Amos, C. L., Manca, E., Ulses, C., Palanques, A., Durrieu de Madron, X., Thompson, C. E. L. (2008). Sediment dispersal from a typical Mediterranean flood: The Têt River, Gulf of Lions. Coninental Shelf Research, 28: 1895–1910.

Canals, M., Puig, P., Durrieu de Madron, X., Heussner, S., Palanques, A., Fabres, J. (2006). Flushing submarine canyons. Nature 444, 354–357. doi:10.1038/nature05271.

Canals M, Danovaro R, Heussner S, Lykousis V, Puig P, et al. (2009). Cascades in Mediterranean submarine grand canyons. Oceanography, 22(1):26–43.

Durán, R., Canals, M., Sanz, J. L., Lastras, G., Amblàs, D., Micallef, A. (2014). Morphology and sediment dynamics of the northern Catalan continental shelf, northwestern Mediterranean Sea. Geomorphology, 204: 1–20. doi:10.1016/j.geomorph.2012.10.004.

Durrieu de Madron, X., Wiberg, P.L., Puig, P. (2008). Sediment dynamics in the Gulf of Lions: the impact of extreme events. Continental Shelf Research, 28:1867–76.

Durrieu de Madron, X., Zervakis, V., Theocharis, A., Georgopoulos D. (2005). Comments on "Cascades of dense water around the world ocean". Progress in Oceanography, 64:83–90.

Ercilla, G. (1992). La sedimentación Cuaternaria en márgenes y cuencas del Mediterráneo noroccidental y suroccidental: Modelos de evolución sedimentaria y factores de control. PhD thesis. 565 pp.

Ercilla, G., Farrán, M., Alonso, B., Díaz, J.I. (1994). Pleistocene progradational growth pattern of the northern Catalonia continental shelf (northwestern Mediterranean). Geo-Marine Letters 14 (4), 264–271. doi:10.1007/BF01274062.

Font, J., Garcia-Ladona, E., Gorriz, E. G. (1995). The seasonality of mesoscale motion in the Northern Current of the western Mediterranean several years of evidence. Oceanologica Acta, 18 (2), 207–219.

Got, H. (1973). Etude des correlations tectonique-sedimentation au cours de l'historie Quaternaire du Precontinent Pyreneo-Catalan.

Ph. Thesis, Université des Sciences et Techniques du Languedoc, France, 294 pp.

Lastras, G., Canals, M., Amblas, D., Lavoie, C., Church, I., De Mol, B., Duran, R., Calafat, A. M., Hughes-Clarke, J. E., Smith, C. J., Heussner, S., and "Euroleón" cruise shipboard party. (2011). Understanding sediment dynamics of two large submarine valleys from seafloor data: Blanes and La Fonera canyons, northwestern Mediterranean Sea. Marine Geology, 280: 20–39.

Liquete, C., Canals, M., De Mol, B., De Batist, M., Trincardi, F. (2008). Quatarnary stratal architecture of the Barcelona prodeltaic continental shelf (NW Mediterranean). Marine Geology 250, 234–250.

Lo Iacono, C., Guillén, J., Puig, P., Ribó, M., Ballesteros, M., Palanques, A., Farrán, M., Acosta, J. (2010). Largescale bedforms along a tideless outer shelf setting in the western Mediterranean. Continental Shelf Research, 30, 1802–1813.

Maillard, A., Mauffret, A.,Watts, A.B., Torné,M., Pascal, G., Buhl, P., Pinet, B. (1992). Tertiary sedimentary history and structure of the Valencia Trough (Western Mediterranean). Tectonophysics 203, 57–76.

Mendoza, E. T., Jimenez, J. A., Mateo, J. (2011). A coastal storm intensity scale for the Catalan Sea (NW Mediterranean). Nat-Hazards Earth Syst. Sci. 11, 2453–2462. doi:10.5194/nhess-11-2453-2011.

Muñoz, J.A., Martínez, A., Vergés, J. (1986). Thrust sequences in the eastern Spanish Pyrenees. Journal of Structural Geology 8, 399–405.

Ogston, A. S., Drexler, T. M., Puig, P. (2008). Sediment delivery, resuspension, and transport in two contrasting canyon environments in the southwest Gulf of Lions. Continental Shelf Research, 28: 2000–2016.

Palanques, A., Durrieu de Madron, X., Puig, P., Fabres, J., Guillén, J., Calafat, A., Canals, M., Heussner, S., Bonnin, J. (2006). Suspended sediment fluxes and transport processes in the Gulf of Lions submarine canyons. The role of storms and dense water cascading. Marine Geology, 234, 43–61

Palanques A., Guillén J., Puig P., Durrieu de Madron X. (2008). Storm-driven shelf-to-canyon suspended sediment transport at the southwestern Gulf of Lions. Continental Shelf Research, 28:1947–56.

Ribó, M., Puig, P., Palanques, A., Lo Iacono, C. (2011). Dense shelf water cascades in the Cap de Creus and Palamós submarine canyons during winters 2007 and 2008. Marine Geology, 284: 175–188.

Roca, E., Sans, M., Cabrera, L., Marzo, M. (1999). Oligocene to Middle Miocene evolution of the Central Catalan margin (northwestern Mediterranean). Tectonophysics 315, 209–233.

Tassone, A., Roca, E., Muñoz, J. A., Cabrera, L., Canals, M. (1994). Evolución del sector septentrional del margen continental catalán durante el Cenozoico. Acta Geológica Hispanica, 29, 3–37.

Ulses, C., Estournel, C., Durrieu de Madron, X., Palanques, A. (2008a). Suspended sediment transport in the Gulf of Lions (NW Mediterranean): Impact of extreme storms and floods. Continental Shelf Research, 28: 2048–2070.

Ulses, C., Estournel, C., Puig, P., Durrieu de Madron, X., Marsaleix, P. (2008b). Dense shelf water cascading in the northwestern Mediterranean during the cold winter 2005: quantification of the export through the Gulf of Lion and the Catalan margin. Geophysical Research Letters, 35, L07610. doi:10.1029/2008GL033257.

Large-Scale Fine-Grained Sediment Waves Over the Gulf of Valencia Continental Slope (NW Mediterranean)

41

Marta Ribó, Pere Puig, Araceli Muñóz, C. Lo Iacono, Pere Masqué, Albert Palanques, Juan Acosta, Jorge Guillén, and María Gómez Ballesteros

M. Ribó (✉) · P. Puig · A. Palanques · J. Guillén
Institut de Ciències del Mar (ICM-CSIC), Barcelona, Spain
e-mail: mribo@icm.csic.es

A. Muñóz
TRAGSATEC-Secretaria General de Pesca, Madrid, Spain

C.L. Iacono
Marine Geoscience National Oceanography Centre (NOC), Southampton, UK

P. Masqué
Universitat Autònoma de Barcelona (UAB), Bellaterra, Spain

P. Masqué
Oceans Institute and School of Physics, The University of Western Australia, 35 Stirling Highway, Crawley, WA 6009, Australia

P. Masqué
School of Natural Sciences and Centre for Marine Ecosystems Research, Edith Cowan University, Joondalup, WA, Australia

J. Acosta · M.G. Ballesteros
Instituto Español de Oceanografía (IEO), Madrid, Spain

J. Guillén et al. (eds.), *Atlas of Bedforms in the Western Mediterranean*,
DOI 10.1007/978-3-319-33940-5_41

Abstract
Recently acquired swath bathymetry, high-resolution seismic profiles and bottom sediment samples have revealed the presence of large-scale fine-grained sediment waves over the Gulf of Valencia continental slope. Like many other deep-water sediment waves, these features were previously attributed to gravitational slope failure related to creep-like deformation, and have now been reinterpreted as sediment wave fields extending from 250 m depth to the continental rise at ~850 m depth. Sediment wave lengths range between 500 and 1000 m and maximum wave heights of up to 50 m are found on the upper slope, decreasing downslope to a minimum height of 2 m. Seismic profiles showed continuous internal reflectors and several sediment wave packages were differentiated, being thicker on the crest of each wave and thinner on the downslope flank, indicating that these sediment waves are upslope-migrating. The sediment wave formation process was inferred from contemporary hydrodynamic observations, and internal wave activity is suggested to be the most probable mechanism for the sediment transport and deposition and subsequent maintenance of the sediment waves over the Gulf of Valencia continental slope.

Keywords
Sediment waves • Continental slope • Gulf of valencia • Internal waves

41.1 Introduction

Sediment waves have been described in many different marine sedimentary environments in the Mediterranean Sea, mostly in the western basin (Marani et al. 1993; Cattaneo et al. 2004; Berndt et al. 2006; Verdicchio and Trincardi 2006; Puig et al. 2007; Fernández-Salas et al. 2007; Bárcenas et al. 2009; Urgeles et al. 2011; among others), with only a few examples in the eastern basin (e.g. Ediger et al. 2002). In the deep waters, sediment waves have been observed on the continental rise, mainly adjacent to submarine canyons (Migeon et al. 2001; Jallet and Giresse 2005; Puig et al. 2014; Babonneau et al. 2012), and also over the continental slope (Marani et al. 1993), but they have not always been interpreted as sediment waves (Rebesco et al. 2009). Over the Gulf of Valencia (GoV) continental slope (NW Mediterranean Sea), similar sediment waves have been described and were also initially related to creep movements, presumably affected by neotectonic fractures which produced mass movements along the continental slope (Díaz del Rio et al. 1986; Herranz et al. 1996; Díaz del Río and Fernández-Salas 2005; Maestro et al. 2005). The data presented in this chapter provide a detailed description of the geometric parameters and internal structure of these features, allowing them to be reinterpreted as fine-grained sediment waves.

The GoV is located in the southwestern part of the Valencia Trough, from the southern end of the Ebro shelf to the Cap La Nao promontory (Fig. 41.1). The GoV margin has an amphitheatre or cul-de-sac shape, with a relatively narrow continental shelf of ~30 km width up to a depth of ~150 m. The general surface circulation in the GoV area is characterized by the convergence of old Atlantic Water (oAW) flowing towards the southwest and the northward intrusions of recent Atlantic Water (rAW) through the Ibiza Channel (Pinot et al. 1995; Salat 1995). The slope circulation is dominated by the presence of mesoscale features such as anticyclonic eddies (Fig. 41.1) with a marked seasonal variability (Font et al. 1988; Pinot et al. 1994, 2002), probably enhanced by instabilities of the general current produced by interaction with the seafloor (Millot 1999).

Recent studies confirmed that the dominant sediment transport on the Ebro margin is along-shelf (Palanques et al. 2002; Puig et al. 2001) and that suspended particles being advected along the GoV shelf are mainly exported to the slope through its southern shelf-edge, near the Cap La Nao promontory (Fig. 41.1), where the margin changes the orientation (Ribó et al. 2013). Moreover, the presence of near-inertial internal waves on the GoV continental slope was confirmed by van Haren et al. (2013) and Ribó et al. (2015), who indicated that this hydrodynamic process plays an important role in the sediment transport and deposition over the continental slope, and in the maintenance of suspended sediment that is advected along- and across-slope.

41.2 Methods

An extended swath bathymetry mosaic covering the continental slope and rise at the southwestern end of the Valencia Trough was built by merging multibeam datasets acquired from 1995 to 2011 (Fig. 41.1). Moreover, a very

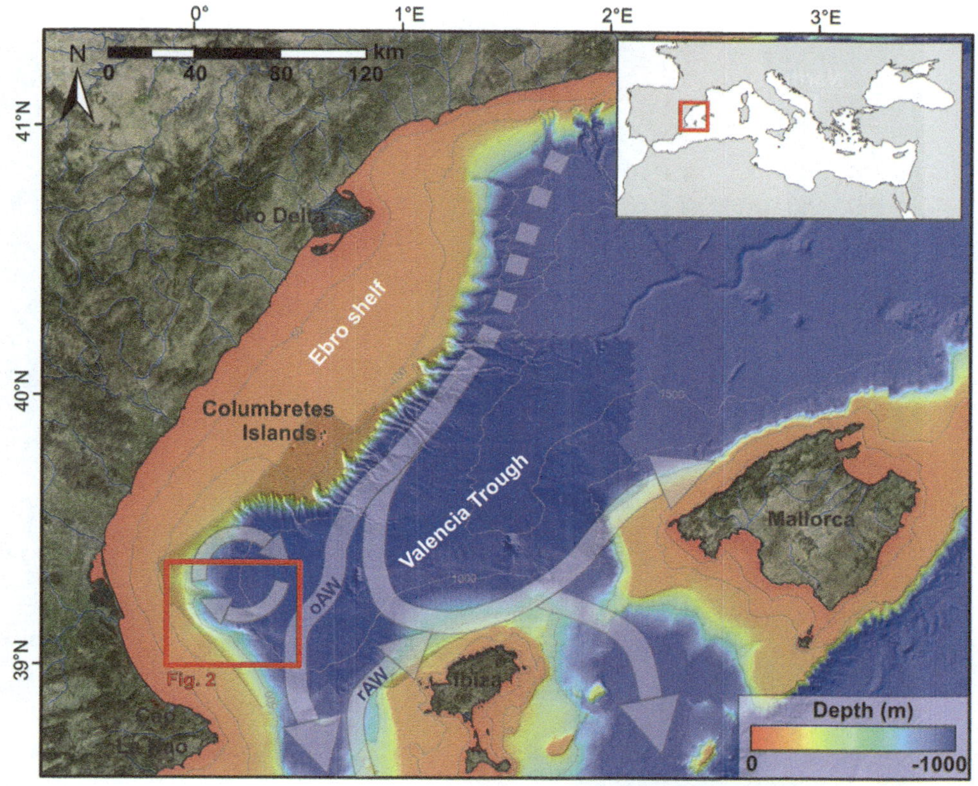

Fig. 41.1 Bathymetric map of the Catalano-Balearic Sea, composition of the GEBCO digital database (IOC et al. 2003) and detailed swath bathymetry mosaic from the Valencia Trough. *Arrows* indicate the regional surface circulation (rAW, recent Atlantic Waters; oAW, old Atlantic Waters). The *red square* indicates the location of the study area

high-resolution swath bathymetry was acquired in April 2011 on board the R/V *Vizconde de Eza* as part of the CAPESME programme, using the Simrad EM300 multibeam echosounder to analyse in detail the morphology of the GoV continental slope (Fig. 41.2). The internal structure was interpreted from 3.5-kHz parametric sub-bottom seismic profiles (TOPAS). Eight sediment cores were collected across the sediment waves using a KC-Denmark Multicorer on board the R/V *García del Cid* during two cruises in the framework of the COSTEM project, in February and June 2011. Grain size analyses were performed, and the concentration of ^{210}Pb was calculated to estimate the sediment accumulation rate (see Sanchez-Cabeza et al. (1998) for further details). Morphological parameters were automatically computed from more than 1000 equidistant transects orthogonal to the sediment wave crests, after removing the noise with a low-pass data filtering step, following the notations of Knappen (2005).

41.3 Results and Discussion

Several undulated features are observed all over the GoV continental margin, the most notable ones being the sediment waves that have developed over the continental slope (Fig. 41.2). Two main sediment wave fields separated by structural highs can be identified, the northern field (~200 km^2) and the central field (~250 km^2), extending from approximately 250 m depth to the continental rise, at 850 m depth. The sediment wave dimensions in these two fields appear to grow upslope, the biggest ones (in height and length) being found on the upper continental slope (Fig. 41.2). The sediment waves have rounded crests and narrow troughs and are parallel to sub-parallel to the bathymetric contours following the margin orientation, and occasionally bifurcating. At some points, it is not clear whether a single wave bifurcates or two crests converge, diminishing their lateral continuity, which ranges from hundreds of metres to tens of kilometres (Fig. 41.2). Grain size analysis from the sediment samples showed that bottom sediments are consistently uniform along the sediment waves, with dominant silt (~60–70 %) and clay (~25 %), and a small portion (~5–15 %) of sand, indicating that they are fine-grained sediment waves and can be classified as mud waves.

The mean sediment wavelength is ~900 m along the continental slope. Near-homogeneous distribution of the sediment wavelengths is observed on the continental slope on both sediment wave fields, but the largest wavelengths are observed on the upper and middle slope, at a water depth of between ~200 and ~600 m, with values ranging between 800 and 1800 m. The sediment wave heights range between <2 m in the deeper areas, increasing upslope to a maximum of 50.5 m, the highest ones being also located on

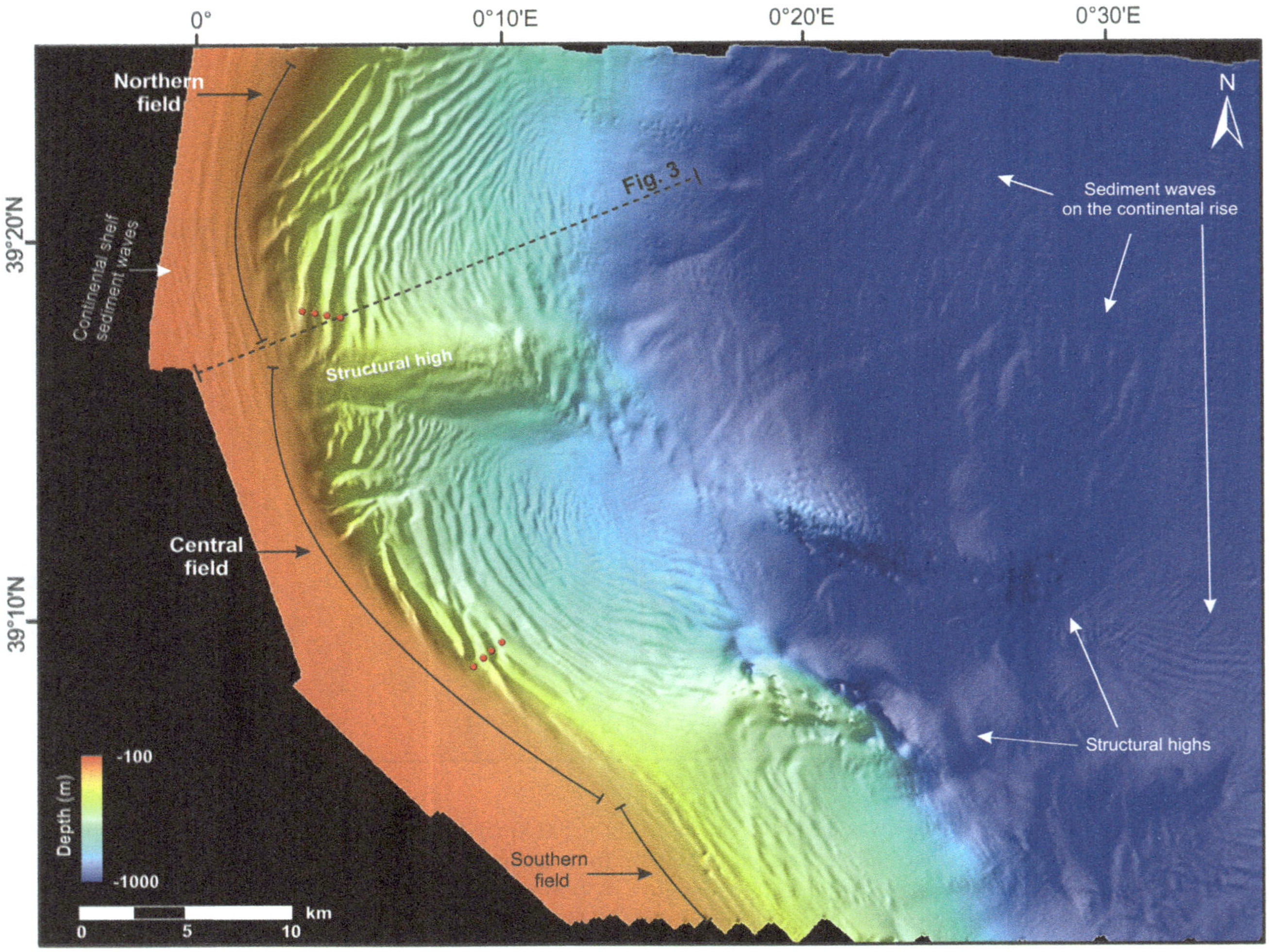

Fig. 41.2 *Shaded* relief image of the Gulf of Valencia digital terrain model (50 × 50 m grid) illustrating the outer continental shelf, continental slope and continental rise morphology. The major morphological features observed are indicated, as well as the location of the sediment samples (*red dots*) and the sub-bottom seismic profiles (*dashed line*)

the upper and middle slope between ~200 and ~600 m depth. In both sediment wave fields, from ~600 m depth to the lowest part of the continental slope, wave heights are clearly smaller, passing from heights of ~15 m to less than 2 m, also coinciding with a clear decrease in the lateral continuity of the sediment waves (Fig. 41.2).

The seismic profile shows that the seafloor displays a concave shape, with sediment waves over the continental slope from ~250 to ~850 m water depth, developing on the foreset region of the margin clinoform and showing a well-layered seismic facies (Fig. 41.3). Across the continental slope, two groups of sediment waves are clearly differentiated: along the lower continental slope region, sediment waves are quasi-stationary "vertically accreting", whereas towards the upper continental slope they become upslope-migrating (Fig. 41.3). This transition coincides with a change in the regional slope gradient across the margin, indicating that the regional slope angle is an important factor controlling the development and shape of these sediment waves. The downslope flank of one wave meets the upslope flank of the adjacent one, within a sharp but traceable zone (masked by diffractions). Although some inflections appear in the troughs, individual reflectors can be continuously traced across-slope throughout the entire seismic profiles (Fig. 41.3). Through the seismic profiles, several sediment wave packages can be differentiated (identified by higher and lower amplitudes, corresponding to clearer and darker shades of grey) and can be followed through the subsurface. These sediment packages are thicker on the crests of each individual sediment wave and thinner on the downslope flank, and their internal layering structure is similar from one wave to the next (see details in the inset in Fig. 41.3a). These observations coincide with the sediment accumulation rates (SARs) computed from the ^{210}Pb analysis performed on the sediment samples collected on the crest and downslope flanks of two consecutive sediment waves. Slightly

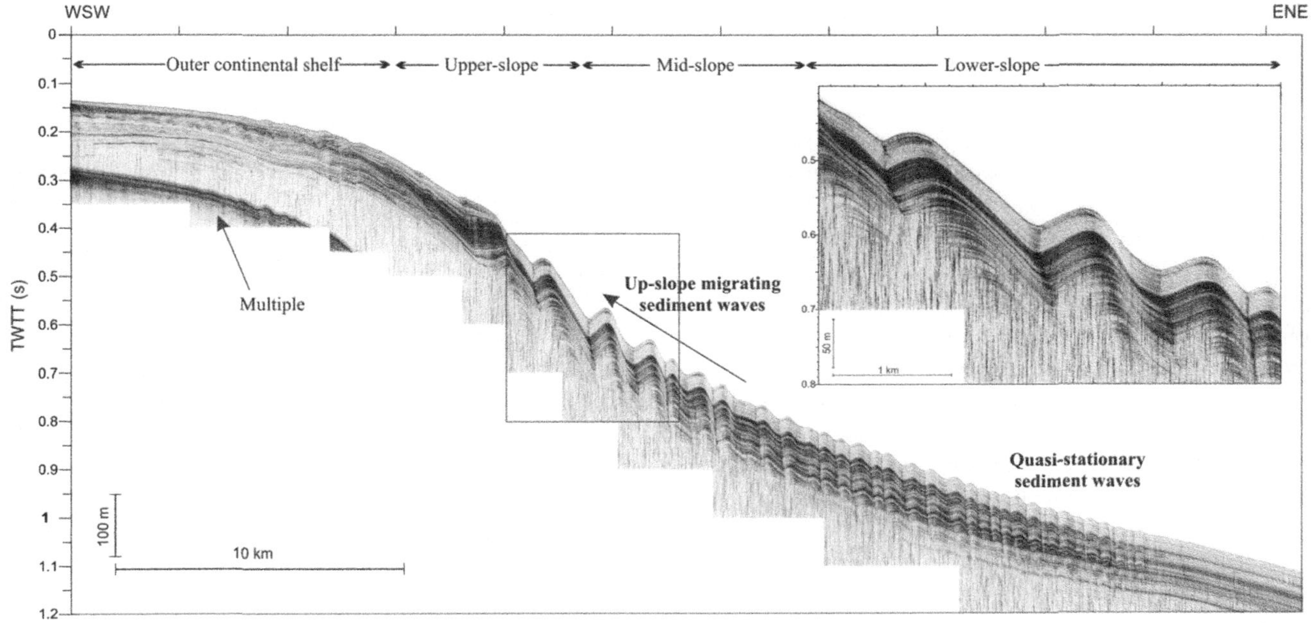

Fig. 41.3 Sub-bottom (TOPAS) seismic profile crossing the sediment waves perpendicularly over the northern field (see location in Fig. 41.2). The different parts of the continental margin and the different types of sediment waves identified on the continental slope are indicated

higher SARs were observed on the crests (from ~8 to ~13 mm $year^{-1}$) of given sediment waves when compared with the adjacent flanks (from ~6 to ~8 mm $year^{-1}$). The differential SARs between sediment wave crests and downslope flanks apparently indicate that the sediment waves over the middle slope are still growing in present conditions (i.e. at a 100-year timescale).

It was observed in the seismic profile that the dipping of the strata is approximately constant within each sediment wave (Fig. 41.3), denoting no synchronous deformation (Lee et al. 2002; Faugères et al. 2002). Furthermore, no headwall scarps on the upper continental slope were observed in the swath bathymetry maps, and at the toe of the sediment wave field there is also a lack of compressional features, which are usually observed in slide deposits (Lee et al. 2002). These observations allowed gravity deformation or creep to be dismissed as the formation mechanism. Three main processes may explain sediment wave formation: turbidity currents, bottom currents and internal waves. On the GoV there is no evidence of the existence of turbidity channels or fan-like depositional features at the lower part of the slope, and there are no major sediment sources near the GoV margin. Moreover, the sediment waves over the GoV continental slope are parallel to the margin bathymetry and to the regional circulation, which makes it unlikely that bottom-currents contribute to their generation. However, internal wave activity has been confirmed in this region (van Haren et al. 2013) and it has been observed that the near-inertial internal waves interacting with the GoV continental slope conditioned the sediment transport and deposition by controlling the across- and along-slope sediment fluxes (van Haren et al. 2013; Ribó et al. 2015). Internal wave motions only contribute to the dispersal of suspended particles, including those passively reaching the continental slope, thus enhancing and modulating the cross-margin component of the flow (Ribó et al. 2015). This depositional pattern is consistent with the perfect layering structures inside the sediment waves, indicating continuous sediment deposition (with relatively low sedimentation rates as derived from the collected sediment cores) and a lack of sediment resuspension causing a disruption of the bed structure (Fig. 41.3).

Taken into account these results, the interaction of internal waves with the seafloor on the GoV continental slope appears to simply modulate the sediment deposition without causing active resuspension, presumably generating regions with enhanced turbulence, and therefore preventing deposition (i.e. on the steep downslope flanks), and calmer regions where particles can preferentially settle (i.e. on the crests or flat flanks). As similarly described in previous studies worldwide (Karl et al. 1986; Faugères et al. 2002; Reeder et al. 2011; Belde et al. 2015), though other generation mechanisms could also be possible, the interaction of internal waves on the GoV continental slope is proposed as a potential formation process for transport and subsequent generation of the observed fine-grained sediment waves.

41.4 Conclusions

The recently acquired swath bathymetry and seismic profiles show large, fine-grained sediment waves over the GoV continental slope, first interpreted as a creep-like deformation, covering an area of 450 km^2 and extending from ~250 m to ~850 m water depth. These sediment waves have wavelengths ranging between 500 and 1000 and wave heights of 2–50 m, decreasing downslope, with rounded crests and narrow troughs. The highest sediment waves were observed over the upper and middle slope regions, with larger wave heights and wavelengths than the ones located on the lower slope, which are more flattened. Quasi-stationary, "vertically accreting" sediment waves are observed along the lower continental slope region, becoming upslope-migrating sediment waves towards the upper part of the continental slope. Slightly higher SARs are found on the crest than on the downslope flanks, indicating that sediment waves are presently growing at 100-year timescales. It is suggested that the interaction of internal waves on the continental slope is a plausible formation process for the sediment transport and deposition, causing the subsequent generation of the fine-grained sediment waves observed over the GoV continental slope.

Acknowledgments This work was funded by the project COSTEM (CMT2009-07806) and FORMED (CGL2012-33989). The authors wish to thank the captain and the crew of the R/V *García del Cid* and the R/V *Vizconde de Eza*, and also the UTM technicians and J. Pozo and M. Lloret from the Instrumental Service, for their assistance. The author M. Ribó was supported by a FPI grant (Ref. BES-2010-029949) from the Spanish Ministry of Economy and Competitiveness. P. Masqué acknowledges the financial support through the ICREA Academia award and a Gledden Visiting Fellowship awarded by the Institute of Advanced Studies at the University of Western Australia.

References

Babonneau, N., Cattaneo, A., Barjavel, B. S. G., Déverchère, J., Yelles, K. (2012). The Kramis deep-sea fan off western Algeria: Role of sediment waves in turbiditic levee growth. In: Applications of the principles of seismic geomorphology to continental-slope and base-of-slope systems: casse studies from seafloor and near-seafloor analogues. SEPM Special Publications, No. 99. ISBN 978-1-56576-304-3, p. 293–308.

Bárcenas, P., Fernández-Salas, L. M., Macías, J., Lobo, F. J., Díaz del Río, V. (2009). Estudio morfométrico comparativo entre las ondulaciones de los prodeltas de los ríos de Andalucía Oriental. Revista de la Sociedad Geológica de España, 22, 43–56.

Belde, J., Back, S., Reuning, L. (2015). Three-dimensional seismic analysis of sediment waves and related geomorphological features on a carbonate shelf exposed to large amplitude internal waves, Browse Basin region, Australia. Sedimentology, 62, 87–109

Berndt, C., Cattaneo, A., Szuman, M., Trincardi, F., Masson, D. (2006). Sedimentary structures offshore Ortona, Adriatic Sea-Deformation or sediment waves? Mar. Geol. 234, 261–270. doi:10.1016/j.margeo.2006.09.016

Cattaneo, A., Correggiari, A., Marsset, T., Thomas, Y., Marsset, B., Trincardi, F. (2004). Seafloor undulation pattern on the Adriatic shelf and comparison to deep-water sediment waves. Mar. Geol. 213, 121–148. doi:10.1016/j.margeo.2004.10.004.

Díaz del Rio, V., Rey, J., Vegas, R. (1986). The Gulf of Valencia continental shelf: Extensional tectonics in Neogene and Quaternary sediments. Mar. Geol. 73, 169–179.

Díaz del Río, V., Fernández-Salas, L.M. (2005). El margen continental del Levante español y las islas Baleares, in: Mapa Geomorfológico de España Y Del Margen Continental. pp. 177–187.

Ediger, V., Velegrakis, A., Evans, G. (2002). Upper slope sediment waves in the Cilician Basin, northeastern Mediterranean. *Marine Geology*, 192, 321–333. doi:10.1016/S0025-3227(02)00562-5.

Faugères, J.C., Gonthier, E., Mulder, T., Kenyon, N., Cirac, P., Griboulard, R., Berné, S., Lesuavé, R. (2002). Multi-process generated sediment waves on the Landes Plateau (Bay of Biscay, North Atlantic). Mar. Geol. 182, 279–302.

Fernández-Salas, L. M., Lobo, F. J., Sanz, J. L., Díaz del Río, V., García, M. C., Moreno, I. (2007). Morphometric analysis and genetic implications of pro-deltaic sea-floor undulations in the northern Alboran Sea margin, western Mediterranean Basin. Marine Geology, 243, 31–56.

Font, J., Salat, J., Tintoré, J. (1988). Permanent features of the circulation in the Catalan Sea. Oceanol. Acta 51–57.

Herranz, P., Acosta, J., Muñoz, A., Palomo, C., Carbó, A., Pardo de Donlebum, M., Sanz, J.L., Uchupi, E. (1996). Resultados Preliminares de la primera campaña en la Zona Económica Exclusiva Española en el Golfo de Valencia y mar Balear. ZEE-95. Geogaceta 20, 347–350.

IOC, IHO, BODC. (2003). Centenary Edition of the GEBCO Digital Atlas, published on CD-ROM on behalf of the Intergovernmental Oceanographic Commission and the International Hydrographic Organization as part of the General Bathymetric Chart of the Oceans, British Oceanographic Data: http://www.gebco.net/

Jallet, L., Giresse, P. (2005). Construction of the Pyreneo-Languedocian Sedimentary Ridge and associated sediment waves in the deep western Gulf of Lions (Western Mediterranean). Marine and Petroleum Geology, 22, 865–888.

Karl, H.A., Cacchione, D.A., Carlson, P.R. (1986). Internal-wave currents as a mechanism to account for large sand waves in Navarinsky canyon head, Bering Sea. J. Sediment. Petrol. 56, 706–714.

Knaapen, M.A.F. (2005). Sandwave migration predictor based on shape information. J. Geophys. Res. 110, F04S11. doi:10.1029/2004JF000195

Urgeles, R., Cattaneo, A., Puig, P., Liquete, C., De Mol, B., Amblàs, D., Sultan, N., Trincardi, F. (2011). A review of undulated sediment features on Mediterranean prodeltas: distinguishing sediment transport structures from sediment deformation. Mar. Geophys. Res. 32, 49–69. doi:10.1007/s11001-011-9125-1

Reeder, D.B., Ma, B.B., Yang, Y.J. (2011). Very large subaqueous sand dunes on the upper continental slope in the South China Sea generated by episodic, shoaling deep-water internal solitary waves. Mar. Geol. 279, 12–18. doi:10.1016/j.margeo.2010.10.009

Lee, H.J., Syvitski, J.P.M., Parker, G., Orange, D., Locat, J., Hutton, E. W.H., Imran, J. (2002). Distinguishing sediment waves from slope failure deposits: field examples, including the "Humboldt slide", and modelling results. Mar. Geol. 192, 79–104.

Maestro, A., Medialdea, T., Llave, E., Somoza, L., Bárcenas, P. (2005). Geomorfología submarina. Explicación de los elementos representados, in: Mapa Geomorfológico de España Y Del Margen Continental. pp. 45–58.

Marani, M., Argnani, A., Roveri, M., Trincardi, F. (1993). Sediment drifts and erosional surfaces in the central Mediterranean: seismic evidence of bottom-current activity. *Sedimentary Geology,* 82, 207–220.

Migeon, S., Savoye, B., Zanella, E., Mulder, T., Faugères, J.C., Weber, O. (2001). Detailed seismic-reflection and sedimentary study of turbidite sediment waves on the Var Sedimentary Ridge (SE France): significance for sediment transport and deposition and for the mechanisms of sediment-wave construction. Marine and Petroleum Geology 18 (2), 179–208.

Millot, C. (1999). Circulation in the Western Mediterranean Sea. J. Mar. Syst. 20, 423–442. doi:10.1016/S0924-7963(98)00078-5Nash, J.D., Toole, J.M., Schmitt, R.W., 2004. Internal tide reflection and turbulent mixing on the continental slope. J. Phys. Oceanogr. 34, 1117–1134.

Palanques, A., Puig, P., Guillén, J., Jiménez, J., Gracia, V., Sánchez-Arcilla, A., Madsen, O. (2002). Near-bottom suspended sediment fluxes on the microtidal low-energy Ebro continental shelf (NW Mediterranean). Cont. Shelf Res. 22, 285–303.

Pinot, J.M., Tintoré, J., Gomis, D. (1994). Quasi-synoptic mesoscale variability in the Balearic Sea. Deep Sea Res. I 41, 897–914.

Pinot, J., Tintoré, J., Gomis, D. (1995). Multivariate analysis of the surface circulation in the Balearic Sea. Prog. Oceanogr. 36, 343–376.

Pinot, J.M., López-Jurado, J., Riera, M. (2002). The CANALES experiment (1996–1998). Interannual, seasonal, and mesoscale variability of the circulation in the Balearic Channels. Prog. Oceanogr. 55, 335–370.

Puig, P., Palanques, A., Guillén, J. (2001). Near-bottom suspended sediment variability caused by storms and near-inertial internal waves on the Ebro mid continental shelf (NW Mediterranean). Mar. Geol. 178, 81–93.

Puig, P., Ogston, A.S., Guillén, J., Fain, A.M.V., Palanques, A. (2007). Sediment transport processes from the topset to the foreset of a crenulated clinoform (Adriatic Sea). *Continental Shelf Research.* 27, 452–474. doi:10.1016/j.csr.2006.11.005.

Puig, P., Palanques, A., Martín, J., Ribó, M., Guillén, J. (2014). Benthic storms in the north-western Mediterranean continental rise caused by deep dense wáter formation. In: 2nd Deep-Water Circulation Congress, 10–12 Sept. 2014, Ghent, Belgium.

Rebesco, M., Neagu, R. C., Cuppari, A., Muto, F., Accettella, D., Dominici, R., Cova, A., Romano, C., Caburlotto, A. (2009). Morphobathymetric analysis and evidence of submarine mass movements in the western Gulf of Taranto (Calabria margin, Ionian Sea). Int. J. Earth Sci (Geol Rundshc), 98, 791–805.

Ribó, M., Puig, P., Salat, J., Palanques, A. (2013). Nepheloid layer distribution in the Gulf of Valencia, northwestern Mediterranean. J. Mar. Syst. 111–112, 130–138. doi:10.1016/j.jmarsys.2012.10.008

Ribó, M., Puig, P., van Haren, H. (2015). Hydrodynamics over the Gulf of Valencia continental slope and their role in sediment transport. Deep Sea Res. I, 95, 54–66. doi:10.1016/j.dsr.2014.10.004

Salat, J. (1995). The interaction between the Catalan and Balearic currents in the southern Catalan Sea. Oceanol. Acta 18, 227–234.

Sanchez-Cabeza, J.A., Masqué, P., Ani-Ragolta, I. (1998). 210Pb and 210Po analysis in sediments and soils by microwave acid digestion. J. Radioanal. Nucl. Chem. 227, 19–22.

van Haren, H., Ribó, M., Puig, P. (2013). (Sub-)inertial wave boundary turbulence in the Gulf of Valencia. J. Geophys. Res. Ocean. 118, 2067–2073. doi:10.1002/jgrc.20168

Verdicchio, G., Trincardi, F. (2006). Short-distance variability in slope bed-forms along the Southwestern Adriatic Margin (Central Mediterranean). *Marine Geology,* 234, 271–292. doi:10.1016/j.margeo.2006.09.007.

Generation of Bedforms by the Mediterranean Outflow Current at the Exit of the Strait of Gibraltar

42

G. Ercilla, D. Casas, F.J. Hernández-Molina, and C. Roque

G. Ercilla (✉)
Institut de Ciències del Mar (ICM-CSIC), Continental Margins Group, P. Marítim de la Barceloneta 37, Barcelona, Spain
e-mail: gemma@icm.csic.es

D. Casas
Instituto Geológico y Minero de España (IGME), Ríos Rosas 23, 08003 Madrid, Spain

F.J. Hernández-Molina
Royal Holloway, University of London, Egham, Surrey TW20 0EX, UK

C. Roque
IDL-Instituto Dom Luíz, University of Lisbon, 1749-016 Lisbon, Portugal
e-mail: cristina.roque@ipma.pt

J. Guillén et al. (eds.), *Atlas of Bedforms in the Western Mediterranean*,
DOI 10.1007/978-3-319-33940-5_42

Abstract

During the MOWER cruise, an ARGUS remotely operated vehicle (ROV) was used to image the effects of the bottom current dynamics of the Mediterranean Outflow Water (MOW) on the Cadiz contourite depositional system (CDS) (Atlantic Sea), at water depths of 500–700 m just downstream of the Strait of Gibraltar. The new images presented here show both longitudinal and transverse bedforms at different scales that are superimposed and coexist throughout the region. Their spatial distribution suggests a general decrease in the MOW velocity westward and the control of the type of bedforms over short distances by sediments and changing seafloor morphology. The superposition of the bedforms also suggests variability in the strength and direction of the bottom current.

Keywords

Longitudinal bedform • Transverse bedform • Mediterranean outflow water • Gulf of cadiz

42.1 Introduction

The Mediterranean outflow water (MOW) results from a combination of light intermediate Mediterranean (western Mediterranean and Levantine waters) and dense deep Mediterranean waters (Thyrrenian deep and western Mediterranean deep waters) (e.g., Millot 2009 and references therein) within the Strait of Gibraltar. The MOW spreads as a westward overflow cascade into the Gulf of Cadiz upon exiting the Mediterranean Sea, veers northwest due to the Coriolis force and flows as a contour current along the middle continental slope of the gulf at water depths between 400 and 1400 m (Baringer and Price 1999; Serra et al. 2010) (Fig. 42.1). The MOW velocity decreases from <240 cm/s just past the Strait of Gibraltar to approximately 80–100 cm/s at the latitude of Cape San Vicente (e.g., Llave et al. 2007 and references therein). The interactions of the MOW with the seafloor have created an impressive contourite depositional system (CDS) that is predominantly studied because clues within the CDS sediment reveal implications of the MOW for oceanography and climate in the northern hemisphere (e.g., Llave et al. 2007; Hernández-Molina et al. 2014a). The general morphosedimentary structure of the CDS, which has kilometre-scale features, and its relationship with the general pattern of the circulation of the MOW have been discussed in several works (e.g., Llave et al. 2007; Hernández-Molina et al. 2014a), whereas the present-day variable bottom current flows, which produce complex patterns of bedforms from decametre to centimetre scales on the seafloor, remain poorly understood (e.g., Nelson et al. 1993; Stow et al. 2013). The effects of the MOW overflow on the CDS were recently filmed just downstream of the Strait of Gibraltar using an ARGUS remotely operated vehicle (ROV) during the MOWER Cruise. The new images presented here at water depths of 500–700 m indicate the presence of a strong near-bottom current and a seabed occupied by sand waves and superimposed ripples, both of which suggest very active near-bottom sediment dynamics generated by the MOW.

42.2 Results and Discussion

The ROV system recorded a large variability in the current-induced bedforms in the domain of the CDS that was recently characterised as a contourite terrace region by Hernández-Molina et al. (2014b). The region primarily comprises channels, though it also contains furrows and drifts (Fig. 42.1). To define the bedforms, we used the criteria (morphology, size and sediment grain size) defined by Stow et al. (2009). Images from the videos presented here show both longitudinal and transverse bedforms at different scales that coexist throughout the region (Figs. 42.2, 42.3 and 42.4). The longitudinal bedforms are represented by surface lineations (Figs. 42.2a; 42.4b), comet scours (Fig. 42.2c), tails, scours around rock fragments with sizes ranging from gravel to large boulders (>1 m) (Fig. 42.2d) and sand and gravel ribbons or irregular patches (Figs. 42.2b, c, e; 42.4d). The transverse forms are exemplified by dunes (Figs. 42.3a, b; 42.5b) with rectilinear crests and sinuous to crescent sand waves (Figs. 42.2f; 42.3a; 42.4c; 42.5a). All of the transverse bedforms show a superposition of primarily asymmetric ripples that are similar in size but have variable linguoid, crescent crestlines (e.g., sinuous or rectilinear) (Figs. 42.2b–f; 42.3a–d; 42.4a–d; 42.5a to d). Benthic organisms such as sponges, which are associated with the tails of clean sand on the lee side of the current flow, have been used at times to deduce the vector of the flow currents (Fig. 42.2e). Sponges are filtering

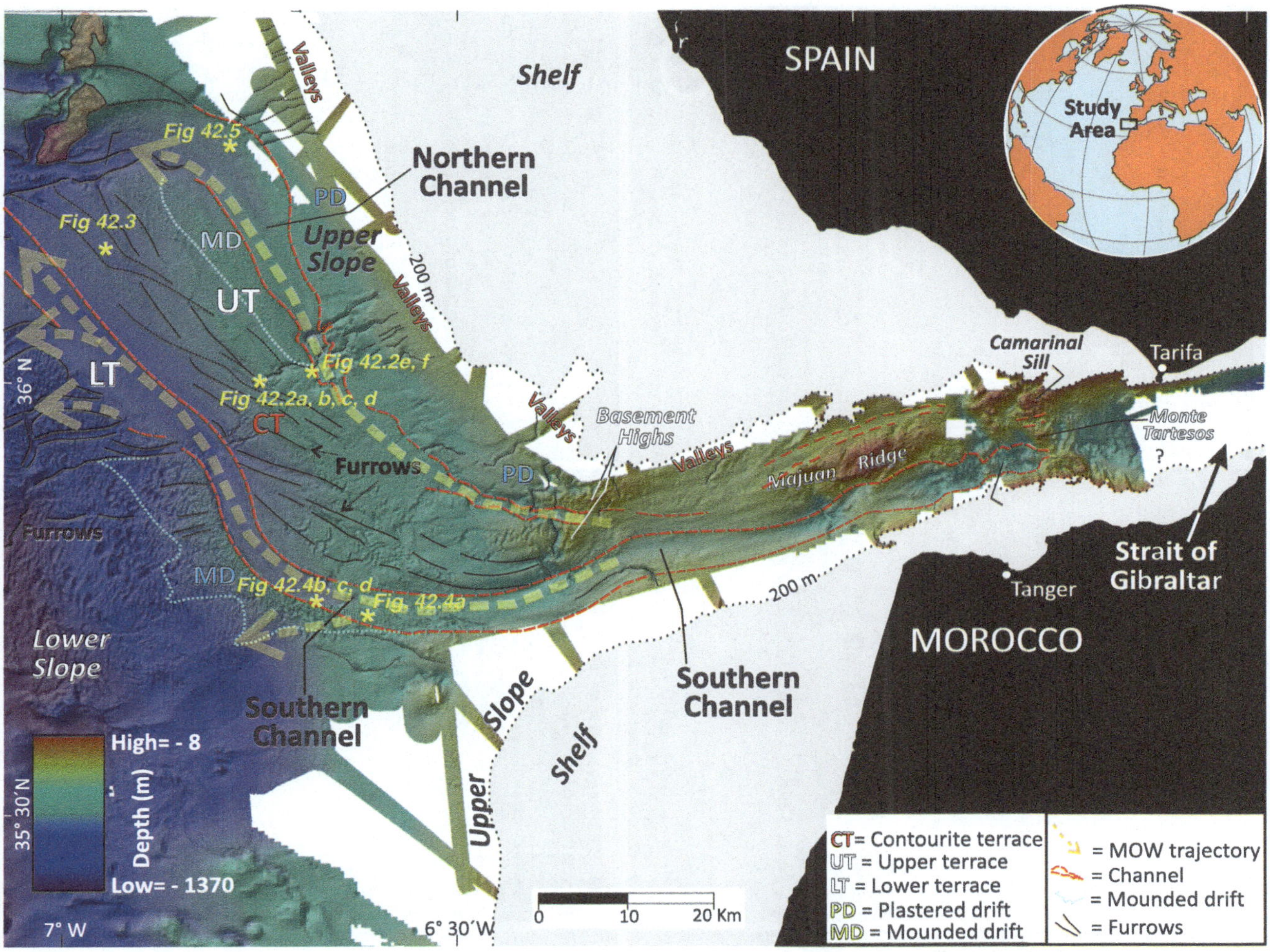

Fig. 42.1 High-resolution multibeam bathymetry at the exit of the Strait of Gibraltar that shows the primary seafloor morphological features, especially the terraces, channels and furrows formed by the MOW (modified from Hernández-Molina et al. 2014a, b). This map also indicates the image location obtained using the ARGUS ROV during the MOWER Cruise

organisms and are oriented perpendicular to the current flow to filter nutrients from the passing flow.

The spatial distribution of bedforms indicates that longitudinal bedforms predominate on the eastern part of the CDS, close to the Strait of Gibraltar, and that the transverse bedforms occur more frequently on the eastern part. This distribution emphasises the generally westward decrease in the intensity of the MOW near-bottom current, as determined by the bedform-velocity matrix for estimating the bottom current velocity from observations (Stow et al. 2009). Additionally, the spatial distribution reflects the control of the type of bedforms by sediments. The bottom currents in the proximal region of the channel are intense, as suggested by the rocky and indurated seafloor (Figs. 42.2b; 42.4a, b), gravelly pavement (Fig. 42.2a, d), relatively coarse particles (gravel to sand) (Figs. 42.2a–f; 42.4d) and scouring of rocky boulders (Fig. 42.2d). We also observed sand in the sand ribbons moving by rolling and suspension. Likewise, the new ROV images indicate the locally erosive and carving effects of the MOW on the mudstone sediments that then form the southern longitudinal contourite channel (Fig. 42.4a). Although erosion and sediment transport by the MOW remain intense toward the west, a recent layer of sediments of relatively fine grain size (silt to sand) and a major presence of sinuous ripples that cover the seafloor (Figs. 42.3 and 42.5) suggest a general decrease in the MOW velocity and/or an increase in sediment deposition. Additionally, the compositions of the proximal and distal textures, including bioclastic, terrigenous and mixed sediments, vary (Figs. 42.2 and 42.4).

The aforementioned bedforms change spatially over short distances because of the seafloor morphology. This change is more evident when the current flow interacts with large isolated boulders (Fig. 42.2c) on the margins of the main

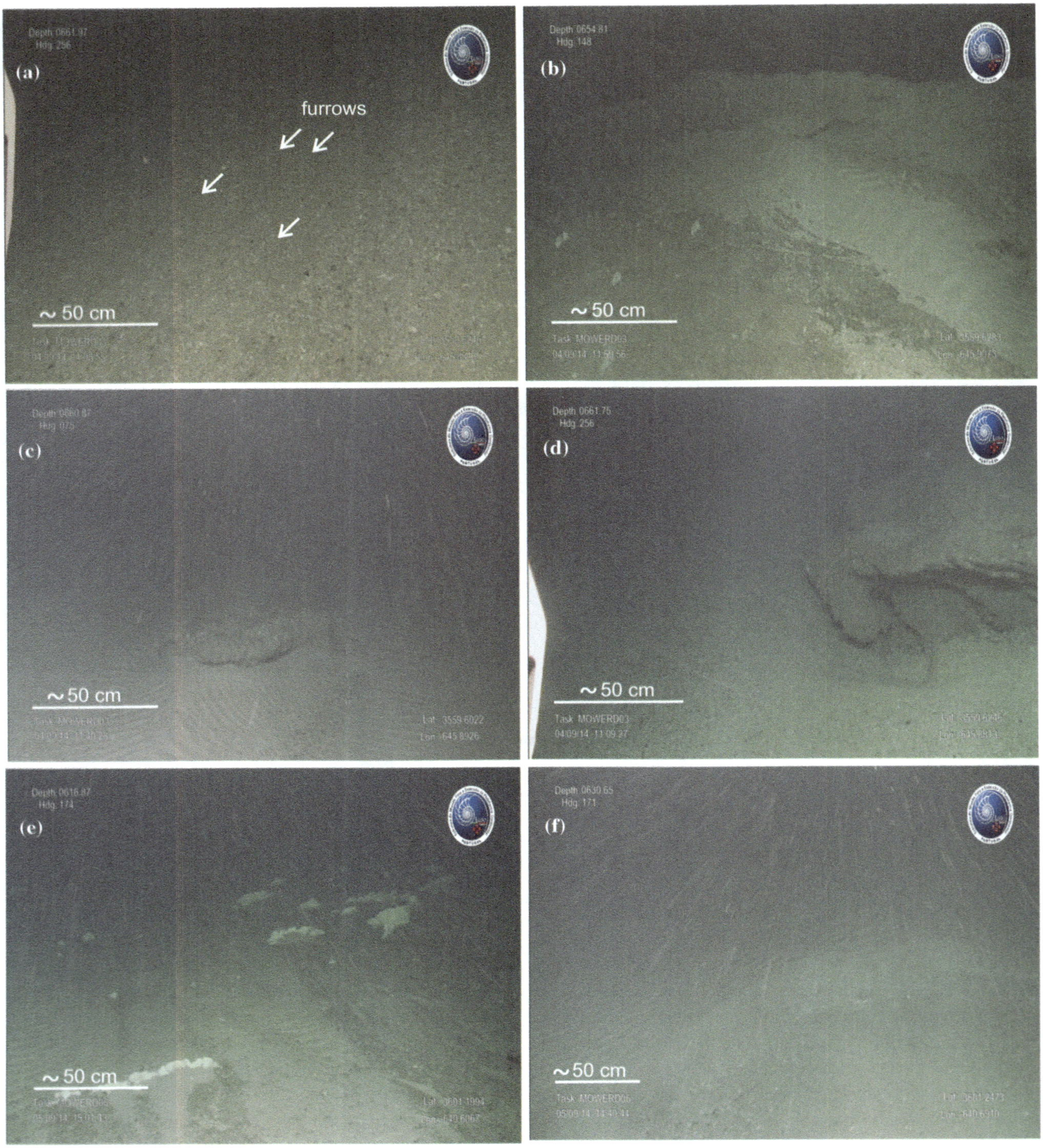
(a)
furrows
~ 50 cm
(b)
~ 50 cm
(c)
~ 50 cm
(d)
~ 50 cm
(e)
~ 50 cm
(f)
~ 50 cm

◀ **Fig. 42.2** Images from the Argus ROV that show bedforms in the proximal furrow area of the upper terrace eroded by the MOW (Fig. 42.1). **a** Small-scale furrows eroding a seafloor with two different textures (*sand and gravel*). The elongate incisions, which are a few centimetres in relief, erode the sandy deposits and are filled by coarser sediments formed by poorly sorted bioclastic fragments (*lighter colour*) and terrigenous pebbles (*darker* and *black* in colour). **b** A small sand ribbon in a rocky area with superimposed smaller regular and asymmetric transverse ripples in centimetres. A visual observation of sediment reveals that its composition is mixed (biogenic and terrigenous); *lighter colours* comprise bioclastic debris located on the lee side of the ripples; **c** Comet scour. The boulder acts as an obstacle to the bottom current, which favours the formation of a longitudinal sandy ribbon superimposed on a small scale and transverse asymmetric ripples similar in form to those in **b**. Note the sandy accumulation on the lee side of the obstacle as well as the two ripple patterns that draw a fishbone feature produced by secondary flows on both sides of the boulder. Additionally, note the seabed armouring by the nearbottom current nearer the ROV. **d** A boulder indicating scouring by secondary helical flows formed from interactions with the current flow. Scouring creates a moat-like feature lateral to the sandy accumulation. **e** Sand ribbons with superimposed ripples resting on the rocky surface that are colonised by sponges. This image highlights obstacle marks formed by the sponges. In contrast with the rocky boulders mentioned in **c**, the sponges produce accumulative bedforms that are represented by the formation of a sand tail. Note that the current direction is responsible for the formation of the tails, sand ribbons and ripples and is coincident with the flow direction inferred from the tracking of the suspended particles in the image. **f** Sinuous sand wave crestline with sinuous to rectilinear ripples on the stoss side and avalanching events on the lee side. The sand wave has developed over a coarser sand and a gravelly seafloor

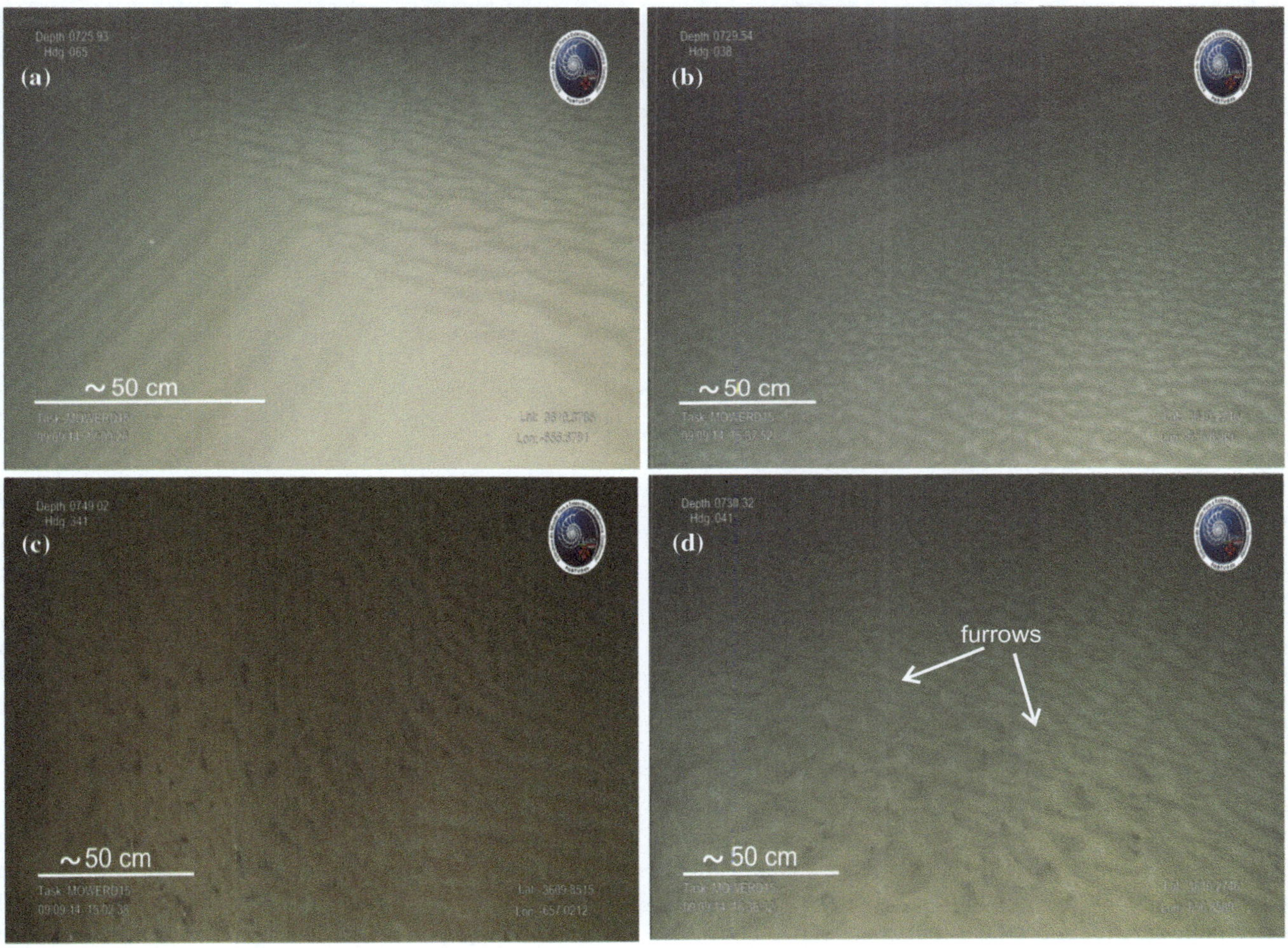

Fig. 42.3 Images from ROV Argus showing bedforms in the distal furrow area of the upper terrace eroded by the MOW (Fig. 42.1). **a** Fine sandy dune showing smaller-scale slightly sinuous ripples superimposed on the stoss side and rectilinear chutes incising the lee side. The sand waves and ripples show coherence with a similar bottom current direction. **b** Section of a sandy dune with superimposed sinuous to linguoid ripples several centimetres in height. The ripples are roughly perpendicular to the dune crestline and flow in different directions. The differences in the sediment colour are explained by different sand fraction compositions, whereby the lighter particles are composed of biogenic debris. **c** Ripple distribution with different crestlines, rectilinear to sinuous. Likewise, note that local variations in the topography produce changes in the orientation of the bedform. The composition of the ripples also appears to be different, as revealed by the colour. **d** Interaction between ripples with coalescing crescent shaped crestlines and small-scale furrows

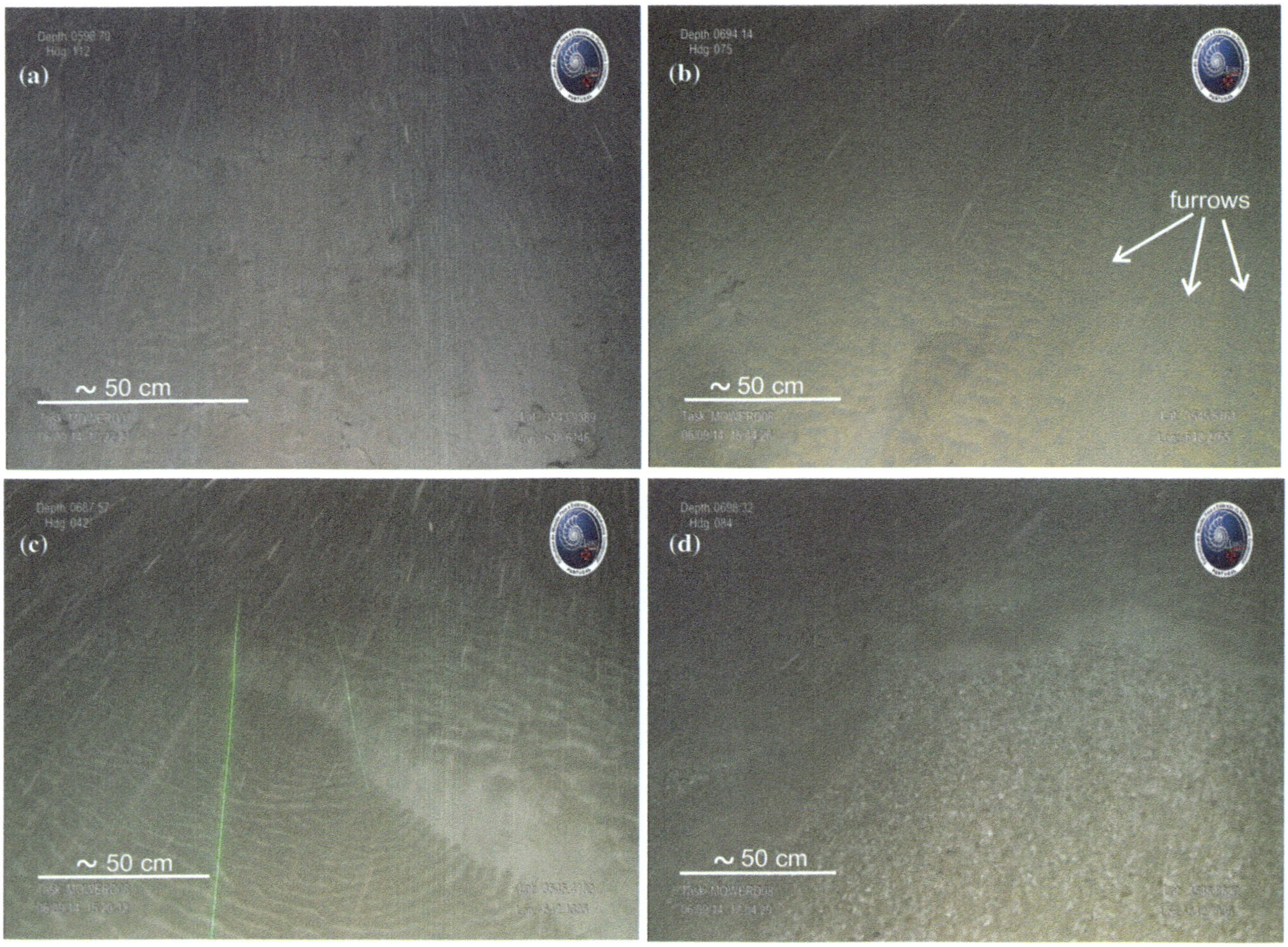

Fig. 42.4 Images from the ROV Argus showing bedforms in the proximal region of the southern channel eroded by the MOW (Fig. 42.1). **a** Indurated sedimentary outcrop, muddy in composition (*grey in colour*), located on the internal margin of the southern contouritic channel. This outcrop indicates intense current flow erosion. The mudstone is covered partially by sand with ripples. **b** Linear erosive furrows excavated on the muddy seafloor channel located adjacent to the previous internal margin. These furrows are covered partially by sandy sediment with starved rectilinear to sinuous asymmetric ripples. This starvation allows for the exposure of the mudstone surface over which the ripples are moving. Note that the furrows and ripples in the bottom current agree with the current direction inferred by tracking the suspended particles. **c** Sinuous sand wave on the channel floor. The sand wave contains superimposed linguoid to sinuous asymmetrical ripples on the stoss side and rectilinear to sinuous ripples in the trough area between the sand waves. The crestlines of the ripples in the trough area are oblique to the sand wave crestline. The wavelength of the ripples in the trough area is <10 cm (the two laser lines are separated 50 cm). **d** Irregular patches of sandy sediment with ripples over a gravelly seafloor

contourite channels (Fig. 42.4) and in starved sediment areas with irregular seafloor topography (Figs. 42.2; 42.4d). However, although long time-series of near-bottom currents are presently unavailable, bedform patterns combined with other indicators (e.g., benthos) reflect the variability in the bottom current direction at a local scale. Similarly, the superposition of bedforms at different scales may imply variability in the current strength, although the temporal scale is unknown. The variable flow direction and strength suggest a complex scenario that is probably characterised by

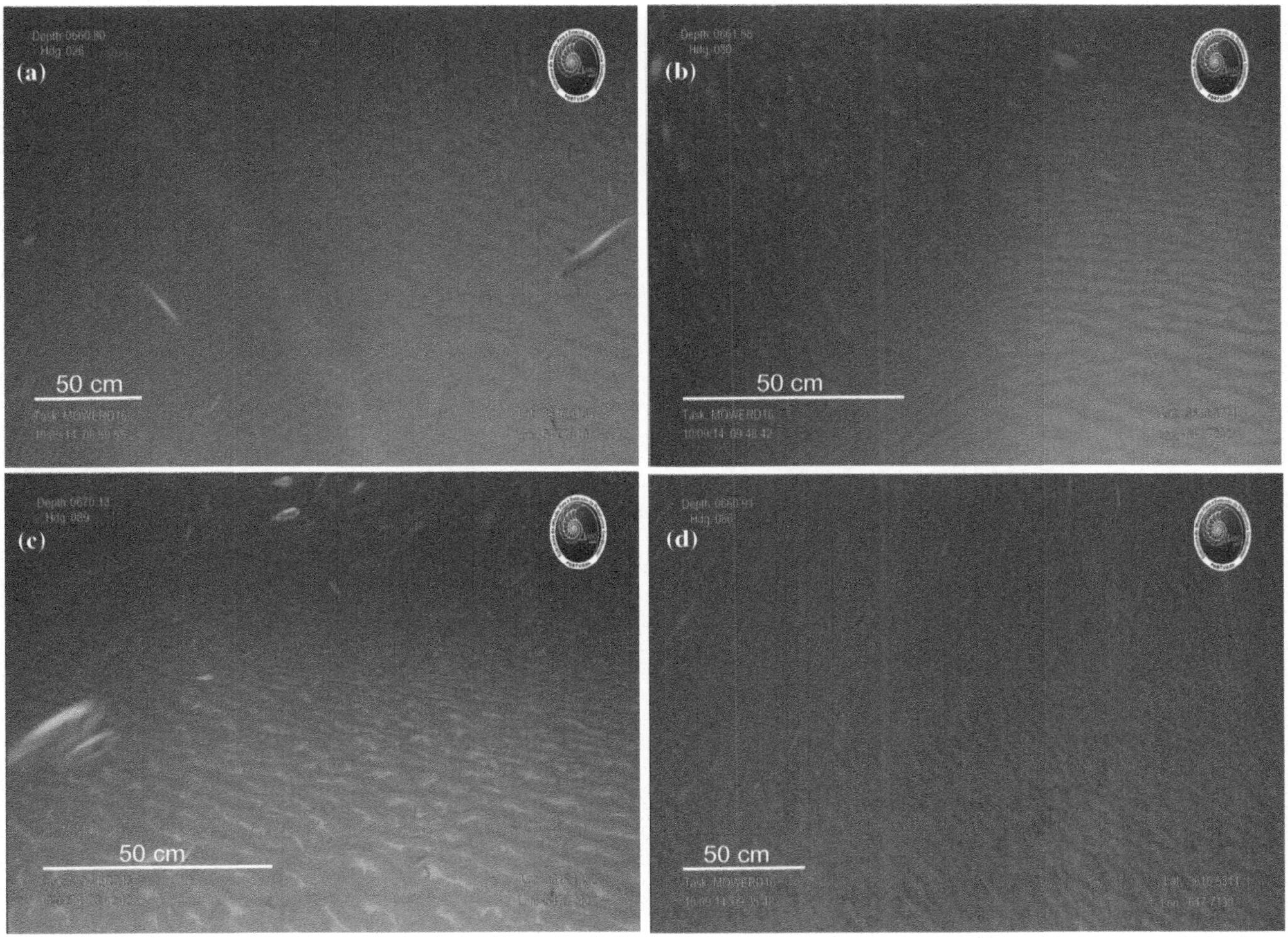

Fig. 42.5 Images from the Argus ROV showing bedforms in the distal part of the northern contouritic channel eroded by the MOW (Fig. 42.1). **a** A sand wave displaying superimposed smaller-scale bedforms. Linguoid ripples with similar orientations to the sand wave appear on the stoss side and chutes on the lee side. Additionally, transverse rectilinear ripples that evolve sharply to larger and oblique sinuous ripples occur in the trough. At this location, the sediment colour highlights that rectilinear ripples are superimposing or reforming some of the oblique sinuous ripples. The bedform distribution suggests that the flow pattern over the sand waves can result in changes to orientation of the small-scale bedforms. **b** A sand wave with small-scale sinuous ripples on the stoss side that are oblique to the dune crestline, which suggests different current flow directions. The rectilinear chutes carve the steeper lee side of the sand wave. **c** Field of sinuous to liguoid asymmetric ripples with bioclastic particles (*white*) deposited at the foot of the lee side. Note that the current direction shown by the tracking of suspended particles is perpendicular to the crestline of the ripples. **d** Field of linguoid ripples that have a trending direction oblique to the current direction at the time of photography

a combination of different hydrodynamic processes and raises questions regarding the contribution of relict or palimpsest bedforms on the observed distribution pattern.

42.3 Concluding Remarks

The films acquired using an ARGUS ROV of the effects of the MOW overflow on the seafloor of the Cadiz CDS show a complex distributional pattern of longitudinal and transverse bedforms at different scales and with superimposed bedforms. Their general spatial distribution throughout the CDS, the type and availability of sediment and the locally irregular seafloor morphology emphasise the general westward decrease in the intensity of the MOW near-bottom current and suggest variable flow directions as well as a combination of different hydrodynamic processes.

Acknowledgments This contribution was supported through the MOWER (CTM 2012-39599-C03) and CONDRIBER (Ref. PTDC/GEO-GEO/4430/2012) projects. We would like to thank the team of ARGUS ROV pilots and technicians from EMEPC (Portugal);

they are a good team who enjoy working with people. We would also like to thank the crew of RV Sarmiento de Gamboa and the UTM technicians for their support during the MOWER cruise.

References

Baringer, M.O., Price, J.F. (1999). A review of the physical oceanography of the Mediterranean Outflow. Marine Geology, 155: 63–82.

Hernández-Molina, F.J., Llave, E., Somoza, L., Fernández-Puga, M.C., Maestro, A., León, R., Medialdea, T., Barnolas, A., García, M., Díaz del Río, V., Fernández-Salas, L.M., Vázquez, J.T., Lobo, F., Alveirinho Dias, J.M., Rodero, J., Gardner, J. (2003). Looking for clues to paleoceanographic imprints: a diagnosis of the Gulf of Cádiz contourite depositional systems. Geology, 31 (1): 19–22.

Hernández-Molina, F.J., Stow, D.A.V., Alvarez-Zarikian, C.A., Acton, G., Bahr, A., Balestra, B., Ducassou, E., Flood, R., Flores, J.A., Furota, S., Grunert, P., Hodell, D., Jimenez-Espejo, F., Kim, J.K., Krissek, L., Kuroda, J., Li, B., Llave, E., Lofi, J., Lourens, L., Miller, M., Nanayama, F., Nishida, N., Richter, C., Roque, C., Pereira, H., Sanchez Goñi, M.F., Sierro, F.J., Singh, A.D., Sloss, C., Takashimizu, Y., Tzanova, A., Voelker, A., Williams, T., Xuan, C. (2014a). Onset of Mediterranean outflow into the North Atlantic. Science 12, 344, 6189, 1244–1250.

Hernández-Molina, F., Llave, E., Preu, B., Ercilla, G., Fontan, A., Bruno, M., Serra, N., Gómiz, J., Brackenridge, R., Sierro, F. (2014b). Contourite processes associated with the Mediterranean Outflow Water after its exit from the Strait of Gibraltar: Global and conceptual implications. Geology 42 (3), 227–230.

Llave, E., Hernández-Molina, F.J., Somoza, L., Stow, D.A.V., Díaz del Río, V. (2007). Quaternary evolution of the contourite depositional system in the Gulf of Cadiz. In: Viana A, Rebesco M (eds) Economic and Paleoceanographic Importance of Contourites. Geol Soc London Sp Publ 276, pp 49–79.

Millot, C. (2009). Another description of the Mediterranean Sea outflow. Progress in Oceanography 82, 101–124.

Nelson, C.H., Baraza, J., Maldonado, A. (1993). Mediterranean undercurrent Sandy contourites, Gulf of Cadiz, Spain. Sedimentary Geology, 82, 103–131.

Serra, N., Ambar, I., Boutov, D. (2010). Surface expression of Mediterranean water dispoles and their contribution to the shelf/slope–open ocean exchange. Ocean Science, 6, 191–209.

Stow, D.A.V., Hernández-Molina, F.J., Llave, E., Sayago, M., Díaz del Río, V., Branson, A. (2009). Bedform-velocity matrix: the estimation of bottom-current velocity form bedform observations. Geology, 37,4, 327–330.

Stow, D.A.V., Hernández-Molina, F.J., Llave, E., Bruno, M., García, M., Díaz del Río, V,. Somoza, L., Brackenridge R.E. (2013). The Cadiz Contourite Channel: Sandy contourites, bedforms and dynamic current interaction. Marine Geology, 343, 99–114.

43 Bedforms on the Distal Valencia Channel and Turbidite System

Albert Palanques

Abstract

The Valencia Fan developed as the distal fill of a deep-sea valley detached from the continental slope and the main sedimentary source. This fan evolves downbasin from channelled to transitional and unchannelled morphological areas. High resolution sidescan sonar records show depositional bedforms on the valley floor of the channelled area and a progressive transversal gradation from depositional to erosional bedforms in the transitional area. This transversal gradation may be due to the effect of the Coriolis force on sediments transported through the Valencia Valley and to the lateral inputs of sediment flows from the Gulf of Lyon continental margin. Bedforms shown in this work include trains of starved ripples and dunes, sand ribbons, and fields of megaflutes.

Keywords

Valencia fan • Channel-lobe transition zone • Dunes • Ripple-like bedforms • NW Mediterranean deep basin

A. Palanques (✉)
Institut de Ciències Del Mar (CSIC), Pg. Marítim de la Barceloneta, 37–49, 08003 Barcelona, Spain
e-mail: albertp@icm.csic.es

J. Guillén et al. (eds.), *Atlas of Bedforms in the Western Mediterranean*,
DOI 10.1007/978-3-319-33940-5_43

43.1 Introduction

The Valencia Fan has developed in the north-western Mediterranean and is located at the end of a large, subsiding deep-sea depression where it fills the end of the Valencia Valley. The Valencia Fan is fed by a complex system of canyons incised into the Iberian margin that merges with the Valencia Channel and is not attached to the continental slope (Palanques and Maldonado 1985; Maldonado et al. 1985a; Alonso et al. 1995). The distal valley is progressively filled and becomes a channel that widens and forms a turbidite system. In the transition zone between the channelled and unchanneled area sanderized material accumulates and erosional features and sandy depositional bedforms develop as a consequence of the increased influence of turbidity currents undergoing a hydraulic jump (Komar 1973; Mutti and Normark 1991). Depressions and scours deeper than 2 m are normally observed on fans inside the area immediately downslope from the channel termination (Mutti and Normark 1991). The Valencia Fan has been active not only during lowering and low stand sea-level periods but also during sea-level rise, and to some extent during the present high-stand period (Palanques et al. 1994). A change in slope on the fan apex and the presence of volcanoes on the upper fan are factors influencing the fan growth pattern (Maldonado et al. 1985b).

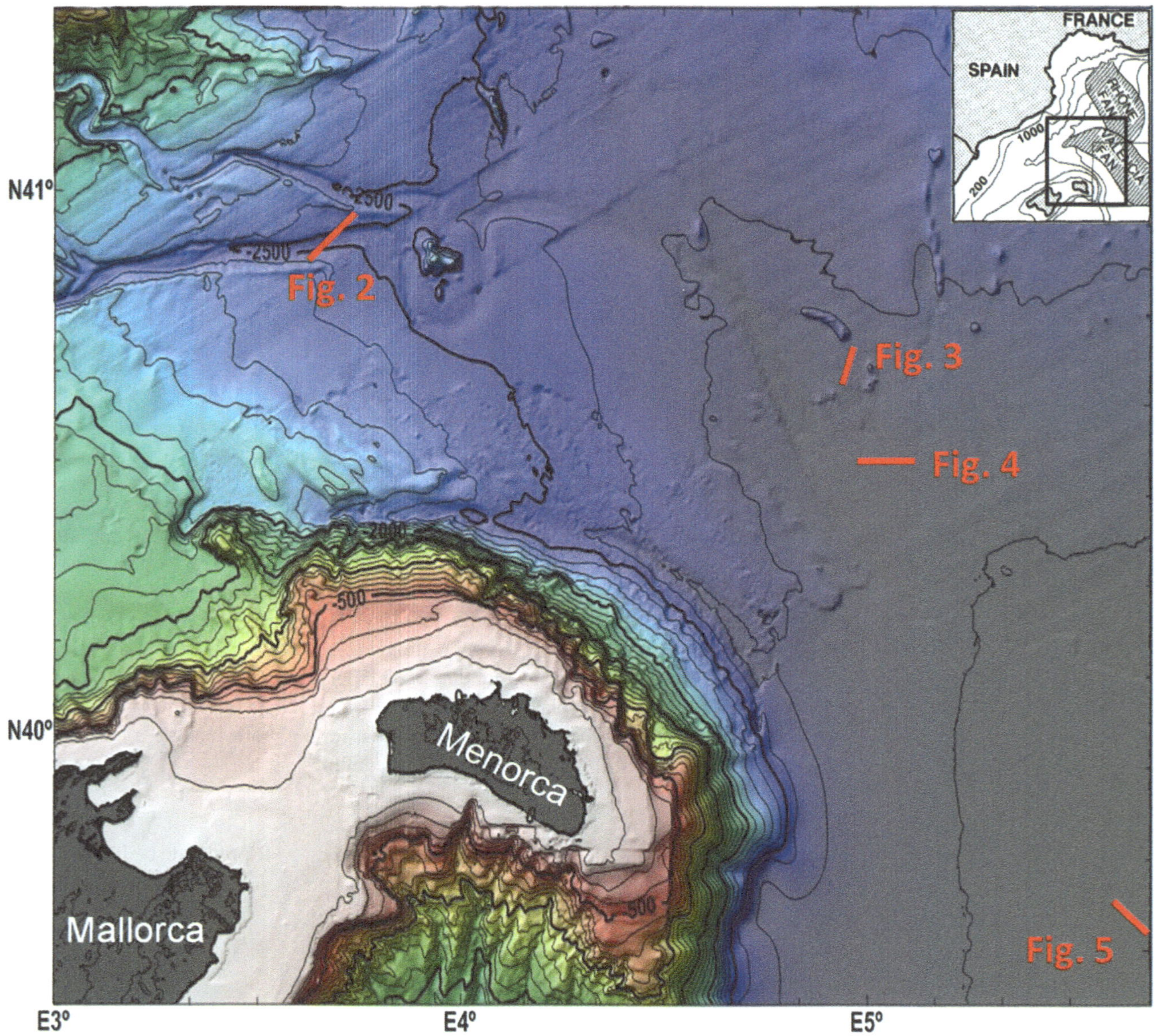

Fig. 43.1 Map of the study area with the location of the Sea MARC and MAK 1 *track lines* used for this work. Contours in metres. *Heavy lines* refer to the position of side scan records Figs. 43.2, 43.3, 43.4 and 43.5

43.2 Methods

This work is based on SeaMARC and MAK I sonographs and sub-bottom profiles (Fig. 43.1). The SeaMARC vehicle with 4.5 kHz sub-bottom profiler was navigated along or slightly oblique to the axis of the main Valencia valley-channel system using a 5-km swath width and recorded the bedforms from the channelled area shown in Fig. 43.1.

The deep towed MAK I sidescan sonar system including a 5 kHz sub-bottom profiler was used at a frequency of 30 kHz and had a swath width of 2 km. Bedforms from Figs. 43.2, 43.3 and 43.4 were recorded by the MAK I System. For more details see Maldonado et al. (1985a) and Palanques et al. (1995).

43.3 Results and Discussion

In the Valencia Valley-Fan there are a large variety of bedforms that show a gradual transition along and across the system. There are depositional bedforms with positive relief and no truncation of sub-bottom reflectors, erosional bedforms with negative relief and truncation of sub-bottom reflectors, and also a third type of bedform which is transitional and shows erosional and depositional characteristics (Maldonado et al. 1985a; Palanques et al. 1995). This work shows some of the most relevant bedforms that have developed in this system.

In the channelled zone of the upper fan system, sand dunes with a barchan shape extend across the valley floor (Fig. 43.2a) in a sector that coincides with the beginning of the fill of the valley by the Valencia Fan. The dune field

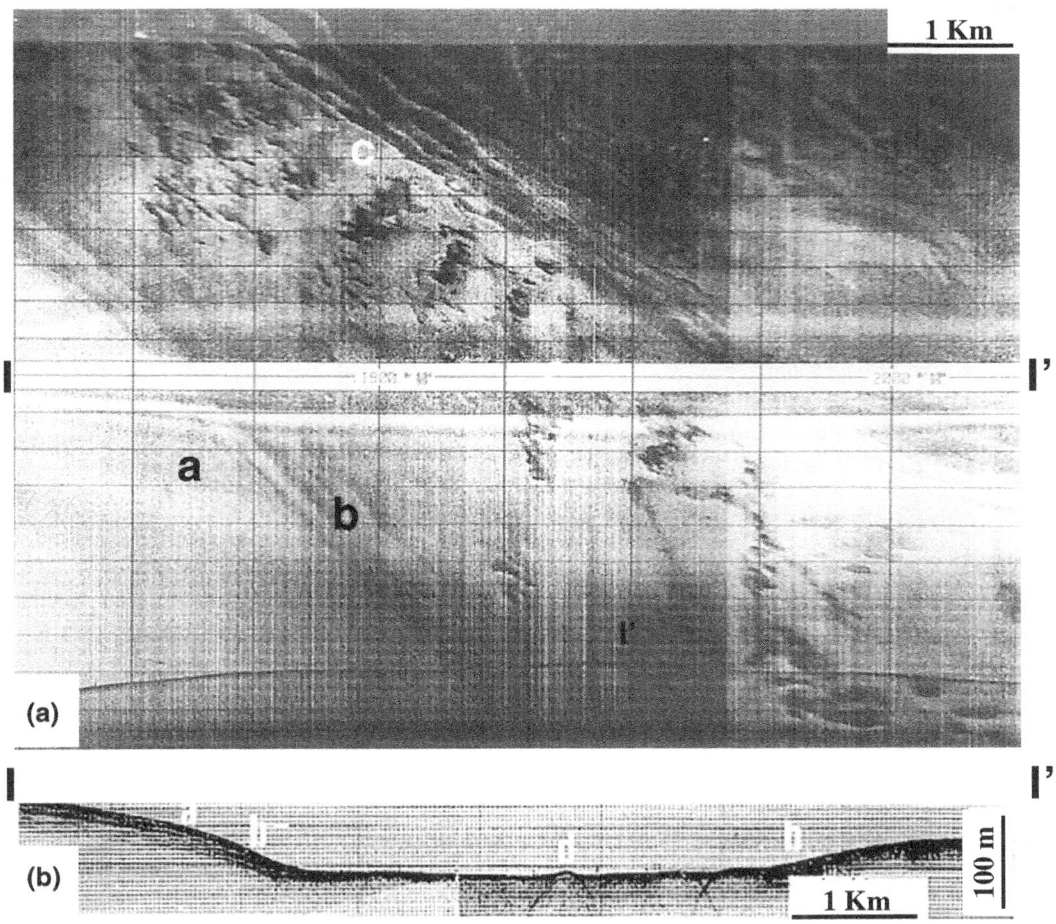

Fig. 43.2 Sonograph image and seismic profile of the upper Valencia Fan at the end of the Valencia Valley. **a**, Processed SeaMARC record oblique to the channel axis. **b**, 4.5-kHz nearbottom profile. The area insonified by the SeaMARC is represented in the record (profile *I-I'*). *a* valley margin; *b* channel wall; *c* barchan sand dunes extending across the entire channel floor; *d* hyperbolic echoes in the 4.5-kHz record attributed to the barchan sand dunes

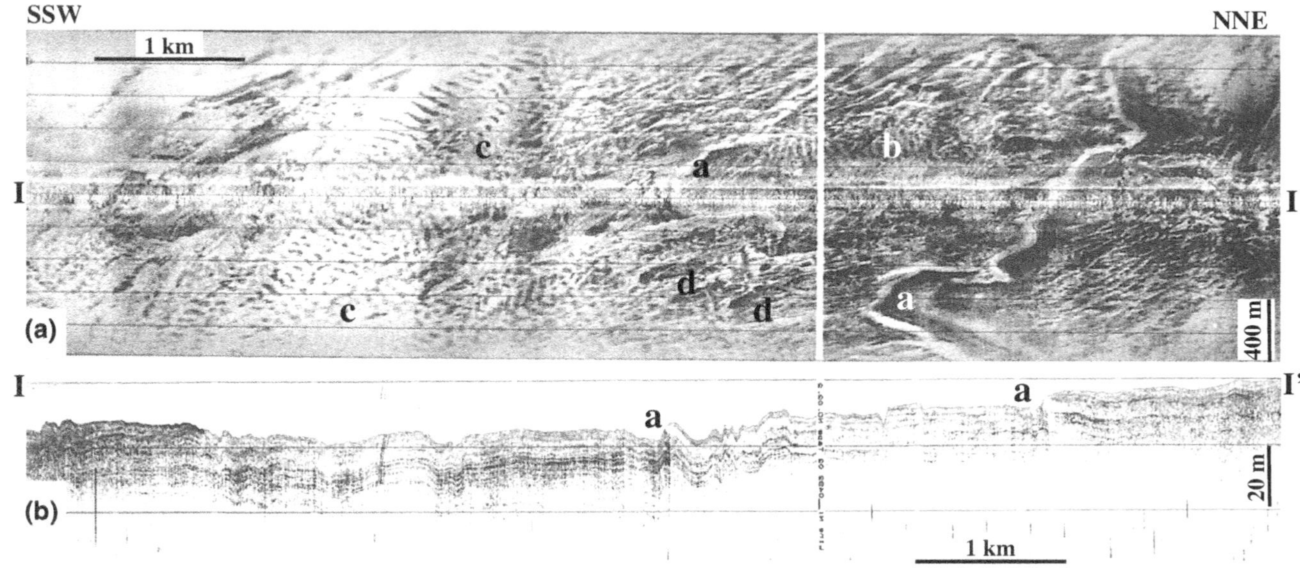

Fig. 43.3 MAK I sonograph (**a**) and 5-kHz seismic profile (**b**) along *lines I-I'* of the northeastern sector of the channelled-unchannelled transition zone on the Valencia Fan. Letters indicate the following features: *a*, linguoid scarps; *b*, small-scale ripple-like bedforms with crestlines oriented E-W; *c*, small-scale ripple-like bedforms with crestlines oriented NE-SW; *d*, ellipsoidal patches

occurs in a zone where the slope of the fan apex changes and channel deposition predominates. These dunes have a length of between 300 and 500 m and from some tens to several hundred meters in width and cause hyperbolic echoes in the 4.5-kHz sub-bottom profiler (Fig. 43.2b). Sediment cores taken in the dune field consist of well-classified channel sand with a median grain size of 3–3.5 phi.

In the transition zone between the channelled and unchanneled area there are small-scale ripple-like bedforms with crest lines oriented E-W and NE-SW (b and c in Fig. 43.3a). These bedforms have a wavelength of about 70 m and a high backscatter level, forming longitudinal streaks up to 1 km long in places. 5-kHz sub-bottom profiles display parallel reflectors of variable thickness with a penetration of about 30 m (Fig. 43.3b). These bedforms have been interpreted as starved ripples and sand ribbons moving as longitudinal bands across less mobile and possibly cohesive sediment (Palanques et al. 1995).

Another relevant bedforms of the transition zone are erosional concavities with a crescentic shape that form bands

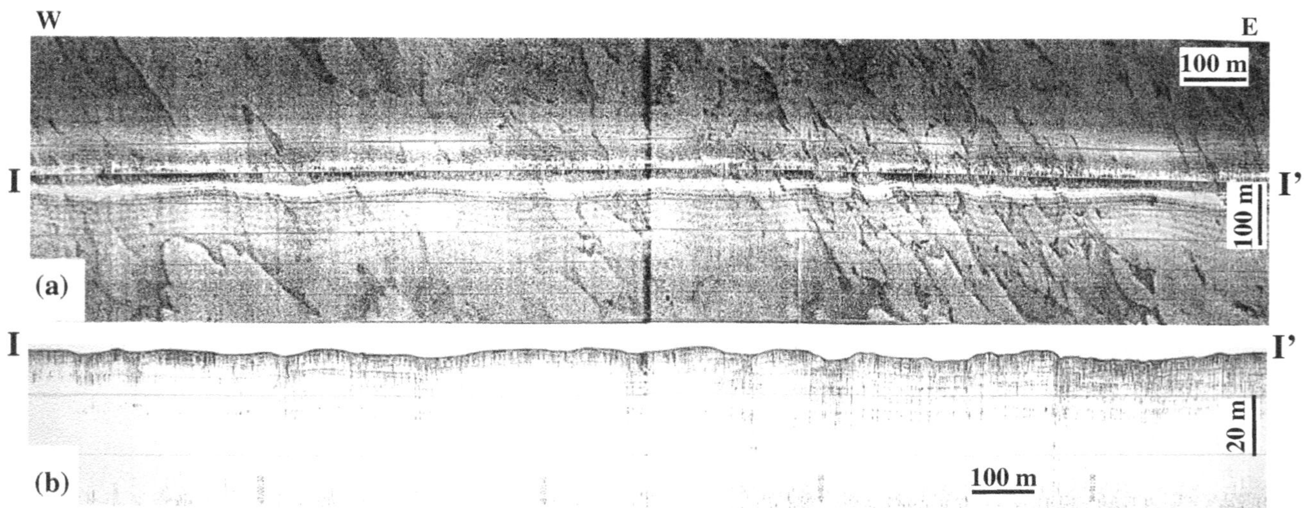

Fig. 43.4 MAK I sonograph (**a**) and 5-kHz seismic profile (**b**) along *lines I-I'* of the southwestern sector of the channelled-unchannelled transition zone on the Valencia Fan, showing bands of ellipsoidal patches believed to be megaflutes

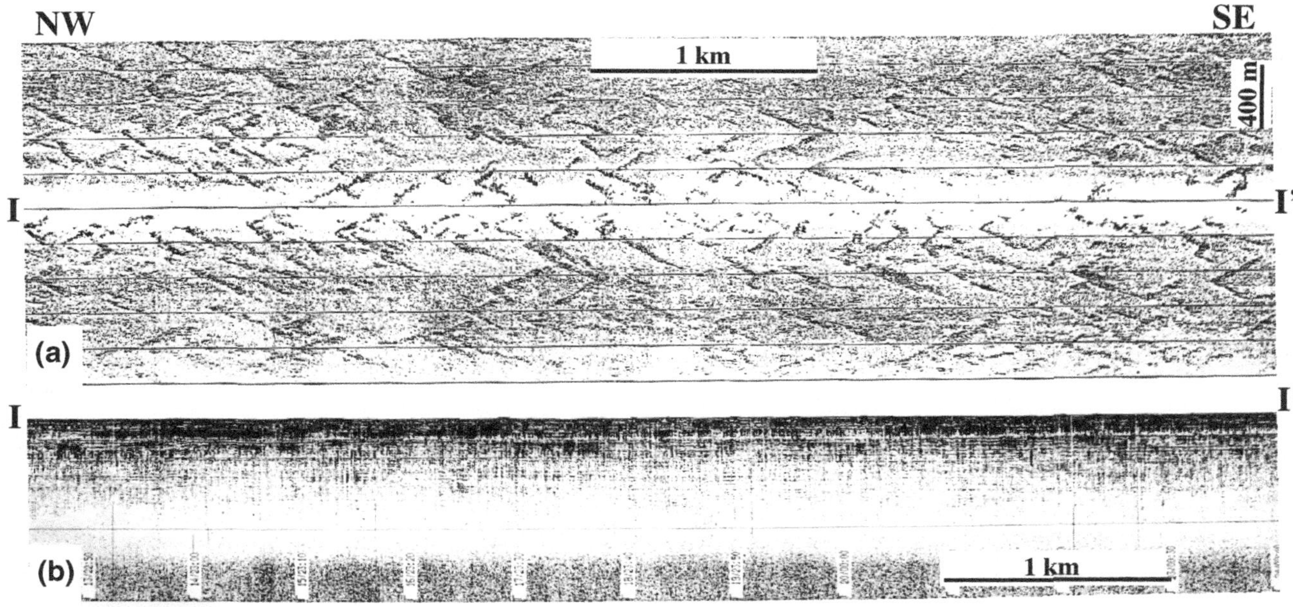

Fig. 43.5 MAK I sonograph (**a**) and 5-kHz seismic profile (**b**) along *lines I-I'* showing a detailed view of V-shaped bedforms from the unchannelled zone on the lower Valencia Fan

of ellipsoidal patches (Fig. 43.4a). These morphologies are oriented basinward. Their width increases with depth from about 50 m at the NE to 200–300 m at the SW and they have a length from 100 to 200 m. The sub-bottom profiles have an acoustic penetration of about 20–30 m that decreases towards the southwest and exhibit an irregular seafloor with 2- to 7-m-deep holes (Fig. 43.4b) that correspond with these ellipse-shaped bedforms. These fields of elongated scours can be identified as megaflutes.

Down-basin, there is a gradual change in the unchannelled zone to V-shaped bedforms (Fig. 43.5a) which size decreases from about 800–300 m in the deeper zone of the system where they are densely grouped. These V-shaped bedforms develop in a mud-dominant area where there is a significant penetration in the 5-kHz sub-bottom profiles (Fig. 43.5b). These bedforms could be starved dunes moving across less mobile cohesive sediment, such as the ripple-like bedforms of the transition zone.

Acknowledgments Data were acquired thanks to various projects. UNESCO and the European Science Foundation (ESF) provided important financial contributions. Funding was also provided by the Joint Committee of Science and Technology of the US Spain Treaty of Friendship, and the U.S. National Science Foundation (grant OCE 81-18069). The officers and crew of the R/V *Gelendzhik* and the R/V *Robert D. Conrad* are gratefully acknowledged. We also thank M.K. Ivanov, W.B.F. Ryan, A. Maldonado and N.H. Kenyon.

References

Alonso, B., Canals, M., Palanques, A., Rehault, J.P. (1995). A deep-sea channel in the northwestern Mediterranean sea: Morpho-sedimentary evolution of the Valencia Channel and its surroundings. Mar. Geophy. Res. 17, 469–484.

Komar, RD. (1973). Continuity of turbidity current flow and systematic variations in deep-sea channel morphology, Geol. Soc. Amer. Bull. 84, 3329–3338.

Maldonado, A., Palanques, A., Alonso, B., Kastens, K.A., Nelson, C. H., O'Connell, S. Ryan, W.B.F. (1985a). Physiography and deposition on a distal deep-sea system: The Valencia Fan (northwestern Mediterranean). Geo-Mar. Lett. 5, 157–164.

Maldonado, A., Got, H., Monaco, A., O'Connell, S. Mirabille, L. (1985b). Valencia Fan (Northwestern Mediterranean): Distal deposition fan variant. Mar. Geol. 62, 295–319.

Mutti, E., Normark, W.R. (1991). An integrated approach to the study of turbidite systems. In: Weimer, P., Link, M.H., (Eds.), Seismic Facies and Sedimentary Processes of Submarine Fans and Turbidite Systems, Springer-Verlag, New York, pp. 75–106

Palanques, A., Maldonado A. (1985). Sedimentology and evolution of the Valencia Valley and Fan (Northwestern Mediterranean). Acta Geol. Hisp. 20, 1–19.

Palanques, A., Alonso, B., Farrán, M. (1994). Progradation and retreat of the Valencia fanlobes controlled by sea level changes during the Plio-Pleistocene (Northwestern Mediterranean). Mar. Geol. 117, 195–205.

Palanques, A., Kenyon, N., Alonso, B., Limonov, A. (1995). Bedforms and depositional patters in the Valencia Channel mouth: An example of a modern channel-lobe transition zone. Mar. Geophys. Res. 17(6), 503–517.

Small-Scale Bedforms Generated by Gravity Flows in the Aeolian Islands

44

D. Casalbore, A. Bosman, C. Romagnoli, and F.L. Chiocci

Abstract

Small-scale crescent-shaped bedforms were identified within flat-bottomed channelized features carving the submarine volcanic flanks of the Stromboli and Salina edifices. The bedforms are downslope-asymmetric, with a sub-horizontal stoss side and a steep lee side. They have a wavelength of tens of metres and a wave height of metres, and their crest-lines are roughly perpendicular to the maximum slope. Based on their size, texture and similarities to other bedforms, they can be interpreted as cyclic steps, i.e. a class of upslope-migrating turbidite sediment waves. Although their genesis must be similar, some differences are observed between the two cases. At Stromboli, repeated multibeam surveys at an 11-year scale showed a significant upslope migration of the bedforms because they lie on the bottom of a channel connected to active coastal dynamics, whose headwall is located at a few metres depth. By contrast, the surveys showed no significant morphological variations at Salina, where the bedforms are larger than at Stromboli and the channel headwall is located at about −100 m, suggesting that these feature are mostly inactive at present.

D. Casalbore (✉) · F.L. Chiocci
Sapienza Università di Roma, Dipartimento Scienze della Terra, Rome, Italy
e-mail: daniele.casalbore@igag.cnr.it

D. Casalbore · A. Bosman · F.L. Chiocci
IGAG-CNR, Istituto di Geologia Ambientale e Geoingegneria
Area della Ricerca di Roma 1, Rome, Italy

C. Romagnoli
Dipartimento di Scienze Biologiche, Geologiche ed Ambientali, Università di Bologna, Bologna, Italy

J. Guillén et al. (eds.), *Atlas of Bedforms in the Western Mediterranean*,
DOI 10.1007/978-3-319-33940-5_44

Keywords
Stromboli • Salina • Repeated bathymetric surveys • Seafloor mapping

44.1 Introduction

The Aeolian Islands are the emerged portion of large strato-volcanoes of Late Quaternary age that extend down to more than −2000 m. These volcanic edifices, together with some seamounts, make up the Aeolian volcanic arc, which spreads between the Calabrian–Peloritan mountain chain and the back-arc Tyrrhenian basin (Fig. 44.1, inset). This geological setting has been related to the NW-trending subduction of the Ionian slab below Calabria (Malinverno and Ryan 1986).

Many studies focus on the geological evolution of the Aeolian Islands (Lucchi et al. 2013 and references therein), whereas only in the last two decades has a big effort been made to extensively map their submarine portions, which account for more than the 80 % of the edifice (Romagnoli et al. 2013 and reference therein). Thanks to these studies a large number of primary volcanic and erosive-depositional features along the submarine flanks have been recognized and mapped (e.g. Gamberi and Marani 1997; Bortoluzzi et al. 2010; Bosman et al. 2014; Casalbore et al. 2011, 2014; Romagnoli et al. 2012). In this paper, we describe the occurrence of small-scale, crescent-shaped bedforms within channelized features carving the insular shelf surrounding the islands of Stromboli (an active volcano with a persistent production of volcaniclastic products) and Salina (where volcanic activity ended 15.6 ka ago, Lucchi et al. 2013). We discuss the possible genesis and development of bedforms on the basis of repeated bathymetric data and single-channel high-resolution seismic and seafloor sampling. Despite their small size, the study of these features is rapidly expanding, because they are ubiquitous in active canyons/channels (e.g. Paull et al. 2010; Conway et al. 2012) and may be considered a proxy of active sedimentary dynamics.

44.2 Data and Methods

High-resolution bathymetry around the Aeolian Islands was collected during several oceanographic cruises onboard R/Vs *Urania* and *Universitatis* and a small launch for shallow water surveys (<−100 m). Details on data acquisition and processing are described in Romagnoli et al. (2013) and Bosman et al. (2015). In the present study, digital elevation models with a cell size varying from a few decimetres in very shallow water (< −20 m) to metres and tens of metres at higher depths were produced. Residual maps, obtained as the difference between successive multibeam bathymetries, were produced to monitor the evolution of the observed morphological features. As one of

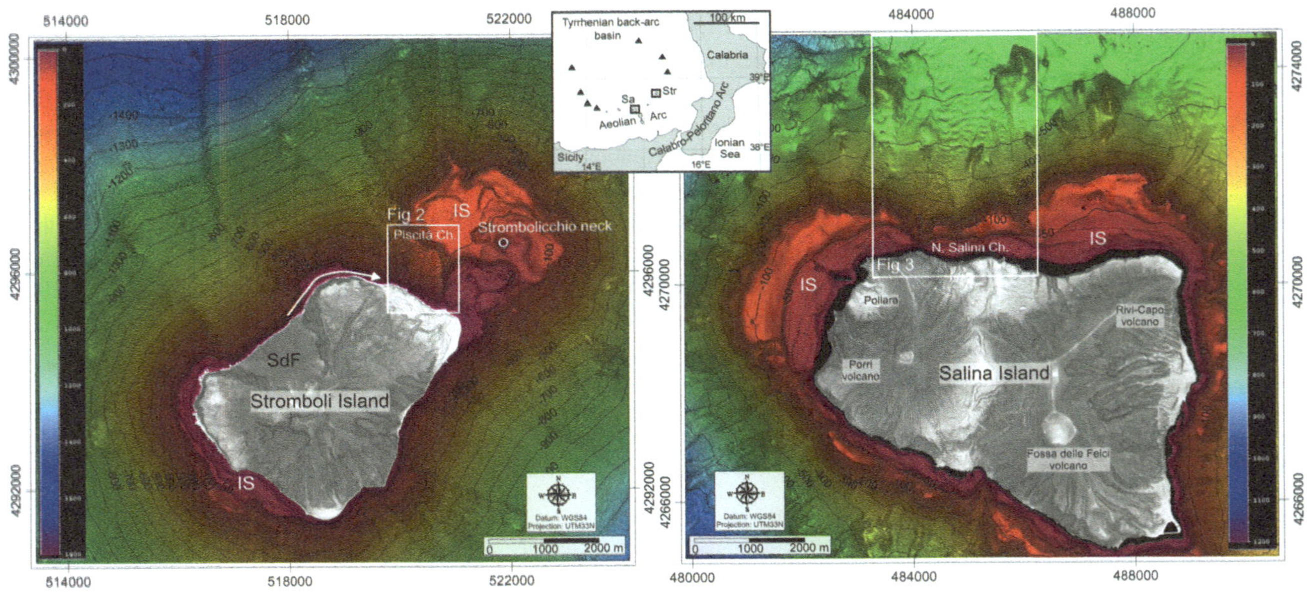

Fig. 44.1 Shaded relief map of the Stromboli (*left*) and Salina (*right*) volcanic edifices, with location of the following figures (*white boxes*); the *white arrow* in Fig. 44.1a indicates the prevailing longshore current around the northern part of Stromboli edifice. In the inset, the regional setting of the area, where triangles represent submarine seamounts also belonging to the Aeolian Arc; Sa, Salina; Str, Stromboli; SdF, Sciara del Fuoco; Piscità Ch., Piscità Channel; N. Salina Ch., North Salina Channel; IS, insular shelf

our target was to detect the migration of the bedforms, we used the stability of the surrounding seafloor to define the accuracy of the residual map and check that the observed variations are an order of magnitude greater than the instrumental noise. Ancillary data used for the interpretation were high-resolution single-channel seismic data acquired with a 0.5–1 kJ Sparker source and Van Veen grabs of seafloor sediment (Casalbore et al. 2013).

44.3 Results and Discussions

Crescent-shaped bedforms with crest-lines trending roughly perpendicular to the slope direction were identified within wide flat-bottomed channels on the northern flank of the Stromboli and Salina edifices (Fig. 44.1).

At Stromboli, a 600-m-wide channel (named the Piscità Channel) carves the saddle between the edifices. The channel head is made up of the coalescence of overlapping slide scars at a hundred-metre scale, and the headwall locally reaches a very shallow depth (−5 m), less than 100 m from the coastline (Fig. 44.2a). Crescent-shaped bedforms are mainly located between −30 and −170 m, where slope gradients become less than 8°–10°. The bedforms have a wavelength of 20–60 m and a wave height of 1–6 m, and their wave dimensions tend to increase downslope. In cross-section, most bedforms are asymmetric downslope (Fig. 44.2d), with a sub-horizontal (or slightly upslope-dipping) stoss side and a steeper lee side (>10°, up to 25°). The multibeam backscatter mosaic shows along- and across-slope variations that can be related to present-day sedimentary dynamics (Fig. 44.2b). A grab from the central

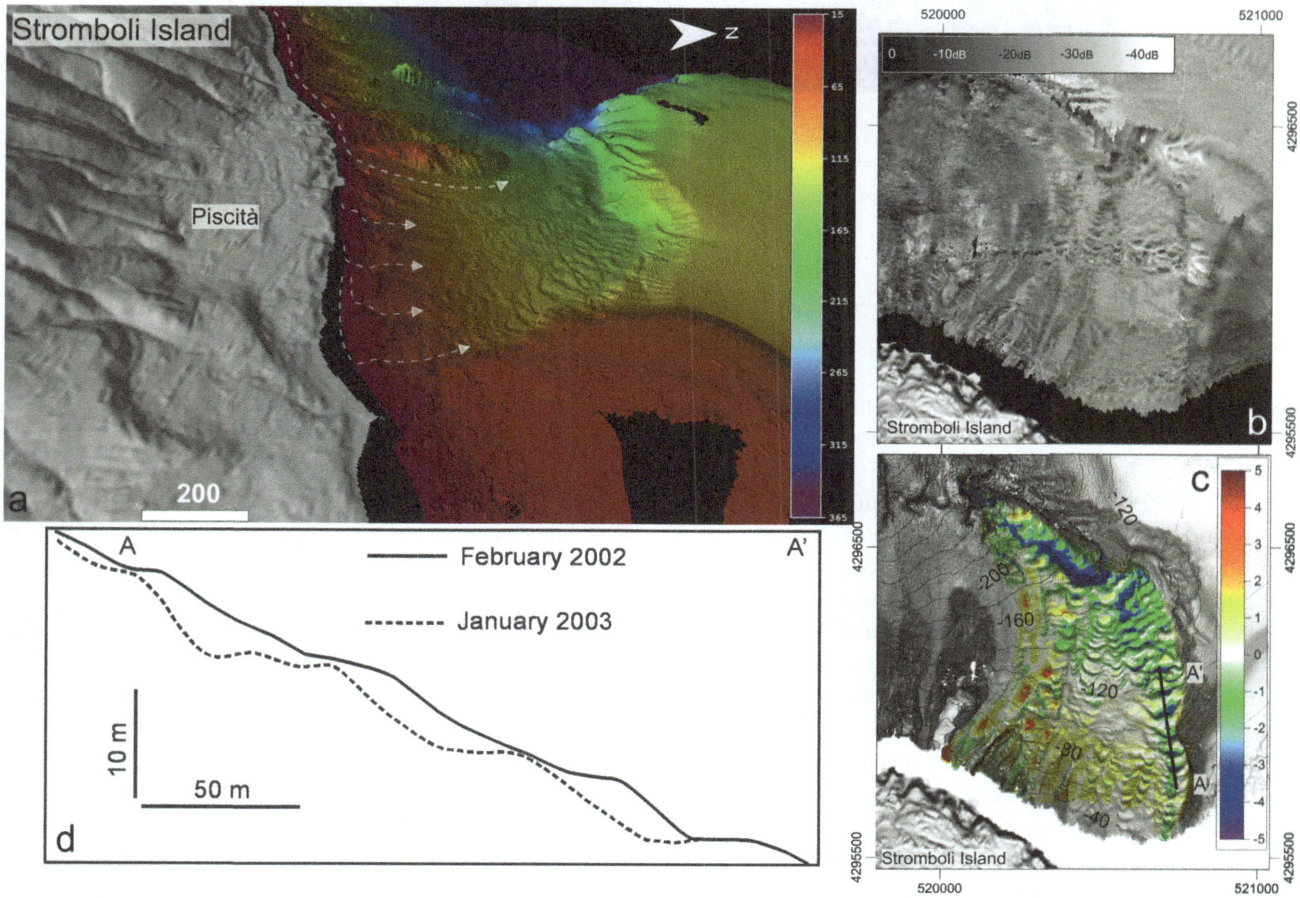

Fig. 44.2 **a** 3-D view of the Piscità Channel on the north flank of Stromboli (location in Fig. 44.1), where crescent-shaped bedforms are evident and the *dashed white arrows* indicate small erosive channels. Scale bar is in metres. **b** Multibeam backscatter map of the Piscità Channel, where backscatter variations suggest active sedimentary processes (*light tones* indicate low backscatter values, dB −40). **c** Residual map obtained from the comparison between the bathymetric surveys of 2002 and 2013 draped over a shaded relief image; scale bar is in metres. **d** 2002 and 2003 bathymetric profiles along an *A–A′* section of Fig. 44.2c show the deepening of the seafloor and the migration of the bedforms in the easternmost part of the Piscità Channel

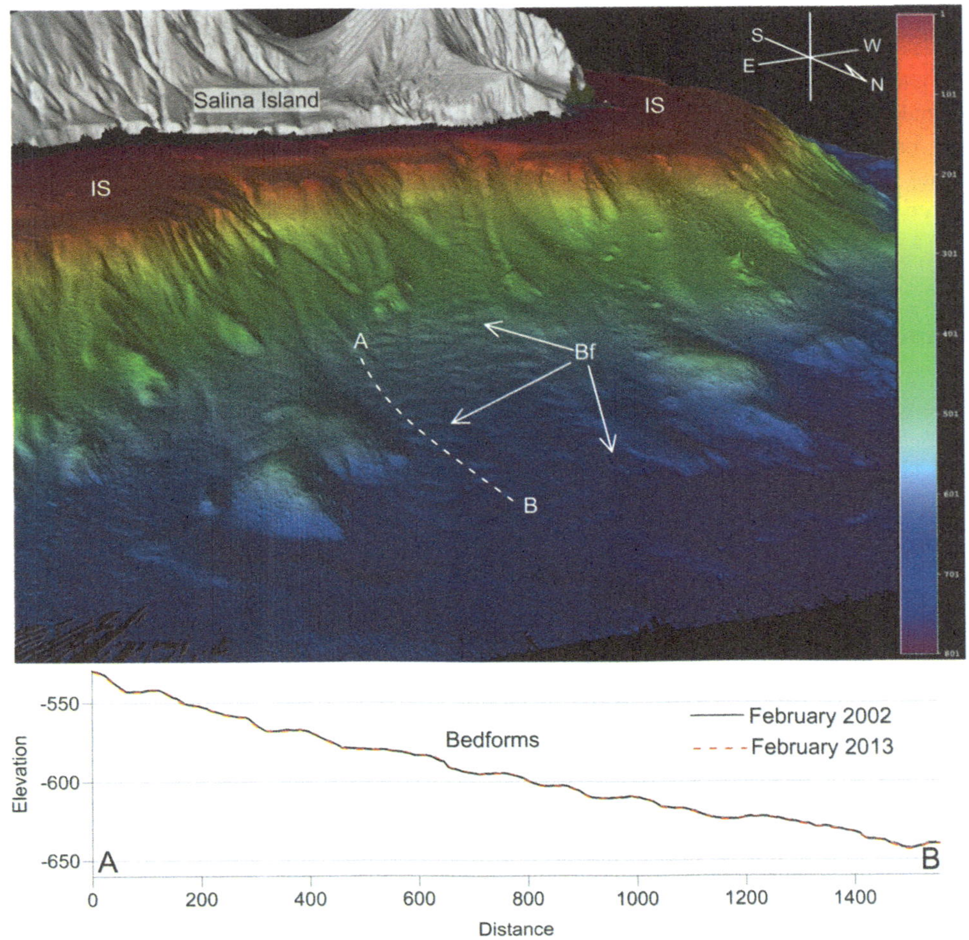

Fig. 44.3 3D view of the North Salina Channel, where the crescent-shaped bedforms (Bf) are evident. Below, 2002 (*black line*) and 2013 (*dashed grey line*) bathymetric profiles along the *A–B* section, where no significant morphological variations are present

part of the Piscità Channel at about −100 recovered poorly-sorted volcanic sand (Casalbore et al. 2013). A residual map, obtained as the difference between the 2002 and 2013 multibeam surveys, shows an overall deepening of the channel floor and an upslope migration of the bedforms of up to tens of metres (Figs. 44.2c). The main variations mostly occurred between 2002 and 2003 in the easternmost part of the Piscità Channel (Fig. 44.2d).

At Salina, a 1-km-wide channel (named the North Salina Channel; Figs. 44.1 and 44.3) is present on the northern submarine flank of the edifice, indenting the outer edge of the insular shelf at depths of −90 to −105 m, 400–600 m from the coastline (Casalbore et al. 2015). This channel actually acts as a main collector for a few smaller (i.e. a few hundred metres wide) channelized features carving the insular shelf edge (Fig. 44.3). Crescent-shaped bedforms are mostly found between −400 and −660 m, where slope gradients markedly decrease to values of less than 9°. They have a wavelength of 70–150 m and a wave height of 3–7 m, and are downslope-asymmetric in cross-section, with the lee side sloping up to 16° (section A–B in Fig. 44.3). Residual maps made from the comparison between the bathymetric surveys of 2002 and 2013 show no significant seafloor variations, as is clearly recognizable in the two overlapping bathymetric sections shown in Fig. 44.3. Three grabs from the North Salina Channel recovered coarse-grained sand in the lower part of the grabs (locally with gravel), passing to fine-sand or sandy silt to silty mud in the upper part (Casalbore et al. 2013).

In both study areas, the crescent-shaped bedforms share many similarities with the bedforms found on the upper reach of active canyons/channels, such as the Monterey Canyon on the Californian margin (Paull et al. 2010) and the Gioia Canyon on the Italian margin (Bosman et al. 2017), where they are interpreted as cyclic steps, i.e. a class of upslope-migrating turbidite sediment waves, in which each downward step is bounded by a hydraulic jump (Kostic 2011 and references therein). In addition to the aforementioned similarities, some differences are observed between the two cases. The bedforms at Salina show both larger wave dimensions and a higher depth range than those recognized at Stromboli. The latter show evidence of migration also at the 1-year scale, whereas no morphological variations are recognized at Salina in a period of 11 years. This evidence can be interpreted as the result of the different erosive processes and states of activity between the two channels. The headwall of the Piscità Channel is located at a few metres depth, so it can directly interact with the coastal dynamics by intercepting the sediment transported by the longshore drift. A recent study by Romagnoli et al. (2006) evidenced a longshore drift of sand and gravel from Sciara del Fuoco in a clockwise direction around the island, due to the prevailing westerly winds, feeding the beaches on the N and NE side of Stromboli island (see white arrows in Fig. 44.1). This transport increases during and after the major eruptive crises that occur at Stromboli, when subaerial lava flows enter the sea and a larger amount of volcaniclastic debris is produced at Sciara del Fuoco. The main morphological variations at the Piscità Channel occurred, in fact, between the 2002 and 2003 multibeam surveys, just after the 2002 tsunamigenic landslide event that affected the Sciara del Fuoco, leading to an increase in volcaniclastic material transported seaward due to the morphological readjustment of the failed subaerial slope of Sciara del Fuoco (Chiocci et al. 2008). In contrast, the headwall of the North Salina Channel is located at higher depths (>90 m), so the link with present-day coastal dynamics is less direct, at least during fair weather conditions. Though the recognition of some fresh-looking scars at the edge of the insular shelf (Figs. 44.1 and 44.3) may suggest the occasional occurrence of slope failures, it is difficult to imagine that these small-scale failures can be the main cause for the bedform formation. Rather, the frequency and energy of processes generating the bedforms might have been higher at the time of the large Pollara explosive eruptions (i.e. 27–17 ka ago), when the sea-level was markedly lower than now (Bintanja et al. 2005) and a more effective morphological continuity between subaerial and submarine processes could be envisaged. These inferences suggest that the bedforms in the North Salina Channel could be considered at present as partially inactive features, consistent with the lack of significant seafloor variations at the 11-year scale and with the fining-upward sedimentary sequence recovered in the grabs, evidencing a decrease in the energy level of sedimentary dynamics.

Acknowledgments This research was funded by the Italian MaGIC (Marine Geohazards along the Italian Coasts) Project and carried out in the framework of the RITMARE (Ricerca ITaliana per il MARE) Project. Crews of R/Vs *Urania* and *Thetis (CNR)*, and *Universitatis* (CoNISMA) are gratefully acknowledged along with the fellow researchers and students who took part in the sea surveys.

References

Bintanja R, Van de Wal RSW, Oerlemans J (2005). Modelled atmospheric temperatures and global sea levels over the past million years. Nature 437:125–128.

Bortoluzzi G, Ligi M, Romagnoli C, Cocchi L, Casalbore D, Sgroi T, Cuffaro M, Caratori Tontini F, D'Oriano F, Ferrante V, Remia A, Riminucci F (2010). Interactions between volcanism and tectonics in the western Aeolian sector, southern Tyrrhenian Sea. Geophysical Journal International 183(1):64–78.

Bosman A, Casalbore D, Romagnoli C, Chiocci FL (2014). Formation of an 'a'ā lava delta: insights from time-lapse multibeam bathymetry and direct observations during the Stromboli 2007 eruption. Bulletin of Volcanology 76(7):1–12.

Bosman A, Casalbore D, Anzidei M, Muccini F, Carmisciano C, Chiocci FL (2015). The first ultra-high resolution Digital Terrain Model of the shallow-water sector around Lipari Island (Aeolian Islands, Italy). Annals of Geophysics, 58(2), p. S0218.

Bosman A, Casalbore D (2017). Cyclic steps in canyon thalweg and in shallow-water slide scars along the Calabrian Margin (Southern Tyrrhenian Sea, Italy). In: J. Guillén et al. (eds) Atlas of Bedforms in the Western Mediterranean. Springer, Heidelberg, pp 287–292.

Conway KW, Barrie JV, Picard K, Bornhold BD (2012). Submarine channel evolution: active channels in fjords, British Columbia, Canada. Geo-Marine Letters 32(4): 301-312.

Casalbore D, Romagnoli C, Bosman A, Chiocci FL (2011). Potential tsunamigenic landslides at Stromboli volcano (Italy): insight from marine DEM analysis. Geomorphology 126: 42–50.

Casalbore D, Bosman A, Romagnoli C, Chiocci FL (2013). Small-scale crescent-shaped bedforms in submarine volcanic setting: examples from Stromboli and Salina island (Italy). GeoActa 12: 37–45.

Casalbore D, Romagnoli C, Bosman A, Chiocci FL (2014). Large-scale seafloor waveforms on the flanks of insular volcanoes (Aeolian Archipelago, Italy), with inferences about their origin. Marine Geology 355: 318–329.

Casalbore D, Bosman A, Romagnoli C, Chiocci FL (2015). Morphological map of Salina offshore. Journal of Map, in press.

Chiocci FL, Romagnoli C, Bosman A (2008). Morphologic resilience and depositional processes due to the rapid evolution of the submerged Sciara del Fuoco (Stromboli island) after the December 2002 submarine slide and tsunami. Geomorphology 100 (3–4): 356–365.

Gamberi F, Marani M, Savelli C (1997). Tectonic, volcanic and hydrothermal features of a submarine portion of the Aeolian arc (Tyrrhenian Sea). Marine Geology, 140(1): 167–181.

Kostic S (2011). Modeling of submarine cyclic steps: controls on their formation, migration and architecture. Geosphere 7: 294–304.

Lucchi F, Keller J, Tranne CA (2013). Regional stratigraphic correlations across the Aeolian archipelago (southern Italy). In: Lucchi, F., Peccerillo, A., Keller, J., Tranne, C.A., Rossi, P.L. (Eds.), The Aeolian Islands Volcanoes. Geological Society, London, Memoirs, 37, pp 55–81.

Malinverno A, Ryan WBF (1986). Extension in the Tyrrhenian sea and shortening in the Apennines as result of arc migration driven by sinking of the lithosphere. Tectonics 5: 227–245.

Paull CK, Ussler III W, Caress DW, Lundsten E, Barry J, Covault JA, Maier KL, Xu JP, Augenstein S (2010). Origins of large crescent-shaped bedforms within the axial channel of Monterey Canyon. Geosphere 6: 755–774.

Romagnoli C, Mancini F, Brunelli R (2006). Historical shoreline changes at an active island volcano: Stromboli, Italy. Journal of Coastal Research 22: 739–749.

Romagnoli C, Casalbore D, Chiocci FL (2012). La Fossa Caldera breaching and submarine erosion (Vulcano island, Italy). Marine Geology 303–306: 87–98.

Romagnoli C, Casalbore D, Bortoluzzi G, Bosman A, Chiocci FL, D'Oriano F, Gamberi F, Ligi M, Marani M (2013). Bathy-morphological setting of the Aeolian islands. In: Lucchi, F., Peccerillo, A., Keller, J., Tranne, C.A., Rossi, P.L. (Eds.), The Aeolian Islands Volcanoes. Geological Society, London, Memoirs, 37, pp. 27–36.

45 Large-Scale Bedforms on Volcaniclastic Aprons Around the Aeolian Islands (Italy)

D. Casalbore, A. Bosman, C. Romagnoli, and F.L. Chiocci

Abstract

Large-scale bedforms, with wavelengths ranging from several tens of metres to over 1 km and wave heights ranging from a few metres to tens/hundreds of metres were observed on multibeam bathymetry in the median and lower part of submarine volcaniclastic aprons surrounding the Aeolian Islands, where a marked decrease in slope gradients below 4°–8° is present. The bedforms are characterized by crest-lines with an arcuate or sinuous shape, trending roughly perpendicular to the maximum slope direction. Based on their size, geometry and texture, they can be interpreted as coarse-grained sediment waves, which are also found in the proximal part of other modern and ancient turbidite systems. In the study areas, they are found within channelized features or on fan-shaped features lying at or close to major sedimentary sources, such as subaerial/submarine depressions left by sector collapses.

D. Casalbore (✉) · F.L. Chiocci
Dipartimento Scienze della Terra, Sapienza Università di Roma, Rome, Italy
e-mail: daniele.casalbore@igag.cnr.it

D. Casalbore · A. Bosman · F.L. Chiocci
IGAG-CNR, Istituto di Geologia Ambientale e Geoingegneria, Area della Ricerca di Roma 1, Monterotondo, Italy

C. Romagnoli
Dipartimento di Scienze Biologiche, Geologiche e Ambientali, Università di Bologna, Piazza di Porta S.Donato 1, Bologna, Italy

J. Guillén et al. (eds.), *Atlas of Bedforms in the Western Mediterranean*,
DOI 10.1007/978-3-319-33940-5_45

Keywords
Sediment gravity flows • Cyclic steps • Morphometric analysis • Coarse-grained sediment waves

45.1 Introduction

Recent advances in seafloor mapping techniques have allowed extensive mapping of the submarine portions of volcanic edifices around the world, greatly enhancing our knowledge of geological processes acting on their flanks. These studies show that morphologies related to mass-wasting and reworking processes generally dominate the architecture of submarine volcaniclastic aprons, covering over 90 % of their surface, as observed at Stromboli and in many other insular/submarine volcanoes (Casalbore et al. 2010 and references therein). In detail, the acquisition of a large dataset of marine geophysical data around the Aeolian Islands highlighted a large number of erosive-depositional features along their submarine flanks, encompassing landslide scars and related deposits, channelized features, fan-shaped features and seafloor bedforms fields (Romagnoli et al. 2013a, b and references therein). The aim of this chapter is to show large-scale bedforms observed on the middle and lower part of the volcaniclastic aprons surrounding the Aeolian Islands, with particular reference to the Stromboli and Vulcano edifices. We also discuss their possible genesis through the integrated analysis of multibeam, side scan sonar and seismic data, and seafloor sampling.

45.2 Data and Methods

Data used for this work were mainly high-resolution multibeam bathymetry collected during several oceanographic cruises carried out in the last ten years around the Aeolian Islands. Details on data acquisition and processing are described in Bosman et al. (2015) and Romagnoli et al. (2013a). In this work, Digital Elevation Models (DEMs) at resolutions varying from 1 m in the first −100 m to 25 m at greater depths (down to −2600 m) were used. Seismic profiles (3.5 kHz SBP) and deep-towed long-range side scan sonar data (TOBI, working at a frequency of 30 kHz) are also available for some areas. Finally, several seafloor dredge, core and grab samples were collected along the submarine flanks of these insular volcanoes (e.g. Casalbore et al. 2014 and reference therein).

45.3 Results and Discussion

In the Aeolian Archipelago, large-scale bedforms were recognized on the submarine flanks of the Stromboli, Panarea, Lipari and Vulcano edifices (Fig. 45.1, Casalbore et al. 2014).

On the whole, these features show wavelengths ranging from several tens of metres to over 1 km, wave heights ranging from a few metres to tens/hundreds of metres, and a lateral extent ranging from hundreds to thousands of metres. Their crest-lines often show an arcuate or sinuous shape and always trend roughly perpendicular to the maximum slope direction, allowing a genesis from geostrophic bottom currents to be ruled out, as bedforms generated by this kind of currents should have crests oblique to the topographic contours and current direction (e.g. Flood et al. 1993). In cross-section, the shape of these bedforms is variable from a stepped morphology, with steep scarps (20°–55°) alternating with relatively flat surfaces, to a more undulating pattern with less steep sides (8°–15°). These bedforms are mostly found within channelized features and/or on fan-shaped features, always associated with a marked decrease in slope gradients to values below 4°–8°.

Below, we describe in detail two examples of these bedforms identified at Stromboli (Fig. 45.2) and Vulcano (Fig. 45.3), respectively. At Stromboli, several coaxial trains of bedforms are recognizable at the foot of the SE flank between −1500 and −1700 m (Fig. 45.2a), as well as on the NW and N flanks below −2000 m (Fig. 45.2b, c, respectively).

The bedforms show an overall low morphological relief, with wavelengths of 60–530 m and wave heights of 4–10 m. In contrast, they are well-recognizable on the side scan sonar image (Fig. 45.2), where they are associated with a marked backscatter variation, with low-backscatter tones on the stoss side and high backscatter tones on the crest-line and lee side. This backscatter zonation can be interpreted as the result of increased flow velocities across the wave crest, leading to the deposition of coarser sediment, and/or the winnowing of sediments downslope, similarly to what has been observed for coarse-grained sediment waves found in the proximal regions of modern turbidite systems (e.g. Wynn et al. 2002).

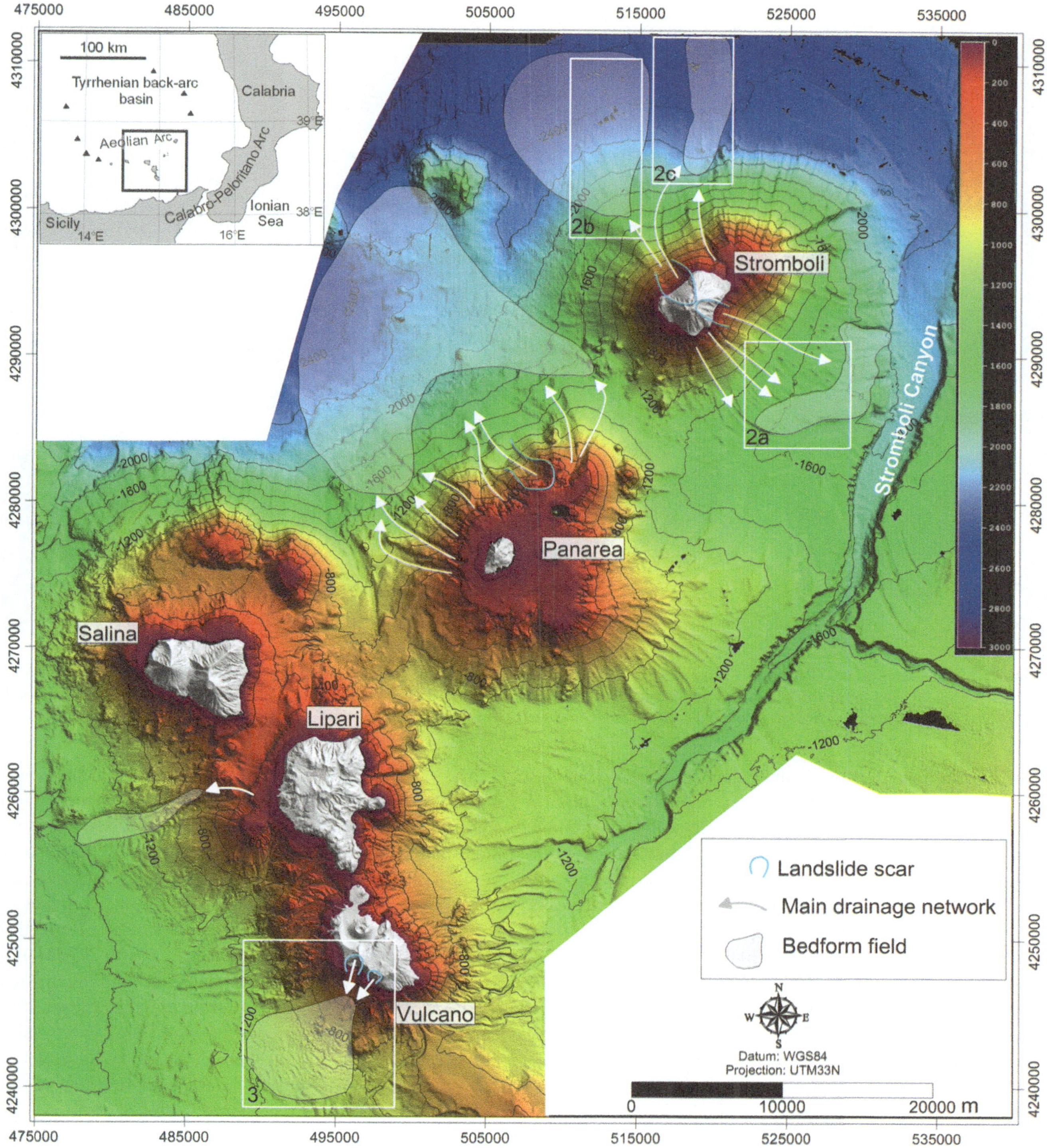

Fig. 45.1 Shaded relief map of central and eastern Aeolian Islands (modified from Casalbore et al. 2014), where the main fields of large-scale bedforms are mapped together with related sediment sources (submarine drainage network and landslide/collapse scars). White boxes correspond to the following figures; scale bar in metres; reference system UTM WGS84 33N. The inset map shows the location of the Aeolian Islands in the southern Tyrrhenian Sea

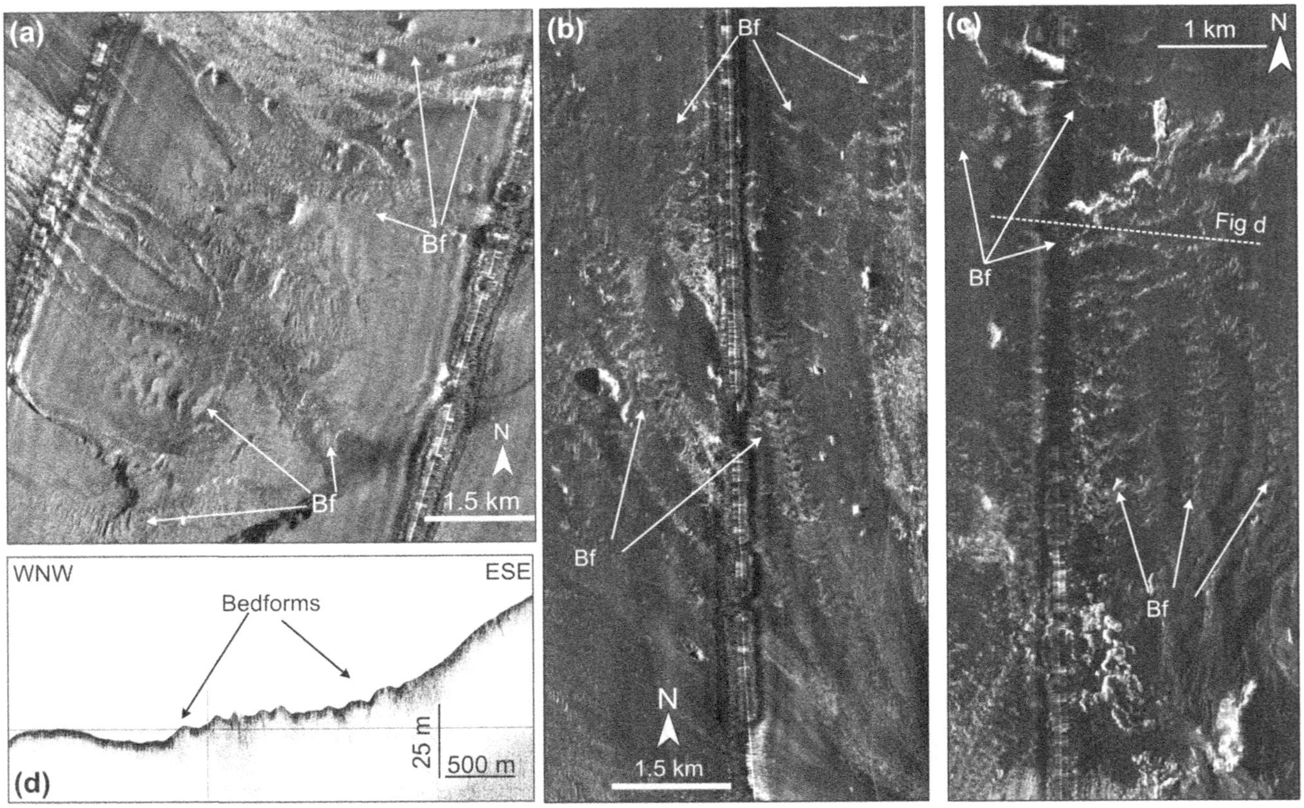

Fig. 45.2 Side scan sonar (TOBI) image of the southeastern (**a**), northwestern (**b**) and northern (**c**) flanks of Stromboli, with the indication of the recognized bedform fields (*Bf*); *light tones* indicate high backscatter values. **d** 3.5 kHz sub-bottom profile across the bedforms (location in Fig. 45.2c), showing a strong seafloor echo and a lack of signal penetration

The coarse-grained nature of these deposits is also witnessed by the lack of penetration on sub-bottom profiles (Fig. 45.2d), as well as by the available seafloor sampling, which recovered pebbles and coarse sand down to −2600 m (Casalbore et al. 2014 and references therein). On the NW and SE side of Stromboli, bedforms mostly occur at the foot of large subaerial-submarine depressions left by multiple collapses affecting the flanks (light-blue scars in Fig. 1, Romagnoli et al. 2009a, b). On the N flank, instead, the bedforms are found within a wide valley at the foot of the saddle separating the Stromboli and Strombolicchio volcanic centres (Figs. 45.1 and 45.2c), where active and strong erosive/depositional processes are observed (Casalbore et al. 2010 and this volume).

At Vulcano, a large field of bedforms is recognizable on a fan-shaped feature extending on the southwestern submarine flank between −800 m and −1200 m (Fig. 45.3). The bedforms have wavelengths of 100–300 m and wave heights of 4–20 m, covering an area of about 30 km^2. Locally, bedforms develop as coaxial trains elongated along the maximum slope, defining a series of small channels on the fan surface.

The fan is fed by two large landslide scars that deeply indent the insular shelf and match a large embayment on the coast, made up of a 350-m high cliff over a coastal tract of more than 2 km (dotted red line in Fig. 45.3). On Sparker profiles, the bedforms develop on top of an acoustically semi-transparent seismic unit (Fig. 45.3b), interpreted as the result of prominent mass-wasting processes (Romagnoli et al. 2013b). Two normally graded, 20- to 30-cm-thick levels of coarse volcaniclastic sands and millimetre-sized scoriae were recovered in a core from the median part of the fan, indicating the occurrence of thick, high-energy sedimentary gravity flows (Casalbore et al. 2014). Several 1- to 2-cm-thick levels of fine sand interbedded in a plastic mud are also recognized, representing the alternation with low-energy sediment flows.

In both case studies, the genesis of the bedforms can be related to the interaction between high-energy sediment gravity flows and the underlying seafloor, because (a) these features are found at or close to major sediment sources, and (b) the wave parameters, geometry and texture of the bedforms show a similarity to those of coarse-grained sediment waves found in the proximal regions of modern and ancient

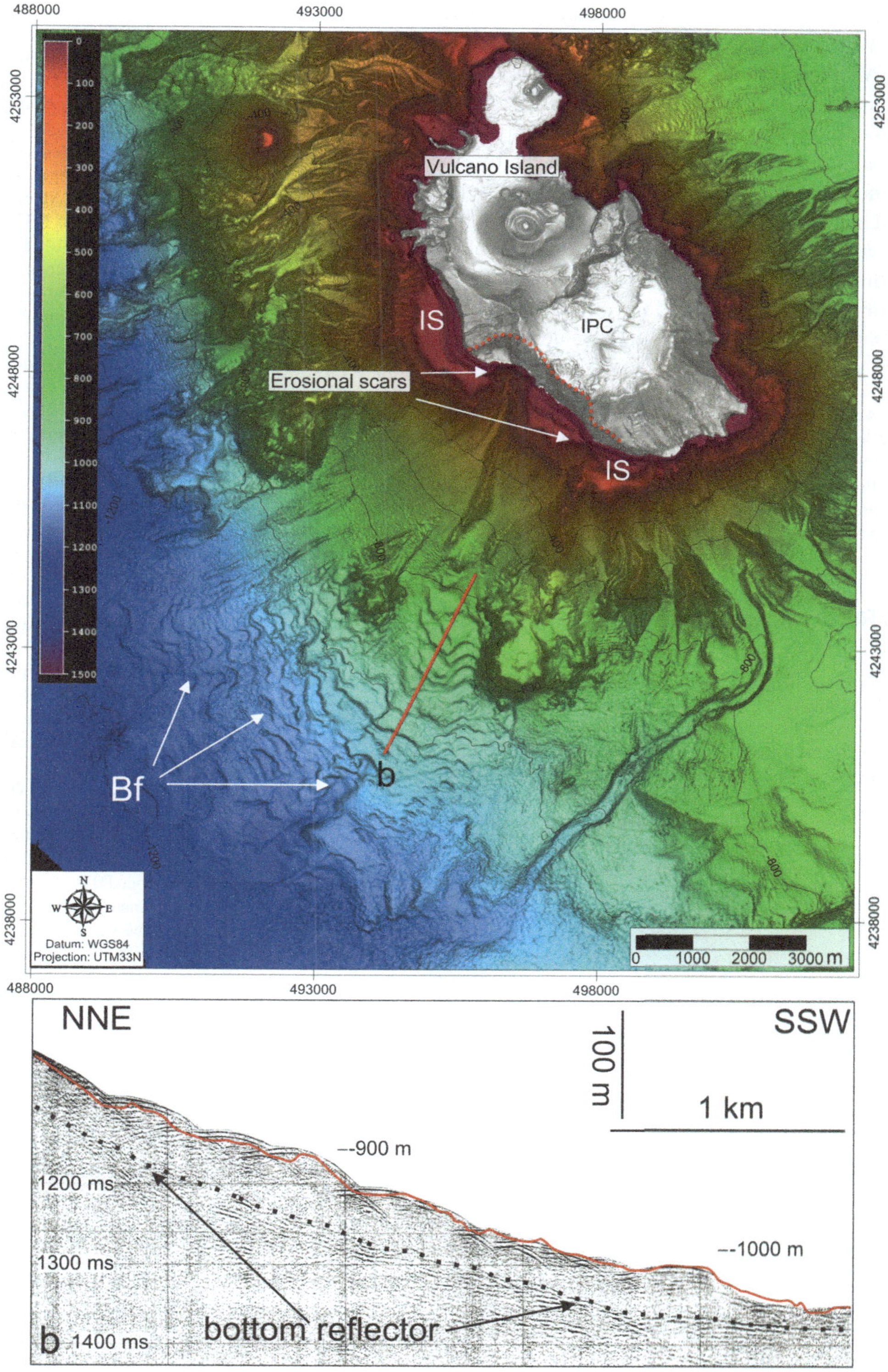

Fig. 45.3 Shaded relief map of the SW flank of Vulcano, where a large bedform field (*Bf*) is present; the *dotted red lines* indicate the wide subaerial erosive scarp affecting the outermost flank of Il Piano Caldera (IPC). *Below* Sparker 4.5 kJ profile on the bedform field (see text for detail, location in the map above). The *red line* on the profile represents the corresponding bathymetric section. A possible bottom reflector of a semi-transparent seismic unit is also indicated as a dotted black line

turbidite systems (e.g. Wynn et al. 2002; Ito 2010). More in detail, the bedforms were recently interpreted as cyclic steps (Casalbore et al. 2014) through comparison with the diagrams proposed by Cartigny et al. (2011). Cyclic steps represent a class of upslope-migrating turbidite sediment waves, in which each downward step is bounded by a hydraulic jump (Kostic 2011 and references therein). In our cases, bedforms start to develop under a marked decrease in the slope gradients, which would force the flow to pass the hydraulic jump at some time (e.g. Postma et al. 2009). Another common characteristics of cyclic steps is their upstream migration. Although we have no direct evidence of bedform migration (i.e. repeated bathymetric surveys), this migration seems to be suggested by the arcuate and downslope-concave shape of the bedform crests over the channel floor (i.e. curving away in downslope direction when approaching the canyon side). This shape might be interpreted as the result of the fastest upslope migration of the waves in the middle of the channel, where the flow velocity reaches the highest values (e.g. Wynn et al. 2002; Cartigny et al. 2011).

Acknowledgments This research was funded by the Italian MaGIC (Marine Geohazards along the Italian Coasts) Project and carried out in the framework of the RITMARE (Ricerca ITaliana per il MARE) Project. Crews of R/Vs *Urania*, *Universitatis* (CoNISMA) and *Thetis* are gratefully acknowledged along with the fellow researchers and students who took part in the sea surveys.

References

Bosman A, Casalbore D, Anzidei M, Muccini F, Carmisciano C, Chiocci FL (2015). The first ultra-high resolution Digital Terrain Model of the shallow-water sector around Lipari Island (Aeolian Islands, Italy). Annals of Geophysics, 58 (2), p. S0218.

Cartigny MJB, Postma G, van den Berg JH, Mastbergen DR (2011). A comparative study of sediment waves and cyclic steps based on geometries, internal structures and numerical modeling. Marine Geology 280: 40–56.

Casalbore D, Romagnoli C, Chiocci F, Frezza V (2010). Morphosedimentary characteristics of the volcanic apron around Stromboli volcano. Marine Geology 269 (3–4): 132–148.

Casalbore D, Romagnoli C, Bosman A, Chiocci FL (2014). Large-scale seafloor waveforms on the flanks of insular volcanoes (Aeolian Archipelago, Italy), with inferences about their origin. Marine Geology 355: 318–329.

Casalbore D, Bosman A, Romagnoli C, Chiocci FL (2017). Small-scale bedforms generated by gravity flows in the Aeolian Islands. J. Guillén et al. (eds.), Atlas of Bedforms in the Western Mediterranean. Springer, Heidelberg. pp 293–298.

Flood RD, Shor AN, Manley PL (1993). Morphology of abyssal mudwaves at project MUDWAVES sites in the Argentine Basin. Deep Sea Research 40: 859–888.

Ito M (2010). Are coarse-grained sediment waves formed as downstream-migrating antidunes? Insight from an early Pleistocene submarine canyon on the Boso Peninsula, Japan. Sedimentary Geology 226: 1–8.

Kostic S (2011). Modeling of submarine cyclic steps: controls on their formation, migration and architecture. Geosphere 7: 294–304.

Postma G, Cartigny M, Kleverlaan K (2009). Structureless, coarse-tail graded Bouma Ta formed by internal hydraulic jump of the turbidity current? Sedimentary Geology 219 (1–4): 1–6.

Romagnoli C, Kokelaar P, Casalbore D, Chiocci FL (2009a). Lateral collapses and active sedimentary processes on the northwestern flank of Stromboli volcano. Marine Geology 265: 101–119.

Romagnoli C, Casalbore D, Chiocci FL, Bosman A (2009b). Offshore evidence of large-scale lateral collapse on the eastern flank of Stromboli, Italy, due to structurally controlled, bi-lateral flank instability. Marine Geology 262: 1–13.

Romagnoli C, Casalbore D, Bortoluzzi G, Bosman A, Chiocci FL, D'Oriano F, Gamberi F, Ligi M, Marani M (2013a). Bathy-morphological setting of the Aeolian islands. In: Lucchi, F., Peccerillo, A., Keller, J., Tranne, C.A., Rossi, P.L. (Eds.), The Aeolian Islands Volcanoes. Geological Society, London, Memoirs, 37: 27–36.

Romagnoli C, Casalbore, D, Bosman, A, Braga, R., Chiocci FL (2013b). Submarine structure of Vulcano volcano (Aeolian Islands) revealed by high-resolution bathymetry and seismo-acoustic data. Marine Geology 338: 30–45.

Wynn RB, Piper DJW, Gee MJR (2002). Generation and migration of coarse-grained sediment waves in turbidity current channels and channel-lobe transition zones. Marine Geology 192: 59–78.

46 Deep-Water, Large-Scale Sediment Waves Bounded by the Valencia Channel

Araceli Muñoz, Elena Elvira, Patricia Jiménez, Juan Acosta, and Laura Pascual

Abstract

Several fields of large sediment waves have developed on the lower continental rise located northward from Mallorca island and southward from the incision of the Valencia Channel. These bedforms have a general E–W orientation and are upslope-migrating morphological features. Because of their similarity to other sediment waves, they are interpreted as being generated by unconfined turbidity flows, probably sourced from the Mallorca northern continental slope.

Keywords

Continental rise • Bedforms • Turbidity currents • Mallorca island

46.1 Introduction and Methods

The Valencia Channel is a mid-oceanic channel incised along the axis of the Valencia Trough and fed by materials that are channelled through the tributary canyons crossing the NW Iberian continental margin and flow into it (Maldonado et al. 1985; Palanques and Maldonado 1985; Alonso et al. 1995, 2000). The middle and lower course of the channel are characterized by high sinuosity, which is partially forced by the presence of volcanic outcrops at the base of the Balearic margin, such as the Valencia Seamount (Fig. 46.1). This chapter describes several fields of large sediment waves that have developed on the lower continental rise located northwards from Mallorca island and are bounded by the Valencia Channel incision.

Multibeam bathymetry in the east and south of the Valencia Channel (Balearic Sea) was acquired aboard the

A. Muñoz (✉) · E. Elvira · P. Jiménez · L. Pascual
Tragsa, C/Valentín Beato, nº6, 28037 Madrid, Spain
e-mail: amur@tragsa.es

J. Acosta
IEO, C/Corazón de María, nº8, 28006 Madrid, Spain

J. Guillén et al. (eds.), *Atlas of Bedforms in the Western Mediterranean*,
DOI 10.1007/978-3-319-33940-5_46

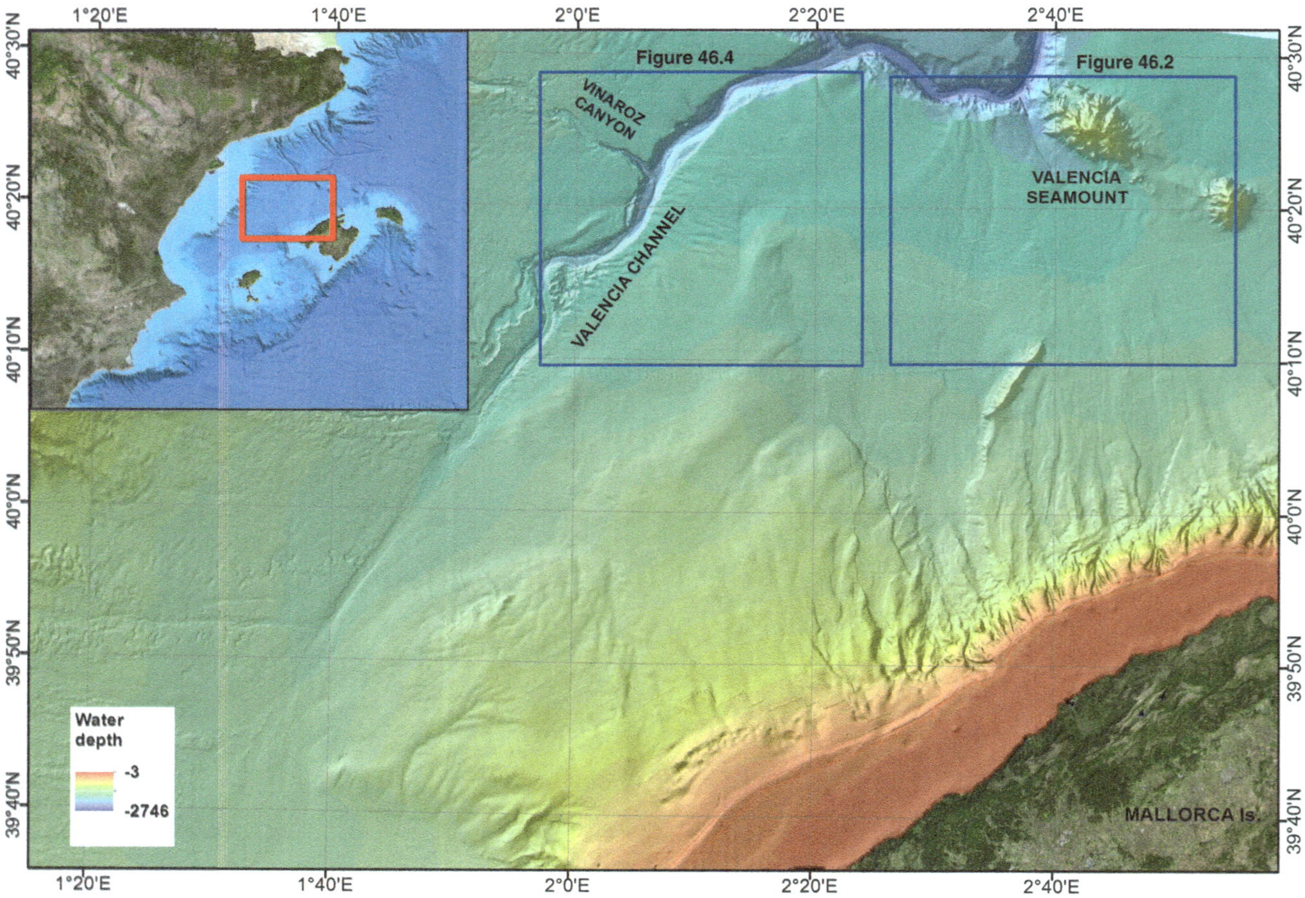

Fig. 46.1 The Valencia Channel and the location of the two fields of sediment waves

R/V *Vizconde de Eza* during the CAPESME-0412 and CAPESME-0213 cruises in March 2012 and February 2013, respectively. These cruises were part of the Fisheries Mapping Programme of the Mediterranean carried out by the Secretaria General de Pesca (MAGRAMA). Swath bathymetric data with a resolution of a few metres were obtained with a Simrad EM-300 system, and tracks were spaced to provide 100 % coverage of the seafloor. The data were logged using specific acquisition software and subsequently processed using the Caris, ArcGIS 9.3 and Fledermause software packages. The final resolution of the data was 25 × 25 m. The seismic lines were obtained with a hull-mounted TOPAS PS 18 high-resolution parametric profiler system with integrated transmitter and receivers. Navigation was via differential GPS Simrad GN33 using satellite corrections from a Fugro system.

46.2 Results and Discussion

Large sedimentary waves are observed on the right side of the Valencia Channel incision. Two fields are identified, one located near the Valencia Seamount and, the other near the confluence between the channel and the Vinaroz Canyon (Fig. 46.1). The deepest boundary of both fields of sediment waves is the Valencia Channel incision.

The region southwest of the Valencia Seamount shows a field composed of 111 sediment waves from 1780 to 1907 m

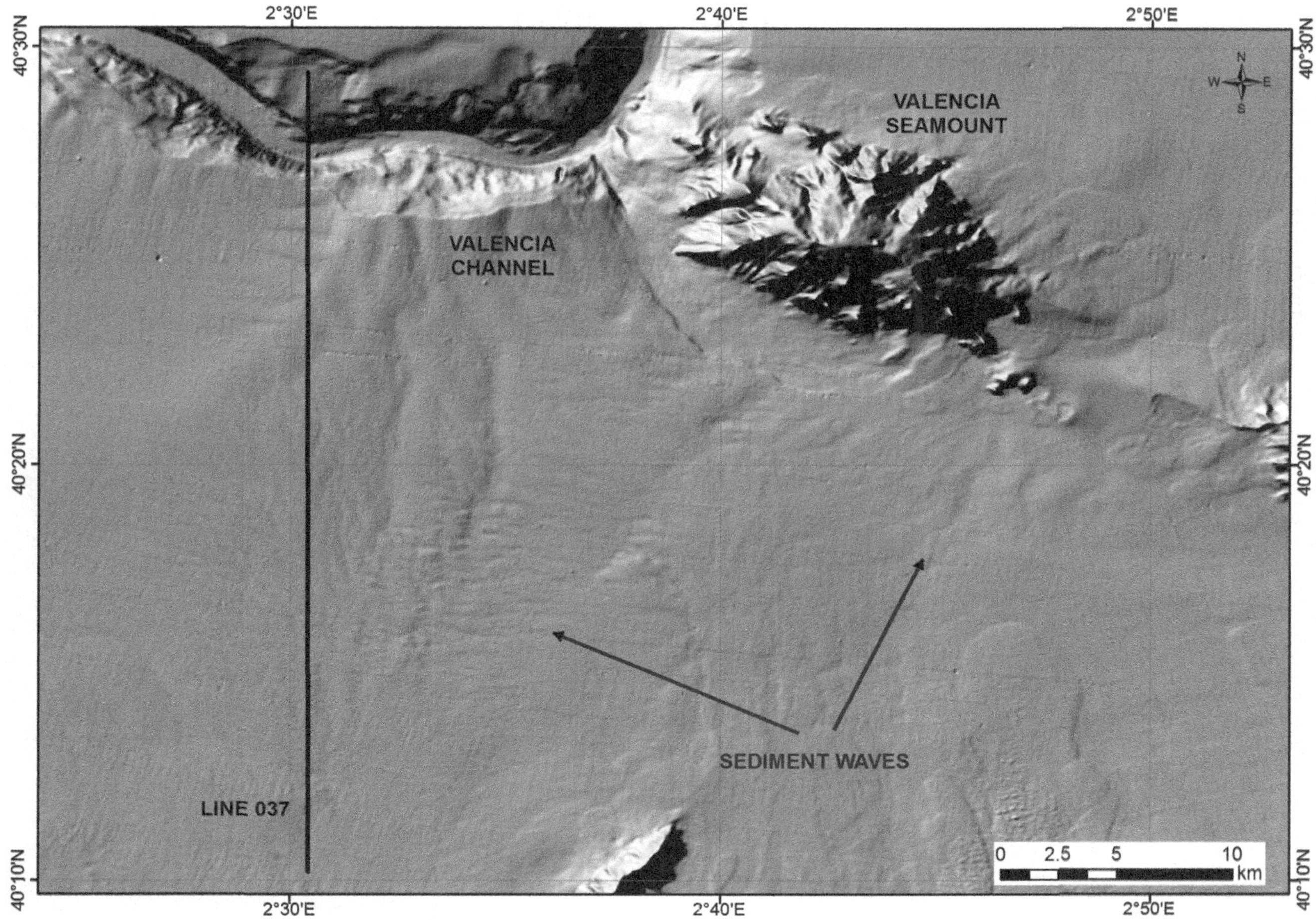

Fig. 46.2 Shaded bathymetric map of the sediment wave field in the vicinity of the Valencia Seamount. *Line 037* indicates the location of the seismic profile in Fig. 46.3

water depth that extends over an area of 418 km^2, (Fig. 46.2). The crestline orientation rotates from NNE–SSW in the eastern sector towards W–E in the western sector. The sediment wave height ranges from 1 to 4 m and the wavelength ranges from 600 to 5000 m, with increasing spacing downslope (Fig. 46.2). These bedforms are interrupted by several slightly incised channels that cross the slope. The seismic profile across the sediment wave field displays parallel, wavy reflectors, showing lateral continuity (Figs. 46.2 and 46.3a). The height and spacing of waves decrease upslope and the migration of the crest is also upslope (Fig. 46.3b, c).

The bedform field located near the confluence of the Vinaroz Canyon but on the right bank of the Valencia Channel (Fig. 46.4) shows 62 sediment waves that have developed between 1765 and 1890 m and cover a surface of 327 km^2 (Fig. 46.4). These bedforms are W–E oriented, their heights range from 2 to 10 m and their spacing ranges from 700 to 1900 m. The sediment waves have a mean length of 3500 m, although the largest can reach up to 10,000 m. Two seismic lines crossing parallel (Figs. 46.4 and 46.5a, b) and perpendicular (Figs. 46.4 and 46.5c) to this field display continuous, parallel wavy reflectors and show the upslope migration of the sediment waves.

Relevant sedimentary mechanisms for generating large sediment waves in deep water include bottom/contour currents and turbidity currents (Wynn and Stow 2002). When the sediment waves are parallel to the slope strikes and migrating upslope, they are usually interpreted as being

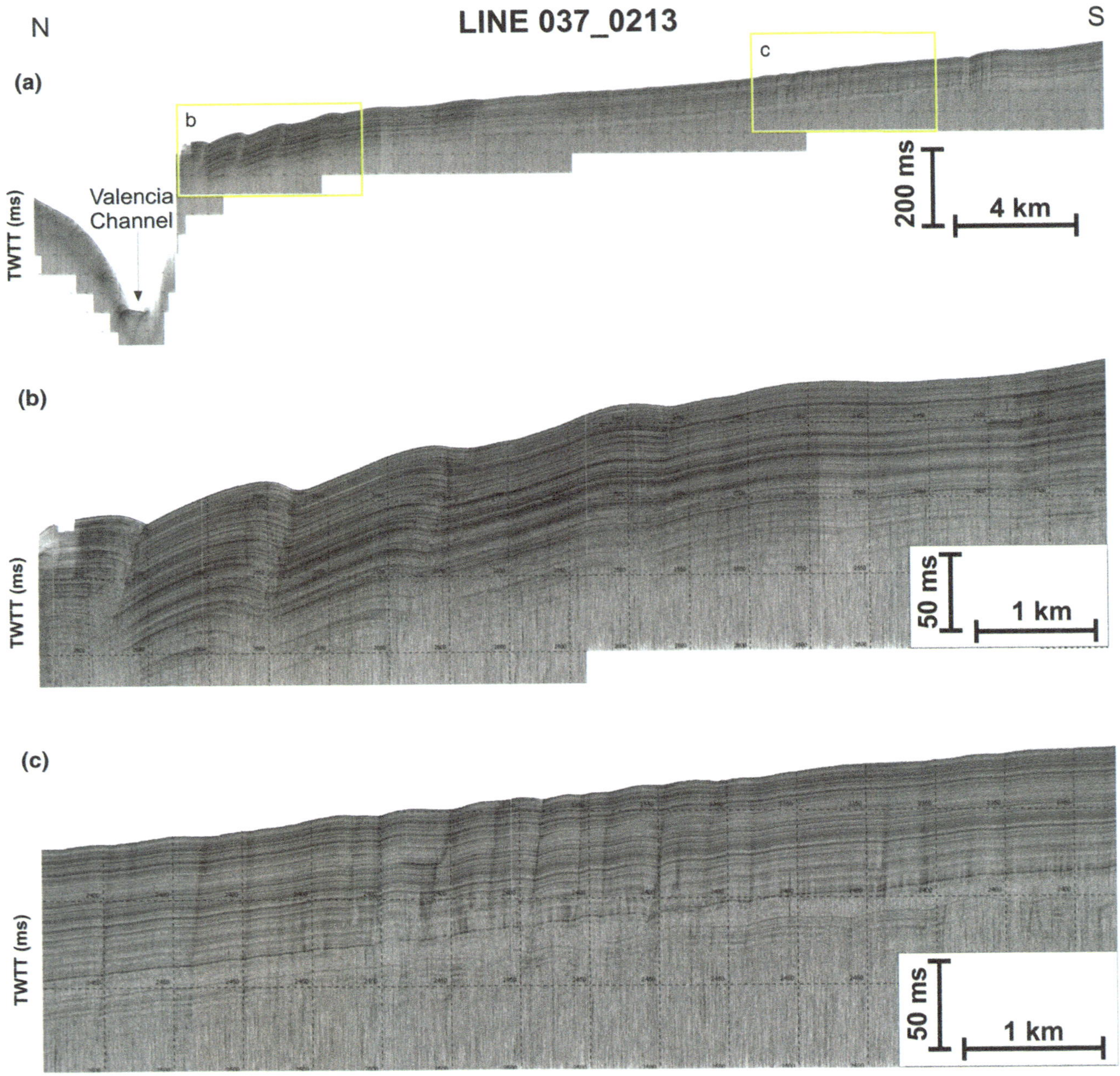

Fig. 46.3 **a** High-resolution TOPAS seismic profile across the Valencia Channel and the sediment wave field in the area southwest of the Valencia Seamount. **b** Detail of up-sloping migration of large-scale sediment waves near the Valencia Channel. **c** Detail of smaller sediment waves in the middle part of the slope (see location of profile in Fig. 46.2)

mainly generated by downslope-flowing unconfined turbidity currents, although low-intensity turbidity/bottom currents could also contribute to the upslope wave migration (Wynn et al. 2000). Studies in different areas revealed that this type of bedform is formed by an alternation of turbidites and pelagic/hemipelagic sediment, and that the provenance of the turbidity currents is the steep slope on the continental margin (Wynn et al. 2000).

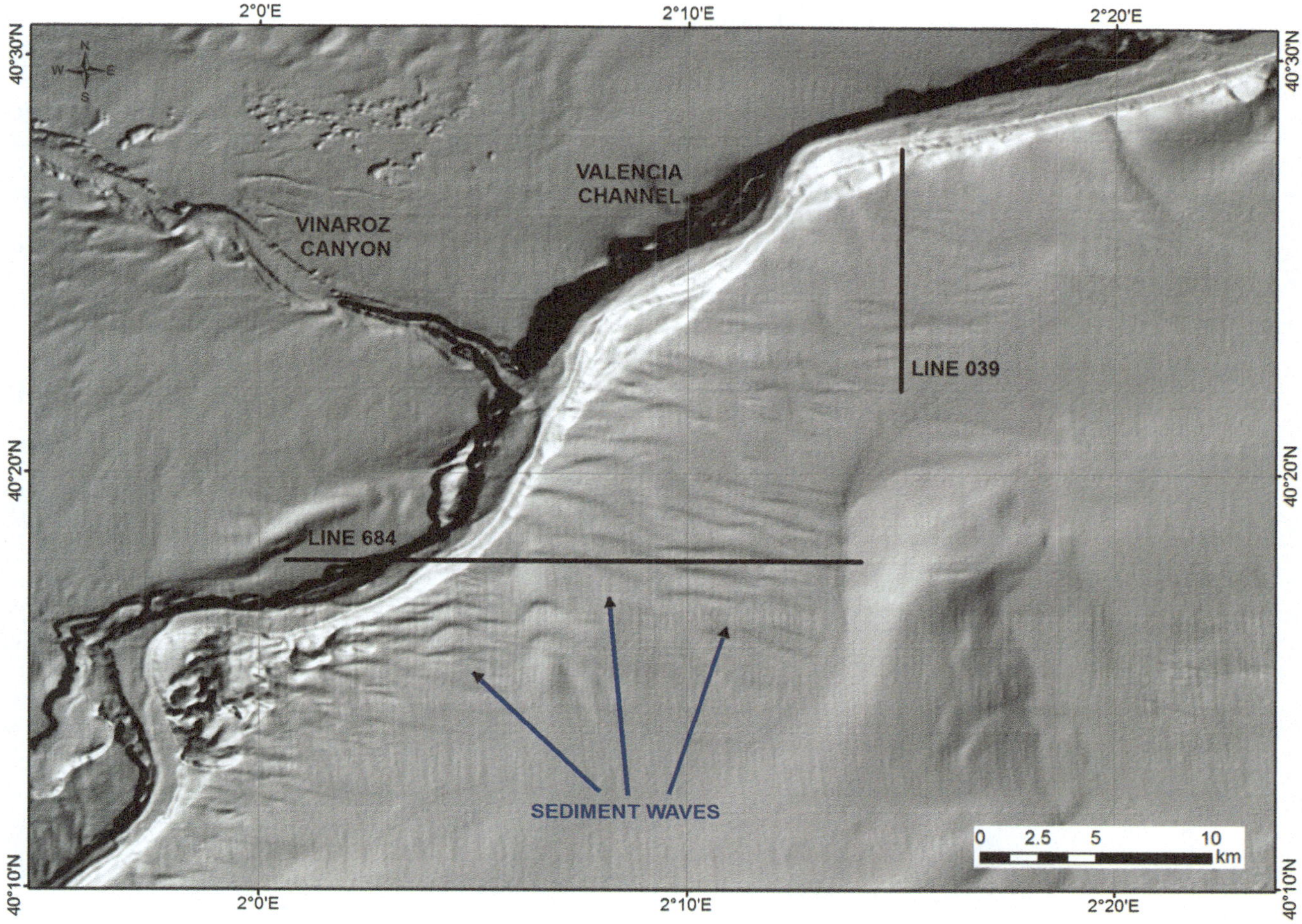

Fig. 46.4 Shaded bathymetric map of the sediment wave field located on the east bank of the Valencia Channel. The bedforms are mainly oriented in a W–E direction. *Lines 039 and 684* indicate the location of the seismic profiles in Fig. 46.5

The location of the sediment waves on the Mallorca continental rise, their orientation parallel to the slope, the continuous parallel reflectors and the upslope migration suggest that their formation may be related to downslope-flowing unconfined turbidity currents, as suggested in other areas (Wynn et al. 2000). In fact, the northern part of Mallorca island is occupied by the Tramuntana Ranges with a height of nearly 1500 m, a relatively narrow continental shelf and a continental slope with the presence of tributaries and channels that are able to supply sediment through turbidity currents (Fig. 46.1). The continuity of seismic reflectors from the sediment waves to the right side

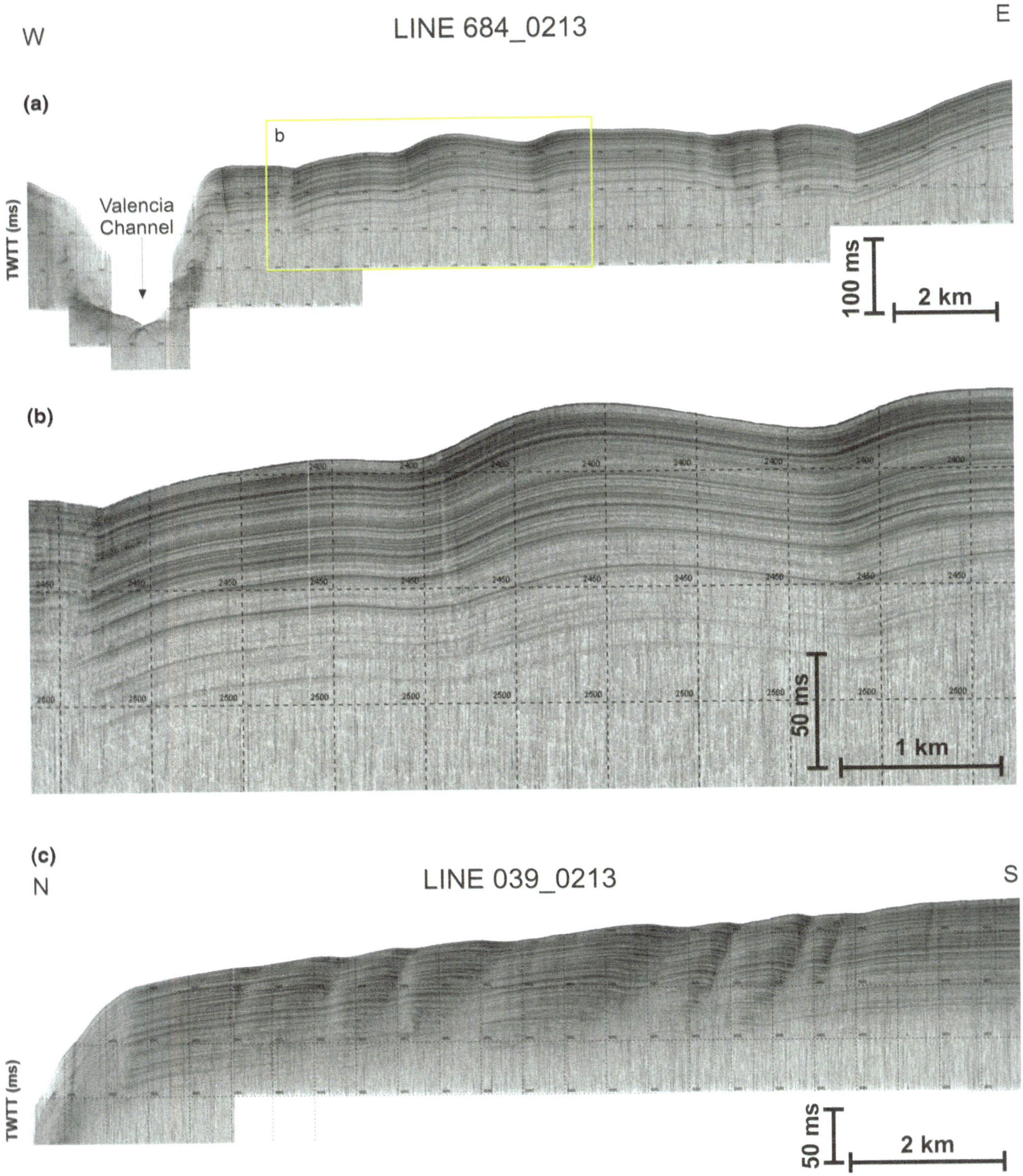

Fig. 46.5 TOPAS profile showing well-developed sediment waves. **a** Seismic profile across the Valencia Channel and parallel to the sediment wave field in the area near the confluence with the Vinaroz Canyon. **b** Detail of the profile showing the large-scale sediment waves composed of parallel, continuous reflectors. **c** Seismic profile almost perpendicular to the sediment wave field. Note the clear upslope migration of the waves and the continuity of reflectors towards the Valencia Channel (see location of profile in Fig. 46.4)

of the Valencia Channel suggests that part of the sediment supplied from the continental margin may reach the bank and axis of the channel.

References

Alonso, B., Canals, M., Palanques, A., Rehault, J.P. (1995). A deep-sea channel in the northwestern Mediterranean sea: morphosedimentary evolution of the Valencia Channel and its surroundings. Marine Geophysical Research. 17, 469–484.

Alonso, B., Canals, M., Palanques, A. (2000). El canal medio-oceánico de Valencia (Mediterráneo noroccidental): procesos sedimentarios y evolución durante el Plio-Cuaternario. In Alonso, B. and Ercilla, G., (eds.). Valles submarinos y sistemas turbidíticos modernos: Consejo Superior de Investigaciones Científicas, p. 229–249.

Maldonado, A., Got, H., Monaco, A., O'Connell, S., Mirabille, L. (1985). Valencia Fan (Northwestern Mediterranean): Distal deposition fan variant. Mar. Geol. 62, 295–319.

Palanques, A., Maldonado A. (1985). Sedimentology and evolution of the Valencia Valley and Fan (Northwestern Mediterranean). Acta Geol. Hisp. 20, 1–19.

Wynn, R.B., Masson, D.G., Stow, D.A., Weaver, P.P.E. (2000). Turbidity current sediment waves on the submarine slopes of the western Canary Islands. Marine Geology, Volume 163, Issues 1–4, 15, 185–198.

Wynn, R.B., Stow, D.A. (2002). Classification and characterisation of deep-water sediment waves. Marine Geology, Volume 192, Issues 1–3, 15 December 2002, 7–22.

List of Reviewers

Giovanni Coco: School of Environment, The University of Auckland, New Zealand.

Ruth Durán: Institut de Ciències del Mar (CSIC), Barcelona, Spain.

Marcel.li Farrán: Institut de Ciències del Mar (CSIC), Barcelona, Spain.

Giovanni de Falco: Istituto per l'Ambiente Marino Costiero, Consiglio Nazionale delle Ricerche, Torregrande —Oristan, Italy.

Albert Falqués: Physics Department, Universitat Politècnica de Catalunya, Barcelona, Catalonia, Spain.

Burghard Flemming: Wilhelmshaven, Germany.

Jordi Font: Institut de Ciències del Mar (CSIC), Barcelona, Spain.

Emilio García: Institut de Ciències del Mar (CSIC), Barcelona, Spain.

John Goff: Institute for Geophysics, The University of Texas at Austin, USA.

Hans van Haren: Royal Netherlands Institute for Sea Research (NIOZ), Den Burg, The Netherlands.

Claudio Lo Iacono: Marine Geosciences, National Oceanography Centre, Southampton, United Kingdom.

Silvia de Juan: Center for Marine Conservation, Las Cruces, Chile.

Aart Kroon: Department of Geosciences and Natural Resource Management, University of Copenhagen, Denmark.

Francisco Lobo: Instituto Andaluz de Ciencias de la Tierra (IACT), CSIC-Universidad de Granada, Spain.

Charles K. Paull: Monterey Bay Aquarium Research Institute, USA.

Pere Puig: Institut de Ciències del Mar (CSIC), Barcelona, Spain.

Pierre-Marie Poulain: Istituto Nazionale di Oceanografia e di Geofisica Sperimentale (OGS), Sgonico (Trieste), Italy.

Francesca Ribas: Physics Department, Universitat Politècnica de Catalunya, Barcelona, Catalonia, Spain.

Gonzalo Simarro: Institut de Ciències del Mar (CSIC), Barcelona, Spain.

Roger Urgelés: Institut de Ciències del Mar (CSIC), Barcelona, Spain.

Álvaro Viudez: Institut de Ciències del Mar (CSIC), Barcelona, Spain.

English editing : **Alan Lounds**, Barcelona, Spain.

Figures support: **Jordi Corbera**, Argentona. Spain.

J. Guillén et al. (eds.), *Atlas of Bedforms in the Western Mediterranean*,
DOI 10.1007/978-3-319-33940-5

The manufacturer's authorised representative in the EU is Springer Nature Customer Service Centre GmbH, Europaplatz 3, 69115 Heidelberg, Germany. If you have any concerns regarding our products, please contact ProductSafety@springernature.com

Printed and bound by CPI Group (UK) Ltd, Croydon, CR0 4YY
30/09/2026
02225516-0008